The Role of Metrics in the Theory of Partial Differential Equations

ADVANCED STUDIES
IN PURE MATHEMATICS 85

Chief Editor of the Series: Ken'ichi Ohshika (Gakushuin University)

The Role of Metrics in the Theory of Partial Differential Equations

Edited by

Yoshikazu Giga (The University of Tokyo)
Nao Hamamuki (Hokkaido University)
Hideo Kubo (Hokkaido University)
Hirotoshi Kuroda (Hokkaido University)
Tohru Ozawa (Waseda University)

Mathematical Society of Japan

This book was typeset by $\mathcal{A}_{\mathcal{M}}\mathcal{S}$-TeX and $\mathcal{A}_{\mathcal{M}}\mathcal{S}$-LaTeX, the TeX macro systems of the American Mathematical Society, together with the style files aspm.sty *and* aspmfm.sty *for $\mathcal{A}_{\mathcal{M}}\mathcal{S}$-TeX written by Dr. Chiaki Tsukamoto and* aspmproc.sty *for $\mathcal{A}_{\mathcal{M}}\mathcal{S}$-LaTeX written by Dr. Akihiro Munemasa and* aspm.cls *for $\mathcal{A}_{\mathcal{M}}\mathcal{S}$-LaTeX provided by Livretech Co., Ltd.*

TeX is a trademark of the American Mathematical Society.

Edited by the Mathematical Society of Japan.

Published by the Mathematical Society of Japan.

Distributed by the Mathematical Society of Japan, the American Mathematical Society and the World Scientific Publishing Co., Ltd.
Distributed exclusively in North America by the American Mathematical Society.
Partially supported by Grant-in-Aid for Publication of Scientific Research Results, Japan Society for the Promotion of Science, Grant Number 17HP1002.

Advanced Studies in Pure Mathematics 85
ISBN 978-4-86497-090-7

PRINTED IN JAPAN
by Livretech Co., Ltd.

2010 Mathematics Subject Classification.
Primary 35-06.
Secondary 35R01, 42B37, 65N12, 83C05.

Preface

This volume consists of peer-reviewed original articles submitted by speakers at the 11th Mathematical Society of Japan, Seasonal Institute (MSJ-SI) "The Role of Metrics in the Theory of Partial Differential Equations", held for the period of July 2–13, 2018 in Hokkaido University, Sapporo. The seasonal institute was supported by the Mathematical Society of Japan and the Department of Mathematics, Hokkaido University. During the period, various kinds of approaches to the analysis of partial differential equations (PDEs) were presented by 6 course lecturers and 46 invited lecturers. These lectures cover a huge variety of fields including PDEs, numerical analysis, harmonic analysis and geometric analysis. The organizing committee members are Y. Giga, H. Kubo, and T. Ozawa.

Sapporo is known to be one of the centers of nonlinear PDEs. The seasonal institute was just held in the year of 150 years anniversary since Hokkaido was founded. The organizing committee organized public lectures in this occasion, in order to involve non-specialists living near Sapporo to the seasonal institute. The aim of these public lectures is to show that mathematics plays an essential role in our real life. There are three lectures by mathematicians working in industry. The lecturers are Dr. Claus Dohmen, Dr. Takashi Takemoto, and Dr. Kouta Usui. In all, we had over 200 participants and fruitful interactions among participants through theses lectures. In this way, we successfully concluded the seasonal institute.

The organizing committee is profoundly grateful to financial support of Sapporo International Communication Plaza Foundation. The organizing committee would also like to express their sincere gratitude to all the colleagues and students who cooperate to manage this seasonal institute running smoothly. Especially, the organizing committee thanks Professors S.-I. Ei, M.-H. Giga, S. Katayama, M. Ohta, H. Takamura, Y. Tonegawa, S. Yamada and Ms. Kuroda, Ms. Ogawa, Ms. Ohno, Ms. Suzuki for their big effort. Without their help, the seasonal institute could be different. Finally, the editorial committee is genuinely grateful to all the contributors for their submission of original manuscripts and to all the referees for their efficient work.

Sapporo, August 2019
Editorial Committee

All papers in this volume have been refereed and are in final form. No version of any of them will be submitted for publication elsewhere.

Program (July 2–6, 2018)

Venue : 学術交流会館 Hokkaido University Conference Hall

July 2, 2018 (Monday)

Course Lecture (Room: 小講堂 Conference Hall)
10:30-10:40 Opening
10:40-11:30 Sergiu Klainerman (Princeton University)
On the mathematical theory of black holes, I
11:30-13:30 お昼休み (Break)
13:30-14:50 Poster Session (Venue: 第1会議室 Conference Room 1)

Session A (Room: 小講堂 Conference Hall)
15:00-15:50 Gerhard Huisken (Universität Tübingen / Mathematisches Forschungsinstitut Oberwolfach)
Mean curvature flow with surgery of embedded mean-convex surfaces in 3-manifolds
16:00-16:50 Mu-Tao Wang (Columbia University)
Quasilocal mass and isometric embedding

Session B (Room: 第1会議室 Conference Room 1)
15:00-15:50 Yi Zhou (Fudan University)
Global Instability of the Multi-dimensional Plane Shocks for the Isothermal Euler Equations
16:00-16:50 Piero D'Ancona (Sapienza - Universitá di Roma)
On the defocusing supercritical wave equation outside a ball

July 3, 2018 (Tuesday)

Course Lecture (Room: 小講堂 Conference Hall)
9:30-10:20 Mihalis Dafermos (Princeton University / University of Cambridge)
Black hole interiors and the strong cosmic censorship conjecture in general relativity, I
10:40-11:30 Sergiu Klainerman (Princeton University)
On the mathematical theory of black holes, II
11:30-13:30 お昼休み (Break)

Session A (Room: 小講堂 Conference Hall)

13:30-14:00	梶ヶ谷徹（東京電機大学）Toru Kajigaya (Tokyo Denki University) Hamiltonian stability for weighted measure and generalized Lagrangian mean curvature flow
14:10-14:40	棚橋典大（九州大学）Norihiro Tanahashi (Kyushu University) Separability of Maxwell equation in rotating black hole spacetime and its geometric aspects
15:00-15:50	Gilbert Weinstein (Ariel University) Harmonic maps with prescribed singularities and applications in general relativity
16:00-16:50	山田澄生（学習院大学）Sumio Yamada (Gakushuin University) Harmonic maps in general relativity

Session B (Room: 第 1 会議室 Conference Room 1)

13:30-14:00	岡本葵（信州大学）Mamoru Okamoto (Shinshu University) Random data Cauchy problem for the energy-critical nonlinear Schrödinger equations
14:10-14:40	戌亥隆恭（大阪大学）Takahisa Inui (Osaka University) Global behavior of solutions to the energy critical nonlinear damped wave equation
15:00-15:50	伊藤健一（東京大学）Kenichi Ito (The University of Tokyo) Asymptotics of generalized eigenfunctions on manifold with Euclidean and/or hyperbolic ends
16:00-16:50	Wei-Min Wang (CNRS / Université de Cergy-Pontoise) Time quasi-periodic solutions to the nonlinear Klein-Gordon equations

July 4, 2018 (Wednesday)

Course Lecture (Room: 小講堂 Conference Hall)

9:30-10:20	Jean-Marc Delort (Université Paris 13) Long time existence results for Hamiltonian non-linear Klein-Gordon equations on some compact manifolds, I
10:40-11:30	Mihalis Dafermos (Princeton University / University of Cambridge) Black hole interiors and the strong cosmic censorship conjecture in general relativity, II
12:00-	Excursion

July 5, 2018 (Thursday)

Course Lecture (Room: 小講堂 Conference Hall)

9:30-10:20 Sergiu Klainerman (Princeton University)
On the mathematical theory of black holes, III

10:40-11:30 Jean-Marc Delort (Université Paris 13)
Long time existence results for Hamiltonian non-linear Klein-Gordon equations on some compact manifolds, II

11:30-13:30 お昼休み (Break)

13:30-14:50 Poster Session (Venue: ホール Hall)

Session A (Room: 小講堂 Conference Hall)

15:00-15:50 Vanessa Styles (University of Sussex)
Stability and error analysis for a diffuse interface approach to an advection-diffusion equation on a moving surface

16:00-16:50 Richard Tsai (The University of Texas at Austin)
Volumetric extensions of variational problems defined on surfaces

Session B (Room: 第 1 会議室 Conference Room 1)

15:00-15:50 Joachim Krieger (École Polytechnique Fédérale de Lausanne)
On stability of blow up solutions for the critical co-rotational wave maps problem

16:00-16:50 片山聡一郎（大阪大学）Soichiro Katayama (Osaka University)
Global existence and the asymptotic behavior for systems of nonlinear wave equations violating the null condition

July 6, 2018 (Friday)

Course Lecture (Room: 小講堂 Conference Hall)

9:30-10:20 Mihalis Dafermos (Princeton University / University of Cambridge)
Black hole interiors and the strong cosmic censorship conjecture in general relativity, III

10:40-11:30 Jean-Marc Delort (Université Paris 13)
Long time existence results for Hamiltonian non-linear Klein-Gordon equations on some compact manifolds, III

11:30-13:30 お昼休み (Break)

Session A (Room: 小講堂 Conference Hall)

13:30-14:00 榊原航也（京都大学）Koya Sakakibara (Kyoto University)
Numerical analysis of one-harmonic equation with values in matrix Lie group

14:10-14:40 三浦達彦（東京大学）Tatsu-Hiko Miura (The University of Tokyo)
Singular limit problem for the Navier-Stokes equations in a curved thin domain

15:00-15:50 小俣正朗（金沢大学）Seiro Omata (Kanazawa University)
Free boundary problems of hyperbolic type with a focus on the free boundary conditions

16:00-16:50 Selim Esedoglu (University of Michigan)
Algorithms for motion of networks by weighted mean curvature

Session B (Room: 第1会議室 Conference Room 1)

13:30-14:00 池田正弘（理化学研究所・慶應義塾大学）Masahiro Ikeda (RIKEN / Keio University)
Blow-up of solutions to semilinear wave equation with a scaling invariant critical damping

14:10-14:40 宮崎隼人（津山工業高等専門学校）Hayato Miyazaki (National Institute of Technology, Tsuyama College)
The initial value problem for the generalized KdV equation with low degree of non-linearity

15:00-15:50 Jacob Sterbenz (University of California San Diego)
Local energy decay and the vector field method

16:00-16:50 Ning-An Lai (Lishui University)
Weighted L^2-L^2 estimate for wave equation and its application

Program (July 9–13, 2018)

Venue : フロンティア応用科学研究棟 Frontier Research in Applied Sciences Building

July 9, 2018 (Monday)

Course Lecture (Room: レクチャーホール Akira Suzuki Hall)

10:40-11:30 Felix Otto
(Max Planck Institute for Mathematics in the Sciences, Leipzig)
Nonlinear partial differential equations and interfacial motions as gradient flow, I

11:30-13:30 お昼休み (Break)

13:30-14:50 Poster Session (Venue: ホワイエ Foyer)

Session A (Room: レクチャーホール Akira Suzuki Hall)

15:00-15:50 二木昭人（清華大学）Akito Futaki (Tsinghua University)
Geometric nonlinear problems and GIT stability through moment maps

16:00-16:50 利根川吉廣（東京工業大学）Yoshihiro Tonegawa
(Tokyo Institute of Technology)
Existence theorem for Brakke flow

Session B (Room: セミナー室2 Seminar Room 2)

15:00-15:50 高村博之（東北大学）Hiroyuki Takamura
(Tohoku University)
Multipliers on the wave-like blow-up for nonlinear damped wave equations

16:00-16:50 関口雄一郎（東邦大学）Yuichiro Sekiguchi
(Toho University)
Numerical relativity: generating space-time on computers

July 10, 2018 (Tuesday)

Course Lecture (Room: レクチャーホール Akira Suzuki Hall)

9:30-10:20 Antonin Chambolle (CNRS / École Polytechnique)
Anisotropic and crystalline mean curvature flows, I

10:40-11:30 Felix Otto
(Max Planck Institute for Mathematics in the Sciences, Leipzig)
Nonlinear partial differential equations and interfacial motions as gradient flow, II

11:30-13:30 お昼休み (Break)

Session A (Room: レクチャーホール Akira Suzuki Hall)

13:30-14:00 本多正平（東北大学）Shouhei Honda (Tohoku University)
Recent developments on metric measure spaces with Ricci bounds from below

14:10-14:40 Kelei Wang (Wuhan University)
Simons cone and saddle solutions of the Allen-Cahn equation

15:00-15:50 芥川和雄（中央大学）Kazuo Akutagawa (Chuo University)
Edge-cone Einstein metrics and Yamabe metrics

16:00-16:50 Matteo Novaga (Università di Pisa)
Evolution by curvature of planar networks

Session B (Room: セミナー室 2 Seminar Room 2)

13:30-14:00 杉山裕介（滋賀県立大学）Yuusuke Sugiyama
(The University of Shiga Prefecture)
Global existence and degeneracy in finite time for 1D quasilinear wave equations

14:10-14:40 佐々木多希子（明治大学）Takiko Sasaki (Meiji University)
The blow-up curve of solutions for semilinear wave equations with Dirichlet boundary conditions in one space dimension

15:00-15:50 Karen Yagdjian (The University of Texas Rio Grande Valley)
A novel integral transform approach to solving hyperbolic equations in the curved spacetimes

16:00-16:50 中村誠（山形大学）Makoto Nakamura
(Yamagata University)
On some effects of background metrics for several partial differential equations

July 11, 2018 (Wednesday)

Course Lecture (Room: レクチャーホール Akira Suzuki Hall)

9:30-10:20	Charles M. Elliott (The University of Warwick) Variational, geometric and surface PDEs in cell biology, I
10:40-11:30	Antonin Chambolle (CNRS / École Polytechnique) Anisotropic and crystalline mean curvature flows, II
12:00-	Excursion

July 12, 2018 (Thursday)

Course Lecture (Room: レクチャーホール Akira Suzuki Hall)

9:30-10:20	Felix Otto (Max Planck Institute for Mathematics in the Sciences, Leipzig) Nonlinear partial differential equations and interfacial motions as gradient flow, III
10:40-11:30	Charles M. Elliott (The University of Warwick) Variational, geometric and surface PDEs in cell biology, II
11:30-13:30	お昼休み (Break)
13:30-14:50	Poster Session (Venue: ホワイエ Foyer)

Session A (Room: レクチャーホール Akira Suzuki Hall)

15:00-15:50	小磯深幸（九州大学）Miyuki Koiso (Kyushu University) Uniqueness problem for closed non-smooth hypersurfaces with constant anisotropic mean curvature
16:00-16:50	Neshan Wickramasekera (University of Cambridge) Regularity of minimal and CMC stable hypersurfaces

Session B (Room: セミナー室2 Seminar Room 2)

15:00-15:50	谷口雅治（岡山大学）Masaharu Taniguchi (Okayama University) Multidimensional traveling fronts in reaction-diffusion equations
16:00-16:50	Elliott Ginder (Meiji University) Approximation methods for oscillatory and constrained interfacial dynamics

July 13, 2018 (Friday)

Course Lecture (Room: レクチャーホール Akira Suzuki Hall)

9:30-10:20	Antonin Chambolle (CNRS / École Polytechnique) Anisotropic and crystalline mean curvature flows, III
10:40-11:30	Charles M. Elliott (The University of Warwick) Variational, geometric and surface PDEs in cell biology, III
11:30-13:30	お昼休み (Break)

Session A (Room: レクチャーホール Akira Suzuki Hall)

13:30-14:00	柳青（福岡大学）Qing Liu (Fukuoka University) Large exponent behavior of power curvature flow and applications
14:10-15:00	高坂良史（神戸大学）Yoshihito Kohsaka (Kobe University) On the dynamics for the volume-preserving geometric flows
15:10-16:00	石井克幸（神戸大学）Katsuyuki Ishii (Kobe University) Convergence of a threshold-type algorithm for mean curvature flow

Session B (Room: セミナー室2 Seminar Room 2)

13:30-14:00	Bronwyn Hajek (University of South Australia) Analytical solutions of nonlinear reaction-diffusion equations in various coordinate systems
14:10-15:00	Petrus van Heijster (Queensland University of Technology) Pinned localised solutions in a heterogeneous three-component FitzHugh-Nagumo model
15:10-16:00	Sanjeeva Balasuriya (The University of Adelaide) Stochastic sensitivity of finite-time Lagrangian motion

16:00-16:10 Closing

The List of Participants [1]

Kazuo Akutagawa	(Chuo University)
Kazuki Aoki	(Osaka University)
Sanjeeva Balasuriya	(University of Adelaide)
Simon Blatt	(University Salzburg)
Lorenzo Cavallina	(Tohoku University)
Ye Sle Cha	(Freie Universität Berlin)
Antonin Chambolle	(CNRS, Mathematical Sciences)
Jui-En Chang	(National Taiwan University)
Haoyang Chen	(Fudan University)
Yan-Hsiou Cheng	(National Taipei University of Education)
Yonggeun Cho	(Chonbuk National University)
Mihalis Dafermos	(Princeton University)
Piero D'Ancona	(Sapienza Università di Roma)
Jean-Marc Delort	(Université Paris 13)
Claus Dohmen	(DB Systel GmbH)
Shin-Ichiro Ei	(Hokkaido University)
Charles M. Elliott	(University of Warwick)
Selim Esedoglu	(University of Michigan)
Yasuhiro Fujita	(University of Toyama)
Yuki Fukui	(The University of Tokyo)
Kei Funano	(Tohoku University)
Akito Futaki	(Tsinghua University)
Anahit Galstyan	(University of Texas Rio Grande Valley)
Mi-Ho Giga	(The University of Tokyo)
Yoshikazu Giga	(The University of Tokyo)
Elliott Ginder	(Meiji University)
Shun'ichi Goto	(Hokkaido University of Education)
Bronwyn Hajek	(University of South Australia)
Nao Hamamuki	(Hokkaido University)
Masaru Hamano	(Saitama University)
Kunio Hidano	(Mie University)
Shouhei Honda	(Tohoku University)
Gerhard Huisken	(Tuebingen University)
Masahiro Ikeda	(Riken, Advanced Intelligence Project)
Takahisa Inui	(Osaka University)
Atsuhide Ishida	(Tokyo University of Science)
Katsuyuki Ishii	(Kobe University)
Aya Ishizeki	(Chiba University)
Kenichi Ito	(The University of Tokyo)
Keisuke Izumi	(Nagoya University)

[1]The participants of the public seminar held on July 7, 2018 and the technical staff are not included in this list. The affiliation written in the list is as of July, 2018.

Takashi Kagaya	(Kyushu University)
Toru Kajigaya	(Tokyo Denki University)
Ryo Kanazawa	(Kyoto University)
Soichiro Katayama	(Osaka University)
Masakazu Kato	(Muroran Institute of Technology)
Daehwan Kim	(Pusan National University)
Kihyun Kim	(Korea Advanced Institute of Science and Technology)
Seokchan Kim	(Changwon National Unversity)
Nobu Kishimot	(Kyoto University)
Jun Kitagawa	(Michigan State University)
Sergiu Klainerman	(Princeton University)
Yoshikazu Kobayashi	(Chuo University)
Yoshihito Kohsaka	(Kobe University)
Naoyuki Koike	(Tokyo University of Science)
Miyuki Koiso	(Kyushu University)
Hiroshi Konno	(Meiji University)
Hideo Kozono	(Waseda University)
Joachim Krieger	(EPFL, Department of Mathematics)
Hideo Kubo	(Hokkaido University)
Keita Kunikawa	(Tohoku University)
Hirotoshi Kuroda	(Hokkaido University)
Masaki Kurokiba	(Muroran Institute of Technology)
Ningan Lai	(Lishui University)
Jihoon Lee	(Yonsei University)
Kuo-Wei Lee	(National Changhua University of Education)
Sanghoon Lee	(Princeton University)
Yangyang Li	(Princeton University)
Mengyun Liu	(Zhejiang University)
Qing Liu	(Fukuoka University)
Shuji Machihara	(Saitama University)
Wechselberger Martin	(University of Sydny)
Jun Masamune	(Hokkaido University)
Yukio Matsumoto	(Gakushuin University)
Ayato Mitsuishi	(Fukuoka University)
Tatsu-Hiko Miura	(The University of Tokyo)
Tatsuya Miura	(The University of Tokyo)
Hayato Miyazaki	(National Institute of Technology, Tsuyama College)
Masashi Mizuno	(Nihon University)
Sang-hyuck Moon	(Korea Advanced Institute of Science and Technology)
Takeyuki Nagasawa	(Saitama University)
Makoto Nakamura	(Yamagata University)
Eungmo Nam	(Pusan National University)
Makoto Narita	(National Institute of Technology, Okinawa College)

Athanasiou Nikolaos (Oxford University)
Yasumasa Nishiura (Tohoku University)
Matteo Novaga (University of Pisa)
Masato Nozawa (Kyoto University)
Takayoshi Ogawa (Tohoku University)
Masaki Ohnuma (Tokushima University)
Masahito Ohta (Tokyo University of Science)
Jihoon Ok (Kyung Hee University)
Jun Okamoto (The University of Tokyo)
Mamoru Okamoto (Shinshu University)
Seiro Omata (Kanazawa University)
Eiji Onodera (Kochi University)
Michiaki Onodera (Tokyo Institute of Technology)
Fumihiko Onoue (The University of Tokyo)
Felix Otto (Max Planck Institute for Mathematics in the Sciences)
Tohru Ozawa (Waseda University)
Sangwoo Park (Pusan National University)
Adam Paxton (Oxford University)
Graeme Pettet (Queensland University of Technology)
Yuji Sagawa (Osaka University)
Koya Sakakibara (Kyoto University)
Daisuke Sakoda (Osaka University)
Takiko Sasaki (Meiji University)
Koh Sato (Otaru University of Commerce)
Yuya Sato (Yamagata University)
Nico Michele Schiavone (University of Bari)
Yukihiro Seki (Kyushu University)
Yuichiro Sekiguchi (Toho University)
Kihoon Seong (Korea Advanced Institute of Science and Technology)
Ikkei Shimizu (Kyoto University)
Toshitada Shintani (retired)
Tetsuya Shiromizu (Nagoya University)
Jacob Sterbenz (University of California San Diego)
Vanessa Styles (University of Sussex)
Yuusuke Sugiyama (The University of Shiga Prefecture)
Kohei Suzuki (Rheinische Friedrich-Wilhelms-Universität Bonn)
Teruo Tada (King Abdullah University of Science and Technology)
Hiroyuki Takamura (Tohoku University)
Keisuke Takasao (Kyoto University)
Norihiro Tanahashi (Kyushu University)
Masaharu Taniguchi (Okayama University)
Grigalius Taujanskas (University of Oxford)
Takashi Teramoto (Asahikawa Medical University)

Yoshihiro Tonegawa	(Tokyo Instituite of Technology)
Richard Tsai	(The University of Texas at Austin)
Kimitoshi Tsutaya	(Hirosaki University)
Petrus J.A. van Heijster	(Queensland University of Technology)
Kyouhei Wakasa	(Tokyo University of Science)
Chengbo Wang	(Zhejiang University)
Kelei Wang	(Wuhan University)
Mu-Tao Wang	(Columbia University)
Wei-Min Wang	(CNRS, Mathematics)
Xuecheng Wang	(Tsinghua University)
Ye-Kai Wang	(National Cheng Kung University)
Gilbert Weinstein	(Ariel University)
Neshan Wickramasekera	(University of Cambridge)
Toshiaki Yachimura	(Tohoku University)
Karen Yagdjian	(University of Texas Rio Grande Valley)
Sumio Yamada	(Gakushuin University)
Hikaru Yamamoto	(Tokyo University of Science)
Shiwu Yang	(Peking Univeristy)
Kazuyoshi Yokoyama	(Hokkaido University of Science)
Borislav Yordanov	(Hokkaido University)
Kogo Yoshikawa	(Hokkaido University)
Qidi Zhang	(East China University of Science and Technology)
Yi Zhou	(Fudan University)
Yi Zhu	(East China University of Science and Technology)

Acknowledgements

The organizing committee would like to extend its sincere gratitude to the following associations and organizations for their valuable support and commitment to make the seasonal institute successful.

♢ Sapporo International Communication Plaza Foundation

♢ Japan Society for the Promotion of Science

- Grant-in-aid for Scientific Research (S) Grant Number JP2622070217
- Grant-in-aid for Scientific Research (S) Grant Number 25220702
- Grant-in-aid for Scientific Research (A) Grant Number 25247008
- Grant-in-aid for Scientific Research (B) Grant Number 24340024
- Grant-in-aid for Scientific Research (C) Grant Number 24540200

♢ Japan Science and Technology Agency, CREST

- Grant Number JPMJCR14D3, Japan

♢ Department of Mathematics, Hokkaido University

♢ The Mathematical Society of Japan

CONTENTS

I. Course Lecturers

II. Invited Lecturers

Advanced Studies in Pure Mathematics 85, 2020
The Role of Metrics in the Theory of Partial Differential Equations
pp. 1–37

Long time existence results for Hamiltonian non-linear Klein-Gordon equations on some compact manifolds

Jean-Marc Delort

Abstract.

Consider a nonlinear Klein-Gordon equation on X, a compact Riemannian manifold without boundary, $(\partial_t^2 - \Delta + m^2)w = N(w)$, where N is a smooth non-linearity. If the non-linearity vanishes at order $n+1$ at zero, and if the Cauchy data are of small size ϵ in a regular enough Sobolev space, the solution exists over time intervals of length c/ϵ^n, and remains of size $O(\epsilon)$ over that interval of time.

In this review paper, we shall discuss the question of getting a similar existence result and $O(\epsilon)$ bounds over an interval of length larger than c/ϵ^n. We shall present first some recent results concerning general manifolds, where one obtains some slight improvement of the lower bound of the time of existence given by local existence theory, when the mass parameter m avoids a subset of zero measure.

Then, we turn to the question of obtaining lower bounds of the time of existence in $c_N\epsilon^{-N}$ for any N. It turns out that this is possible if one works on very special manifolds (Zoll manifolds), with Hamiltonian non linearities, and again takes the parameter m of the equation outside a set of zero measure.

We shall focus first on the semi-linear case, where ideas of proof may be described easily. We shall then discuss with less details the more involved case of quasi-linear equations.

Received October 11, 2018.

Revised February 14, 2019.

2010 *Mathematics Subject Classification.* Primary 35L72; Secondary 37K45, 35S50, 58J45.

Key words and phrases. Long time existence, Nonlinear Klein-Gordon equation, Normal forms.

The author is partially supported by the ANR project 13-BS01-0010-02 “Analyse asymptotique des équations aux dérivées partielles d’évolution”.

§1. Introduction

The aim of this survey paper is to review results that have been obtained during the last ten years about long time existence for nonlinear hyperbolic equations with small smooth data on compact manifolds. We shall be interested mainly on nonlinear Klein-Gordon equations.

Denote by (X, g) a compact Riemannian manifold of dimension $d \geq 1$, by Δ the Laplace-Beltrami operator on X. Let m be given in $]0, +\infty[$ and let n be in $\mathbb{N}^*$. We consider the following problem

$$\begin{aligned} (\partial_t^2 - \Delta + m^2)w &= w^{n+1} \\ w|_{t=0} &= \epsilon w_0 \\ \partial_t w|_{t=0} &= \epsilon w_1 \end{aligned} \tag{1.1}$$

where $\epsilon > 0$ is a small parameter and (w_0, w_1) is in the unit ball of the product of Sobolev spaces $H^{s+1}(X, \mathbb{R}) \times H^s(X, \mathbb{R})$ for some large enough s. Our general goal is to obtain lower bounds for the length of the interval $] - T_\epsilon, T_\epsilon[$ over which the solution exists and satisfies $\|w(t, \cdot)\|_{H^{s+1}} + \|\partial_t w(t, \cdot)\|_{H^s} \leq K\epsilon$ for some constant K.

Let us recall that a trivial lower bound for T_ϵ, of the form $T_\epsilon \geq c\epsilon^{-n}$, is provided for free by local existence theory and the energy inequality associated to the equation. In fact, define the Sobolev energy

$$E_s(w, t) = \|w(t, \cdot)\|^2_{H^{s+1}} + \|\partial_t w(t, \cdot)\|^2_{H^s} \tag{1.2}$$

and consider the following statement:

There are constants $A_0 > 0$, $A_1 > 0$, $\epsilon_0 > 0$, such that, if $T \in [0, T_\epsilon[$ and if the solution of (1.1) exists on $[0, T]$ and satisfies for any $t \in [0, T]$,

$$\begin{aligned} E_s(w, t) &\leq A_0 \epsilon^2 \\ \|w(t, \cdot)\|_{L^\infty} &\leq A_1 \epsilon, \end{aligned} \tag{1.3}$$

then, actually, for t in $[0, T]$ and $\epsilon \in]0, \epsilon_0[$, one has

$$\begin{aligned} E_s(w, t) &\leq \frac{1}{2} A_0 \epsilon^2 \\ \|w(t, \cdot)\|_{L^\infty} &\leq \frac{1}{2} A_1 \epsilon. \end{aligned} \tag{1.4}$$

Assume that the preceding statement has been proved for a value of s which is large enough so that local existence and uniqueness holds for initial data in $H^{s+1} \times H^s$ (say $s > \frac{d}{2}$). Then the solution may be extended up to time T_ϵ. Actually, one has just to consider the supremum

$\tilde{T}_\epsilon$ of those $T \in [0, T_\epsilon[$ such that the solution exists on $[0, T]$ and satisfies (1.3) on $[0, T]$. Then, we deduce from the implication (1.3) $\Rightarrow$ (1.4) that the solution of (1.1) satisfies (1.4) on $[0, T]$. By local existence theory, w may be extended to some slightly larger interval $[0, T + \delta]$, and since the left hand side of (1.4) is a continuous function of time, if (1.4) holds on $[0, T]$, then (1.3) will be satisfied on $[0, T + \delta]$, for some δ independent of $T < \tilde{T}_\epsilon$. This shows that $\tilde{T}_\epsilon$ cannot be smaller than T_ϵ.

In order to get lower bounds for T_ϵ, we thus just have to establish (1.3) $\Rightarrow$ (1.4) for convenient values of this T_ϵ. The trivial lower bound $T_\epsilon \geq c\epsilon^{-n}$ is obtained in that way: Computing the time derivative of the energy (1.2) and using the equation, we get for any $t \in [0, T_\epsilon[$ the estimate

$$E_s(w,t) \leq E_s(w,0) + C \int_0^t \|w(\tau,\cdot)\|_{H^s}^{n+2}\, d\tau. \tag{1.5}$$

If we bound $\|w\|_{H^s}^{n+2} \leq C\|w\|_{L^\infty}^n \|w\|_{H^s}^2$ and use (1.3), we get

$$E_s(w,t) \leq E_s(w,0) + CtA_1^n A_0 \epsilon^{n+2}. \tag{1.6}$$

If A_0 has been chosen large enough relatively to the Cauchy data, we may assume that $E_s(w,0) \leq \frac{A_0}{4}\epsilon^2$, so that, if we take $t \leq T_\epsilon = c\epsilon^{-n}$ with c small enough, we obtain from (1.6) that the first inequality (1.4) holds true. The fact that the second equation (1.4) is satisfies follows just from Sobolev embedding (since $s > \frac{d}{2}$) and a convenient choice of A_1 in terms of A_0.

The question we want to address is the following one:

Can we extend the solution of (1.1), together with a bound of the form (1.3), over a longer time interval than $c\epsilon^{-n}$? More precisely, can we get for T_ϵ a lower bound $T_\epsilon \geq c\epsilon^{-An}$ for some $A > 1$? Under convenient assumptions, could we eventually get such a lower bound for any $A > 1$?

Remark. Let us stress here the difference between the problem we consider and the corresponding one on $X = \mathbb{R}^d$, with small, smooth and *decaying* Cauchy data. In that framework, because of the dispersive properties of the equation, solutions are expected to decay in L^∞ norm: $\|w(t,\cdot)\|_{L^\infty} = O(t^{-\frac{d}{2}})$ when t goes to infinity. This decay allows for much better lower bounds for the time of existence. For instance, in dimension larger or equal to three, one gets that for any $n \geq 1$, solutions of (1.1) with small smooth decaying initial data are global. The proof relies on bounds of the form (1.3), (1.4) with A_1 replaced by $A_1 t^{-\frac{d}{2}}$. The new ingredient that is needed, in addition to energy inequality, is a way to exploit the dispersion in order to obtain the L^∞ bounds in (1.3), (1.4).

We refer to the papers of Klainerman [18] and Shatah [19] where these results are proved (including for more general non-linearities than those of (1.1)) and to the recent paper of Stingo [21] for up to date references on that dispersive case.

On the other hand, when X is a compact manifold, we no longer have any dispersion, and one must rely on other properties in order to extend solutions beyond the time $c\epsilon^{-n}$ given by local existence theory. The key tool will be normal forms. We shall first describe this normal forms method on a simple example in next section.

§2. The normal forms method

The normal forms method, that is very classical for ordinary differential equations, has been introduced for PDEs by Shatah [19] and Simon-Taflin [20]. To apply it to equation (1.1), let us introduce the following notation. Define by functional calculus $\Lambda_m = \sqrt{-\Delta + m^2}$, set $D_t = \frac{1}{i}\frac{\partial}{\partial t}$ and introduce

$$w_+ = (D_t + \Lambda_m)w, \ w_- = (D_t - \Lambda_m)w = -\bar{w}_+$$

so that (1.1) may be written

$$(D_t - \Lambda_m)w_+ = F(w_+, w_-) \tag{2.1}$$

with

$$F(w_+, w_-) = \Big(\frac{1}{2}\Lambda_m^{-1}(w_+ - w_-)\Big)^{n+1}. \tag{2.2}$$

Setting $u = w_+$, we see that the non-linearity in (2.1) may be written as a sum for $p = 0, \dots, n+1$ of expressions $M(u, \dots, u, \bar{u}, \dots, \bar{u})$ with p arguments u and $n+1-p$ arguments $\bar{u}$, $M(u_1, \dots, u_{n+1})$ being $(n+1)$-linear. Consider as a model the case of the non-linearity given by just one of these contributions, namely the equation

$$(D_t - \Lambda_m)u = M(\underbrace{u, \dots, u}_{p}, \underbrace{\bar{u}, \dots, \bar{u}}_{n+1-p}) \tag{2.3}$$

for some $0 \le p \le n+1$. The basic idea of normal forms is to find another multilinear map $\tilde{M}(u_1, \dots, u_{p+1})$, chosen in such a way that if $v = u - \tilde{M}(u, \dots, u, \bar{u}, \dots, \bar{u})$, then

(2.4) $u \to v$ is a local diffeomorphism at zero in H^s (for $s \gg 1$),

(2.5) v solves an equation $(D_t - \Lambda_m)v = O\big(\|v\|_{H^s}^{n'+1}\big)$

for some $n' > n$. If (2.4), (2.5) hold, one will obtain an energy inequality

$$\|v(t,\cdot)\|_{H^s}^2 \leq \|v(0,\cdot)\|_{H^s}^2 + C\int_0^t \|v(\tau,\cdot)\|_{H^s}^{n'+2}\,d\tau \tag{2.6}$$

so that, by the same reasoning as in the introduction, one will get that v exists and stays of norm $O(\epsilon)$ in H^s over an interval of length $c\epsilon^{-n'}$. By (2.4), the solution u of (2.3) satisfies as well such a bound. Since $n' > n$, one has thus obtained an improved lower bound for the time of existence.

The case of a quadratic non-linearity on the circle

Let us explain how an $\tilde{M}$ such that (2.4), (2.5) hold may be constructed in the case when the non-linearity M in (2.3) is quadratic and $X = \mathbb{S}^1$ the unit circle, following essentially Bourgain [6] (See also the papers of Bambusi [1] and Bambusi-Grébert [3]). For k in $\mathbb{N}$, denote by Π_k the spectral projector associated to the eigenvalue k of $\sqrt{-\Delta}$ i.e. if u is in $L^2(\mathbb{S}^1,\mathbb{C})$, set

$$\Pi_0 u = \frac{\hat{u}(0)}{\sqrt{2\pi}},\quad \Pi_k u = \hat{u}(k)\frac{e^{ikx}}{\sqrt{2\pi}} + \hat{u}(-k)\frac{e^{-ikx}}{\sqrt{2\pi}} \tag{2.7}$$

where $\hat{u}(\xi) = \frac{1}{\sqrt{2\pi}}\int_{-\pi}^{\pi} e^{-ix\xi}u(x)\,dx$ for ξ in $\mathbb{Z}$. We have thus

$$\mathrm{Id} = \sum_{k=0}^{+\infty}\Pi_k. \tag{2.8}$$

Let u be a solution of (2.3) and let us look for a multilinear expression

$$(u_1,\dots,u_{n+1}) \to \tilde{M}(u_1,\dots,u_{n+1}) \tag{2.9}$$

such that, if we set

$$v = u - \tilde{M}(\underbrace{u,\dots,u}_{p},\underbrace{\bar{u},\dots,\bar{u}}_{n+1-p}), \tag{2.10}$$

we obtain for v an equation of the form (2.5). Compute

$$\begin{aligned}(D_t - \Lambda_m)v = (D_t - \Lambda_m)u - \sum_{j=1}^{p}\tilde{M}(u,\dots,D_t u,\dots,u,\bar{u},\dots,\bar{u})\\ + \sum_{j=p+1}^{n+1}\tilde{M}(u,\dots,u,\bar{u},\dots,\overline{D_t u},\dots,\bar{u}) + \Lambda_m\tilde{M}(u,\dots,u,\bar{u},\dots,\bar{u}).\end{aligned} \tag{2.11}$$

Using equation (2.3), we may rewrite the right hand side of (2.11) as a term homogeneous of degree $n+1$, namely

$$\tag{2.12} M(u,\dots,u,\bar u,\dots,\bar u) - \sum_{j=1}^{p} \tilde M(u,\dots,\Lambda_m u,\dots,u,\bar u,\dots,\bar u) \\ + \sum_{j=p+1}^{n+1} \tilde M(u,\dots,u,\bar u,\dots,\Lambda_m \bar u,\dots,\bar u) + \Lambda_m \tilde M(u,\dots,u,\bar u,\dots,\bar u)$$

and contributions homogeneous of degree $n' = 2n+1 > n+1$, that will eventually be $O(\|u\|_{H^s}^{n'+1})$ in H^s if we get good enough bounds on $\tilde M$. We thus have to find $\tilde M$ so that (2.12) vanishes, in other words, we need to solve the equation between $n+1$-linear maps

$$\tag{2.13} \sum_{j=1}^{p} \tilde M(u_1,\dots,\Lambda_m u_j,\dots,u_{n+1}) - \sum_{j=p+1}^{n+1} \tilde M(u_1,\dots,\Lambda_m u_j,\dots,u_{n+1}) \\ - \Lambda_m \tilde M(u_1,\dots,u_{n+1}) = M(u_1,\dots,u_{n+1}).$$

In (2.13), replace u_j by $\Pi_{k_j} u_j$ for $j = 1,\dots,n+1$, where k_j is in $\mathbb{N}$. Then, by definition of Π_{k_j}, $\Lambda_m \Pi_{k_j} u_j = \sqrt{m^2 + k_j^2}\,\Pi_{k_j} u_j$, so that, if we make act also on the left on (2.13) another spectral projector $\Pi_{k_{n+2}}$, we reduce (2.13) to

$$\tag{2.14} \mathcal{G}_m^p(k_1,\dots,k_{n+2})\Pi_{k_{n+2}} \tilde M(\Pi_{k_1}u_1,\dots,\Pi_{k_{n+1}}u_{n+1}) \\ = \Pi_{k_{n+2}} M(\Pi_{k_1}u_1,\dots,\Pi_{k_{n+1}}u_{n+1})$$

where

$$\tag{2.15} \mathcal{G}_m^p(k_1,\dots,k_{n+2}) = \sum_{j=1}^{p} \sqrt{m^2+k_j^2} - \sum_{j=p+1}^{n+2} \sqrt{m^2+k_j^2}.$$

The construction of $\tilde M$ is thus reduced to the division problem (2.14) and so to the fact that one may obtain for (2.15) lower bounds that are good enough so that (2.4) will hold true.

Let us consider first the case $n = 1, p = 2$, which corresponds to (2.3) with a quadratic non-linearity of the form $M(u,u)$. Notice also that by the definition (2.7) of Π_k and the fact that $e^{ik_1x}e^{ik_2x} = e^{i(k_1+k_2)x}$, both sides of (2.14) vanish identically unless for some choice of $\epsilon_j \in \{-1,1\}$

$$\tag{2.16} \sum_{j=1}^{n+2} \epsilon_j k_j = 0,$$

so that we have only to get lower bounds for (2.15) under the restriction (2.16). In the case $n = 1$, $p = 2$, we have

$$\begin{aligned} (2.17) \qquad |\mathcal{G}_m^2(k_1,k_2,k_3)| &= \left|\sqrt{m^2+k_1^2}+\sqrt{m^2+k_2^2}-\sqrt{m^2+k_3^2}\right| \\ &\geq c\big(1+\min(k_1,k_2,k_3)\big)^{-1} \end{aligned}$$

under the restriction (2.16), as follows from elementary computations. Consequently, we may define $\tilde{M}$ from (2.14) dividing by $\mathcal{G}_m^2$, and we get a bound

$$\begin{aligned} (2.18) \quad & \|\Pi_{k_3}\tilde{M}(\Pi_{k_1}u_1,\Pi_{k_2}u_2)\|_{L^2} \\ & \leq C\big(1+\min(k_1,k_2,k_3)\big)\|\Pi_{k_3}M(\Pi_{k_1}u_1,\Pi_{k_2}u_2)\|_{L^2}. \end{aligned}$$

Assume that M satisfies

$$(2.19) \quad \|\Pi_{k_3}M(\Pi_{k_1}u_1,\Pi_{k_2}u_2)\|_{L^2} \leq C\big(1+\min(k_1,k_2,k_3)\big)^{\nu}\prod_{j=1}^{2}\|u_j\|_{L^2}$$

for some ν (This is the case if $M(u_1,u_2) = u_1u_2$ for instance, with $\nu = 0$). Then a similar estimate is satisfied by $\tilde{M}$ according to (2.18), with ν replaced by $\nu+1$, and one easily deduces from this and from condition (2.16) that $\tilde{M}$ is bounded from $H^s \times H^s$ to H^s, if s is large enough relatively to ν. It thus follows that (2.4), (2.5) hold with $n' = 2$ in the case of a quadratic non-linearity, so that an equation like (2.3) with right hand side u^2 and initial data of size ϵ in $H^s(\mathbb{S}^1)$, with s large enough, has a solution defined on an interval of length at least $c\epsilon^{-2}$. The same holds for the non-linearities $\bar{u}^2$ and $|u|^2$.

We would like to obtain a similar improvement of the time of existence when the right hand side of (2.3) is homogeneous of degree $n+1$, with $n \geq 2$. We find immediately a difficulty: if $n \geq 2$, expression (2.15) for arguments $(k_1,\dots,k_{n+2}) \in \mathbb{N}^{n+2}$ satisfying (2.16) may vanish for some values of m. The idea is thus to exclude some values of m, i.e. to show that there is a subset $\mathcal{N}$ of $]0,+\infty[$ of zero measure such that, if m is in $]0,+\infty[-\mathcal{N}$, there are $c_0 > 0$, $N_0 \in \mathbb{N}$, such that

$$(2.20) \qquad |\mathcal{G}_m^p(k_1,\dots,k_{n+2})| \geq c\big(1+\max\nolimits_3(k_1,\dots,k_{n+2})\big)^{-N_0}$$

for any $(k_1,\dots,k_{n+2})$ satisfying (2.16), $\max_3(k_1,\dots,k_{n+2})$ being the third largest among these integers. If an inequality of the form (2.20) holds, one may repeat the above reasoning, and construct from (2.14) an $\tilde{M}$ such that (2.4), (2.5) are satisfied. One then extends the solution of (2.3) at least over a time interval of length $c\epsilon^{-2n}$.

We shall see below that one may obtain an estimate of the form (2.20) when n is odd (i.e. for non-linearities homogeneous of even degree) if $X = \mathbb{S}^1$ or more generally a Zoll manifold. On the other hand, if n is even, there are some k_j's for which the left hand side of (2.20) vanishes identically (when in (2.15), $p = \frac{n}{2} + 1$ and $\{k_1, \ldots, k_p\} = \{k_{p+1}, \ldots, k_{n+2}\}$). We shall have to identify then a special structure of the non-linearity that ensures that the corresponding contributions – that cannot be eliminated – do not make grow the Sobolev energy.

In the case of a general manifold, (2.20) does not hold true in general. Nevertheless, we shall see that in this case, a weaker property may be obtained. For equations like (1.1), this property will be sufficient to perform a (partial) normal form and obtain a slight improvement of the time of existence. We shall describe that in next section.

§3. Semi-linear equations on a compact manifold

We consider in this section (X, g) a compact Riemannian manifold. We still denote by $-\Delta$ the positive Laplace operator on X. Our main result asserts that, for solutions of an equation of the form (1.1), we may find a lower bound for the time of existence in $c\epsilon^{-An}$, for some constant $A > 1$. More precisely, we have the following statement, obtained in joint work with Rafik Imekraz [11] (see also [8, 14] in the case of the torus):

Theorem 1. *Let (X, g) be a compact Riemannian manifold. There is $\mathcal{N} \subset]0, +\infty[$, a subset of zero measure, and for any m in $]0, +\infty[-\mathcal{N}$, there are $A > 1$, $s_0 > 0$, and for any $s > s_0$, there are $c > 0$, $K > 0$, $\epsilon_0 > 0$, such that, for any (w_0, w_1) in the unit ball for $H^{s+1}(X, \mathbb{R}) \times H^s(X, \mathbb{R})$, any $\epsilon \in]0, \epsilon_0[$, equation (1.1) with data $w|_{t=0} = \epsilon w_0$, $\partial_t w|_{t=0} = \epsilon w_1$, has a unique solution w in $C^0(]-T_\epsilon, T_\epsilon[, H^{s+1}) \cap C^1(]-T_\epsilon, T_\epsilon[, H^s)$ with $T_\epsilon \geq c\epsilon^{-An}$. Moreover for any $t \in]-T_\epsilon, T_\epsilon[$, one has*

$$\|w(t, \cdot)\|_{H^{s+1}} + \|\partial_t w(t, \cdot)\|_{H^s} \leq K\epsilon. \tag{3.1}$$

The idea of proof will be to try to adapt the normal forms method to the general situation of compact manifolds. Several difficulties are in order. The first one is that we cannot expect to use a decomposition of the identity like (2.8), in terms of the spectral projectors associated to the individual eigenvalues. Actually, in order to get lower bounds for (2.15) (with k_j replaced by some eigenvalue ξ_{k_j}), we need to have some lower estimate on the gap between consecutive distinct eigenvalues. But on a general manifold, one has no control on such a gap. Our idea will be to replace in (2.8) spectral projectors associated to *one* eigenvalue by spectral projectors on an interval of the spectrum, showing that we may obtain a convenient lower bound for the gap between two consecutive intervals. More precisely, we have the following proposition:

Proposition 2. *Denote by $(\mu_j)_{j\in\mathbb{N}}$ the increasing sequence of distinct eigenvalues of $\sqrt{-\Delta}$ on X. There are a constant $\alpha \geq 1$ and a strictly increasing sequence of integers $(j_k)_{k\in\mathbb{N}}$, such that the following properties hold for the intervals $I_0 = [0, \mu_{j_0}]$ and $I_k = [\mu_{j_{k-1}+1}, \mu_{j_k}]$ $(k \geq 1)$, when k goes to infinity:*

$$
\begin{aligned}
&\mathrm{dist}\,(I_k, 0) = \mu_{j_{k-1}+1} = 3\alpha k + O(1)\\
&\mathrm{Length}\,(I_k) = \mu_{j_k} - \mu_{j_{k-1}+1} = O(1)\\
&\mathrm{dist}\,(I_k, I_{k+1}) = \mu_{j_k+1} - \mu_{j_k} \geq c\mu_{j_k}^{-(d-1)} \sim ck^{-(d-1)}\\
&\mathrm{Card}\,(I_k) \cap \mathrm{Spect}\,(\sqrt{-\Delta}) = O(k^{d-1}).
\end{aligned}
\tag{3.2}
$$

The geometric situation described by the above proposition may be illustrated by the following picture:

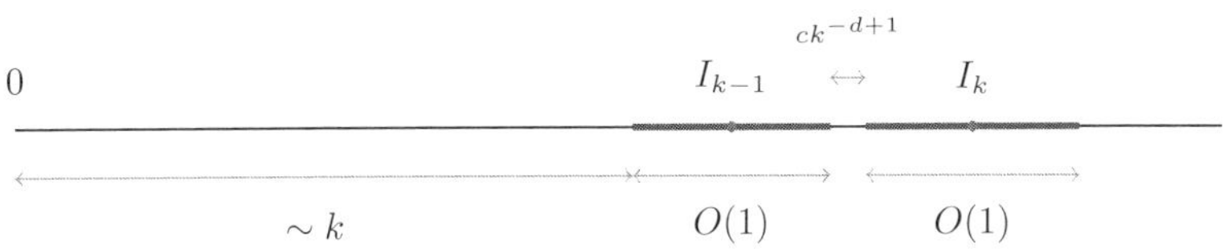

Fig. 1. Spectrum of $\sqrt{-\Delta}$

We refer to [11] for the proof of Proposition 2. The fact that it holds on any compact manifold is a consequence of the Weyl formula, asserting that if $(\lambda_j)_j$ denotes the sequence of eigenvalues of $\sqrt{-\Delta}$ counted with multiplicities, then

$$\mathrm{Card}\,\{j \in \mathbb{N}; \lambda_j \leq \lambda\} = c(d)\mathrm{Vol}\,(X)\lambda^d + O(\lambda^{d-1}),\ \lambda \to +\infty,$$

for some constant $c(d)$ depending only on the dimension. One deduces from that that there are constants $\alpha \geq 1$, $C > 0$, such that, for any k in $\mathbb{N}$

$$C^{-1}(k+1)^{d-1} \leq \mathrm{Card}\,\big\{j \in \mathbb{N}; \lambda_j \in]\alpha k, \alpha(k+1)]\big\} \leq C(k+1)^{d-1}.$$

One then uses this to construct an increasing sequence $(j_k)_k$ of integers such that the eigenvalues $(\mu_j)_j$ of $\sqrt{-\Delta}$, counted without multiplicities, satisfy, for this subsequence of integers,

$$3\alpha k \le \mu_{j_k} \le 3\alpha(k+1),\ \mu_{j_k+1} - \mu_{j_k} \ge \frac{1}{C\mu_{j_k}^{d-1}},$$

which gives the proposition.

We define now

$$\Pi_k = \mathbf{1}_{I_k}(\sqrt{-\Delta}) \tag{3.3}$$

so that $\mathrm{Id} = \sum_{k=0}^{+\infty} \Pi_k$. We would like to prove Theorem 1 using the same ideas as in Section 2. We have seen there, in the case of equation (1.1) on the circle, that we could perform a normal forms method as soon as we have a lower bound on small divisors of the form (2.17) or (2.20). We would thus need here a control of a similar quantity. It turns out that, for a general compact manifold, we cannot expect as good a lower bound as in (2.20). We shall be able to obtain only the following weaker statement on small divisors.

Proposition 3. *There is a subset $\mathcal{N}$ of $]0,+\infty[$ and for every m in $]0,+\infty[-\mathcal{N}$, there are $c>0$, $N_0 \in \mathbb{N}$, such that, for any $p \in [0, n+2]$, any $(k_1,\dots,k_{n+2}) \in \mathbb{N}^{n+2}$, any $(\xi_{k_1},\dots,\xi_{k_{n+2}}) \in \prod_{j=1}^{n+2}(I_{k_j} \cap \mathrm{Spec}\sqrt{-\Delta})$, one has*

$$\left|\sum_{j=1}^{p}\sqrt{m^2+\xi_{k_j}^2} - \sum_{j=p+1}^{n+2}\sqrt{m^2+\xi_{k_j}^2}\right| \ge c\big(1+\max(\xi_{k_1},\dots,\xi_{k_{n+2}})\big)^{-N_0} \tag{3.4}$$

except if n is even, $p = \frac{n}{2}+1$ and

$$\{k_1,\dots,k_p\} = \{k_{p+1},\dots,k_{n+2}\}. \tag{3.5}$$

The proof of the proposition follows from [12]. The idea is the following. For fixed $(k_1,\dots,k_{n+2})$ that do not satisfy (3.5) when n is even and $p = \frac{n}{2}+1$, the analytic function

$$m \to \mathcal{G}_m^p(\xi_{k_1},\dots,\xi_{k_{n+2}}) = \sum_{j=1}^{p}\sqrt{m^2+\xi_{k_j}^2} - \sum_{j=p+1}^{n+2}\sqrt{m^2+\xi_{k_j}^2} \tag{3.6}$$

is not identically zero. Consequently, it has on any compact interval $[a,b] \subset [0,+\infty[$ only finitely many zeros, and these zeros have bounded order. It follows (see next picture) that there are $\rho>0, C>0$, such that for any $\alpha \in]0,1[$,

$$\mathrm{meas}\left\{m \in [a,b];\, |\mathcal{G}_m^p(\xi_{k_1},\dots,\xi_{k_{n+2}})| < \alpha\right\} \le C\alpha^\rho. \tag{3.7}$$

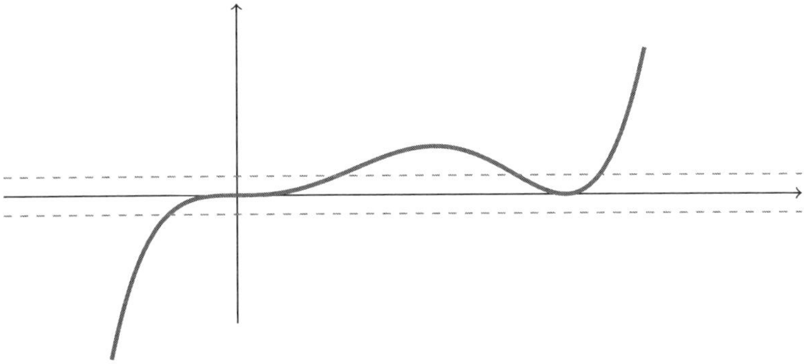

Of course, C and ρ in (3.7) may depend on the k_j's and the key of the proof of Proposition 3 is to show that ρ may be taken independent of $(k_1, \dots, k_{n+2})$ and $C = O(\max(k_1, \dots, k_{n+2})^{N_1})$ for some N_1. Taking $N_0 \gg N_1$ and α of the form $\delta(1 + \max(k_1, \dots, k_{n+2}))^{-N_0}$, one deduces from (3.7) that (3.4) holds true, except if m is in a subset of measure $O(\delta^\rho)$.

Idea of proof of Theorem 1. Let us explain the idea of the proof on a model of the form (2.3):

$$(3.8) \qquad \begin{aligned} (D_t - \Lambda_m)u &= (\Lambda_m^{-1}u)^p(\Lambda_m^{-1}\bar{u})^{n+1-p} \\ &= M\big(\underbrace{\Lambda_m^{-1}u, \dots, \Lambda_m^{-1}u}_{p}, \underbrace{\Lambda_m^{-1}\bar{u}, \dots, \Lambda_m^{-1}\bar{u}}_{n+1-p}\big). \end{aligned}$$

An important feature, that we shall use below, is the presence of the operator of degree -1 Λ_m^{-1} acting on each factor in the right hand side of (3.8). Assume also to simplify the argument that n is odd, and let us also pretend (which is of course not true in general!) that for any k, $I_k \cap \operatorname{Spec}(\sqrt{-\Delta})$ is reduced to a single (multiple) eigenvalue ξ_k. The general case may be obtained by a technical modification of the arguments we outline below. Define also the Sobolev norm on the manifold X by

$$(3.9) \qquad \Theta_s^0(u) = \|u\|_{H^s}^2 = \sum_{k=0}^{+\infty}(k+1)^{2s}\|\Pi_k u\|_{L^2}^2$$

with the spectral projectors Π_k defined in (3.3). Using the equation, a direct computation shows that, for s large enough,

$$(3.10) \qquad \frac{d}{dt}\Theta_s^0(u(t,\cdot)) = O\big(\|u(t,\cdot)\|_{H^s}^{n+2}\big),$$

so that one gets an estimate of the form (1.5) for $t \geq 0$:

$$\Theta_s^0(u(t,\cdot)) \leq \Theta_s^0(u(0,\cdot)) + C\int_0^t \|u(\tau,\cdot)\|_{H^s}^n \Theta_s^0(u(\tau,\cdot))\,d\tau. \tag{3.11}$$

This implies, as in (1.6), that a bound $\Theta_s^0(u(t,\cdot)) = O(\epsilon^2)$ holds true over an interval of the form $[0, c\epsilon^{-n}]$, if the initial data is of size ϵ in H^s. In order to prove the theorem, we need to improve the power n in the integral in the right hand side of (3.11) to An for some $A > 1$. Let us notice first that in estimate (3.10), we did not use the presence of order -1 operators in the right hand side of (3.8). Actually, we have

$$\frac{d}{dt}\Theta_s^0(u(t,\cdot)) = 2\mathrm{Re}\,\langle u, M(\Lambda_m^{-1}u,\ldots,\Lambda_m^{-1}\bar{u})\rangle_{H^s}. \tag{3.12}$$

In the right hand side of (3.12), let us decompose

$$u = u_L + u_H,\ u_L = \sum_{k;k<\epsilon^{-\delta}} \Pi_k u \tag{3.13}$$

for some small $\delta > 0$ to be chosen. Consider first the contribution to the right hand side of (3.12) of $M(\Lambda_m^{-1}u_1,\ldots,\Lambda_m^{-1}u_{n+1})$ where at least one of the u_j's has been replaced by u_H. Since

$$\|\Lambda_m^{-1}u_H\|_{H^s} \leq C\epsilon^{\delta}\|u\|_{H^s}, \tag{3.14}$$

we shall get from (3.12)

$$\frac{d}{dt}\Theta_s^0(u(t,\cdot)) = 2\mathrm{Re}\,\langle u, M(\Lambda_m^{-1}u_L,\ldots,\Lambda_m^{-1}\bar{u}_L)\rangle_{H^s} + O\big(\epsilon^{\delta}\|u\|_{H^s}^{n+2}\big) \tag{3.15}$$

and since ϵ is the size of the initial data, the last term in (3.15) is essentially of the form $O(\|u\|_{H^s}^{n'+2})$ with $n' = n+\delta > n$. Consequently, if we ignore the first term in the right hand side of (3.15), we would get a bound of the form (3.11) with the power n replaced by n', whence a lower estimate for the time of existence in $c\epsilon^{-n'}$. In order to get the conclusion of the theorem, we have thus to show that we may get rid of the first term in the right hand side of (3.15) by normal forms. To do so, we shall look for a perturbation $\tilde{\Theta}_s(u)$ of $\Theta_s^0(u)$ such that

$$\Theta_s^0(u) - \tilde{\Theta}_s(u) \sim \Theta_s^0(u) \text{ for small } u \tag{3.16}$$

and

$$\frac{d}{dt}\big(\Theta_s^0(u(t,\cdot)) - \tilde{\Theta}_s(u(t,\cdot))\big) = O\big(\epsilon^{\delta'}\|u(\tau,\cdot)\|^{n+2}\big) \tag{3.17}$$

for some $\delta' > 0$. If we may achieve (3.16) and (3.17), we shall get by integration for $t \geq 0$

$$\begin{aligned}\|u(t,\cdot)\|_{H^s}^2 &= \Theta_s^0(u(t,\cdot)) \sim \Theta_s^0(u(t,\cdot)) - \tilde{\Theta}_s(u(t,\cdot)) \\ &\leq C\left[\|u(0,\cdot)\|_{H^s}^2 + \epsilon^{\delta'}\int_0^t \|u(\tau,\cdot)\|_{H^s}^{n+2}\,d\tau\right]\end{aligned} \tag{3.18}$$

which implies existence of the solution up to time of magnitude $\epsilon^{-(n+\delta')}$ and the conclusion.

In view of (3.15), property (3.17) will hold true if we construct $\tilde{\Theta}_s$ so that

(3.19)
$$\frac{d}{dt}\tilde{\Theta}_s(u(t,\cdot)) = 2\mathrm{Re}\,\langle u, M(\Lambda_m^{-1}u_L,\dots,\Lambda_m^{-1}\bar{u}_L)\rangle_{H^s} + O\big(\epsilon^{\delta'}\|u\|_{H^s}^{n+2}\big).$$

We look for $\tilde{\Theta}_s$ as

$$\begin{aligned}\tilde{\Theta}_s(u) = -2\mathrm{Re}\sum_{k_1}\cdots\sum_{k_{n+2}}(k_{n+2}+1)^{2s} i\mathcal{G}_m^p(\xi_{k_1},\dots,\xi_{k_{n+2}})^{-1} \\ \times\int \Pi_{k_{n+2}}\bar{u}M(\Lambda_m^{-1}\Pi_{k_1}u_L,\dots,\Lambda_m^{-1}\Pi_{k_{n+1}}\bar{u}_L)\,dx,\end{aligned} \tag{3.20}$$

where ξ_{k_j} is the unique point in $I_{k_j}\cap\mathrm{Spec}\,(\sqrt{-\Delta})$ and $\mathcal{G}_m^p$ is given by (3.6). We notice first that (3.16) will hold true if δ in (3.13) is chosen small enough. Actually, by (3.4) and the fact that all frequencies ξ_{k_j} are smaller than $\epsilon^{-\delta}$ because of the definition of u_L, $|(\mathcal{G}_m^p)^{-1}| \leq C\epsilon^{-\delta N_0}$, so that $\tilde{\Theta}_s(u) = O(\epsilon^{-\delta N_0}\|u\|_{H^s}^{n+2}) = O(\epsilon^{n-\delta N_0}\Theta_s^0(u))$ as long as $\|u\|_{H^s} = O(\epsilon)$. If $\delta N_0 \ll n$, we do get a small perturbation to $\Theta_s^0(u)$.

On the other hand, when we compute $\frac{d}{dt}\tilde{\Theta}_s(u(t,\cdot))$ and make act the time derivative on each argument in the right hand side of (3.20), we get using equation (3.8):

• Terms where $\frac{du}{dt}$ is replaced by $i\Lambda_m u$. As $\Lambda_m\Pi_{k_j} = \sqrt{m^2+\xi_{k_j}^2}\,\Pi_{k_j}$ we obtain, because of the definition of $\mathcal{G}_m^p$, a contribution equal to the first term in the right hand side of (3.19).

• Terms where $\frac{du}{dt}$ is replaced by the non-linearity in (3.8). These terms are bounded, using that $|\mathcal{G}_m^p|^{-1} \leq C\epsilon^{-\delta N_0}$ and that $\|u\|_{H^s} = O(\epsilon)$, by

$$C\|u\|_{H^s}^{2n+2}\epsilon^{-\delta N_0} = O\big(\epsilon^{n-\delta N_0}\|u\|_{H^s}^{n+2}\big),$$

so that they contribute to the remainder in (3.19), with $\delta' = n - \delta N_0 > 0$, if δ is small enough. This concludes the proof. Q.E.D.

Remark.

• The proof we outlined above works for n odd. When n is even and $p = \frac{n}{2} + 1$, the lower bound (3.4) does not hold for indices such that $\{k_1, \ldots, k_p\} = \{k_{p+1}, \ldots, k_{n+2}\}$. One has then to check that these modes give a zero contribution when one computes (3.15). We shall come back to this issue in next section.

• Clearly, the proof of Theorem 1 uses in an essential way the fact that in equation (1.1), the non-linearity does not contain derivatives of w. This is what makes appear the operator Λ_m^{-1} in the non-linearity of (3.8), these operators playing a seminal role in the proof of Theorem 1. In particular, the proof does not extend to non-linearities of the form $(\partial_x u)^{n+1}$ and even less to quasi-linear non-linearities, like $u^n(\partial_x^2 u)$.

Nevertheless, let us mention a very recent result of Ionescu and Pusateri [16] of long time existence for solutions of capillary-gravity wave equations on $\mathbb{T}^2$, for almost all values of the gravity-surface tension parameters. This equations is quasi-linear (actually even fully nonlinear). Though, it is possible to apply a normal form procedure in order to obtain a lower bound for the time of existence in $\epsilon^{-\frac{5}{3}}$, instead of the ϵ^{-1} that would be given by local existence theory. The key point is that the structure of the non-linearity allows one to gain derivatives, when eliminating quadratic contributions to the non-linearity that are spectrally localized in phase space regions giving rise to small divisors.

§4. Almost global existence on Zoll manifolds

In the preceding section, Theorem 1 provided, for an equation of the form (1.1), a slight improvement of the time of existence of solutions with data of size ϵ, through a normal forms method that was replacing a non-linearity vanishing at some order at zero by another one, vanishing at an higher order. A natural idea, in order to get a better improvement of the time of existence, and eventually reaching "almost global existence", i.e. existence over a time interval of length $c_N\epsilon^{-N}$ for any N, would be to iterate the normal forms construction, in order to reduce oneself to higher and higher order vanishing non-linearities. Clearly, one cannot expect to do so in the general framework of Section 3: we remarked that we used at the first normal forms step the fact that the non-linearity in equation (3.8) involves the action of operators of degree -1 on each factor, and this property is lost on the higher order contributions generated by the reduction. Nevertheless, in dimension one, Bourgain [6] constructed iterated normal forms that allow one to get almost global existence of solutions. Our goal here is to obtain similar results in higher dimensions. We shall be able to do so only on a very restricted class of compact manifolds. Actually, in one space dimension, one uses several special facts: first, the property that the product of two eigenfunctions of the Laplace operator on the circle is still an eigenfunction: $e^{ikx}e^{ik'x} = e^{i(k+k')x}$. It turns out that this may be replaced by an inequality that holds true on any compact manifold, so that this property is not essential for the proof of long time existence. Second, one uses an estimate of the form (2.20) that gives a control of small divisors in terms of a negative power of the *third* largest of their arguments. On a general manifold, we have only the lower bound (3.4), that involves a negative power of the largest frequency. This bound would provoke losses of derivatives in a normal forms procedure, and thus prevents one from using the method in general. We shall thus limit ourselves to the case of manifolds whose spectrum has special separation properties, so that a lower bound of type (2.20) will hold true. This is the class of Zoll manifolds, whose definition is recalled below.

Definition 4. *A Zoll manifold is a compact Riemannian manifold* (X, g) *on which the geodesic flow is periodic.*

Spheres are obviously Zoll manifolds, but that class contains other examples (see Besse [5]). For our purposes, the key point will be a characterization of Zoll manifolds in terms of their spectral properties due to Weinstein [22], Colin de Verdière [7] and Guillemin [15].

Theorem 5. *Let* X *be a compact Riemannian manifold. There is equivalence between:*

(i) X *is Zoll.*

(ii) *There are* $\tau > 0$ *(the period of the geodesic flow),* $\delta > 0$, $c_0 > 0$, $\alpha \in \mathbb{R}$ *such that* $\operatorname{Spec}(\sqrt{-\Delta}) \subset \bigcup_{n=0}^{+\infty} I_k$, *where the union is disjoint,* I_0 *is some compact interval and for* $k \geq 1$,

$$I_k = \left[\frac{2\pi}{\tau}k + \alpha - \frac{c_0}{k^\delta}, \frac{2\pi}{\tau}k + \alpha + \frac{c_0}{k^\delta}\right]. \tag{4.1}$$

Let us illustrate statement (ii) on a picture:

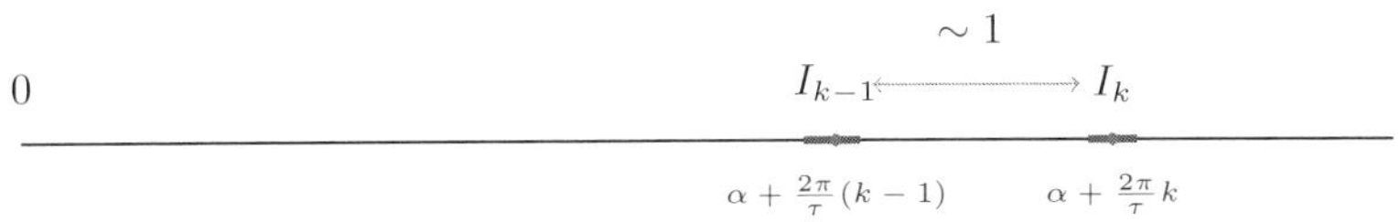

Fig. 2. Spectrum of $\sqrt{-\Delta}$ on a Zoll manifold

The spectrum of $\sqrt{-\Delta}$ is contained in the union of intervals, centered at an arithmetic progression, whose length goes to zero as the energy increases, so that two consecutive I_k's are separated by a gap bounded from below by a constant. Notice that this is in sharp contrast with the situation described by the figure on page 9, for a general manifold, where we could just include the spectrum in intervals separated by a shrinking gap. We may exploit this separation property in order to prove the following:

Proposition 6. *Assume X is Zoll. There is $\mathcal{N} \subset]0,+\infty[$ of zero measure such that, for any $n \in \mathbb{N}^*$, any $m \in]0,+\infty[-\mathcal{N}$, there are $c > 0$, $N_0 \in \mathbb{N}$, so that, for any $k_1,\dots,k_{n+2}$ in $\mathbb{N}$, for any $\xi_{k_j} \in \operatorname{Spec}(\sqrt{-\Delta}) \cap I_{k_j}$, any $0 \le p \le n+2$, one has, using notation* (3.6)

$$|\mathcal{G}^p_m(\xi_{k_1},\dots,\xi_{k_{n+2}})| \ge c\big(1+\max{}_3(k_1,\dots,k_{n+2})\big)^{-N_0} \tag{4.2}$$

($\max_3(k_1,\dots,k_{n+2})$ *standing for the third largest among* $(k_1,\dots,k_{n+2})$), *unless*

$$n \text{ is even, } p = \frac{n+2}{2} \text{ and } \{k_1,\dots,k_p\} = \{k_{p+1},\dots,k_{n+2}\}. \tag{4.3}$$

Idea of proof. By Proposition 3, we know that a lower bound for the left hand side of (4.2) in terms of $c(1+\max(k_1,\dots,k_{n+2}))^{-N_0}$ holds true, since $\xi_{k_j} \sim k_j$ for any ξ_{k_j} in interval I_{k_j}. To get the better lower bound (4.2), let us assume for instance that $k_1, k_{n+2} \gg \max(k_2,\dots,k_{n+1})$ and write (when $1 < p < n+2$, which is the only serious case),

(4.4)
$$\mathcal{G}^p_m(\xi_{k_1},\dots,\xi_{k_{n+2}}) = \sqrt{m^2+\xi_{k_1}^2} - \sqrt{m^2+\xi_{k_{n+2}}^2} + \tilde{\mathcal{G}}^p_m(\xi_{k_2},\dots,\xi_{k_{n+1}})$$

with

$$\tilde{\mathcal{G}}^p_m(\xi_{k_2},\dots,\xi_{k_{n+1}}) = \sum_{j=2}^{p} \sqrt{m^2+\xi_{k_j}^2} - \sum_{j=p+1}^{n+1} \sqrt{m^2+\xi_{k_j}^2}. \tag{4.5}$$

Since $\xi_{k_j} \in I_{k_j}$, interval centered at $\alpha + \frac{2\pi}{\tau}k_j$ according to (4.1), and since the length of these intervals is $O(k_j^{-\delta})$, we may write

$$\begin{aligned}\sqrt{m^2+\xi_{k_1}^2} - \sqrt{m^2+\xi_{k_{n+2}}^2} &= \Big(\alpha + \frac{2\pi}{\tau}k_1\Big) - \Big(\alpha + \frac{2\pi}{\tau}k_{n+2}\Big)\\ &\quad + O(k_1^{-\delta}) + O(k_{n+2}^{-\delta})\end{aligned} \tag{4.6}$$

from which we can conclude

(4.7)
$$\mathcal{G}^p_m(\xi_{k_1},\dots,\xi_{k_{n+2}}) = \frac{2\pi}{\tau}q + \tilde{\mathcal{G}}^p_m(\xi_{k_2},\dots,\xi_{k_{n+1}}) + O(k_1^{-\delta}) + O(k_{n+2}^{-\delta})$$

where q belongs to $\mathbb{Z}$. Because of that, and of the fact that the arguments of $\tilde{\mathcal{G}}_m^p$ are all $O(\max_3(k_1,\dots,k_{n+2}))$, a modification of the proof of Proposition 3 shows that, for any q in $\mathbb{Z}$, any $\xi_{k_j} \in \operatorname{Spec}(\sqrt{-\Delta}) \cap I_{k_j}$, any m outside a subset $\mathcal{N}$ of zero measure,

$$\left|\frac{2\pi}{\tau}q + \tilde{\mathcal{G}}_m^p(\xi_{k_2},\dots,\xi_{k_{n+1}})\right| \geq c\big(1+\max_3(k_1,\dots,k_{n+2})\big)^{-N_0} \tag{4.8}$$

for some N_0. As we may assume that

$$k_1, k_{n+2} \geq \max_3(k_1,\dots,k_{n+2})^{N_1}$$

for a large enough N_1 (since otherwise (4.2) follows from (3.4)), one sees that the perturbations of the right hand side of (4.7) are small enough so that (4.8) implies (4.2). Q.E.D.

According to Proposition 6, we have an improved bound (4.2) for small divisors in the case of Zoll manifolds. This will be the key point that will allow us to prove the following theorem due to Bambusi, Delort, Grébert and Szeftel [2]:

Theorem 7. *Let X be a Zoll manifold. Let $f : (x,w) \to f(x,w)$ from $X \times \mathbb{R}$ to $\mathbb{R}$ be smooth, vanishing at order three at $w = 0$. There is $\mathcal{N} \subset]0,+\infty[$, of zero measure and, for any $m \in]0,+\infty[-\mathcal{N}$, any $N \in \mathbb{N}$, there is $s_0 > 0$ and for any $s > s_0$, there are $\epsilon_0 > 0$, $c > 0$, such that, for all $\epsilon \in]0,\epsilon_0[$, any (w_0,w_1) in the unit ball of $H^{s+\frac12}(X,\mathbb{R}) \times H^{s-\frac12}(X,\mathbb{R})$,*

$$\begin{aligned}(\partial_t^2 - \Delta + m^2)w &= \partial_w f(x,w)\\ w|_{t=0} &= \epsilon w_0\\ \partial_t w|_{t=0} &= \epsilon w_1\end{aligned} \tag{4.9}$$

has a unique solution

$$w \in C^0(]-T_\epsilon,T_\epsilon[, H^{s+\frac12}(X,\mathbb{R})) \cap C^1(]-T_\epsilon,T_\epsilon[, H^{s-\frac12}(X,\mathbb{R}))$$

with $T_\epsilon \geq c\epsilon^{-N}$. Moreover $(w(t,\cdot),\partial_t w(t,\cdot))$ is $O(\epsilon)$ in $H^{s+\frac12}(X,\mathbb{R}) \times H^{s-\frac12}(X,\mathbb{R})$ for t in $]-T_\epsilon,T_\epsilon[$.

The idea of proof of the above theorem will be to apply a sequence of normal forms, in order to reduce the non-linearity to terms of higher and higher order of homogeneity. It turns out that it is not possible to eliminate all contributions of lower degree: this is related to the fact that the small divisors estimate (4.2) can be achieved except when (4.3) holds. One has thus to check that the corresponding terms in the non-linearity do not make grow the Sobolev energy. For the first step in a normal forms procedure, this may be verified by hand. But along the iteration, the new nonlinear contributions induced by successive reductions become more and more complicated, so that the cancellation properties cannot be checked directly. Though, a way of achieving that in a systematic way is to put the problem in an Hamiltonian framework, in order to be able to use a Birkhoff normal forms method. Then terms corresponding to indices satisfying (4.3) are those contributions to the Hamiltonian that depend only on the actions, so that the fact that they do not make grow the energy becomes trivial. We introduce this Hamiltonian formulation in next section.

§5. Hamiltonian formulation and Birkhoff normal forms

Denote by $\langle V, V' \rangle = \int_X V \cdot V' \, dx$ the usual scalar product on the space $L^2(X, \mathbb{R}^2)$ (where dx stands for the Riemannian volume element). Define a symplectic form by

$$\omega(V, V') = \langle {}^t JV, V' \rangle \tag{5.1}$$

where $J = \begin{bmatrix} 0 & -1 \\ 1 & 0 \end{bmatrix}$. If $V \to G(V)$ is a C^1 function on an open subset of $H^s(X, \mathbb{R}^2)$, for some s, we denote by $\nabla G(V)$ the unique element of $H^{-s}(X, \mathbb{R}^2)$ such that, for any vector H in $H^s(X, \mathbb{R}^2)$,

$$DG(V) \cdot H = \langle \nabla G(V), H \rangle$$

and the symplectic gradient $X_G(V)$ by

$$X_G(V) = J \nabla G(V). \tag{5.2}$$

In the sequel, we shall always take s large enough and consider only functions G such that when V is in H^s, then $\nabla G(V)$ or $X_G(V)$ belongs to H^s (or H^{s-1}), like polynomials in V for instance. The Hamiltonian equation associated to G is

$$\begin{aligned} \partial_t V &= X_G(V) \\ V|_{t=0} &= V_0 \in H^s(X, \mathbb{R}^2). \end{aligned} \tag{5.3}$$

Let us write the Klein-Gordon equation (4.9) under the form (5.3). Recall that we set

$$\Lambda_m = \sqrt{-\Delta + m^2}$$

and if we are given a function of t in some interval $[0, T]$, with values in $H^{s+\frac{1}{2}}$, such that

$$w \in C^0\big([0, T], H^{s+\frac{1}{2}}(X, \mathbb{R})\big) \cap C^1\big([0, T], H^{s-\frac{1}{2}}(X, \mathbb{R})\big),$$

define

$$V = \begin{bmatrix} V^1 \\ V^2 \end{bmatrix} = \begin{bmatrix} \Lambda_m^{-\frac{1}{2}} \partial_t w \\ \Lambda_m^{\frac{1}{2}} w \end{bmatrix} \in C^0\big([0,T], H^s(X,\mathbb{R}^2)\big). \tag{5.4}$$

Let $(x,w) \to f(x,w)$ be a smooth function on $X \times \mathbb{R}$ with values in $\mathbb{R}$. Define

$$F(w) = \int_X f(x, w(x))\, dx. \tag{5.5}$$

This is a smooth function on $H^s(X,\mathbb{R})$ for s large enough, such that $\nabla F(w) = \frac{\partial f}{\partial w}(x, w(x))$ belongs to $H^s(X,\mathbb{R})$. Introduce

$$G_0(V) = \frac{1}{2}\int_X V \cdot (\Lambda_m V)\, dx,\ G(V) = G_0(V) - F\big(\Lambda_m^{-\frac{1}{2}} V^2\big) \tag{5.6}$$

using notation (5.4). Then the linear Klein-Gordon equation $(\partial_t^2 - \Delta + m^2)w = 0$ is equivalent to $\partial_t V = X_{G_0}(V)$ and equation (4.9) is equivalent to $\partial_t V = X_G(V)$. For simplification, assume that f is a polynomial in w independent of x. Then our system (4.9) may be written

$$\begin{aligned} \partial_t V &= X_G(V) \\ V|_{t=0} &= \epsilon V_0 \end{aligned} \tag{5.7}$$

with V_0 given in the unit ball of $H^s(X,\mathbb{R}^2)$ and $G(V)$ a polynomial $G(V) = \sum_{n=0}^{N'} G_n(V)$, with

$$\begin{aligned} G_0(V) &= \frac{1}{2}\int_X \big(\Lambda_m^{\frac{1}{2}} V\big) \cdot \big(\Lambda_m^{\frac{1}{2}} V\big)\, dx \\ G_n(V) &= \int_X P_n\big(\Lambda_m^{-\frac{1}{2}} V\big)\, dx \end{aligned} \tag{5.8}$$

where P_n is an homogeneous polynomial of order $n+2$.

Let us rewrite equation (5.7) in complex coordinates, introducing

$$u = \frac{\sqrt{2}}{2}\big(V^1 + iV^2\big) \tag{5.9}$$

and

$$\nabla_u = \frac{\sqrt{2}}{2}\big(\nabla_{V^1} - i\nabla_{V^2}\big),\ \nabla_{\bar{u}} = \frac{\sqrt{2}}{2}\big(\nabla_{V^1} + i\nabla_{V^2}\big). \tag{5.10}$$

Then equation (5.7) may be rewritten in terms of u as

$$\begin{aligned} \partial_t u &= i\nabla_{\bar{u}} G(u,\bar{u}) \\ u|_{t=0} &= \epsilon u_0 \end{aligned} \tag{5.11}$$

with $u_0 \in H^s(X,\mathbb{C})$ and where, if we set $U = (u,\bar{u})$, the Hamiltonian G may be rewritten from (5.8) as

$$G = \sum_{n=0}^{N'} G_n(U), \tag{5.12}$$

with

$$\begin{aligned} G_0(U) &= \int_X (\Lambda_m u)\bar{u}\,dx \\ G_n(U) &= \sum_{p=0}^{n+2} \int_X P_{n,p}(u,\dots,u,\bar{u},\dots,\bar{u}) \end{aligned} \tag{5.13}$$

where $P_{n,p}$ is a $(n+2)$-linear expression, depending on p arguments u and $n+2-p$ arguments $\bar{u}$. We did not make appear in the expression of $P_{n,p}$ the operators $\Lambda_m^{-\frac{1}{2}}$ as they will play no role.

The strategy of proof of Theorem 7 will be the same as the one described in above sections 1 and 2, except that we shall iterate the normal forms construction, and will rely for that on the Hamiltonian framework. Denote

$$\Theta_s^0(U) = \frac{1}{2}\int_X (\Lambda_m^s u)(\overline{\Lambda_m^s u})\,dx \sim \|u\|_{H^s}^2 \tag{5.14}$$

the Sobolev energy. We shall take s large enough and look for another function $\Theta_s(U)$ that satisfies

$$\Theta_s(U) \sim \|u\|_{H^s}^2 \text{ if } \|u\|_{H^s} \ll 1 \tag{5.15}$$

and

$$\frac{d}{dt}\Theta_s(U(t,\cdot)) = O\big(\|u(t,\cdot)\|_{H^s}^{N+2}\big),\ U \to 0 \tag{5.16}$$

if U is a solution of (5.11), and N is an arbitrary given integer. As already seen, (5.15) and (5.16) imply that the solution may be extended up to a time in $c\epsilon^{-N}$, and that the Sobolev norm of the solution remains of size $O(\epsilon)$ over such an interval.

Birkhoff normal forms method

We look for Θ_s as

$$\Theta_s(U) = \Theta_s^0 \circ \chi(U) \tag{5.17}$$

where χ is a local diffeomorphism going from a neighborhood of zero in $H^s(X, \mathbb{R}^2) \simeq H^s(X, \mathbb{C})$ to itself, sending zero to zero, that is assumed moreover *symplectic*. For F, G two smooth maps on an open set of H^s, we define their Poisson bracket by

$$\{F, G\}(U) = DG(U) \cdot X_F(U) \tag{5.18}$$

(Assuming that $X_F(U)$ is sufficiently regular so that the action of $DG(U)$ on it makes sense, which will be the case in our applications below). In complex coordinates, one gets

$$\{F, G\}(U) = i\big[D_u G(U) \cdot \nabla_{\bar{u}} F(U) - D_{\bar{u}} G(U) \cdot \nabla_u F(U)\big]. \tag{5.19}$$

Moreover, if χ is symplectic, we get

$$\{F \circ \chi, G\} = \big\{F, G \circ \chi^{-1}\big\} \circ \chi(U). \tag{5.20}$$

The Birkhoff method relies on the construction of the symplectic diffeomorphism χ from an auxiliary given function F, C^1 on an open subset of $H^s(X, \mathbb{R}^2) \simeq H^s(X, \mathbb{C})$, such that $U \to X_F(U)$ is also C^1. Actually, the Cauchy-Lipschitz theorem allows one to solve the ODE

$$\begin{aligned} \frac{\partial \Phi}{\partial \sigma}(\sigma, U) &= X_F(\Phi(\sigma, U)) \\ \Phi(0, U) &= U, \end{aligned} \tag{5.21}$$

and for small enough initial data U, the solution is defined for σ in $[-1, 1]$. If one sets

$$\chi(U) = \Phi(-1, U) \tag{5.22}$$

then χ is symplectic by construction. We may then state:

Lemma 8. *Let G be given as in* (5.12), (5.13). *Assume that one may find a polynomial function $F(U) = \sum_{n \geq 1} F_n(U)$, where F_n is homogeneous of order $n+2$, such that, if we define*

$$\mathrm{Ad}_F G = \{F, G\},\ \mathrm{Ad}_F^k G = \mathrm{Ad}_F \mathrm{Ad}_F^{k-1} G,\ k \geq 1, \tag{5.23}$$

then

$$\left\{\Theta_s^0, \sum_{k=0}^{N-1} \frac{1}{k!} \mathrm{Ad}_F^k G\right\} = O\big(\|U\|_{H^s}^{N+2}\big),\ U \to 0. \tag{5.24}$$

Then Theorem 7 holds.

Proof. This is just the basic computation of the Birkhoff normal forms method. If U is a solution to (5.21) and Θ_s is defined by (5.17), with χ given by (5.22), one has by (5.18) and (5.20)

$$\begin{aligned}\frac{d}{dt}\Theta_s(U(t,\cdot)) &= D\Theta_s(U(t,\cdot))\cdot\frac{dU}{dt} = D\Theta_s(U)\cdot X_G(U)\\ &= -\{\Theta_s,G\} = -\{\Theta_s^0\circ\chi,G\} = -\{\Theta_s^0,G\circ\chi^{-1}\}\circ\chi.\end{aligned}\tag{5.25}$$

In order to prove the theorem, we just have to check that, if F satisfies (5.24), then (5.16) holds ((5.15) will follow from the fact that χ is a local diffeomorphism with $\chi(0)=0$). By the definition (5.22) of χ and Taylor formula,

$$\begin{aligned}G\circ\chi^{-1}(U) &= G\circ\Phi(1,U)\\ &= \sum_{k=0}^{N-1}\frac{1}{k!}\frac{d^k}{d\sigma^k}(G\circ\Phi(\sigma,\cdot))|_{\sigma=0} + O(\|U\|_{H^s}^{N+2}).\end{aligned}\tag{5.26}$$

But, according to (5.21), (5.18),

$$\frac{d}{d\sigma}G\circ\Phi(\sigma,\cdot) = DG\cdot\frac{d\Phi}{d\sigma} = DG\cdot X_F(\Phi) = \{F,G\}(\Phi(\sigma,\cdot))$$

and more generally, $\frac{d^k}{d\sigma^k}G\circ\Phi(\sigma,\cdot) = \mathrm{Ad}_F^k G(\Phi(\sigma,\cdot))$. We thus get from (5.26)

$$G\circ\chi^{-1}(U) = \sum_{k=0}^{N-1}\frac{1}{k!}\mathrm{Ad}_F^k\cdot G(U) + O(\|U\|_{H^s}^{N+2}).\tag{5.27}$$

Plugging this equality in (5.25) and using assumption (5.24), we do get, since $\chi(0)=0$, that $\frac{d}{dt}\Theta_s(U(t,\cdot)) = O(\|U\|_{H^s}^{N+2})$ i.e. property (5.16). This concludes the proof. Q.E.D.

§6. Resolution of the homological equation

In Lemma 8 of Section 5, we have reduced the proof of Theorem 7 to the resolution of the homological equation (5.24). The goal of this section is to perform this last step, i.e. to prove the following proposition.

Proposition 9. *Let $G_0(U) = \int_X(\Lambda_m u)\bar u\,dx$ and for $n\geq 1$ define*

$$G_n(U) = \sum_{p=0}^{n+2}\int_X P_{n,p}(u,\dots,u,\bar u,\dots,\bar u),\tag{6.1}$$

where $P_{n,p}$ is defined after (5.13). *Then for any N, there is a polynomial $F(U) = \sum_{n\geq 1} F_n(U)$, with F_n homogeneous of degree $n+2$, such that the homological equation*

$$\left\{\Theta_s^0, \sum_{k=0}^{N-1} \frac{1}{k!}\mathrm{Ad}_F^k G\right\} = O(\|U\|_{H^s}^{N+2}),\ U \to 0 \tag{6.2}$$

holds.

We shall make clear in the proof what we mean by "polynomial function F". Actually, the key point will be to define a class of homogeneous functions inside which we may solve (6.2). More precisely, we shall prove the following lemma:

Lemma 10. *Assume given for any n in $\mathbb{N}^*$, any $\nu > 0$, a class $\mathcal{L}_n^\nu$ of functions homogeneous of degree $n+2$ from H^s (with s large enough) to $\mathbb{R}$, such that the following four properties hold true:*

(i) *For any $n \geq 1$, the function G_n defined in* (5.13) *belongs to $\mathcal{L}_n^\nu$ for some ν.*

(ii) *For any ν, n, n', there is ν' such that $\{\mathcal{L}_n^\nu, \mathcal{L}_{n'}^\nu\} \subset \mathcal{L}_{n+n'}^{\nu'}$.*

(iii) *If L is in $\mathcal{L}_n^\nu$ and s is large enough, $\{\Theta_s^0, L\}(U) = O(\|U\|_{H^s}^{n+2})$.*

(iv) *If H_n is in $\mathcal{L}_n^\nu$, there are some ν' and F_n in $\mathcal{L}_n^{\nu'}$ so that $\{F_n, G_0\}$ is in $\mathcal{L}_n^\nu$ and $\{\Theta_s^0, \{F_n, G_0\} - H_n\} = 0$.*

Then Proposition 9 *holds.*

Idea of proof of the lemma. According to (5.23), $\mathrm{Ad}_F^k G$ is a sum of contributions of the form

$$\{F_{n_1}, \{F_{n_2}, \{\cdots \{F_{n_k}, G_{n_0}\}\}\}\} \tag{6.3}$$

that belong to $\mathcal{L}_{n_0+\cdots+n_k}^{\nu'}$ for some ν'. This follows from (i) and the stability property (ii), together with the fact that if $n_0 = 0$, then $\{F_{n_k}, G_0\}$ is in $\mathcal{L}_{n_k}^\nu$ by (iv). As $G = \sum_{n=0}^{N'} G_n$, we may write the left hand side of (6.2) as

$$\left\{\Theta_s^0, \sum_{k=0}^{N-1} \frac{1}{k!}\mathrm{Ad}_F^k G\right\} = \sum_{n\geq 1} \{\Theta_s^0, \{F_n, G_0\} - H_n\}, \tag{6.4}$$

where the sum in the right hand side is finite and where H_n is computed from expressions of the form (6.3) with $n_0 + \cdots + n_k = n$ and $n_j < n$ for all $j = 1, \ldots, k$, since the only term where one of these n_j is equal to n is the $\{F_n, G_0\}$ contribution in the right hand side of (6.4). By what we said after (6.3), H_n is in $\mathcal{L}_n^{\nu'}$, and since it depends only on

$F_{n'}$ for $n' < n$, we may use (iv) in order to find successively F_n solving $\{\Theta_s^0, \{F_n, G_0\} - H_n\} = 0$ for $n = 1, \dots, N-1$. In that way, the only remaining terms in the right hand side of (6.4) are of the form $\{\Theta_s^0, L\}$ for elements L of $\mathcal{L}_n^{\nu'}$ for some ν' and $n \geq N$. According to (iii), they provide contributions to the right hand side of (6.2). This concludes the proof. Q.E.D.

We have thus reduced the proof of Theorem 7 to the construction of classes $\mathcal{L}_n^\nu$ such that properties (i) to (iv) hold. In order to do so, we define first classes of multilinear maps.

Definition 11. *Let n in $\mathbb{N}^*$ and $\nu > 0$. For k in $\mathbb{N}$, denote by Π_k the spectral projector associated to the interval I_k introduced in Theorem 5. We define $\tilde{\mathcal{L}}_n^\nu$ as the space of $(n+2)$-linear forms $(u_1, \dots, u_{n+2}) \to L(u_1, \dots, u_{n+2})$ from $H^s \times \cdots \times H^s$ to $\mathbb{C}$ such that, for any N in $\mathbb{N}$, there is $C_N > 0$ and for any $k_1, \dots, k_{n+2}$ in $\mathbb{N}$,*

$$|L(\Pi_{k_1} u_1, \dots, \Pi_{k_{n+2}} u_{n+2})| \leq C_N \frac{(k_3^*)^{\nu+N}}{(k_1^* - k_2^* + k_3^*)^N} \prod_{j=1}^{n+2} \|u_j\|_{L^2} \tag{6.5}$$

where

$$k_1^* = \max(k_1, \dots, k_{n+2}) + 1 \geq k_2^* = \max{}_2(k_1, \dots, k_{n+2}) + 1 \geq k_3^* = \max{}_3(k_1, \dots, k_{n+2}) + 1,$$

where we have denoted by $\max_j(k_1, \dots, k_{n+2})$ the j-th largest among all integers $k_1, \dots, k_{n+2}$.

The basic example of an element of $\tilde{\mathcal{L}}_n^\nu$ is given by

$$L(u_1, \dots, u_{n+2}) = \int_X u_1 \cdots u_{n+2}\, dx. \tag{6.6}$$

Let us show this in the special case when $X = \mathbb{S}^d$, so that each interval I_k is reduced to a single point $\{\xi_k\}$. If for instance $k_1 \geq k_2 \geq k_3 \geq \cdots \geq k_{n+2}$, so that $k_j^* = k_j + 1$, estimate first the left hand side of (6.5) by

$$|L(\Pi_{k_1} u_1, \dots, \Pi_{k_{n+2}} u_{n+2})| \leq \|\Pi_{k_1} u_1\|_{L^2} \|\Pi_{k_2} u_2\|_{L^2} \prod_{j=3}^{n+2} \|\Pi_{k_j} u_j\|_{L^\infty}$$

and use Sobolev injection to estimate the L^∞ norms by

$$\prod_{j=3}^{n+2} \|\Pi_{k_j} u_j\|_{L^\infty} \leq (k_3^*)^\nu \prod_{j=3}^{n+2} \|\Pi_{k_j} u_j\|_{L^2}$$

for some ν. We thus obtain (6.5) in the special case $k_1^* - k_2^* \le Ck_3^*$. On the other hand, if $k_1^* - k_2^* \gg k_3^*$, we may write, setting $a(x) = \prod_{j=3}^{n+2} \Pi_{k_j} u_j$,

$$\begin{aligned}(\xi_{k_1}^2 - \xi_{k_2}^2) &\int_X (\Pi_{k_1} u_1)(\Pi_{k_2} u_2) a(x)\, dx \\ &= \int \Big[-(\Delta\Pi_{k_1} u_1)(\Pi_{k_2} u_2) + (\Pi_{k_1} u_1)(\Delta\Pi_{k_2} u_2)\Big] a(x)\, dx \\ &= \int_X (\Pi_{k_1} u_2)[a, \Delta](\Pi_{k_2} u_2)\, dx.\end{aligned}$$

It follows, as the $\mathcal{L}(H^1, L^2)$ norm of the commutator $[a, \Delta]$ may be bounded from $\|\nabla a\|_{L^\infty}$, that

$$(\xi_{k_1}^2 - \xi_{k_2}^2)|L(\Pi_{k_1} u_1, \dots, \Pi_{k_{n+2}} u_{n+2})| \le C(k_3^*)^{\nu+1} k_2^* \prod_{j=3}^{n+2} \|\Pi_{k_j} u_j\|_{L^2}.$$

As $\xi_{k_j} \sim k_j$, this implies (6.5) with $N = 1$. Iterating the process, one gets the result.

Remark.

• We gave the proof of estimate (6.5) for the example (6.6) in the case when X is the sphere. Actually, the above proof may be adapted to an arbitrary compact Riemannian manifold (with the Π_{k_j} defined from the intervals I_{k_j} of Proposition 2). In other words, we do not use any geometric assumption in the proof of the fact that (6.6) belongs to the class $\tilde{\mathcal{L}}_n^\nu$.

• The same proof shows that $\int_X \prod_{j=1}^{n+2} (\Lambda_m^{-\frac{1}{2}} u_j)\, dx$ belongs to $\tilde{\mathcal{L}}_n^\nu$.

Let us define now the classes $\mathcal{L}_n^\nu$ that will enjoy the properties of Lemma 10.

Definition 12. *One says that $U \to H_n(U)$ is in $\mathcal{L}_n^\nu$ if there are elements $H_{n,p}$ of $\tilde{\mathcal{L}}_n^\nu$ for $p = 0, \dots, n+2$ such that*

$$H_n(U) = \sum_{p=0}^{n+2} H_{n,p}(\underbrace{u, \dots, u}_{p}, \underbrace{\bar{u}, \dots, \bar{u}}_{n+2-p}). \tag{6.7}$$

We have to check that properties (i) to (iv) of Lemma 10 hold true.

• Property (i) follows from the above example. Actually, by (5.8), G_n may be written as an expression (6.7), where $H_{n,p}$ is given by a quantity of the form (6.6) with each factor u_j replaced by $\Lambda_m^{-\frac{1}{2}} u_j$, that provides an element of $\mathcal{L}_n^\nu$ as already remarked.

• Property (ii), the stability under Poisson brackets, may be reduced to the proof of the fact that if L is in $\tilde{\mathcal{L}}_n^\nu$ and if M is a $(n'+1)$-linear map $(u_1,\dots,u_{n'+1}) \to M(u_1,\dots,u_{n'+1})$, satisfying estimates of the form

$$\|\Pi_{k_{n'+2}} M(\Pi_{k_1} u_1,\dots,\Pi_{k_{n'+1}} u_{n'+1})\|_{L^2} \leq C_N \frac{(k_3^*)^{\nu+N}}{(k_1^*-k_2^*+k_3^*)^N} \prod_{j=1}^{n'+1} \|u_j\|_{L^2} \tag{6.8}$$

then

$$L\big(M(u_1,\dots,u_{n'+1}),u_{n'+2},\dots,u_{n+n'+2}\big)$$

is in $\tilde{\mathcal{L}}_{n+n'}^{\nu'}$ for some ν'. This follows from the combination of estimates (6.5) and (6.8) and some elementary computations.

• Property (iii) may be deduced as well from the bounds (6.5) if one takes s large enough relatively to ν.

Finally, we have reduced the proof of Theorem 7 to the fact that we may solve the homological equation of (iv) of Lemme 10 i.e. prove the following proposition.

Proposition 13. *There is a zero measure subset $\mathcal{N}$ of $]0,+\infty[$ such that, if m belongs to $]0,+\infty[$, for any $\nu>0$, $n\in\mathbb{N}^*$, any H_n in $\mathcal{L}_n^\nu$, one may find F_n in $\mathcal{L}_n^{\nu'}$, with some ν' depending only on ν, m, such that*

$$\big\{\Theta_s^0, \{F_n, G_0\} - H_n\big\} = 0. \tag{6.9}$$

We shall proceed as in Section 3, except that we make the reasoning on linear forms like H_n, F_n instead of linear maps, in order to exploit the Hamiltonian structure. We prove first the proposition in the simpler case of odd n, where we may actually solve a stronger equation than (6.9).

Lemma 14. *Assume that n is odd. Then, with the notation of Proposition 13, we may find for any H_n in $\mathcal{L}_n^\nu$ some F_n in $\mathcal{L}_n^{\nu'}$ solving*

$$\{F_n, G_0\} - H_n = 0. \tag{6.10}$$

Proof. We give the proof in the special case $X=\mathbb{S}^d$ so that each interval I_k is reduced to the point $\{\xi_k\}$. By Definition 12, we may write H_n under the form (6.7), and we look for F_n as $F_n(U) = \sum_{p=0}^{n+2} F_{n,p}(u,\dots,u,\bar{u},\dots,\bar{u})$ with $F_{n,p}$ in $\tilde{\mathcal{L}}_n^{\nu'}$. To solve (6.10) we have to find $F_{n,p}$ so that

(6.11) $\{F_{n,p}(u,\dots,u,\bar u,\dots,\bar u),G_0\} = H_{n,p}(u,\dots,u,\bar u,\dots,\bar u).$

Recall from (5.19) that

$$\{F_{n,p},G_0\} = -i[D_uF_{n,p}\nabla_{\bar u}G_0 - D_{\bar u}F_{n,p}\nabla_u G_0]$$

so that, since by (5.13) $\nabla_{\bar u}G_0 = \Lambda_m u$, $\nabla_u G_0 = \Lambda_m\bar u$, equality (6.11) will follow from

$$(6.12)\quad \sum_{j=1}^{p} F_{n,p}(u_1,\dots,\Lambda_m u_j,\dots,u_{n+2}) - \sum_{j=p+1}^{n+2} F_{n,p}(u_1,\dots,\Lambda_m u_j,\dots,u_{n+2}) = iH_{n,p}(u_1,\dots,u_{n+2}).$$

We replace in (6.12) u_j by $\Pi_{k_j}u_j$. Since we assume that $I_{k_j}=\{\xi_{k_j}\}$, we have $\Lambda_m\Pi_{k_j}u_j = \sqrt{m^2+\xi_{k_j}^2}\,\Pi_{k_j}u_j$, so that (6.12) may be written as

$$(6.13)\quad \mathcal{G}_m^p(\xi_{k_1},\dots,\xi_{k_{n+2}})F_{n,p}(\Pi_{k_1}u_1,\dots,\Pi_{k_{n+2}}u_{n+2}) = iH_{n,p}(\Pi_{k_1}u_1,\dots,\Pi_{k_{n+2}}u_{n+2})$$

using notation (3.6). If n is odd, assumption (4.3) of Proposition 6 holds, so that by (4.2) we may divide (6.13) by $\mathcal{G}_m^p(\xi_{k_1},\dots,\xi_{k_{n+2}})$ and get that if $H_{n,p}$ satisfies bounds of the form (6.5), so does $F_{n,p}$ with ν replaced by $\nu'=\nu+N_0$.

This concludes the proof when $X=\mathbb{S}^d$. When X is a general Zoll manifold, each interval I_k contains several eigenvalues. But, since the length of I_k is $O(k^{-\delta})$, they are close enough to each other so that one may still solve (6.13) by a perturbative argument (see [12, 2]). Q.E.D.

End of proof of Proposition 13. We have to complement the conclusion of the preceding lemma treating the case of even n, for which condition (4.3) is not always satisfied. Actually, it still holds except if n is even, $p=\frac{n}{2}+1$ and $\{k_1,\dots,k_p\}=\{k_{p+1},\dots,k_{n+2}\}$, in which case $\mathcal{G}_m^p$ vanishes identically. But notice that we have to solve (6.9) and not (6.10), and by definition of Θ_s^0, if $\{k_1,\dots,k_p\}=\{k_{p+1},\dots,k_{n+2}\}$,

$$\begin{aligned}\{\Theta_s^0, H_{n,p}(\Pi_{k_1}u_1,\dots,\Pi_{k_p}u_p,\Pi_{k_1}\bar u_1,\dots,\Pi_{k_p}\bar u_p)\} &= -i\left(\sum_{j=1}^{p}\langle\xi_{k_j}\rangle^{2s} - \sum_{j=p+1}^{n+2}\langle\xi_{k_j}\rangle^{2s}\right)\\ &\quad\times H_{n,p}(\Pi_{k_1}u_1,\dots,\Pi_{k_p}u_p,\Pi_{k_1}\bar u_1,\dots,\Pi_{k_p}\bar u_p)=0\end{aligned}$$

so that, for these values of the indices, we may just define

$$F_{n,p}(\Pi_{k_1}u_1,\dots,\Pi_{k_{n+2}}u_{n+2})\equiv 0.$$

This conclusion the proof. Q.E.D.

§7. Quasi-linear equations

Up to now, we have discussed long time existence problems for semi-linear equations of the form

$$(\partial_t^2-\Delta+m^2)w=N(w) \tag{7.1}$$

where the non-linearity $N(w)$ vanishes at least at order two at zero and *does not* depend on derivatives of w. We would like now to discuss non-linearities depending also on derivatives of w. We cannot hope for a statement as Theorem 7, even on $X=\mathbb{S}^1$, for *any* semi-linear non-linearity depending on derivatives of u. Actually, examples of Yordanov [23] and Keel and Tao [17] show that, in general, one cannot expect existence over a time interval of length larger than the one given by local existence theory. This shows that, if in the right hand side of (7.1) $N(w)$ is replaced by an expression containing first order derivatives of w, the normal form procedure we used in the preceding sections will not always work. The key point is that some contributions to the non-linearity that may not be eliminated by normal forms will induce a growth of the Sobolev energy. On the contrary, if the non-linearity depends only on w, the Hamiltonian formulation that we introduced in Section 5 showed that all non eliminable contributions do not make grow the energy, as they depend only on the actions, so that, in an homological equation of the form (6.9), their Poisson bracket with Θ_s^0 vanishes.

If we want to study non-linearities depending also on derivatives, we thus have to restrict ourselves to classes of nonlinear expressions that possess some special properties that ensure that non eliminable terms in the normal form procedure do not make grow energy. Moreover, this special structure has to be preserved under normal forms reduction, so that one is able to take profit of it at any step of the construction.

One possibility in order to achieve that is to consider "reversible" equations. We refer to the paper [13] of Fang, Han and Zhang, devoted to semi-linear equations on the circle, with non-linearities depending on a reversible way of $\partial_t w,\partial_x w$, i.e. verifying the condition

$$N(-w,\partial_t w,-\partial_x w)=-N(w,\partial_t w,\partial_x w).$$

For odd or even small initial data, they get almost global solutions, when the mass parameter stays outside some subset of zero measure.

Another option, that we shall describe below, is to stay in the Hamiltonian framework, but allowing the Hamiltonian (5.5) to depend not only on w, but also on $\partial_x w$, so that, in (5.8), $G(V)$ would be given in terms of an integral of $P(\Lambda_m^{\frac{1}{2}}V)$ instead of $P(\Lambda_m^{-\frac{1}{2}}V)$. In this case, the evolution equation is necessarily a quasi-linear one, and not just semi-linear. In fact, one may check that if a non-linearity is both semi-linear and Hamiltonian, then it has to depend only on w and not its derivatives, so that one gets just equation (7.1). In this section, we shall thus state a theorem of almost global existence for *quasi-linear* Hamiltonian Klein-Gordon equations on spheres.

Let us fix some notation. From now on, X will stand for the sphere $\mathbb{S}^d$. Let $f : \mathbb{R} \times T^*X \to \mathbb{R}$ be a smooth map $(z, \rho) \to f(z, \rho)$, polynomial in the fiber variable of ρ, and vanishing at least at order 3 at $(z = 0, \rho \in X)$, where we identify X to the zero section of T^*X. We consider a non-linearity $P[w]$ given in local coordinates by

$$\begin{aligned} P[w] &= \frac{\partial f}{\partial z}\Big(w, x, \frac{\partial w}{\partial x}\Big) \\ &\quad - \sum_{j=1}^{d} (\det g)^{-1/2} \frac{\partial}{\partial x_j}\Big[(\det g)^{1/2}\Big(\frac{\partial f}{\partial \xi_j}\Big)\Big(w, x, \frac{\partial w}{\partial x}\Big)\Big], \end{aligned} \tag{7.2}$$

(where g is the metric of the sphere) and the quasi-linear Klein-Gordon equation

$$(\partial_t^2 - \Delta + m^2)w = P[w]. \tag{7.3}$$

This is a Hamiltonian equation associated to the Hamiltonian

$$\int_X \Big[\frac{1}{2}[(\partial_t w)^2 + (\partial_x w)^2 + m^2 w^2] - f(w, dw)\Big]\, dx. \tag{7.4}$$

If we set

$$F(w) = \int_X f(w, dw)\, dx, \tag{7.5}$$

and define V from $(w, \partial_t w)$ as in (5.4), equation (7.3) may be rewritten as

$$\partial_t V = X_G(V) \tag{7.6}$$

with

$$G(V) = G_0(V) - F\big(\Lambda_m^{-\frac{1}{2}} V^2\big),\ G_0(V) = \frac{1}{2}\int_X \big(\Lambda_m^{\frac{1}{2}} V\big)\cdot\big(\Lambda_m^{\frac{1}{2}} V\big)\, dx \tag{7.7}$$

where, because of the first order derivative in (7.5), $F(\Lambda_m^{-\frac{1}{2}}V^2)$ is actually of order $\frac{1}{2}$ in V^2. Defining as before $u = \frac{\sqrt{2}}{2}(V^1 + iV^2)$, $U = (u, \bar{u})$ and writing the expression of the Hamiltonian in terms of U, namely

$$G(U) = \int_X (\Lambda_m u)\bar{u}\, dx - F\Big(-i\frac{\sqrt{2}}{2}\Lambda_m^{-\frac{1}{2}}(u - \bar{u})\Big) \tag{7.8}$$

the equation may be written again

$$\begin{aligned} \dot{u} &= i\nabla_{\bar{u}} G(U) \\ u|_{t=0} &= \epsilon u_0. \end{aligned} \tag{7.9}$$

The new feature is that F in (7.8) being of the form (7.5), one should think of $G(U)$ as $\sum_{n=0}^{N'} G_n(U)$, with $G_0(U)$ given by the first formula in (5.13), but where $G_n(U)$ is no longer given by the second formula (5.13), but instead by an expression of the form

$$G_n(U) = \sum_{p=0}^{n+2} \int_X P_{n,p}(\Lambda_m^{\frac{1}{2}}u, \dots, \Lambda_m^{\frac{1}{2}}u, \Lambda_m^{\frac{1}{2}}\bar{u}, \dots, \Lambda_m^{\frac{1}{2}}\bar{u}) \tag{7.10}$$

with $P_{n,p}$ a $(n+2)$-linear form. Actually, the expression one gets is more general than (7.10), since it involves operators of order $\frac{1}{2}$ that are not necessarily the $\Lambda_m^{\frac{1}{2}}$'s.

The main theorem is as follows:

Theorem 15. *With the above notation, there is a zero measure subset $\mathcal{N}$ of $]0, +\infty[$ and for any $m \in]0, +\infty[- \mathcal{N}$, any $N \in \mathbb{N}$, there is $s_0 > 0$ and for any $s > s_0$, there are $\epsilon_0 > 0$, $c > 0$, $C > 0$, such that, for any $\epsilon \in]0, \epsilon_0[$, any (w_0, w_1) in the unit ball of $H^{s+\frac{1}{2}}(X) \times H^{s-\frac{1}{2}}(X)$, one may find a unique w in*

$$C^0(] - T_\epsilon, T_\epsilon[, H^{s+\frac{1}{2}}(X)) \cap C^1(] - T_\epsilon, T_\epsilon[, H^{s-\frac{1}{2}}(X))$$

solution of

$$\begin{aligned} (\partial_t^2 - \Delta + m^2)w &= P[w] \\ w|_{t=0} &= \epsilon w_0 \\ \partial_t w|_{t=0} &= \epsilon w_1, \end{aligned} \tag{7.11}$$

where $T_\epsilon \geq c\epsilon^{-N}$ and

$$\sup_{]-T_\epsilon, T_\epsilon[} \left[\|w(t,\cdot)\|_{H^{s+\frac{1}{2}}} + \|\partial_t w(t,\cdot)\|_{H^{s-\frac{1}{2}}}\right] \leq C\epsilon. \tag{7.12}$$

Remark. In the above theorem, we limit ourselves to the case of spheres, while for semi-linear equations we proved a similar result more generally on Zoll manifolds. Very likely, Theorem 15 holds true as well in this Zoll setting. However, the proof of this more general statement would involve extra technical difficulties. That's why we restricted ourselves in [10] (see also [9] for the case $X = \mathbb{S}^1$) to spheres.

We shall discuss some ideas of the proof in the model case (that already contains all difficulties) given by (7.9) with $G(U) = \sum_{n=0}^{N'} G_n(U)$, where $G_0(U) = \int_X (\Lambda_m u)\bar{u}\,dx$ and G_n is given by (7.10) for $n \geq 1$.

Let us explain the main new ideas that have to be introduced. In the semi-linear case, we have instead of (7.10) an expression of G_n given by (5.13) i.e. by generalization of quantities of the form

$$\int_X u \cdots u\bar{u} \cdots \bar{u}\,dx. \tag{7.13}$$

We may write (7.13) in terms of expressions

$$\int_X (A(U)u)\bar{u}\,dx, \quad \int_X (A(U)u)u\,dx, \quad \int_X (A(U)\bar{u})\bar{u}\,dx \tag{7.14}$$

where $A(U)$ is a quantity of the from $u^{n'}\bar{u}^{n''}$ with $n'+n'' = n$. The operator $A(U)$ is thus of order zero, so that it does not lose any derivative. In the homological equation (6.9), we could have written the function H_n under the form (7.14) as well. The solution F_n we have constructed in the proof of Proposition 13 has the same form as H_n. Moreover, the contributions to $H_{n+1}, H_{n+2}, \ldots$ generated by F_n through Poisson brackets in (6.3) have also structure (7.14). Actually, the Poisson bracket between two functions of the form (7.14) involving operators $A'(U), A''(U)$ may be written as (7.14) with $A(U)$ replaced by $A'(U) \circ A''(U)$ or $A''(U) \circ A'(U)$. As A', A'' are of order zero, so is A i.e. the structure (7.14) is preserved along the normal forms procedure.

For quasi-linear equations, this is no longer the case at first glance. If instead of (7.13) we start from a formula similar to (7.10), namely

$$\int_X \big(\Lambda_m^{\frac{1}{2}} u\big) \cdots \big(\Lambda_m^{\frac{1}{2}} u\big)\big(\Lambda_m^{\frac{1}{2}} \bar{u}\big) \cdots \big(\Lambda_m^{\frac{1}{2}} \bar{u}\big)\,dx \tag{7.15}$$

we get in (7.14) operators of *order one*. Then the functions F_n obtained solving the homological equation are expected to be also given by (7.14), with operators $A(U)$ of order one, so that the contributions induced by Poisson bracket between F_n and $H_{n'}$ are expected to contain compositions of operators of order one, that is operators of order two. Clearly, each new step would generate operators of higher and higher order, so that the procedure would lead to derivative losses.

In order to overcome these difficulties, one has to show that the structure of the Poisson bracket allows one to avoid these increasing losses. Actually, one may check easily that when computing

$$\Big\{\int_X (A'(U)u)\bar{u}\,dx, \int_X (A''(U)u)\bar{u}\,dx\Big\} \tag{7.16}$$

one gets a main contribution of the form

$$\int_X ([A'(U), A''(U)]u)\bar{u}\,dx \tag{7.17}$$

plus other terms having still the structure (7.14) with $A(U)$ of order one. We thus need to know that $[A'(U), A''(U)]$ is also of order one, so that we stay in the same classes of integral quantities. In other words, we need to know that $A'(U), A''(U)$ belong to a class of operators enjoying some "symbolic calculus". Examples of such classes are pseudo-differential operators. Since in our problem, the coefficients of the operators depend on the unknown U, that has limited regularity, it is actually better to use para-differential operators. We shall not recall here the definition of the latter. Let us just say that one may define for any m in $\mathbb{R}$, n in $\mathbb{N}$, families of operators $H\Psi^\ell_n$, that are homogeneous polynomials of degree n in U, such that for any fixed U, the operator $V \to A(U)V$ satisfies the following:

- If A is in $H\Psi^\ell_n$, then for any s in $\mathbb{R}$, $V \to A(U)V$ sends $H^s(X)$ to $H^{s-\ell}(X)$ (and its operator norm depends only on $\|U\|_{H^\sigma}$ for some σ independent of s).
- If A is in $H\Psi^\ell_n$ and B is in $H\Psi^{\ell'}_{n'}$, then $A \circ B$ is in $H\Psi^{\ell+\ell'}_{n+n'}$ and $[A, B]$ is in $H\Psi^{\ell+\ell'-1}_{n+n'}$.

One may show that equation (7.11) may be written equivalently as

$$\begin{aligned} \dot{u} &= i\nabla_{\bar{u}} G(U) \\ u|_{t=0} &= \epsilon u_0 \end{aligned} \tag{7.18}$$

where $G(U) = \sum_{n=0}^{N'} G_n(U)$ with essentially

$$G_n(U) = \operatorname{Re} \int_X (A^1_n(U)u)\bar{u}\,dx + \operatorname{Re}\int_X (A^2_n(U)u)u\,dx \tag{7.19}$$

with A^1_n, A^2_n in $H\Psi^1_n$. Let us consider the model case $A^2_n = 0$ and prove:

Proposition 16. *Let A_n be in $H\Psi^1_n$ for $n = 1, \dots, N'$ and define for $n \geq 1$, $G_n(U) = \int_X \big(A_n(U)u\big)\bar{u}\,dx$ and $G(U) = \sum_{n=0}^{N'} G_n(U)$. Then, for any N in $\mathbb{N}$, if m is outside a subset $\mathcal{N}$ of zero measure of $]0, +\infty[$, there is $s_0 > 0$ and for any $s > s_0$, there are $\epsilon_0 > 0$, $c > 0$, $C > 0$ such that, if $\epsilon \in]0, \epsilon_0[$, u_0 is in the unit ball of $H^s(X, \mathbb{C})$, system (7.18) has a unique solution u in $C^0(] - T_\epsilon, T_\epsilon[, H^s(X))$ with $T_\epsilon \geq c\epsilon^{-N}$ and $\sup_{]-T_\epsilon,T_\epsilon[} \|u(t,\cdot)\|_{H^s} \leq C\epsilon$.*

Strategy of proof. The strategy of proof is similar to the one we used in the semi-linear case, except that we have to exploit the expression of the G_n in terms of para-differential operators in order to avoid derivative losses.

First step: Construction of modified energy

We have seen that the first step in the proof of long time existence is the construction of a modified energy Θ_s of the form (5.17). In the semi-linear case, Θ_s was defined using the canonical transformation χ given by (5.22) from the solution Φ of the ODE (5.21). In the quasi-linear framework, (5.21) is no longer an ODE, as the right hand side X_F will have to be defined in terms of a function F given by

$$\int_X \big(B(U)u\big)\bar{u}\,dx \tag{7.20}$$

with $B(U)$ of order one. Thus, (5.21) is then a PDE involving in its right hand side the action of the operator of order one $B(U)$. Though, it would be still possible to solve the corresponding evolution equation for Φ, as the PDE one gets is an hyperbolic one. Nevertheless, it is not even necessary to do so, because one just needs a finite Taylor expansion of $G \circ \chi^{-1}$ (see (5.26)), so that it is enough to define the action of composition of G with χ^{-1} at finite order, i.e. to give a meaning to the sum in the right hand side of (5.27). The general term of that sum is made of Poisson brackets like

$$\Big\{\int_X \big(B_{n'}(U)u\big)\bar{u}\,dx, \int_X \big(A_{n''}(U)u\big)\bar{u}\,dx\Big\} \tag{7.21}$$

with $A_{n''}, B_{n'}$ para-differential operators of order one. According to (7.17) and to the fact that commutators $[B_{n'}, A_{n''}]$ are still in $H\Psi^1_{n'+n''}$, one sees that the sum in (5.27) will be made essentially of terms of the form (7.20) with operators $B(U)$ of order one.

Second step: Resolution of the homological equation

As in the semi-linear case, the first step reduces the proof of Proposition 16 to an homological equation of the form (6.9). For the sake of simplicity, let us consider only odd degrees of homogeneity n, so that one can solve instead of (6.9)

$$\{F_n, G_0\} - H_n = 0 \tag{7.22}$$

where $H_n(U) = \int_X \big(A_n(U)u\big)\bar{u}\,dx$ for some self-adjoint operator A_n in $H\Psi^1_n$, and look for F_n as $\int_X \big(B_n(U)u\big)\bar{u}\,dx$, with B_n to be determined also in $H\Psi^1_n$. Looking for A_n, B_n under the form

$$\begin{aligned} A_n(U) &= \sum_{p=0}^{n} A_n^p(\underbrace{u,\dots,u}_{p},\underbrace{\bar{u},\dots,\bar{u}}_{n-p}),\\ B_n(U) &= \sum_{p=0}^{n} B_n^p(\underbrace{u,\dots,u}_{p},\underbrace{\bar{u},\dots,\bar{u}}_{n-p}) \end{aligned} \tag{7.23}$$

where $A_n^p(u_1,\dots,u_n)$, $B_n^p(u_1,\dots,u_n)$ are first order para-differential operators that are n-linear in $(u_1,\dots,u_n)$, one may rewrite (7.22) as an equation of the form

$$\begin{aligned} &[B_n^p(\mathcal{U}),\Lambda_m] + \sum_{j=1}^{p} B_n^p(u_1,\dots,\Lambda_m u_j,\dots,u_n)\\ &\quad - \sum_{j=p+1}^{n} B_n^p(u_1,\dots,\Lambda_m u_j,\dots,u_n) = -iA_n^p(\mathcal{U}) + \text{remainders} \end{aligned} \tag{7.24}$$

where $\mathcal{U} = (u_1,\dots,u_n)$. This equation replaces (6.12) in the quasi-linear setting. In order to solve it, one first shows that one may replace in (7.24) $[B_n^p,\Lambda_m]$ by $[B_n^p,\widetilde{\Lambda}]$, where $\widetilde{\Lambda} = \sqrt{-\Delta + \big(\frac{d-1}{2}\big)^2}$, up to contributions that are better in terms of the number of derivatives involved in the operator. One is thus essentially reduced to an equation of the form

$$\begin{aligned} &[B_n^p(\mathcal{U}),\widetilde{\Lambda}] + \sum_{j=1}^{p} B_n^p(u_1,\dots,\Lambda_m u_j,\dots,u_n)\\ &\qquad - \sum_{j=p+1}^{n} B_n^p(u_1,\dots,\Lambda_m u_j,\dots,u_n) = -iA_n^p(\mathcal{U}). \end{aligned} \tag{7.25}$$

One then replaces each u_j by $\Pi_{k_j}u_j$ which, denoting

$$\Pi_k\mathcal{U} = (\Pi_{k_1}u_1,\dots,\Pi_{k_n}u_n),$$

brings (7.25) to

$$[B_n^p(\Pi_k\mathcal{U}),\widetilde{\Lambda}] + \tilde{\mathcal{G}}_m^p(\xi_{k_1},\dots,\xi_{k_n})B_n^p(\Pi_k\mathcal{U}) = -iA_n^p(\Pi_k\mathcal{U}) \tag{7.26}$$

with $\tilde{\mathcal{G}}^p_m$ given by (4.5), the ξ_{k_j} being the eigenvalues of the square root of the laplacian on the sphere. Taking θ in $C_0^\infty(\mathbb{R})$, equal to one close to zero, one can check that if we set

$$B^p_n(\Pi_k\mathcal{U}) = -\int_0^{+\infty} e^{-it\widetilde{\Lambda}}A(\Pi_k\mathcal{U})e^{it\widetilde{\Lambda}}e^{i\tilde{\mathcal{G}}^p_m(\xi_{k_1},\dots,\xi_{k_n})}\theta(\epsilon t)\,dt, \tag{7.27}$$

one gets a solution to (7.25), up to a remainder of the form $\epsilon^N R$, where R is again a para-differential operator of order one. The contribution of this last term to the homological equation (6.9) is $O(\epsilon^{N+2})$, i.e. is $O(\|u\|^{N+2}_{H^s})$, as the right hand side of (5.16), since ϵ is essentially the size of $\|u\|_{H^s}$.

Finally, the last step of the proof is to show that if A in the right hand side of (7.27) is a para-differential operator of order one, so is B. This may be proved using that $t \to e^{it\widetilde{\Lambda}}$ is 2π-periodic (this is the place where one uses that X is a sphere, as this implies that the eigenvalues of $\widetilde{\Lambda}$ are integers) and using that $\tilde{\mathcal{G}}^p_m(\xi_{k_1},\dots,\xi_{k_n})$ satisfies small divisor bounds of the form (3.4). Notice that here, because $A^p_n(\Pi_k\mathcal{U})$ is a para-differential operator, the frequencies $k_1,\dots,k_n$ that appear in the argument of $\tilde{\mathcal{G}}^p_m$ are actually the *small* frequencies, so that a lower bound in $(1+\max(k_1,\dots,k_n))^{-N_0}$ is the analogous of the lower bound (4.2) in $(1+\max_3(k_1,\dots,k_{n+2}))^{-N_0}$. This concludes the sketch of proof of the proposition. Q.E.D.

To conclude this section, let us mention that recent results of almost global existence for quasi-linear equations have also been obtained by Berti and Delort in [4] for the capillary-gravity water waves equation on the circle. In this case, the situation is different from the one we considered here, as the linear part of the system is given by an operator of order $\frac{3}{2}$ instead of one. The eigenvalues of that operator are then separated by growing gaps, that allow one to perform a reduction to constant coefficients.

References

[1] D. Bambusi, Birkhoff normal form for some nonlinear PDEs, *Comm. Math. Physics*, 234:253–285, 2003.

[2] D. Bambusi, J.-M. Delort, B. Grébert and J. Szeftel, Almost global existence for Hamiltonian semilinear Klein-Gordon equations with small Cauchy data on Zoll manifolds, *Comm. Pure Appl. Math.*, 60(11):1665–1690, 2007.

[3] D. Bambusi and B. Grébert, Birkhoff normal form for PDEs with tame modulus, *Duke Math. J.*, 135:507–567, 2006.

[4] M. Berti and J.-M. Delort, *Almost global solutions of capillary-gravity water waves equations on the circle*, Volume 24 of *Lecture Notes of the Unione Matematica Italiana*, Springer, Cham; Unione Matematica Italiana, [Bologna], 2018.

[5] A. L. Besse, *Manifolds all of whose geodesics are closed*, volume 93 of *Ergebnisse der Mathematik und ihrer Grenzgebiete*, Springer-Verlag, Berlin-New York, 1978, With appendices by D. B. A. Epstein, J.-P. Bourguignon, L. Bérard-Bergery, M. Berger and J. L. Kazdan.

[6] J. Bourgain, Construction of approximative and almost periodic solutions of perturbed linear Schrödinger and wave equations, *Geom. Funct. Anal.*, 6(2): 201–230, 1996.

[7] Y. Colin de Verdière, Sur le spectre des opérateurs elliptiques à bicaractéristiques toutes périodiques, *Comment. Math. Helv.*, 54(3): 508–522, 1979.

[8] J.-M. Delort, Temps d'existence pour l'équation de Klein-Gordon semi-linéaire à données petites périodiques, *American Journal of Mathematics*, pages 663–689, 1998.

[9] J.-M. Delort, A quasi-linear Birkhoff normal forms method. Application to the quasi-linear Klein-Gordon equation on $\mathbb{S}^1$, *Astérisque*, (341):vi+113, 2012.

[10] J.-M. Delort, Quasi-linear perturbations of Hamiltonian Klein-Gordon equations on spheres, *Mem. Amer. Math. Soc.*, 234 (1103):vi+80, 2015.

[11] J.-M. Delort and R. Imekraz, Long-time existence for the semilinear Klein-Gordon equation on a compact boundary-less Riemannian manifold, *Comm. Partial Differential Equations*, 42(3):388–416, 2017.

[12] J.-M. Delort and J. Szeftel, Long-time existence for small data nonlinear Klein-Gordon equations on tori and spheres, *Int. Math. Res. Not.*, (37):1897–1966, 2004.

[13] D. Fang, Z. Han and Q. Zhang, Almost global existence for the semi-linear Klein-Gordon equation on the circle, *J. Differential Equations*, 262(9): 4610–4634, 2017.

[14] D. Fang and Q. Zhang, Long-time existence for semi-linear Klein–Gordon equations on tori, *Journal of Differential Equations*, 249(1): 151–179, 2010.

[15] V. Guillemin, Lectures on spectral theory of elliptic operators, *Duke Math. J.*, 44(3):485–517, 1977.

[16] A. D. Ionescu and F. Pusateri, Long-time existence for multi-dimensional periodic water waves, *Geom. Funct. Anal.*, 29(3): 811–870, 2019.

[17] M. Keel and T. Tao, Small data blow-up for semilinear Klein-Gordon equations, *Amer. J. Math.*, 121(3): 629–669, 1999.

[18] S. Klainerman, Global existence of small amplitude solutions to nonlinear Klein-Gordon equations in four space-time dimensions, *Comm. Pure Appl. Math.*, 38(5): 631–641, 1985.

[19] J. Shatah, Normal forms and quadratic nonlinear Klein-Gordon equations, *Comm. Pure Appl. Math.*, 38(5): 685–696, 1985.

[20] J. C. H. Simon and E. Taflin, The Cauchy problem for nonlinear Klein-Gordon equations, *Comm. Math. Phys.*, 152(3): 433–478, 1993.

[21] A. Stingo, Global existence and asymptotics for quasi-linear one-dimensional Klein-Gordon equations with mildly decaying Cauchy data, *Bull. Soc. Math. France*, 146(1): 155–213, 2018.

[22] A. Weinstein, Asymptotics of eigenvalue clusters for the Laplacian plus a potential, *Duke Math. J.*, 44(4): 883–892, 1977.

[23] B. Yordanov, Blow-up for the one dimensional Klein-Gordon equation with a cubic nonlinearity, Preprint, 1996.

Université Sorbonne Paris-Nord,
LAGA, CNRS, UMR 7539,
99, Avenue J.-B. Clément,
F-93430 Villetaneuse, France
E-mail address: delort@math.univ-paris13.fr

Advanced Studies in Pure Mathematics 85, 2020
The Role of Metrics in the Theory of Partial Differential Equations
pp. 39–61

Small deformations of spherical biomembranes

Charles M. Elliott, Luke Hatcher and Philip J. Herbert

Abstract.

In this contribution to the proceedings of the 11th Mathematical Society of Japan (MSJ) Seasonal Institute (July 2018) we give an overview of some recent work on a mathematical model for small deformations of a spherical membrane. The idea is to consider perturbations to minimisers of a surface geometric energy. The model is obtained from consideration of second order approximations to a perturbed energy. In particular, the considered problems involve particle constraints and surface phase field energies.

§1. Introduction

Mathematical models in cell biology involving biomembranes are of burgeoning interest within biophysical applications, analysis of geometric partial differential equations and numerical analysis. Biomembranes are thin lipid bilayers which surround cells and some organelles providing a barrier between their interior and exterior. They are principally composed of lipid molecules whose tails are hydrophobic and thus forming a bilayer in which the hydrophilic lipid heads point out of the bilayer. Within the bilayer and its local environment there are a large number of proteins. These proteins have a great variety of forms and purposes. They may be attached to a surface of the bilayer, embedded into it or span the entire bilayer causing local deformations. The great variety of proteins and their possible interactions with the membrane mean there are many models for membrane deformation. They are known to regulate cell morphology and a variety of cellular functions see, for example, [31].

Received May 31, 2018.
Revised July 25, 2019.
2010 *Mathematics Subject Classification.* Primary 35Q80; Secondary 65M60, 92BXX.
Key words and phrases. spherical biomembranes, Canham-Helfrich, phase separation, embedded proteins, phase field, point constraints.

In biophysics, established continuum models treat the biomembrane as a deformable fluid surface of infinitesimal thickness whose deformation is described by an energy functional that depends on the curvature of the surface. Our focus is on equilibrium configurations for biomembranes associated with the interaction of proteins and the curvature of the lipid bilayers. Proteins can induce curvature by local shaping but also elastic forces may organize membrane proteins.

In this contribution we consider two models for small perturbations of the surface energy. First we observe that a large area of study is the effect of embedded inclusions which deform the membrane by their shape [14, 27, 31]. The cross-section of membrane proteins and the membrane surface usually differs dramatically in scale. Here we consider individual proteins as imposing point constraints or forces on the membrane. This leads to fourth order equations with point constraints. Secondly we consider varying concentrations of proteins. The impact on the membrane morphology is then modelled by concentration-dependent material properties, phase-dependent spontaneous curvature, and a gradient energy. For this regime we consider continuous distributions of particles which separate the membrane into two phases with differing properties.

§2. Canham-Helfrich models

A classical energy for the biomembrane is the energy functional [7, 20],

$$\mathcal{E}_B(\Gamma) := \int_\Gamma \left(\frac{1}{2}\kappa(H - H^s)^2 + \sigma + \kappa_G K \right) \mathrm{d}\Gamma. \tag{2.1}$$

Here Γ is a two dimensional hypersurface in $\mathbb{R}^3$ modelling the biomembrane. The interior of the biomembrane is Ω, an open bounded subset of $\mathbb{R}^3$, whose boundary $\partial\Omega = \Gamma$ has outward pointing unit normal ν. We denote by $\mathcal{A}(\Gamma)$ and $\mathcal{V}(\Gamma)$ the area and volume of the surface and enclosed bulk domain which satisfy the isoperimetric inequality

$$\mathcal{A}^3(\Gamma) \geq 36\pi\mathcal{V}^2(\Gamma).$$

The mean and Gaussian curvatures are denoted by H and K. We use the notation ∇_Γ to denote the tangential gradient, and $\mathcal{H} = \nabla_\Gamma \nu$ to denote the extended Weingarten map with H equal to the sum and K being the product of the two principal curvatures which are the eigenvalues of $\mathcal{H}$ associated with two tangential eigenvectors. Note that this definition of the mean curvature differs from the more common definition being the arithmetic mean. The constants $\kappa, \kappa_G > 0$ and $\sigma \geq 0$ measure the bending rigidity and the surface tension of the membrane, respectively. The remaining parameter H^s is the spontaneous curvature, that is the preferred curvature of the membrane. This may be due to asymmetry in the configuration of the lipid bilayer. It may vary over the membrane. Without the surface tension term this energy functional is called the Canham-Helfrich functional. This was introduced by Canham [7] in the case $H^s = 0$. The spontaneous curvature H^s was introduced by Helfrich [20].

The biological context has led to the minimisation problem for these energy functionals being studied with respect to various constraints. Some typical examples would be: taking into account osmotic pressure effects, which leads to a fixed volume constraint; taking the membrane to be incompressible, which leads to an area constraint; supposing that molecules cannot be exchanged between the two layers of the bilayer, which leads to a constraint on the curvature [36]. In the presence of proteins and varying lipid concentrations these equations are typically coupled with nonlinear partial differential equations describing the evolution and location of interfaces, concentrations and particles on the surface.

Difficulties in studying these minimisation problems arise since the Euler-Lagrange equations associated with these surface energy functionals usually are fourth-order, nonlinear (geometric) partial differential equations. For example, some minimisation problems of the Willmore energy were considered in [4, 39] and some minimisation problems for the Helfrich energy were considered in [8, 9, 21].

In the case that κ_G is constant we have by the Gauss-Bonnet theorem

$$\kappa_G \left(\int_\Gamma K + \int_{\partial\Gamma} H_g \right) = \kappa_G 2\pi\chi(\Gamma),$$

where H_g is the geodesic curvature of $\partial\Gamma$ and $\chi(\Gamma)$ is the Euler characteristic. Note that if $\partial\Gamma = \emptyset$ then the corresponding term in the formula vanishes. If either κ_G is zero or the integral of the geodesic curvature over $\partial\Gamma$ is constant then we may neglect the Gaussian curvature term in the energy. The Willmore functional

$$\mathcal{W}(\Gamma) := \int_\Gamma \frac{1}{2} H^2 \, \mathrm{d}\Gamma, \tag{2.2}$$

is a special case, well studied in differential geometry, [42].

Our base functional in this discussion is obtained by setting $H^s = 0$ and supposing the Gaussian curvature term is constant. The resulting energy is

$$\mathcal{E}(\Gamma) = \int_\Gamma \left(\frac{\kappa}{2} H^2 + \sigma\right) \mathrm{d}\Gamma. \tag{2.3}$$

Let Γ_0 be a fixed reference compact hypersurface over which we seek a map $X : \Gamma_0 \to \Gamma \subset \mathbb{R}^3$ that is

$$\Gamma = \{x \in \mathbb{R}^3 \mid x = X(p), p \in \Gamma_0\}.$$

The reference hypersurface Γ_0 might be chosen for convenience of the problem under consideration. In this article we wish to use the perturbation approach of [12] which takes advantage of the evaluation of the first variation of the energy functional at Γ_0 being zero in order to derive quadratic energy functional approximations. This allows the study of approximate models which have a simpler variational structure. In the following sections we choose Γ_0 to be a sphere $\mathbb{S}^2(R)$ of radius R with enclosed volume $V(R)$.

We conclude this section by setting out briefly several problems. In each we constrain the volume enclosed by the surface. We write, with an abuse of notation.

$$\mathcal{E}(X) := \int_{\Gamma_0} \left(\frac{1}{2}\kappa H^2 + \sigma\right) \mathrm{d}\mu$$

where $\mathrm{d}\mu$ denotes the surface area element of Γ associated with the map X.

2.1. Point constraints

We begin with point constraints which model the interaction of a biomembrane with proteins. To begin we suppose that there is an interaction between particles external to the membrane with a quadratic energy.

Problem 2.1. *Let $\{z_j\}_{j=1}^L$ be a collection of disjoint points in $\mathbb{R}^3$. Find $X^* \in K := \{X \in C^2(\Gamma_0) \mid \mathcal{V}(X(\Gamma_0)) = V_0\}$ such that*

$$\mathcal{E}(X^*) + \sum_{j=1}^L \frac{1}{2\delta_j} d(X^*(\Gamma_0), z_j)^2 = \inf_{X \in K} \left(\mathcal{E}(X) + \sum_{j=1}^L \frac{1}{2\delta_j} d(X(\Gamma_0), z_j)^2 \right),$$

where $\delta_j > 0$ might represent the reciprocal of an elastic spring constant associated to the particle attachment.

Here $d\colon \mathbb{R}^3 \times \mathbb{R}^3 \to \mathbb{R}^+$ is the standard metric on $\mathbb{R}^3$, with the usual extension to sets,

$$d(X(\Gamma_0), z) := \inf_{y \in X(\Gamma_0)} d(y, z).$$

In the case of vanishing δ_j we obtain point Dirichlet constraints.

Problem 2.2. *Let* $\{z_j\}_{j=1}^{L}$ *be a collection of disjoint points. Find* $X^* \in K(\Gamma) := \{X \in C^2(\Gamma_0) \mid \exists p_j \in \Gamma_0 : X(p_j) = z_j, j = 1,\ldots,L, \mathcal{V}(X(\Gamma_0)) = V_0\}$ *such that*

$$\mathcal{E}(X^*) = \inf_{X \in K} \mathcal{E}(X).$$

The point constraints in Problem 2.2 are said to be hard constraints. The energy of Problem 2.1 may be viewed as involving soft constraints corresponding to the penalisation of the hard constraints.

2.2. Two phase surfaces

We will consider the case that some intercalated particles within the membrane induce a spontaneous curvature. To model this we introduce an order parameter $\phi : \Gamma \to \mathbb{R}$ to represent the difference of two relative concentrations, which we assume satisfies the constraint $⨍_\Gamma \phi = \alpha$ ($⨍_\Gamma := \frac{1}{|\Gamma|}\int_\Gamma$). Two phases of the membrane can then be distinguished; a particle rich phase and a particle deficient one. We also introduce a Ginzburg-Landau energy functional to approximate the line tension between the differing phases and obtain the energy functional

$$\mathcal{E}_{JL}^{GL}(X,\phi) = \int_{\Gamma_0} \left(\frac{\kappa}{2}\left(H - H^s(\phi)\right)^2 + \sigma + \frac{b\epsilon}{2}|\nabla_\Gamma \phi|^2 + \frac{b}{\epsilon}W(\phi) \right) \mathrm{d}\mu. \tag{2.4}$$

Here $W(\phi)$ is a double well potential, for example given by $W(\phi) = \frac{1}{4}(\phi^2 - 1)^2$, which favours the two distinct values $\phi = \pm 1$ generating phases, $\epsilon \in \mathbb{R}$ is a smallness parameter commensurate with the width of a diffuse interface, $b \in \mathbb{R}$ is a line tension coefficient and as before $\mathrm{d}\mu$ denotes the surface area element of Γ associated with the map X. If we take a phase-dependent spontaneous curvature of the form

$$H^s(\phi) = \Lambda\phi,$$

where $\Lambda \in \mathbb{R}$ is a constant, then we obtain the energy first introduced in [29]. See also [40]. We define the set, K^{GL}, of admissable maps associated with (2.4) by

$$K^{GL} = \left\{ \begin{array}{r} (X,\phi) \in C^2(\Gamma_0;\mathbb{R}^3) \times H^1(\Gamma;\mathbb{R}) : \mathcal{V}(X(\Gamma_0)) = V_0 \\ \text{and } ⨍_{\Gamma_0} \phi \,\mathrm{d}\mu = \alpha \end{array} \right\} \tag{2.5}$$

and consider the minimisation problem:

Problem 2.3. *Find* $(X^*, \phi^*) \in K^{GL}$, *such that*

$$\mathcal{E}_{JL}^{GL}(X^*, \phi^*) = \inf_{(X,\phi) \in K^{GL}} \mathcal{E}_{JL}^{GL}(X, \phi).$$

This minimisation problem for a diffuse interface energy can be understood, for small ϵ, as an approximation to a minimisation problem of a sharp interface energy, [15, 16, 17], introduced in [24, 25]. A three dimensional phase field model was proposed in [41] in which the surface energy was also approximated by a bulk phase field energy.

To write down the sharp interface energy we first let γ_0 denote a smooth curve on Γ_0 that carries line energy and divides Γ_0 into two non-empty regions Γ_0^1 and Γ_0^2, that is $\gamma_0 = \partial\Gamma_0^1 = \partial\Gamma_0^2$ and $\Gamma_0 = \Gamma_0^1 \cup \gamma_0 \cup \Gamma_0^2$. Then given a mapping $X : \Gamma_0 \to \Gamma$ we suppose that

$$\Gamma = \Gamma^1 \cup \gamma \cup \Gamma^2$$

where Γ^1 and Γ^2 are defined by

$$\Gamma^i = \left\{ x \in \mathbb{R}^3 \mid x = X(p),\, p \in \Gamma_0^i \right\}$$

for $i \in \{1, 2\}$ and γ is defined by

$$\gamma = \left\{ x \in \mathbb{R}^3 \mid x = X(p),\, p \in \gamma_0 \right\}.$$

The sharp interface energy is obtained from a generalised Canham-Helfrich energy functional by adding a line energy, [24, 25],

$$\mathcal{E}_{JL}(X) := \sum_{i=1}^{2} \int_{\Gamma_0^i} \left(\frac{\kappa_i}{2}(H - H_i^s)^2 + \sigma \right) \mathrm{d}\mu + \int_{\gamma} \sigma_\gamma \, \mathrm{d}\gamma. \tag{2.6}$$

Here $\sigma_\gamma := C_W b \in \mathbb{R}$ is a line tension coefficient related to b by the constant C_W depending only on the double well. We suppose that $\alpha \in (-1, 1)$ is given. The corresponding sharp interface minimisation problem is then given by

Problem 2.4. *Find* $X^* \in K(X)$,

(2.7)

$$K(X) = \left\{ \begin{array}{r} X \in C^1(\Gamma_0; \mathbb{R}^3) : X|_{\Gamma_0^i} \in C^2(\Gamma_0^i; \mathbb{R}^3), \mathcal{V}(X(\Gamma_0)) = V_0, \\ \textit{and } |\Gamma^1| - |\Gamma^2| = \alpha|\Gamma| \end{array} \right\}$$

such that

$$\mathcal{E}_{JL}(X^*) = \inf_{X \in K} \mathcal{E}_{JL}(X).$$

We anticipate that as $\epsilon \to 0$, Problem 2.4 may be obtained as a limit of Problem 2.3.

§3. Perturbation models of critical points

A standard approach in biophysics is to consider small deformations of flat surfaces written as graphs over the flat domain. This leads to variational problems for quadratic energy functionals involving the height of the graph together with appropriate boundary conditions and constraints modelling the interaction with individual particles, [14]. These ideas were extended to critical points of the surface energy by [12]. In particular the resulting quadratic energies were explicitly computed for the sphere and Clifford torus. In the following, after considering flat domains and the Monge gauge, we describe the calculations of [12].

3.1. Flat domains and the Monge gauge

In the Monge gauge, one first assumes that the surface is nearly flat in the sense that it can be written as a graph

$$\Gamma = \{(x_1, x_2, u(x_1, x_2)) | (x_1, x_2) \in \Omega\} \tag{3.1}$$

over a two-dimensional reference domain $\Omega \subset \mathbb{R}^2$. Then, the mean curvature H and the Gaussian curvature K of the membrane Γ are given by

$$H = -\nabla \cdot \frac{\nabla u}{(1+|\nabla u|^2)^{1/2}},$$
$$K = \left(\frac{\partial^2 u}{\partial x_1^2} \frac{\partial^2 u}{\partial x_2^2} - \left(\frac{\partial^2 u}{\partial x_1 \partial x_2} \right)^2 \right) \Big/ (1+|\nabla u|^2)^{1/2} .$$

A common approach to derive an approximate model is to assume that the displacement of the membrane from the (x_1, x_2) plane produced by the particles is small, i.e. $|\nabla u| \ll 1$. This leads to the well known geometric linearisation

$$\begin{gathered} (1+|\nabla u|^2)^{1/2} \rightsquigarrow 1 + \frac{1}{2}|\nabla u|^2, \\ H \rightsquigarrow -\Delta u, \quad K \rightsquigarrow \frac{\partial^2 u}{\partial x_1^2} \frac{\partial^2 u}{\partial x_2^2} - \left(\frac{\partial^2 u}{\partial x_1 \partial x_2} \right)^2, \end{gathered} \tag{3.2}$$

modelling perturbations from a flat surface yielding, up to a constant term, the quadratic energy

$$\mathcal{E}(u) = \int_\Omega \left(\tfrac{1}{2}\kappa(\Delta u)^2 + \kappa_G \left(\tfrac{\partial^2 u}{\partial x_1^2} \tfrac{\partial^2 u}{\partial x_2^2} - \left(\tfrac{\partial^2 u}{\partial x_1 \partial x_2} \right)^2 \right) + \tfrac{1}{2}\sigma |\nabla u|^2 \right) \mathrm{d}x.$$

Finally, ignoring Gaussian curvature as discussed above, we obtain a quadratic approximation of the energy taking the form

$$\mathcal{E}(u) = \int_\Omega \left(\frac{1}{2}\kappa(\Delta u)^2 + \frac{1}{2}\sigma|\nabla u|^2 \right) \mathrm{d}x. \tag{3.3}$$

The deformation of the membrane from the flat configuration may be in response to small forces or deformation constraints. The minimization of the resulting energy leads to fourth order equations involving the biharmonic operator. These are considerably simpler than the fourth order nonlinear equations arising from minimising the original energy.

Remark 3.1. *The energy* (3.3) *with appropriate perturbations is systematically studied in* [14] *in the case of point constraints arising from attachments of the membrane to proteins modelled as points. A phase field perturbation is studied in* [18].

3.2. First and second variations of geometric quantities on spheres

We now outline the derivation of the approximating quadratic energy functional in the case that the membrane is a small displacement of a sphere. Following [12], the idea is to take advantage of the fact that the sphere is a critical point of the Willmore energy.

3.2.1. *Variations of functionals* It is convenient to define the following Lagrangian

$$\mathcal{L}(\Gamma, \lambda) := \kappa\mathcal{W}(\Gamma) + \sigma\mathcal{A}(\Gamma) + \lambda\left(\mathcal{V}(\Gamma) - V_0\right), \tag{3.4}$$

where λ is the Lagrange multiplier associated with the volume constraint.

Remark 3.2. *λ can be interpreted as a hydrostatic pressure.*

We consider graph like surfaces of the form,

$$\Gamma_\rho(u) := \{x + \rho u(x)\nu(x) : x \in \Gamma_0\} \tag{3.5}$$

where $u \in C^2(\Gamma_0)$, ν is the outward pointing unit normal to $\Gamma_0 = \mathbb{S}^2(R)$ and ρ is a constant. We will assume $\rho \ll 1$ and hence only consider small perturbations. To compute the required derivatives of a functional J we use the following formulae relating them to variations of J.

$$J'(\Gamma_0)[u\nu] := \left.\frac{\mathrm{d}J(\Gamma_\rho)}{\mathrm{d}\rho}\right|_{\rho=0}$$

$$J''(\Gamma_0)[u\nu, u\nu] := \left.\frac{\mathrm{d}^2 J(\Gamma_\rho)}{\mathrm{d}\rho^2}\right|_{\rho=0}.$$

For our case, the second variation will coincide with the variation of the first variation (see Remark 3.2 in [12]) given by

$$J''(\Gamma_0)[g\nu, u\nu] := \left.\frac{\mathrm{d}J'(\Gamma_\rho)[g\nu_\rho]}{\mathrm{d}\rho}\right|_{\rho=0}, \tag{3.6}$$

where ν_ρ is the outward unit normal to Γ_ρ. Hence we will use (3.6) below to calculate the second variation.

The first and second variations of the Willmore functional on a closed surface Γ_0 are

$$\mathcal{W}'(\Gamma_0)[u\nu] = \int_{\Gamma_0} \left(-\Delta_{\Gamma_0} H - H|\mathcal{H}|^2 + \frac{1}{2}H^3\right) u \,\mathrm{d}\Gamma_0. \tag{3.7}$$

$$\begin{aligned}\mathcal{W}''(\Gamma_0)[u\nu, g\nu] &= \int_{\Gamma_0} (\Delta_{\Gamma_0} u + |\mathcal{H}|^2 u)(\Delta_{\Gamma_0} g + |\mathcal{H}|^2 g) \,\mathrm{d}\Gamma_0 \\ &+ \int_{\Gamma_0} 2H\mathcal{H} : (g\nabla_{\Gamma_0}\nabla_{\Gamma_0} u + u\nabla_{\Gamma_0}\nabla_{\Gamma_0} g) \,\mathrm{d}\Gamma_0 \\ &+ \int_{\Gamma_0} \Big(2H\mathcal{H}\nabla_{\Gamma_0} u \cdot \nabla_{\Gamma_0} g - \frac{3}{2}H^2 \nabla_{\Gamma_0} u \cdot \nabla_{\Gamma_0} g \\ &\qquad\qquad - \frac{3}{2}H^2(u\Delta_{\Gamma_0} g + g\Delta_{\Gamma_0} u)\Big) \,\mathrm{d}\Gamma_0 \\ &+ \int_{\Gamma_0} \left(2HTr(\mathcal{H}^3) - \frac{5}{2}H^2|\mathcal{H}|^2 + \frac{1}{2}H^4\right) gu \,\mathrm{d}\Gamma_0\end{aligned} \tag{3.8}$$

The first and second variation of the area functional $\mathcal{A}(\Gamma_0)$ are given by

$$\mathcal{A}'(\Gamma_0)[u\nu] = \int_{\Gamma_0} uH \,\mathrm{d}\Gamma_0, \tag{3.9}$$

$$\mathcal{A}''(\Gamma_0)[u\nu, g\nu] = \int_{\Gamma_0} \nabla_{\Gamma_0} u \cdot \nabla_{\Gamma_0} g + (H^2 - |\mathcal{H}|^2)ug \,\mathrm{d}\Gamma_0. \tag{3.10}$$

The first and second variation of the volume functional $V(\Gamma_0)$ are given by

$$\mathcal{V}'(\Gamma_0)[u\nu] = \int_{\Gamma_0} u \,\mathrm{d}\Gamma_0, \tag{3.11}$$

$$\mathcal{V}''(\Gamma_0)[u\nu, g\nu] = \int_{\Gamma_0} Hgu \,\mathrm{d}\Gamma_0. \tag{3.12}$$

Writing $\lambda_\rho = \lambda_0 + \rho\mu$ we may expand

$$\mathcal{L}(\Gamma_\rho, \lambda_\rho) = \mathcal{L}(\Gamma_0, \lambda_0) + \rho \frac{\mathrm{d}\mathcal{L}(\Gamma_\rho, \lambda_\rho)}{\mathrm{d}\rho}\bigg|_{\rho=0} + \frac{\rho^2}{2}\frac{\mathrm{d}^2\mathcal{L}(\Gamma_\rho, \lambda_\rho)}{\mathrm{d}\rho^2}\bigg|_{\rho=0} + O(\rho^3). \tag{3.13}$$

We follow [12] where the idea is to exploit the vanishing of the first variation of the Lagrangian at (Γ_0, λ_0) and neglect the $O(\rho^3)$ term yielding an approximation based on a quadratic Lagrangian.

Definition 3.1. *For the surface $\Gamma_0 \subset \mathbb{R}^3$ with associated Lagrange multiplier $\lambda_0 \in \mathbb{R}$, the quadratic Lagrangian $L : H^2(\Gamma_0) \times \mathbb{R} \to \mathbb{R}$ is given by*

$$\begin{aligned} L(u, \mu) :&= \frac{1}{2}\frac{\mathrm{d}^2\mathcal{L}(\Gamma_\rho(u), \lambda(\rho))}{\mathrm{d}\rho^2}\bigg|_{\rho=0} \\ &= \frac{1}{2}\mathcal{E}''(\Gamma_0)[u\nu, u\nu] + \frac{1}{2}\lambda_0\mathcal{V}''(\Gamma_0)[u\nu, u\nu] + \mu\mathcal{V}'(\Gamma_0)[u\nu]. \end{aligned}$$

3.2.2. *Variations of functionals on sphere* Since a sphere has zero first variation of the Willmore functional, the first variation of the Lagrangian with respect to Γ evaluated on $\mathbb{S}^2(R)$ in the normal direction is

$$\sigma\frac{2}{R} + \lambda.$$

Setting $\lambda_0 = -2\sigma/R$ we observe that the first variation vanishes when $\lambda = \lambda_0$ so (Γ_0, λ_0) is a critical point of the Lagrangian $\mathcal{L}(\cdot, \cdot)$ with $V_0 = V(R)$.

The second variations on $\Gamma_0 = \mathbb{S}^2(R)$ may be calculated to be

$$\begin{aligned} \mathcal{W}''(\Gamma_0)[u\nu, g\nu] &= \int_{\Gamma_0}\left(\Delta_{\Gamma_0}u\Delta_{\Gamma_0}g - \frac{2}{R^2}\nabla_{\Gamma_0}u\cdot\nabla_{\Gamma_0}g\right)\mathrm{d}\Gamma_0, \\ \mathcal{A}''(\Gamma_0)[u\nu, g\nu] &= \int_{\Gamma_0}\left(\nabla_{\Gamma_0}u\cdot\nabla_{\Gamma_0}g + \frac{2}{R^2}ug\right)\mathrm{d}\Gamma_0, \\ \mathcal{V}''(\Gamma_0)[u\nu, g\nu] &= \frac{2}{R}\int_{\Gamma_0} ug\,\mathrm{d}\Gamma_0. \end{aligned}$$

It follows that

$$\begin{aligned} L(u, \mu) = \frac{\kappa}{2}\mathcal{W}''(\Gamma_0)[u\nu, u\nu] + \frac{\sigma}{2}\mathcal{A}''(\Gamma_0)[u\nu, u\nu] \\ - \frac{\sigma}{R}\mathcal{V}''(\Gamma_0)[u\nu, u\nu] + \mu\mathcal{V}'(\Gamma_0)[u\nu] \end{aligned}$$

yielding

(3.14)

$$L(u,\mu)=\int_{\Gamma_0}\left(\frac{\kappa}{2}|\Delta_{\Gamma_0}u|^2+\frac{1}{2}\left(\sigma-\frac{2\kappa}{R^2}\right)|\nabla_{\Gamma_0}u|^2-\frac{\sigma}{R^2}u^2+\mu u\right)\mathrm{d}\Gamma_0.$$

so that

$$L(u,\mu)=\frac{1}{2}a(u,u)+\mu(u,1)_{L^2(\Gamma_0)}. \tag{3.15}$$

Here we have defined the bilinear form associated to the quadratic part of $L(\cdot,\cdot)$, $a(\cdot,\cdot)\colon H^2(\Gamma_0)\times H^2(\Gamma_0)\to\mathbb{R}$ as follows,

(3.16)

$$a(u,v)=\int_{\Gamma_0}\left(\kappa\Delta_{\Gamma_0}u\Delta_{\Gamma_0}v+\left(\sigma-\frac{2\kappa}{R^2}\right)\nabla_{\Gamma_0}u\cdot\nabla_{\Gamma_0}v-\frac{2\sigma}{R^2}uv\right)\mathrm{d}\Gamma_0.$$

It can be shown that $a(\cdot,\cdot)$ is coercive over the space $\{1,\nu_1,\nu_2,\nu_3\}^\perp$ [12], where $\perp$ is meant in the sense of $H^2(\Gamma_0)$. However, since these functions are positive eigenfunctions of $-\Delta_{\Gamma_0}$ it is equivalent to take the $L^2(\Gamma_0)$ orthogonality. Note on $\mathbb{S}^2(R)$ that $\nu=x/R$. The orthogonality constraint

$$(1,u)=0$$

may be interpreted as enforcing the volume constraint and

$$(x_i,u)=0,\ \ i=1,2,3$$

as ensuring the surface has a fixed center of mass.

Remark 3.3. *The Monge-Gauge approximation* (3.3) *for near flat membranes is recovered by formally taking $R\to\infty$ in* (3.16).

3.2.3. *Perturbed energy functional* We wish to consider critical points of the perturbed Lagrangian

$$\mathcal{L}_\rho(\Gamma_\rho(u),\lambda_\rho)=\mathcal{L}(\Gamma_\rho(u),\lambda_\rho)+\mathcal{F}_\rho(\Gamma_\rho(u)). \tag{3.17}$$

Here $\mathcal{F}_\rho(\cdot)$ is a perturbative energy whose precise form depends on the type of deformation being considered. Here we supose that it may be expanded as

$$\mathcal{F}_\rho(\Gamma_\rho)=\rho\mathcal{F}_0(\Gamma_0)+\rho^2\mathcal{F}_1(\Gamma_0)[u\nu]+\mathcal{O}(\rho^3). \tag{3.18}$$

We write the Taylor expansion of (3.17) with $\Gamma_\rho(u)$ set as in (3.5) and $\lambda_\rho = \lambda_0 + \mu\rho$ up to third order in ρ as

$$
\begin{aligned}
\mathcal{L}_\rho(\Gamma_\rho(u), \lambda_\rho) &= \mathcal{L}(\Gamma_0, \lambda_0) \\
&+ \rho \left(\mathcal{E}'(\Gamma_0)[u\nu] + \lambda_0 \mathcal{V}'(\Gamma_0)[u\nu] + \mu(\mathcal{V}(\Gamma_0) - V_0) + \mathcal{F}_0(\Gamma_0) \right) \\
&+ \frac{\rho^2}{2} \left(\mathcal{E}''(\Gamma_0)[u\nu, u\nu] + \lambda_0 \mathcal{V}''(\Gamma_0)[u\nu, u\nu] + 2\mu\mathcal{V}'[u\nu] + 2\mathcal{F}_1(\Gamma_0)[u\nu] \right) \\
&+ \mathcal{O}(\rho^3).
\end{aligned}
$$

Evaluating on the critical point $(\mathbb{S}^2(R), \lambda_0 = \frac{-2\sigma}{R})$ of $\mathcal{L}(\cdot, \cdot)$ we find

$$
\begin{aligned}
\mathcal{L}_\rho(\Gamma_\rho(u), \lambda(\rho)) = \mathcal{L}(\Gamma_0, \lambda_0) &+ \rho \mathcal{F}_0(\Gamma_0) \\
&+ \frac{\rho^2}{2} \left(L(u, \mu) + 2\mathcal{F}_1(\Gamma_0)[u\nu] \right) + \mathcal{O}(\rho^3).
\end{aligned}
$$

It follows that up to third order in ρ a critical point of (3.17) may be approximated by finding a critical point (u, μ) of

$$
\mathcal{J}(u, \mu) := L(u, \mu) + 2F(u) \tag{3.19}
$$

where $F(u) := \mathcal{F}_1(\Gamma_0)[u\nu]$.

§4. Point constraints

Here we further explore problems where we wish to minimise the energy (2.3) subject to point constraints as described in Problems 2.1 and 2.2. From the physical perspective this models the interaction of proteins with a biomembrane. To include such mechanisms, we wish to allow for groups of point constraints to move rigidly. A group of point constraints may be considered to be representative of a large biological structure, for example a BAR protein [22], which are believed to attach to the membrane at a finite number of points. We consider N such groupings of identical particles, each of which has L points of attachment. Following [27], for $q \in \mathbb{R}^6$, we write

$$
\phi(q, x) := R_x(q_1) R_y(q_2) R_z(q_3) x + (q_4, q_5, q_6)^T \tag{4.1}
$$

to be the map describing the rigid transformation, where R_x, R_y and R_z are the rotations in the x, y and z axes respectively. This then allows us to pose the Problems 2.1 and 2.2 in the setting of the surface being a graph with rigid transformations of particles.

Problem 4.1. *Find* u, $\{q^i\}_{i=1}^N$ *minimising*

$$\int_{\Gamma_\rho(u)} \Big(\frac{\kappa}{2}H^2 + \sigma\Big)\, \mathrm{d}\Gamma_\rho(u) + \frac{1}{2\delta}\sum_{i=1}^{N}\sum_{j=1}^{L} d(\Gamma_\rho(u), \phi(q^i, z_j))^2 \tag{4.2}$$

subject to $\mathcal{V}(\Gamma_\rho(u)) = V_0$.

Problem 4.2. *Find* u, $\{q^i\}_{i=1}^N$ *minimising* (2.3) *subject to* $\phi(q^i, z_j) \in \Gamma_\rho(u)$ *for* $j = 1, \ldots, L$, $i = 1, \ldots, N$ *and* $\mathcal{V}(\Gamma_\rho(u)) = V_0$.

Remark 4.1. *Something we have neglected to consider is that the protein is not only a collection of points which are attached to the membrane, but also may be inside or outside the interior of the surface. This may be modelled with inequality constraints.*

We see that these problems are highly non-linear and should they be solvable, numerical methods will be challenging. We wish to exploit ρ as a small parameter writing

$$\mathcal{F}_\rho(\Gamma_\rho) = \frac{1}{2\delta}\sum_{i=1}^{N}\sum_{j=1}^{L} d\big(\Gamma_\rho(u), \phi(q^i, z_j(\rho))\big)^2$$

in order to simplify the problem as discussed in sub-section 3.2.3. Here as we assume the deformations are small, we write the points, z_j, as a graph over Γ_0, specifically,

$$z_j = p_j + \rho Z_j \nu(p_j) \tag{4.3}$$

for each $j = 1, \ldots, L$, where $Z_j \in \mathbb{R}$ and the $p_j \in \Gamma_0$ are distinct.

We now limit ourselves to a single group of constraints ($N = 1$) and we fix q so that we discuss only the geometric minimisation for a fixed location of a single particle.

As in [12], we calculate

$$d(\Gamma_\rho(u), p_j + \rho Z_j \nu(p_j))|_{\rho=0} = d(\Gamma_0, p_j) = 0,$$

with first derivative term,

$$\begin{aligned}\frac{\mathrm{d}}{\mathrm{d}\rho} d(\Gamma_\rho(u), p_j + \rho Z_j\nu(p_j))|_{\rho=0} &= \nabla d(\Gamma_0, p_j)\cdot \frac{\mathrm{d}}{\mathrm{d}\rho}(p_j + \rho Z_j \nu(p_j))|_{\rho=0} \\ &\quad + \dot{\partial} d(\Gamma_\rho(u), p_j)|_{\rho=0} \\ &= Z_j - u(p_j),\end{aligned}$$

which results in

$$d(\Gamma_\rho(u), p_j + \rho Z_j \nu(p_j))^2 = \rho^2 (Z_j - u(p_j))^2 + \mathcal{O}(\rho^3) \tag{4.4}$$

and

$$F(u) = \frac{1}{2\delta}\sum_{j=1}^{L} |u(p_j) - Z_j|^2. \tag{4.5}$$

The Lagrangian (3.19) becomes

$$\mathcal{J}(u,\mu) = \frac{1}{2}a(u,u) + \mu(1,u) + \frac{1}{2\delta}\sum_{j=1}^{L} |u(p_j) - Z_j|^2. \tag{4.6}$$

By considering

$$U_0 := \{v \in H^2(\Gamma_0) \mid (1,v) = 0\}$$

we may now consider finding a minimiser of the quadratic functional

$$\mathcal{Q}(u) := \frac{1}{2}a(u,u) + \frac{1}{2\delta}\sum_{j=1}^{L} |u(p_j) - Z_j|^2 \tag{4.7}$$

over U_0. In [23] it is seen that the largest subspace of U_0 where a is coercive is

$$U_\nu := \left\{v \in H^2(\Gamma_0) : v \in (\mathrm{Sp}\{1,\nu_1,\nu_2,\nu_3\})^\perp\right\}.$$

In the case of $L \geq 4$ and there is j_1, j_2, j_3, j_4 with $\{p_{j_k}\}_{k=1}^4$ non-coplanar, it may be seen that Q is coercive over U_0. We now have the following existence and uniqueness results, which follow from the Lax-Milgram theorem.

Theorem 4.1. *Given distinct $\{p_i\}_{i=1}^L \subset \Gamma_0$ and $\{Z_i\}_{i=1}^L \subset \mathbb{R}$, with penalty parameter $\delta > 0$, there is a unique $u^\delta \in U_\nu$ such that for any $v \in U_\nu$,*

$$a(u^\delta, v) + \frac{1}{\delta}\sum_{j=1}^{L} u^\delta(p_j)v(p_j) = \frac{1}{\delta}\sum_{j=1}^{L} Z_j v(p_j).$$

Theorem 4.2. *Given disjoint $\{p_i\}_{i=1}^L \subset \Gamma_0$, and $\{Z_i\}_{i=1}^L \subset \mathbb{R}$, there is a unique $u \in U_\nu$ such that $u(p_i) = Z_i$ for $i = 1, \ldots, L$ and for any $v \in U_\nu$ with $v(p_i) = 0$ for $i = 1, \ldots, L$, it holds*

$$a(u,v) = 0.$$

Viewing u^δ as a penalised approximation to u, an interesting question to now ask is, as $\delta \to 0^+$, does $u^\delta \to u$? If so, in what sense does it converge (i.e. what topology)? If it converges in a norm, can we show a reasonable error estimate? These questions lead to the following convergence result.

Theorem 4.3. *Let $u^\delta \in U_\nu$ be as in Theorem 4.1, let $u \in U_\nu$ be as in Theorem 4.2, with the constraints $\{p_j\}_{j=1}^L \subset \Gamma_0$ and $\{Z_j\}_{j=1}^L \subset \mathbb{R}$. Then there is $C > 0$ independent of δ such that*

$$\|u - u^\delta\|_{H^2(\Gamma_0)} \leq C\sqrt{\delta}.$$

We have previously fixed q for convenience since finding a minimiser for fixed q is a standard variational problem, whereas the problem for minimising q is not variational. The problem for finding a minimising q is developed as a gradient descent in [27], which deals with a near flat membrane and also that of a near cylindrical membrane.

The methods for numerically solving the penalty problems are based on splitting the fourth order problem into coupled second order problems [13]. The methods for extending this to the case of the hard constraint problem are to be found in a work in preparation by Elliott and Herbert.

The simulations illustrated in Figures 4.1, 4.2 and 4.3 are implemented using the DUNE framework [3], in particular the ALU-Grid module [1].

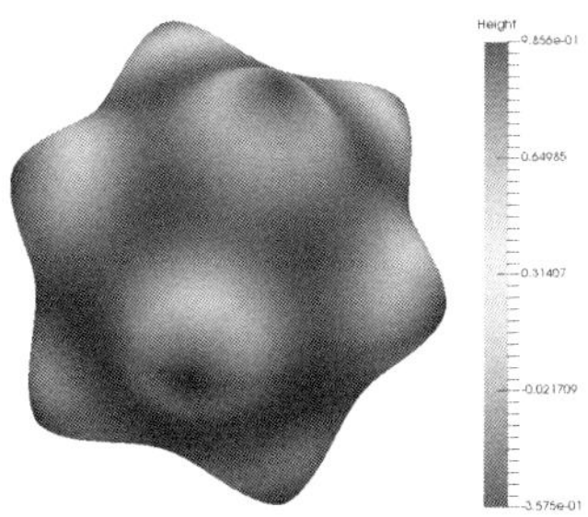

Fig. 4.1. Example numerical simulation with $\delta = 10^{-4}$ and 12 points X_i positioned on the vertices of a regular icosahedron with heights 1, $\rho = 0.2$.

§5. Phase field interaction

The second perturbation model we review takes into account phase separation between two differing lipid compositions within the membrane. Phase separation has been observed in artificial membranes such as in giant unilamellar vesicles (GUVs) or supported lipid bilayers (SLBs). Experimental results involving GUVs and SLBs suggest the curvature of the biological membrane plays an important role in the

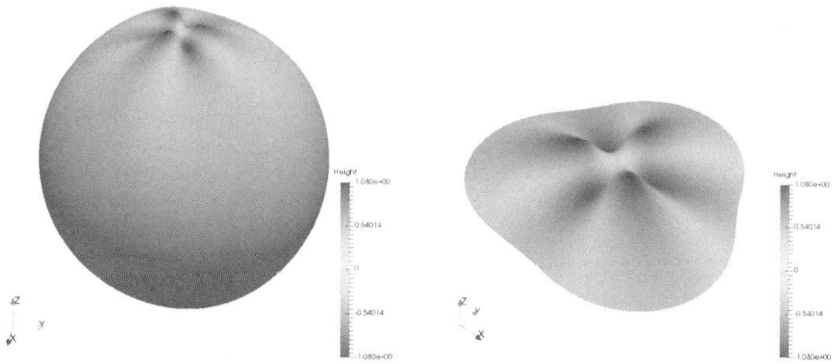

Fig. 4.2. Example simulation with $\delta = 10^{-6}$, two particles positioned around each of the North and South poles with 6 points of attachment, placed with heights alternating between -1 and 1, antisymmetrically, $\rho = 0.05$. Left: whole sphere; Right: zoom in of North pole.

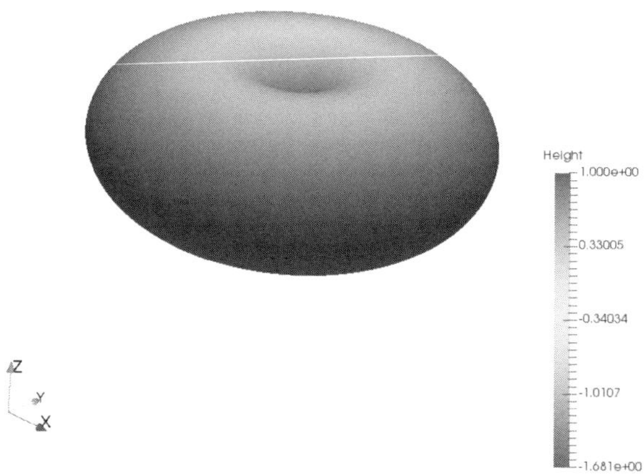

Fig. 4.3. Example simulation with $\delta = 10^{-6}$, with 10 points of attachment equally spaced on the equator, with height 1, $\rho = 0.6$.

formation of these domains [5, 6, 32, 34, 35]. Lipid rafts are small (10-200nm) domains which compartmentalise cellular functions such as signal transduction, protein sorting and host-pathogen interactions [33, 38].

Here we introduce a concentration ϕ taking the values ± 1 in each of two phases and consider the energy

$$\mathcal{E}(\Gamma, \phi) = \kappa \mathcal{W}(\Gamma) + \sigma \mathcal{A}(\Gamma) - \kappa \Lambda \mathcal{I}(\Gamma, \phi) + b \mathcal{P}(\Gamma, \phi) \tag{5.1}$$

subject to the following concentration and volume constraints

$$\mathcal{R}(\Gamma, \phi) := \fint_{\Gamma} \phi = \alpha, \qquad \mathcal{V}(\Gamma) = V(R). \tag{5.2}$$

Here $\mathcal{I}(\cdot,\cdot)$ and $\mathcal{P}(\cdot,\cdot)$ are given by

$$\mathcal{I}(\Gamma, \phi) := \int_{\Gamma} H\phi \, \mathrm{d}\Gamma, \tag{5.3}$$

$$\mathcal{P}(\Gamma, \phi) := \int_{\Gamma} \left(\frac{\epsilon}{2} |\nabla_{\Gamma}\phi|^2 + \frac{1}{\epsilon} f(\phi) \right) \mathrm{d}\Gamma \tag{5.4}$$

and where

$$f(\phi) := W(\phi) + \frac{\epsilon\kappa\Lambda^2\phi^2}{2b}. \tag{5.5}$$

This leads to the consideration of the Lagrangian

$$\mathcal{L}(\Gamma, \phi, \lambda, \eta) := \mathcal{E}(\Gamma, \phi) + \lambda(\mathcal{V}(\Gamma) - V_0) + \eta(\mathcal{R}(\Gamma, \phi) - \alpha) \tag{5.6}$$

where λ and η are the Lagrange multipliers for the constraints.

The energy (5.1) may be rewritten to be in the form (2.4) in which the protein molecules induce a spontaneous curvature of the form

$$H^s(\phi) = \Lambda\phi.$$

This model was proposed in [29]. Similar models have been considered in [15, 16, 17, 19, 41] where phase-dependent bending rigidities were considered.

We now suppose that the spontaneous curvature is small leading to a perturbation of the underlying equilibrium. It is reasonable to expect the deformations to be of the same order, and hence we scale both of these terms by ρ. In [28] it was calculated that the line tension for lipid rafts should depend quadratically on the spontaneous curvature, so we scale this term by ρ^2. This leads to a perturbation model for which we have the scaling

$$\Gamma \rightsquigarrow \Gamma_\rho, \qquad \Lambda \rightsquigarrow \rho\Lambda, \qquad b \rightsquigarrow \rho^2 b.$$

Carrying out calculations similar to those given earlier leads to an energy

$$E(u, \phi) := \frac{1}{2} a(u, u) + I(u, \phi) + G(\phi) \tag{5.7}$$

where

$$I(u,\phi) := \int_{\Gamma_0} \left(\kappa\Lambda\phi\Delta_{\Gamma_0} u + \frac{2\kappa\Lambda u\phi}{R^2} \right) d\Gamma_0, \tag{5.8}$$

$$G(\phi) := b \int_{\Gamma_0} \left(\frac{\epsilon}{2}|\nabla_{\Gamma_0}\phi|^2 + \frac{1}{\epsilon} f(\phi) \right) d\Gamma_0. \tag{5.9}$$

The perturbed equations for the volume and mean concentration constraints become

$$\int_{\Gamma_0} u = 0 \qquad \text{and} \qquad \fint_{\Gamma_0} \phi = \alpha. \tag{5.10}$$

Formally considering $R \to \infty$ we obtain the approximation in the case Γ_0 is a plane. This is in agreement with the linearisation outlined in [14, 18] and the energy used in [2, 26, 29, 30, 37].

Observe that for components of the normal ν_i we have that

$$E(u + \tau\nu_i, \phi) = E(u,\phi) \text{ for all } \tau \in \mathbb{R} \text{ and } i \in \{1,2,3\}. \tag{5.11}$$

This is reasonable, since this corresponds to order ρ translations of the centre of mass, for which we would expect the energy to be invariant. We can therefore further constrain (5.7) by requiring that

$$\int_{\Gamma_0} u\nu_i = 0 \qquad \text{for } i \in \{1,2,3\} \tag{5.12}$$

to remove this redundancy. This has the added benefit that $E(\cdot,\cdot)$ is coercive over this subspace. By considering $u = \tau\nu_1$ and $\phi = \alpha$, it follows that (5.12) is neccessary for coercivity since

$$\lim_{\tau\to\infty} \left(\|\tau\nu_1\|_{H^2(\Gamma_0)} + \|\alpha\|_{H^1(\Gamma_0)} \right) = \infty$$

but

$$E(\tau\nu_1, \alpha) = \frac{b}{\epsilon}\int_{\Gamma_0} f(\alpha) \leq C$$

where C is some constant independent of τ. In fact by defining

$$\mathcal{K} := \left\{ (u,\phi) \in H^2(\Gamma_0) \times H^1(\Gamma_0) : u \in \mathrm{Sp}\{1,\nu_1,\nu_2,\nu_3\}^{\perp} \text{ and } \fint_{\Gamma_0} \phi = \alpha \right\},$$

it follows that this is sufficient since we can prove there exists positive constants C_1 and C_2 such that the energy functional $E(\cdot,\cdot) : \mathcal{K} \to \mathbb{R}$ satisifies the following inequality

$$E(u,\phi) \geq C_1 \left(\|\phi\|^2_{H^1(\Gamma_0)} + \|u\|^2_{H^2(\Gamma_0)} \right) - C_2 \qquad \forall (\phi, u) \in \mathcal{K}.$$

Hence it can be shown using the direct method of calculus of variations that there exists a solution to the following problem.

Problem 5.1. *Find* $(u^*, \phi^*) \in \mathcal{K}$ *such that*

$$E(u^*, \phi^*) = \inf_{(u,\phi)\in\mathcal{K}} E(u, \phi).$$

Finally, in order to calculate critical points of $E(\cdot, \cdot)$ over $\mathcal{K}$ it is convenient to investigate these numerically using a conserved L^2-gradient flow. This yields the following system

(5.13)
$$\begin{cases} \alpha_1 \phi_t + \frac{b}{\epsilon} f'(\phi) - b\epsilon \Delta_{\Gamma_0} \phi + \kappa \Lambda \Delta_{\Gamma_0} u + \frac{2\kappa\Lambda u}{R^2} + \lambda_\phi = 0 \\ \alpha_2 u_t - \left(\sigma - \frac{2\kappa}{R^2}\right) \Delta_{\Gamma_0} u + \kappa \Delta^2_{\Gamma_0} u - \frac{2\sigma u}{R^2} + \kappa\Lambda \Delta_{\Gamma_0} \phi + \frac{2\kappa\Lambda\phi}{R^2} + \lambda_u = 0 \end{cases}$$

subject to some initial conditions which satisfy the constraints (5.10) and (5.12). These constraints are preserved by the evolution when values of the Lagrange multipliers corresponding to the constraints (5.10) have the following explicit form

$$\lambda_\phi = -\frac{b}{\epsilon} \fint_{\Gamma_0} f'(\phi), \tag{5.14}$$

$$\lambda_u = -\frac{2\kappa\Lambda\alpha}{R^2}. \tag{5.15}$$

DUNE software with a PYTHON module (c.f. [10]) was used to implemement a surface FEM method, [11], for (5.13). Some examples of almost stationary solutions are given in Figure 5.1. Additional details and results for this model will be given in a work in preparation by Elliott, Hatcher and Stinner.

§6. Outlook

Well posedness and numerical analysis of the problems described in Sections 4 and 5 with detailed proofs are the subjects of works in preparation by the authors. These works also include geometric motion forced by surface deformation for the sharp interface limit of the phase field problem and the movement of point particles on the surface.

§ Acknowledgements

The research of C. M. Elliott was partially supported by the Royal Society via a Wolfson Research Merit Award. L. Hatcher and P. Herbert

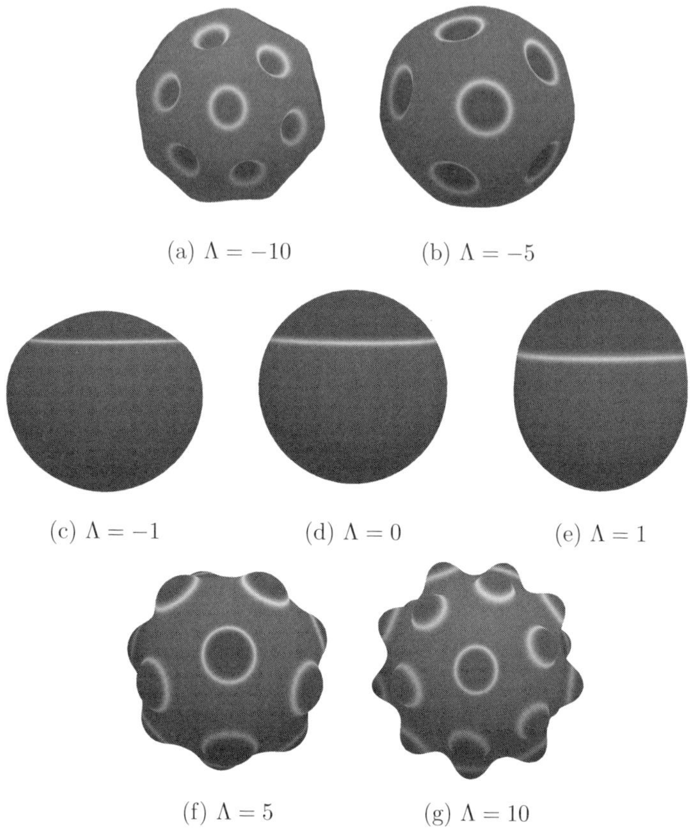

Fig. 5.1. Effect of varying the coupling coefficient Λ. The simulations show the two phases of the order parameter ϕ. In each case the surface is deformed according to u. For illustration purposes $\rho = 1$, whereas in reality it would be much smaller.

were supported by the UK Engineering and Physical Sciences Research Council (EPSRC) Grant EP/H023364/1 within the MASDOC Centre for Doctoral Training.

References

[1] M. Alkämper, A. Dedner, R. Klöfkorn and M. Nolte, *The DUNE-ALUGrid Module*, Archive of Numerical Software, 4 (2016), pp. 1–28.

[2] D. T. Andelman, T. Kawakatsu and K. Kawasaki, *Equilibrium shape of two-component unilamellar membranes and vesicles*, Europhys. Lett., 19 (1992), pp. 57–62.

[3] P. Bastian, M. Blatt, A. Dedner, C. Engwer, R. Klofkorn, M. Oldberger and O. Sander, *A Generic grid interface for parallel and adaptive scientific computing. part* II: *Impelementation and tests in DUNE*, Computing, 82, (2008), pp. 121–138.

[4] M. Bauer and E. Kuwert, *Existence of minimizing Willmore surfaces of prescribed genus*, International Mathematics Research Notices, 2003 (2003), pp. 553–576.

[5] T. Baumgart, S. Das, W. W. Webb and J. T. Jenkins, *Membrane elasticity in giant vesicles with fluid phase coexistence*, Biophysical Journal, 89 (2005), pp. 1067–1080.

[6] T. Baumgart, S. Hess and W. W. Webb, *Imaging coexisting fluid domains in biomembrane models coupling curvature and line tension*, Nature, 425 (2003), pp. 821–824.

[7] P. Canham, *The minimum energy of bending as a possible explanation of the biconcave shape of the red blood cell*, J. Theor. Biol., 26 (1970), pp. 61–81.

[8] R. Choksi, M. Morandotti and M. Veneroni, *Global minimizers for axisymmetric multiphase membranes*, ESAIM: Control, Optimisation and Calculus of Variations, 19 (2013), pp. 1014–1029.

[9] R. Choksi and M. Veneroni, *Global minimizers for the doubly-constrained Helfrich energy: the axisymmetric case*, Calculus of Variations and Partial Differential Equations, 48 (2013), pp. 337–366.

[10] A. Dedner and M. Nolte, *The DUNE PYTHON module*, arxiv preprint, arxiv:1807.05252, (2018).

[11] G. Dziuk and C. M. Elliott, *Finite element methods for surface partial differential equations*, Acta Numerica, 22 (2013), pp. 289–396.

[12] C. M. Elliott, H. Fritz and G. Hobbs, *Small deformations of Helfrich energy minimising surfaces with applications to biomembranes*, Math. Mod. Meth. Appl. Sci., 27 (2017), pp. 1547–1586.

[13] ———, *Second order splitting for a class of fourth order problems*, Math. Comp., 88 (2019), 2635–2664.

[14] C. M. Elliott, C. Gräser, G. Hobbs, R. Kornhuber and M.-W. Wolf, *A variational approach to particles in lipid membranes*, Archive for Rational Mechanics and Analysis, 222 (2016), pp. 1011–1075.

[15] C. M. Elliott and B. Stinner, *Modeling and computation of two phase geometric biomembranes using surface finite elements*, J. Comput. Phys., 229 (2010), pp. 6585–6612.

[16] ———, *A surface phase field model for two-phase biological membranes*, SIAM Journal on Applied Mathematics, 70 (2010), pp. 2904–2928.

[17] ———, *Computations of two-phase biomembranes with phase dependent material parameters using surface finite elements*, Communications in Computational Physics, 13 (2012), pp. 325–360.

[18] I. Fonseca, G. Hayrapetyan, G. Leoni and B. Zwicknagl, *Domain formation in membranes near the onset of instability*, Journal of Nonlinear Science, 26 (2016), pp. 1191–1225.

[19] T. J. Healey and S. Dharmavaram, *Symmetry-breaking global bifurcation in a surface continuum phase field model for lipid bilayer vesicles*, SIAM Journal on Mathematical Analysis, 49 (2017), pp. 1027–1059.

[20] W. Helfrich, *Elastic properties of lipid bilayers: theory and possible experiments*, Z. Naturforsch, C 28 (1973), pp. 693–703.

[21] M. Helmers, *Convergence of an approximation for rotationally symmetric two-phase lipid bilayer membranes*, Quarterly Journal of Mathematics, 66 (2015), pp. 143–170.

[22] W. M. Henne, H. M. Kent, M. G. Ford, B. G. Hegde, O. Daumke, P. J. G. Butler, R. Mittal, R. Langen, P. R. Evans and H. T. McMahon, *Structure and analysis of FCHo2 F-BAR domain: a dimerizing and membrane recruitment module that effects membrane curvature*, Structure, 15 (2007), pp. 839–852.

[23] G. Hobbs, *Particles and biomembranes: a variational PDE approach*, PhD thesis, The University of Warwick, 2016.

[24] F. Jülicher and R. Lipowsky, *Domain-induced budding of vesicles*, Phys. Rev. Lett., 70 (1993), pp. 2964–2967.

[25] ———, *Shape transformations of vesicles with intramembrane domains*, Phys. Rev. E, 53 (1996), pp. 2670–2683.

[26] T. Kawakatsu, D. Andelman, K. Kawasaki and T. Taniguchi, *Phase transitions and shapes of two component membranes and vesicles* I: *strong segregation limit*, J. Phys. II (France), 3 (1993), pp. 971–997.

[27] T. Kies, *Gradient methods for membrane-mediated particle interactions*, PhD thesis, Institut fur Mathematik, Frei Universitat Berlin, 2019.

[28] P. I. Kuzmin, S. A. Akimov, Y. A. Chizmadzhev, J. Zimmerberg and F. S. Cohen, *Line tension and interaction energies of membrane rafts calculated from lipid splay and tilt*, Biophysical journal, 88 (2005), pp. 1120–1133.

[29] S. Leibler, *Curvature instability in membranes*, Journal de Physique, 47 (1986), pp. 507–516.

[30] S. Leibler and D. Andelman, *Ordered and curved meso-structures in membranes and amphiphilic films*, Journal de physique, 48 (1987), pp. 2013–2018.

[31] H. T. McMahon and J. L. Gallop, *Membrane curvature and mechanisms of dynamic cell membrane remodelling*, Nature, 438 (2005), pp. 590–596.

[32] R. Parthasarathy, C.-H. Yu and J. T. Groves, *Curvature-modulated phase separation in lipid bilayer membranes*, Langmuir, 22 (2006), pp. 5095–5099.

[33] L. J. Pike, *Rafts defined: a report on the keystone symposium on lipid rafts and cell function*, Journal of lipid research, 47 (2006), pp. 1597–1598.

[34] M. Rinaldin, P. Fonda, L. Giomi and D. J. Kraft, *Geometric pinning and antimixing in scaffolded lipid vesicles*, arXiv preprint, arXiv:1804.08596, (2018).

[35] S. Rozovsky, Y. Kaizuka and J. T. Groves, *Formation and spatio-temporal evolution of periodic structures in lipid bilayers*, Journal of the American Chemical Society, 127 (2005), pp. 36–37.

[36] U. Seifert, *Configurations of fluid membranes and vesicles*, Adv. Phys., 46 (1997), pp. 1–137.

[37] M. Seul and D. Andelman, *Domain shapes and patterns: the phenomenology of modulated phases*, Science, 267 (1995), p. 476.

[38] E. Sezgin, I. Levental, S. Mayor and C. Eggeling, *The mystery of membrane organization: composition, regulation and roles of lipid rafts*, Nature Reviews Molecular Cell Biology, 18 (2017), p. 361.

[39] L. Simon, *Existence of surfaces minimizing the Willmore functional*, Communications in Analysis and Geometry, 1 (1993), pp. 281–326.

[40] T. Taniguchi, *Shape deformation and phase separation dynamics of two-component vesicles*, Phys. Rev. Lett., 76 (1996), pp. 4444–4447.

[41] X. Wang and Q. Du, *Modelling and simulations of multi-component lipid membranes and open membranes via diffuse interface approaches*, J. Math. Biol., 56 (2008), pp. 347–371.

[42] T. J. Willmore, *Riemannian Geometry*, Clarendon Press, Oxford, 1993.

Mathematics Institute, University of Warwick
E-mail address: c.m.elliott@warwick.ac.uk
E-mail address: l.hatcher@warwick.ac.uk
E-mail address: p.j.herbert@warwick.ac.uk

Advanced Studies in Pure Mathematics 85, 2020
The Role of Metrics in the Theory of Partial Differential Equations
pp. 63–93

The thresholding scheme for mean curvature flow and de Giorgi's ideas for minimizing movements

Tim Laux and Felix Otto

Abstract.

We consider the thresholding scheme and explore its connection to De Giorgi's ideas on gradient flows in metric spaces; here applied to mean curvature flow as the steepest descent of the interfacial area. The basis of our analysis is the observation by Esedoğlu and the second author that thresholding can be interpreted as a minimizing movements scheme for an energy that approximates the interfacial area. De Giorgi's framework provides an optimal energy dissipation relation for the scheme in which we pass to the limit to derive a dissipation-based weak formulation of mean curvature flow. Although applicable in the general setting of arbitrary networks, here we restrict ourselves to the case of a single interface, which allows for a compact, self-contained presentation.

§1. Introduction and context

The purpose of these notes is to draw a connection between De Giorgi's tools for minimizing movements, that is, gradient flows in metric spaces on the one hand, and the very popular thresholding scheme for flow of a hyper-surface by its mean curvature on the other hand. While we have developed this connection in the case of multiple phases with surface energies and mobilities depending on the pair of phases, as is relevant for grain growth in polycrystals, and when the notion of viscosity solution is not available, we present our results here in the simplest setting of two phases. Our presentation is essentially self-contained.

Received May 6, 2019.
Revised July 9, 2019.
2010 *Mathematics Subject Classification.* Primary 35A15; Secondary 53E10, 65M12, 74N20.
Key words and phrases. thresholding, diffusion generated motion, mean curvature flow, gradient flows, minimizing movements.

What makes the evolution of the boundary $\partial\Omega$ of a set Ω by its mean curvature H valuable for modeling in materials science is that it is driven by the reduction of the (total) interfacial area of $\partial\Omega$, which relies on the mean curvature H, the sum of the principal curvatures, being the first variation of the interfacial area. There is a more intimate connection between mean curvature flow (MCF) and the functional E of interfacial area of a configuration: MCF can formally be understood as a gradient flow of E. We stress that a dynamical system that can be written as a gradient flow, that is, a steepest descent in an energy landscape, does not just rely on the height function E, but also on a notion of distance on configuration space, which is typically described by a metric tensor g in the sense of Riemannian geometry. In case of MCF, the tangent space in some configuration Ω should be thought of as consisting of all normal velocities V, i.e., functions on $\partial\Omega$, while the configuration-dependent metric tensor g_Ω is given by the L^2-inner product on $\partial\Omega$.

Still formally, any gradient flow allows for a natural discretization in time. Every step of the discretization comes in form of a variational problem, just involving the functional E and the induced distance d, cf. (9), but not the metric tensor g and the differential of E – it thus relies rather on the "metric", but not the differential structure. Following De Giorgi, we call such a scheme a minimizing movements scheme. We recall that, as in elementary differential geometry, the induced distance d on a Riemannian manifold $(\mathcal{M}, g)$ is defined via $d^2(\chi_0, \chi_1) := \inf\{\int_0^1 g_{\chi_s}(\frac{d\chi}{ds}, \frac{d\chi}{ds})ds\}$, where the infimum is taken over all curves $[0,1] \ni s \mapsto \chi_s$ connecting χ_0 to χ_1 (we use the letter χ because we think of a characteristic function describing the configuration). We note that in the Euclidean case, the Euler-Lagrange equation of (9) turns into the implicit Euler scheme for $\frac{d\chi}{dt} = -\operatorname{grad} E|_\chi$.

However, this infinite-dimensional Riemannian structure making MCF a gradient flow leads to a degenerate induced metric (i.e. $d \equiv 0$): It can be seen that the infimum of $\int_0^1 \int_{\partial\Omega_s} V_s^2 ds$ over all curves of configurations $[0,1] \ni s \mapsto \Omega_s$ with normal velocity $[0,1] \ni s \mapsto V_s$, connecting some given configurations Ω_0 and Ω_1, vanishes [9].

Nonetheless, a minimizing movements scheme (before the latter) for MCF has been formulated by Almgren et.al. [1], with $E(\Omega)$ being the surface area of $\partial\Omega$ and with $d^2(\Omega_1, \Omega_0) = 4\int_{\Omega_1 \triangle \Omega_0} \operatorname{dist}(\cdot, \partial\Omega_0)$. Luckhaus et.al. [7] have established a (long-time) convergence result for this scheme. This convergence result is conditional in the sense that a condition like in (5) has to be imposed.

Thresholding, cf. (1), is a very well performing and widely used numerical scheme for MCF, introduced by Osher et.al. [8]. Also the convolution step, which after spatial discretization can be carried out by the Fast Fourier Transform, is of low complexity. Right from the beginning, thresholding has attracted the attention of analysts; since it obviously conserves the comparison principle for MCF, it has been shown to converge to MCF in the sense of viscosity solution in the two-phase case [4].

Esedoğlu and the second author [3] realized that thresholding also respects the gradient-flow structure of MCF, in the sense that it can be interpreted as a minimizing movements scheme, cf. Lemma 2. This was used in the multi-phase case to extend thresholding to surface tensions and mobilities [10] that depend on the pair of grains, while keeping its low complexity. It was also used by the present authors to provide several types of convergence results; presently, all of them are conditional in the sense of assumption (5), in the tradition of [7].

The first result [5] provided the same limiting notion of solution for MCF as in [7]. However, this weak notion of solution does not imply the dissipation inequality natural to a gradient flow. It is Brakke's weak notion of solution for MCF that is based on a localization of the dissipation inequality; in [6], we establish a (still conditional) convergence result towards this inequality-based notion of solution.

For any gradient flow in a Riemannian context $(\mathcal{M}, g, E)$, there is yet another notion of weak solution based on a single inequality, namely $E(\chi(T)) + \int_0^T \frac{1}{2} g_\chi(\frac{d\chi}{dt}, \frac{d\chi}{dt}) + \frac{1}{2}|\mathrm{grad} E_{|\chi}|^2 dt \leq E(\chi(0))$. This elementary observation is credited to De Giorgi; its appeal lies in the fact that it is potentially more stable in limiting procedures because only lower semi-continuity is needed (as provided by Propositions 1 and 2). The main result of this paper, Theorem 1, precisely establishes this inequality in the case of MCF, cf. (8).

One advantage of a minimizing movements scheme, cf. (9), lies in the fact that it automatically comes with the a priori estimate $E(\chi^N) + \sum_{n=1}^N \frac{1}{2h} d^2(\chi^n, \chi^{n-1}) \leq E(\chi^0)$, which is obtained by using χ^{n-1} as a competitor in (9). In the limit $h \downarrow 0$, this inequality formally turns into $E(\chi(T)) + \int_0^T \frac{1}{2} g_\chi(\frac{d\chi}{dt}, \frac{d\chi}{dt}) dt \leq E(\chi(0))$, which misses the formally correct identity by a factor of 2. On the level of the metric structure, De Giorgi provides tools to capture the missing term $\int_0^T \frac{1}{2}|\mathrm{grad} E_{|\chi}|^2 dt$, see Lemma 1. We take the proof from the monograph [2].

As a consequence of these notions and tools of De Giorgi, our (conditional) convergence proof for the thresholding scheme in fact is rather "soft", softer than [5] which relied on the notion of tilt excess and the fine structure of Caccioppoli sets, and certainly softer than [7] which relied on regularity theory for minimal surfaces. We believe that these tools have a wider potential for geometric evolutions or non-linear PDE of gradient-flow type. For the broader context and more references, we refer to [5].

§2. Main result and structure of proof

Given an initial configuration, as described by its (Lebesgue-measurable) characteristic function $\chi^0 \colon \mathbb{R}^d \to \{0,1\}$, and a time step size $h > 0$, the thresholding scheme iteratively produces configurations at time steps $n = 1, 2, \cdots$, encoded by their characteristic functions χ^n, via convolution and "thresholding":

$$\chi^n := \left\{ \begin{array}{ll} 1 & \text{where } G_h * \chi^{n-1} > \frac{1}{2} \\ 0 & \text{else} \end{array} \right\}, \tag{1}$$

where G_h denotes the heat kernel at time $\frac{h}{2}$, that is,

$$G_h(z) := \frac{1}{\sqrt{2\pi h}^d} \exp\left(-\frac{|z|^2}{2h}\right) \tag{2}$$

(like in stochastic analysis, we take $\frac{h}{2}$ so that G_1 is the standard Gaussian). We interpolate piecewise constant in time:

$$\chi_h(t) = \chi^n \quad \text{for } \ t \in [nh, (n+1)h), \qquad \chi_h(t) = \chi^0 \quad \text{for } \ t \leq 0. \tag{3}$$

For simplicity, we pass from the whole space $\mathbb{R}^d$ to a torus as the spatial domain; by rescaling, we may w.l.o.g. take the unit torus $[0,1)^d$. We also restrict to the finite time horizon $T < \infty$ and $h \leq 1$.

Our main result is the following convergence result, which is only a conditional one since assumption (5) on the energies E_h defined below in (15) presumably cannot be verified. It is the opposite direction, $c_0 \int_{(0,T)\times[0,1)^d} |\nabla\chi| dt \leq \liminf_{h\downarrow 0} \int_0^T E_h(\chi_h(t)) dt$, that follows from (4), see (19) in Lemma 3. Here and in the sequel, $|\nabla\chi| dt$ denotes the total variation of the distribution $\nabla\chi$ in $(0,T) \times [0,1)^d$, provided the latter is a bounded measure. This notation is justified since in the present case, $|\nabla\chi| dt$ is (Lebesgue) equi-integrable in t and thus admits a density $|\nabla\chi|$. In the sequel, $\nu \in L^1(|\nabla\chi| dt)$ denotes the measure-theoretic normal (characterized through the polar factorization $\nabla\chi = \nu|\nabla\chi| dt$ and $|\nu| = 1$ $|\nabla\chi| dt$-almost everywhere).

Theorem 1. *Given χ^0 as above and such that $\nabla\chi^0$ is a bounded measure, and a sequence $h \downarrow 0$; let χ_h be defined by (1) and (3). Suppose that there exists a $\chi\colon (0,T)\times[0,1)^d \to [0,1]$ such that*

$$\chi_h \rightharpoonup \chi \quad \textit{in } L^1\big((0,T)\times[0,1)^d\big). \tag{4}$$

Then we have $\chi \in \{0,1\}$ (Lebesgue)-a.e. and $\nabla\chi$ is a bounded measure which is equi-integrable in t. If we assume in addition

$$\limsup_{h\downarrow 0} \int_0^T E_h(\chi_h(t))\,dt \le c_0 \int_{(0,T)\times[0,1)^d} |\nabla\chi|\,dt, \tag{5}$$

where $c_0 := \frac{1}{\sqrt{2\pi}}$, then there exists $H \in L^2(|\nabla\chi|dt)$ with

$$\int_{(0,T)\times[0,1)^d} (\nabla\cdot\xi - \nu\cdot\nabla\xi\nu)|\nabla\chi|\,dt = -\int_{(0,T)\times[0,1)^d} H\nu\cdot\xi|\nabla\chi|\,dt \tag{6}$$

for all $\xi \in C_0^\infty((0,T)\times[0,1)^d)^d$, and $V \in L^2$ $(|\nabla\chi|dt)$ with

$$\begin{aligned}\int_{[0,1)^d} \zeta(t=0)\chi^0\,dx &+ \int_{(0,T)\times[0,1)^d} \partial_t\zeta\chi\,dxdt \\ &+ \int_{(0,T)\times[0,1)^d} \zeta V|\nabla\chi|\,dt = 0\end{aligned} \tag{7}$$

for all $\zeta \in C_0^\infty([0,T)\times[0,1)^d)$, such that

$$\begin{aligned}\limsup_{\tau\downarrow 0} \frac{1}{\tau} &\int_{(T-\tau,T)\times[0,1)^d} |\nabla\chi|\,dt \\ &+ \int_{(0,T)\times[0,1)^d} \left(V^2 + \left(\frac{H}{2}\right)^2\right)|\nabla\chi|\,dt \le \int_{[0,1)^d} |\nabla\chi^0|.\end{aligned} \tag{8}$$

We note that in case of $\{\chi = 1\}$ being smooth in time-space $[0,T]\times[0,1)^d$, $|\nabla\chi|$ coincides with the surface measure, ν with the (inner) normal, and H and V coincide with the mean curvature (with the convention that convex sets have positively curved boundary) and normal velocity (with the convention that growing sets have positive velocity), respectively. In addition, (7) yields that $\chi(t=0) = \chi^0$. Moreover, expanding the square $V^2 + (\frac{H}{2})^2 = (V+\frac{H}{2})^2 - VH$, and appealing to the classical formula $\frac{d}{dt}\int_{[0,1)^d}|\nabla\chi| = \int_{[0,1)^d} VH|\nabla\chi|$ (which relies on the fact that mean curvature describes the first variation of the surface area), we see that (8) turns into $\int_{(0,T)\times[0,1)^d}(V+\frac{H}{2})^2|\nabla\chi|dt \le 0$, and thus MCF in form of $V = -\frac{H}{2}$ (the factor $\frac{1}{2}$ stems from the normalization in (2)). Therefore, the inequality (8) may be considered a weak notion of MCF.

In the sequel, we omit writing the time-space domain $(0,T)\times[0,1)^d$ when integrating the Lebesgue measure $dxdt$ or the limiting surface measure $|\nabla\chi|dt$. However, the convolution $*$, for which we reserve the z-variable, is always w.r.t. $\mathbb{R}^d$.

The next elementary lemma provides the necessary notions and results on abstract minimizing movements schemes.

Lemma 1. *Let $(\mathcal{M}, d)$ be a compact metric space and $E\colon \mathcal{M}\to\mathbb{R}$ be continuous. Given $\chi^0\in\mathcal{M}$ and $h>0$ consider a sequence $\{\chi^n\}_{n\in\mathbb{N}}$ satisfying*

$$\chi^n \quad \text{minimizes} \quad \frac{1}{2h}d^2(u,\chi^{n-1}) + E(u) \quad \text{among all} \quad u\in\mathcal{M}. \tag{9}$$

Then we have for all $t\in\mathbb{N}h$

$$E(\chi(t)) + \frac{1}{2}\int_0^t \left(\frac{1}{h^2}d^2(\chi(s+h),\chi(s)) + |\partial E(u(s))|^2\right) ds \le E(\chi^0). \tag{10}$$

Here $\chi(t)$ is the piecewise constant interpolation, cf. (3), $u(t)$ is another interpolation (the "variational interpolation") satisfying

$$\int_0^\infty \frac{1}{2h^2} d^2(u(t),\chi(t))\,dt \le E(\chi^0), \tag{11}$$

$$E(u(t)) \le E(\chi(t)) \quad \text{for all}\, t\ge 0, \tag{12}$$

and $|\partial E(u)|$ is the "metric slope" defined through

$$|\partial E(u)| := \limsup_{v:d(v,u)\to 0} \frac{(E(u)-E(v))_+}{d(v,u)} \in [0,\infty]. \tag{13}$$

The next elementary but crucial lemma establishes that the thresholding scheme is a minimizing movements scheme.

Lemma 2. *Expression (1) satisfies (9) provided we define*

$$\mathcal{M} := \{u\colon [0,1)^d\to[0,1]\ \text{measurable}\}, \tag{14}$$

$$E_h(u) := \frac{1}{\sqrt{h}}\int (1-u)\, G_h * u\, dx, \tag{15}$$

$$d_h(u,u') := \left(2\sqrt{h}\int |G_{\frac{h}{2}} * (u-u')|^2\,dx\right)^{\frac{1}{2}}. \tag{16}$$

Furthermore, $(\mathcal{M}, d_h)$ is a compact metric space and E_h continuous.

We will mostly use (16) in form of

$$\frac{1}{2h} d_h^2(u,u') = \frac{1}{\sqrt{h}} \int |G_{\frac{h}{2}} * (u-u')|^2\,dx. \tag{17}$$

The first part of the next lemma provides compactness. The second part contains the (only) way we use the convergence assumption (5); loosely speaking, it ensures convergence of the (oriented) normal down to (spatial) scales of $O(\sqrt{h})$. In particular, it rules out ghost interfaces. Since it will also be used for the variational interpolation, cf. Lemma 1, it is formulated for a $[0,1]$-valued sequence $\{u_h\}_{h\downarrow 0}$.

Lemma 3. i) *Consider a sequence $\{\chi_h\}_{h\downarrow 0}$ of $\{0,1\}$-valued functions on $(0,T)\times[0,1)^d$ that satisfies*

$$\operatorname{esssup}_{t\in(0,T)} E_h(\chi_h(t)) + \int_0^T \frac{1}{2h^2} d_h^2\big(\chi_h(t),\chi_h(t-h)\big)\,dt \quad \text{stays bounded as } h\downarrow 0, \tag{18}$$

and that is piecewise constant in the sense of (3). Such a sequence is compact in $L^1((0,T)\times[0,1)^d)$; any (weak) limit χ is such that $\nabla\chi$ is a bounded measure, equi-integrable in t, with

$$c_0 \int |\nabla\chi|\,dt \le \liminf_{h\downarrow 0} \int_0^T E_h(\chi_h(t))\,dt. \tag{19}$$

ii) *Consider a sequence $\{u_h\}_{h\downarrow 0}$ of $[0,1]$-valued functions on $(0,T)\times[0,1)^d$ and a $\{0,1\}$-valued function χ on $(0,T)\times[0,1)^d$ that satisfies (4) and (5) (with χ_h replaced by u_h) and*

$$\operatorname{ess\,sup}_{t\in(0,T)} E_h(u_h(t)) \quad \text{stays bounded as } h\downarrow 0. \tag{20}$$

Then, as measures on (z,t,x)-space, we have the weak convergences

$$G_1(z)\frac{1}{\sqrt{h}} u_h(t,x)(1-u_h)(t,x-\sqrt{h}z)\,dxdtdz \rightharpoonup G_1(z)(\nu\cdot z)_+|\nabla\chi|\,dtdz, \tag{21}$$

$$G_1(z)\frac{1}{\sqrt{h}} (1-u_h)(t,x)u_h(t,x-\sqrt{h}z)\,dxdtdz \rightharpoonup G_1(z)(\nu\cdot z)_-|\nabla\chi|\,dtdz. \tag{22}$$

The test functions may even have polynomial growth in z.

The next two propositions are at the core and provide the link between (10) and (8). Proposition 1 ensures that the metric d_h, cf. (16), is strong enough to control the right notion of distance of curves in configuration space. Proposition 2 makes sure that it is not too strong so that the metric slope $|\partial E_h|$, cf. (13), controls the gradient of the limiting functional.

Proposition 1. *Suppose that (4) and the conclusion of Lemma 3 hold (with u_h replaced by χ_h). Provided the l.h.s. of (24) is finite, there exists $V \in L^2(|\nabla\chi|dt)$ that is the normal velocity in the sense of*

$$\partial_t \chi = V|\nabla\chi| \quad \textit{distributionally,} \tag{23}$$

and that is dominated in the sense of

$$\liminf_{h\downarrow 0} \int_0^T \frac{1}{2h^2}\, d_h^2(\chi_h(t+h), \chi_h(t))dt \geq c_0 \int V^2|\nabla\chi|\, dt. \tag{24}$$

Proposition 2. *Suppose that the conclusions of Lemma 3 ii) hold. Then there exists $H \in L^2(|\nabla\chi|dt)$ that is the mean curvature in the sense of (6) and that is dominated in the sense of*

$$\liminf_{h\downarrow 0} \int_0^T \frac{1}{2}|\partial E_h(u_h(t))|^2\, dt \geq c_0 \int \left(\frac{H}{2}\right)^2 |\nabla\chi|\, dt. \tag{25}$$

§3. Proofs

We will repeatedly use the (parabolic) scaling of G_h, cf. (2),

$$G_h(z) = \frac{1}{\sqrt{h}^d} G_1\left(\frac{z}{\sqrt{h}}\right) \tag{26}$$

and its semi-group property in form of

$$G_h * G_{h'} = G_{h+h'} \quad \text{in particular} \quad G_{\frac{h}{2}} * G_{\frac{h}{2}} = G_h. \tag{27}$$

The constant $c_0 = \frac{1}{\sqrt{2\pi}}$ appears because of the identity

$$\int G_1(z)(z_1)_+ dz = G_1^{d=1}(0) = c_0, \tag{28}$$

where $G_1^{d=1}(z_1) := \frac{1}{\sqrt{2\pi}}\exp(-\frac{z_1^2}{2})$ denotes the standard Gaussian in a single variable. Indeed, by the factorization of the d-dimensional standard Gaussian into $G_1^{d=1}$ and the $(d-1)$-dimensional one, and by the normalization of the latter, the integral in (28) reduces to $\int_0^\infty G_1^{d=1} z_1 dz_1$. The formula then follows from writing $z_1 G_1^{d=1} = -\frac{d}{dz_1} G_1^{d=1}$.

PROOF OF THEOREM 1.

Note that Lemma 2 allows to make use of Lemma 1, so that we have (10) with (E, d, χ, u) replaced by (E_h, d_h, χ_h, u_h). We start with the l.h.s. of (10), for which we plainly have

$$E_h(\chi^0) \le c_0 \int |\nabla \chi^0|. \tag{29}$$

Indeed, dropping the index 0, this follows by making the l.h.s. explicit $\frac{1}{\sqrt{h}} \int G_h(z)(1-\chi)\chi(\cdot - z)dxdz$, which by (26) and $\chi \in \{0,1\}$ coincides with $\int G_1(z) \frac{1}{\sqrt{h}}(\chi - \chi(\cdot - \sqrt{h}z))_- dxdz$. It remains to appeal to the mean-value inequality $\int \frac{1}{\sqrt{h}}(\chi - \chi(\cdot - \sqrt{h}z))_- dx \le \int (z \cdot \nu)_- |\nabla \chi|$ and to (28).

Note that because of (10) and (29), (18) is satisfied. Hence we may apply Lemma 3 i), which yields $\chi \in \{0,1\}$ a.e. and that $\nabla \chi$ is a bounded measure which is equi-integrable in t. By Lemma 3 ii), in view of the theorem's assumption (5), we obtain (21) and (22) with u_h replaced by χ_h, so that we may apply Proposition 1. We now argue that (11) and (12) imply that (4) and (5) hold with χ_h replaced by u_h, so that we may use Proposition 2 also for u_h. Indeed, (5) for u_h follows immediately from (5) for χ_h and (12). We now turn to (4); because it is $[0,1]$-valued, the sequence $\{u_h\}_{h\downarrow 0}$ always admits a subsequence that has a weak limit u, so that it remains to argue that $u = \chi$, w.l.o.g. assuming that the entire sequence converges. We momentarily fix $h_0 > 0$ and note that by (27) together with Jensen's inequality we have for all $h \le 2h_0$ that

$$\begin{aligned}\int |G_{h_0} * (u_h - \chi_h)|^2\, dxdt &\le \int |G_{\frac{h}{2}} * (u_h - \chi_h)|^2\, dxdt \\ &\overset{(17)}{=} \frac{1}{2\sqrt{h}} \int_0^T d_h^2(u_h(t), \chi_h(t))\, dt \overset{(11)}{\le} h\sqrt{h} E_h(\chi^0) \overset{(29)}{\le} c_0 h\sqrt{h} \int |\nabla \chi^0|,\end{aligned}$$

so that by lower-semi continuity of the l.h.s. under weak convergence we obtain $\int |G_{h_0} * (u - \chi)|^2 dxdt = 0$. From letting h_0 tend to zero we obtain the desired $u = \chi$.

Momentarily setting $\rho(t) := E_h(\chi_h(t)) + \frac{1}{2}\int_0^t \big(\frac{1}{h^2} d_h^2(\chi_h(s), \chi_h(s - h)) + |\partial E_h(u_h(s))|^2\big)ds$, we note that by definition $\rho(t) = \rho(nh) + \delta(t)$, for $t \in \big[nh, (n+1)h\big)$, where $\delta(t) := \frac{1}{2}\int_{nh}^t \big(\frac{1}{h^2} d_h^2(\chi_h(s), \chi_h(s-h)) +$

$|\partial E_h(u_h(s))|^2)ds$. By (10), $\int_0^T \delta(t)dt \le hE_h(\chi^0)$. Hence if we multiply (10) in form of $\rho(nh) \le E_h(\chi^0)$ with $\eta(nh) - \eta((n+1)h)$ for some non-increasing $\eta \in C_0^\infty([0,T))$, we obtain $\int_0^\infty(-\frac{d\eta}{dt})\rho dt \le \big(\eta(0) + h\sup|\frac{d\eta}{dt}|\big)E_h(\chi^0)$. By an integration by parts and with the choice $\eta(t) = \max\{\min\{\frac{T-t}{\tau},1\},0\}$, this turns into

$$\frac{1}{\tau}\int_{T-\tau}^{T} E_h(\chi_h(t))dt + \frac{1}{2}\int_0^{T-\tau}\big(\tfrac{1}{h^2}d_h^2(\chi_h(t),\chi_h(t-h)) + |\partial E_h(u_h(t))|^2\big)\,dt \le \big(1+\tfrac{h}{\tau}\big)E_h(\chi^0).$$

Passing to the limit $h \downarrow 0$, for the first l.h.s. term, we appeal to (19) in Lemma 3 with $(0,T)$ replaced by $(T-\tau,T)$. For the second l.h.s. term, we apply (24) (note that we may extend the integral down to 0 because of the second item in (3)) and (25), both with $(0,T)$ replaced by $(0,T-\tau)$. For the r.h.s. term, we use (29). Summing up, we obtain

$$\frac{c_0}{\tau}\int_{(T-\tau,T)\times[0,1)^d}|\nabla\chi|\,dt + c_0\int_{(0,T-\tau)\times[0,1)^d}\left(V^2+\left(\frac{H}{2}\right)^2\right)|\nabla\chi|\,dt \le c_0\int|\nabla\chi^0|.$$

Dividing by c_0 and letting $\tau \downarrow 0$ yields (8).

Finally, we argue why (23) is sufficient to infer (7). Indeed, by the trivial extension of χ_h to $t \le 0$, cf. (3), the assumptions (4) and (5) extend to $(-T,T)$, where for (5) we appeal to (29). Likewise, the l.h.s. integral in (24) extends to $(-T,T)$. Hence (23) holds distributionally on $(-T,T)\times[0,1)^d$, which turns into (7) because of $\chi = \chi^0$ for $t < 0$.

Proof of Lemma 1.
We reproduce the proof of [2, Theorem 3.1.4 and Lemma 3.1.3]. We start with the definition of the variational interpolation u. Since by assumption, $(\mathcal{M},d)$ is compact and E continuous, for any $n \in \mathbb{N}$ and any $t \in ((n-1)h, nh]$, there exists $u(t)$ that minimizes

$$\frac{d^2(u,\chi^{n-1})}{2(t-(n-1)h)} + E(u) \quad \text{among} \quad u \in \mathcal{M}. \tag{30}$$

W.l.o.g. we may assume that $u(nh) = \chi^n$, cf. (9), so that u is indeed an interpolation of $\{\chi^n\}_{n\in\mathbb{N}}$. Since by comparison with $u = \chi^{n-1}$ we have $E(u(t)) \le E(\chi^{n-1})$, (12) follows immediately from the way we defined the piecewise linear interpolation, cf. (3).

Fixing $n \in \mathbb{N}$ and introducing

$$(31)\qquad e(t) := \min_{u\in\mathcal{M}} \left(\frac{d^2(u,\chi^{n-1})}{2(t-(n-1)h)} + E(u) \right), \quad (n-1)h < t \le nh,$$

we now establish the two crucial inequalities

$$(32)\qquad \begin{aligned} \frac{d^2(u(s),\chi^{n-1})}{2(s-(n-1)h)(t-(n-1)h)} &\le \frac{e(s)-e(t)}{t-s} \\ &\le \frac{d^2(u(t),\chi^{n-1})}{2(s-(n-1)h)(t-(n-1)h)}, \quad (n-1)h < s < t \le nh. \end{aligned}$$

For notational simplicity we consider the case $n = 1$; for any $s, t > 0$ we have by definitions (30) and (31)

$$\begin{aligned} e(s) &= \frac{1}{2s} d^2(u(s),\chi^0) + E(u(s)) \\ &\le \frac{1}{2s} d^2(u(t),\chi^0) + E(u(t)) = \left(\frac{1}{2s} - \frac{1}{2t}\right) d^2(u(t),\chi^0) + e(t). \end{aligned}$$

Writing $\frac{1}{2s} - \frac{1}{2t} = \frac{t-s}{2st}$ this gives the upper bound in (32) after division by $t - s > 0$. Exchanging the roles of s and t, we likewise get the lower one.

We now argue that

$$(33)\qquad |\partial E(u(t))| \le \frac{d(u(t),\chi^{n-1})}{t-(n-1)h} \quad \text{for } t \in ((n-1)h, nh].$$

Again, for notational simplicity we consider $n = 1$ and give ourselves a $v \in \mathcal{M}$. By the characterizing property (30) of $u(t)$ we have $\frac{1}{2t}d^2(u(t),\chi^0)$ $+E(u(t)) \le \frac{1}{2t}d^2(v,\chi^0) + E(v)$, so that $E(u(t)) - E(v) \le \frac{1}{2t}(d(v,\chi^0) - d(u(t),\chi^0))(d(v,\chi^0) + d(u(t),\chi^0))$. By the triangle inequality, this implies

$$E(u(t)) - E(v) \le d(v,u(t))\frac{1}{t}\left(d(u(t),\chi^0) + \frac{1}{2}d(v,u(t)) \right),$$

so that (33) follows from definition (13) of the metric slope.

We now may conclude on (10). By telescoping and according to the piecewise constant interpolation, it is sufficient to establish

$$E(\chi^n) + \frac{1}{2h}\, d^2(\chi^n,\chi^{n-1}) + \int_{(n-1)h}^{nh} \frac{1}{2}\, |\partial E(u(s))|^2\, ds \le E(\chi^{n-1}),$$

which according to (33) follows from

$$E(\chi^n) + \frac{1}{2h}\, d^2(\chi^n, \chi^{n-1}) + \int_{(n-1)h}^{nh} \frac{d^2(u(s), \chi^{n-1})}{2(s-(n-1)h)^2}\, ds \le E(\chi^{n-1}),$$

and with help of (31) may be rewritten as

$$e(nh) + \int_{(n-1)h}^{nh} \frac{d^2(u(s), \chi^{n-1})}{2(s-(n-1)h)^2}\, ds \le E(\chi^{n-1}). \tag{34}$$

Here comes the argument for (34): We first learn from (32) that

$$((n-1)h, nh] \ni s \mapsto d^2(u(s), \chi^{n-1}) \quad \text{is monotone increasing} \tag{35}$$

and thus continuous outside of a countable set of s's. We then learn that e is locally Lipschitz continuous on $((n-1)h, nh]$ and differentiable where (35) is continuous. In particular, we have in those (Lebesgue) almost every time points s, $\frac{de}{dt}(s) = -\frac{1}{2s^2} d^2(u(s), \chi^{n-1})$. Integrating this relationship from some $t \in ((n-1)h, nh]$ to nh we obtain $e(nh) + \int_t^{nh} \frac{1}{2s^2} d^2(u(s), \chi^{n-1}) ds \le e(t)$. Using the obvious $e(t) \le E(\chi^{n-1})$, cf. (31), and letting $t \downarrow (n-1)h$ we obtain (34) by monotone convergence.

We finally turn to (11). According to (35) we have $d^2(u(t), \chi^{n-1}) \le d^2(\chi^n, \chi^{n-1})$ for $t \in ((n-1)h, nh]$ and therefore $\int_0^\infty d^2(u(t), \chi(t)) dt \le \int_0^\infty d^2(\chi(t), \chi(t-h)) dt$, so that (11) follows from (10).

Proof of Lemma 2.

By the definitions (15) and (17), the latter in conjunction with (27), we have

$$\frac{1}{2h}\, d_h^2(u, \chi^{n-1}) + E_h(u) = \langle u - \chi^{n-1}, u - \chi^{n-1}\rangle + \langle 1-u, u\rangle,$$

where we momentarily introduced the bilinear form $\langle u, u'\rangle := \frac{1}{\sqrt{h}} \int u\, G_h * u' dx$. Since this form is symmetric, we may rewrite the r.h.s. as $\langle u, 1 - 2\chi^{n-1}\rangle + \langle \chi^{n-1}, \chi^{n-1}\rangle$, so that

$$\frac{1}{2h}\, d_h^2(u, \chi^{n-1}) + E_h(u) = \frac{1}{\sqrt{h}} \int u(1 - 2G_h * \chi^{n-1}) + C,$$

where $C := \langle \chi^{n-1}, \chi^{n-1}\rangle$ does not depend on u. It is now obvious that (1) minimizes $\frac{1}{2h} d_h^2(u, \chi^{n-1}) + E_h(u)$ among all $u \in \mathcal{M}$, cf. (14).

It remains to argue that the metric space $(\mathcal{M}, d_h)$ is compact and E_h continuous. Both follows from the fact that d_h metrizes weak convergence on $\mathcal{M} \subset L^2([0,1)^d)$. The latter can be seen as follows: In terms of Fourier series, we have $\frac{2}{\sqrt{h}}d_h^2(u,u') = \sum_{k\in 2\pi\mathbb{Z}^d}\exp(-\frac{|hk|^2}{2})$ $|\mathcal{F}(u-u')|^2(k)$, and note $|\mathcal{F}(u-u')|^2(k) \le \int (u-u')^2dx \le 1$. Hence by dominated convergence, $d_h(u_n,u)\to 0$ is equivalent to $\mathcal{F}(u_n-u)(k)\to 0$ for all $k\in 2\pi\mathbb{Z}^d$, which by the L^2-boundedness of $\{u_n\}_{n\uparrow\infty}\subset\mathcal{M}$ is equivalent to weak convergence.

PROOF OF LEMMA 3.

STEP 1. Some useful inequalities on E_h and d_h. We claim for any $[0,1]$-valued function u of space:

$$\int |u - G_h * u|\,dx \le 2\sqrt{h}E_h(u), \tag{36}$$

$$E_{h_0}(u) \le E_h(u) \quad \text{for } h_0\in\mathbb{N}^2 h, \tag{37}$$

which we claim combine to

$$\int (u - G_{h_0} * u)^2\,dx \le 4\sqrt{h_0}E_h(u) \quad \text{for all } h_0 \ge h. \tag{38}$$

We also claim for any pair of $\{0,1\}$-valued functions χ,χ' of space

$$\int |\chi-\chi'|\,dx \le \frac{1}{2\sqrt{h}}\,d_h^2(\chi,\chi') + 2\sqrt{h}\big(E_h(\chi)+E_h(\chi')\big). \tag{39}$$

We first tackle (36); by Jensen's inequality in form of $|u-G_h*u|(x) \le \int G_h(z)|u(x)-u(x-z)|dz$, and by the definition (15) of E_h which together with the symmetry of G_h* yields $2\sqrt{h}E_h(u) = \int((1-u)G_h*u$ $+uG_h*(1-u))dx$, (36) follows from the elementary inequality

$$|u-u'| \le (1-u)u' + u(1-u') \quad \text{for } u,u'\in[0,1]. \tag{40}$$

We now turn to (37) which we (iteratively) establish in the more general form of

$$\sqrt{h_0}E_{h_0} \le \sqrt{h}E_h + \sqrt{h'}E_{h'} \quad \text{provided } \sqrt{h_0} = \sqrt{h}+\sqrt{h'}. \tag{41}$$

Indeed, by definition (15) and the scaling (26) we have $\sqrt{h}E_h(u) = \int G_1(z)(1-u)(x)u(x-\sqrt{h}z)dxdz$ and by a change of variables in x, $\sqrt{h'}E_{h'}(u) = \int G_1(z)(1-u)(x-\sqrt{h}z)u(x-(\sqrt{h}+\sqrt{h'})z)dxdz$. Hence (41) follows from the elementary inequality

$$(1-u)u'' \le (1-u)u' + (1-u')u'' \quad \text{for } u,u',u''\in[0,1].$$

This is equivalent to $u'(u+u''-1)\le uu''$, which because of $u'\in[0,1]$ and $uu''\ge 0$ follows from $u+u''-1\le uu''$. The latter is equivalent to $u''(1-u)\le 1-u$, which holds because of $u,u''\in[0,1]$.

We now turn to the upgrade (38). We first observe that the l.h.s. is monotone increasing in h_0, as can be seen by the Fourier representation $\sum_k(1-\exp(-\frac{|h_0 k|^2}{2}))|\mathcal{F}u(k)|^2$, where $\mathcal{F}u(k)$, $k\in 2\pi\mathbb{Z}^d$, denotes the Fourier series of u and $\exp(-\frac{|h_0 k|^2}{2})$ is the Fourier transform of G_{h_0}. Now given $h_0\geq h$, we write $\sqrt{h_0}=N\sqrt{h}-s$ with $N\in\mathbb{N}$ and $s\in[0,\sqrt{h})$. By the above monotonicity we have $\int(u-G_{h_0}*u)^2\leq\int(u-G_{N^2h}*u)^2\leq\int|u-G_{N^2h}*u|$, to which we first apply (36) with h replaced by N^2h and second apply (37) with N^2h playing the role of h_0. Hence we end up with $\int(u-G_{h_0}*u)^2\leq 2N\sqrt{h}E_h(u)$, which because of $N\sqrt{h}\leq\sqrt{h_0}+\sqrt{h}$ turns into (38).

We finally address (39), which according to the definitions (15) and (16) of E_h and d_h and the simple estimate (36) follows from integrating the following inequality in x, appealing to the symmetry of G_h*,

$$|\chi-\chi'|\leq(\chi-\chi')G_h*(\chi-\chi')+|\chi-G_h*\chi|+|\chi'-G_h*\chi'|.$$

Writing $|\chi-\chi'|=(\chi-\chi')^2=(\chi-\chi')G_h*(\chi-\chi')+(\chi-\chi')(\chi-G_h*\chi)+(\chi'-\chi)(\chi'-G_h*\chi')$, we see that the inequality relies on $(\chi-\chi')(\chi-G_h*\chi)\leq|\chi-G_h*\chi|$ and on the same inequality with the roles of χ and χ' exchanged.

STEP 2. Modulus of continuity in time: We claim that for every $\{0,1\}$-valued function χ of time-space that is piecewise constant in time, cf. (3), and any time shift $s\in[0,1]$ we have

$$I(s)\leq C_0\left\{\begin{array}{ccc}\frac{s}{\sqrt{h}} & \text{for} & s\leq h\\ 2\sqrt{h} & \text{for} & h\leq s\leq\sqrt{h}\\ 4s & \text{for} & \sqrt{h}\leq s\end{array}\right\}\leq 4C_0\sqrt{s}, \tag{42}$$

where in this step of the proof, we use the abbreviation

$$I(s):=\int_{(s,T)\times[0,1)^d}|\chi(t,x)-\chi(t-s,x)|\,dxdt,$$

$$C_0:=\int_h^T\frac{1}{2h^2}\,d_h^2(\chi(t),\chi(t-h))\,dt+4\int_0^T E_h(\chi(t))\,dt. \tag{43}$$

Indeed, for $s\geq h$, we use (39) with $(\chi,\chi')=(\chi(t),\chi(t-s))$ and integrate in $t\in(s,T)$ to obtain

$$I(s)\leq\int_s^T\frac{1}{2\sqrt{h}}\,d_h^2(\chi,\chi(\cdot-s))\,dt+4\sqrt{h}\int_0^T E_h(\chi)\,dt, \tag{44}$$

where we write $\chi(\cdot-s)$ for the time-shifted function $(t,x)\mapsto\chi(t-s,x)$.

We first use this to treat (42) in case of $s \leq h$: By the piecewise constant interpolation, cf. (3), we have $I(s) \leq \frac{s}{h} I(h)$, into which we insert (44) for $s = h$ in form of $I(h) \leq C_0\sqrt{h}$. We now treat (42) in case of $h \leq s \leq \sqrt{h}$, and first restrict ourselves to $s = Nh$ with $N \in \mathbb{N}$ in order to use the triangle inequality for d_h in form of $d_h^2(\chi, \chi(\cdot - s)) \leq N \sum_{n=1}^N d_h^2(\chi(\cdot - (n-1)h), \chi(\cdot - nh))$ so that we obtain from (44)

$$
\begin{aligned}
(45) \qquad I(s) &\leq \left(\frac{s}{h}\right)^2 \int_h^T \frac{1}{2\sqrt{h}} d_h^2(\chi, \chi(\cdot - h))\, dt + 4\sqrt{h} \int_0^T E_h(\chi)\, dt \\
&\overset{(43)}{\leq} C_0 \max\left\{\frac{s^2}{\sqrt{h}}, \sqrt{h}\right\} = C_0\sqrt{h}.
\end{aligned}
$$

For the unrestricted range of $h \leq s \leq \sqrt{h}$, we write $s = Nh + s'$ with $N \in \mathbb{N}$ and $s' \in [0, h)$, use $I(s) \leq I(Nh) + I(s')$, and appeal to (45) for the first contribution and to (42) in the previously treated case of $s' \leq h$ for the second contribution. Finally, for (42) in the remaining case of $s \geq \sqrt{h}$, we write $s = N\sqrt{h} + s'$ with $N \in \mathbb{N}$ and $s' \in [0, \sqrt{h})$, use $I(s) \leq NI(\sqrt{h}) + I(s')$, and appeal to the previously treated case of (42) for both terms.

STEP 3. Proof of the compactness statement. From (38) and thanks to the first part of our assumed bound (18), we learn that $G_{h_0} * \chi_h$ is close to χ_h in $L^\infty((0,T), L^2([0,1)^d))$ (and thus in $L^1((0,T) \times [0,1)^d)$), as $h_0 \downarrow 0$, uniformly in $h \downarrow 0$. Hence it remains to argue for fixed $h_0 > 0$ that $\{G_{h_0} * \chi_h\}_{h\downarrow 0}$ is compact in L^1. Because of the convolution in space, and of the equi-integrability following from $G_{h_0} * \chi_h \in [0,1]$, this follows from a modulus of continuity in time in L^1 that is uniform in $h \downarrow 0$. Thanks to our assumption (18), this holds for χ_h itself by Step 2; it transmits to $G_{h_0} * \chi_h$ by Jensen's inequality.

STEP 4. Before establishing the exact inequality (19), which will be done at the end of Step 6, it is convenient to first argue that $\nabla\chi$ is a bounded measure, equi-integrable in t, under the mere assumption (20). We focus on $\partial_1\chi$, give ourselves a $\zeta \in C_0^\infty((0,T) \times [0,1)^d)$ and note that as a consequence of (28) we have

$$
\begin{aligned}
-c_0\partial_1\zeta &= \lim_{h\downarrow 0} \frac{1}{\sqrt{h}} \int_{z_1>0} \big(\zeta - \zeta(\cdot + \sqrt{h}z)\big) G_1(z)\, dz \\
&\overset{(26)}{=} \lim_{h\downarrow 0} \frac{1}{\sqrt{h}} \int_{z_1>0} (\zeta - \zeta(\cdot + z)) G_h(z)\, dz
\end{aligned}
$$

uniformly in time-space, where we write $\zeta(\cdot+z)$ for the space-shifted function $(t,x)\mapsto\zeta(t,x+z)$. Hence it follows from (4) that

$$-c_0\int\partial_1\zeta\chi\,dxdt=\lim_{h\downarrow 0}\int_{z_1>0}\zeta\frac{1}{\sqrt{h}}(\chi_h-\chi_h(\cdot-z))G_h(z)\,dxdtdz$$

and thus

$$c_0\left|\int\partial_1\zeta\chi\,dxdt\right|\le\int_0^T\sup_x|\zeta|\,dt\liminf_{h\downarrow 0}\operatorname*{ess\,sup}_t\frac{1}{\sqrt{h}}\int|\chi_h-G_h*\chi_h|\,dx.$$

According to (36), the second r.h.s. factor is estimated by $2\operatorname{ess\,sup}_t E_h(\chi_h)$, the $\liminf_{h\downarrow 0}$ of which is bounded by our assumption (20).

Step 5. Turning to (21) and (22), it is convenient to have the following equi-integrability of the non-negative

$$\rho_h(z,t,x):=G_1(z)\frac{1}{\sqrt{h}}((1-u_h)(t,x)u_h(t,x-\sqrt{h}z)\\ +u_h(t,x)(1-u_h)(t,x-\sqrt{h}z))$$

in the (non-compact) variables t and z in the sense of

$$\int\exp\left(\frac{3|z|^2}{8}\right)\rho_h(z,t,x)\,dxdz\le 2^{d+2}E_h(u_h(t)),$$

cf. (20). Indeed, we first observe that $G_4(z)=\frac{1}{2^d}\exp(\frac{3|z|^2}{8})G_1(z)$, so that by the scaling (26) we obtain $\int\exp(\frac{3|z|^2}{8})\rho_h dxdz=\frac{2^d}{\sqrt{h}}\int((1-u_h)\,G_{4h}*u_h+u_hG_{4h}*(1-u_h))dx$, so that by symmetry of G_{4h} and the definition (15) of E_{4h}, $\int\exp(\frac{3|z|^2}{4})\rho_h dxdz=2^{d+2}E_{4h}(u_h)$, so that it remains to appeal to (37).

Step 6. Proof of (21) and (22). We focus on (21); according to Step 5, it is sufficient to treat bounded continuous test functions $\zeta(z,t,x)$, which by linearity and splitting into positive and negative part we may assume to be $[0,1]$-valued. The statement splits into the local lower bound

$$\liminf_{h\downarrow 0}\int\zeta G_1(z)\frac{1}{\sqrt{h}}u_h(1-u_h)(\cdot-\sqrt{h}z)\,dxdtdz\\ \ge\int\zeta G_1(z)(\nu\cdot z)_+|\nabla\chi|\,dtdz \tag{46}$$

and the global upper bound

$$\limsup_{h\downarrow 0}\int G_1(z)\frac{1}{\sqrt{h}}u_h(1-u_h)(\cdot-\sqrt{h}z)\,dxdtdz \le \int G_1(z)(\nu\cdot z)_+|\nabla\chi|\,dtdz. \tag{47}$$

Indeed, splitting a given test function $\zeta = 1-(1-\zeta)$, appealing to linearity and using (46) with ζ replaced by $1-\zeta\in[0,1]$, we obtain also the local upper bound.

We note that the global upper bound (47) is nothing else than our assumption (5) (with χ_h replaced by u_h): For the l.h.s. this follows from the scaling (26), the symmetry of G_h*, and the definition (15) of E_h. For the r.h.s. this follows from (28).

Hence it remains to establish (46) by a typical l.s.c. argument: By Fatou's Lemma, it is enough to establish for fixed z

$$\liminf_{h\downarrow 0}\int \zeta\frac{1}{\sqrt{h}}u_h(1-u_h)(\cdot-\sqrt{h}z)\,dxdt \ge \int \zeta(\nu\cdot z)_+|\nabla\chi|\,dt.$$

Since by Step 4, we already know that $\nabla\chi$ is a bounded measure, we have $(\nu\cdot z)_+|\nabla\chi| = (\partial_z\chi)_+$. Hence by the definition of the positive part of a measure, and by the equi-integrability in t established in Step 5, it is enough to establish for any non-negative $\zeta\in C_0^\infty((0,T)\times[0,1)^d)$

$$\liminf_{h\downarrow 0}\int \zeta\frac{1}{\sqrt{h}}u_h(1-u_h)(\cdot-\sqrt{h}z)\,dxdt \ge -\int \partial_z\zeta\chi\,dxdt.$$

Since by assumption (4), the r.h.s. is the limit of

$$\int \frac{1}{\sqrt{h}}(\zeta-\zeta(\cdot+\sqrt{h}z))u_h = \int \zeta\frac{1}{\sqrt{h}}(u_h-u_h(\cdot-\sqrt{h}z)),$$

the statement follows from the elementary inequality $u-u'\le u(1-u')$ that is valid for any $u,u'\in[0,1]$.

We note that this argument for (46) did not involve the extra assumption (5) and thus applies also under the assumptions of part i) of the lemma. Statement (46) applied to $\zeta=1$ yields (19) (for the same reason, given above, that (47) is a mere reformulation of (5)).

PROOF OF PROPOSITION 1.
Note that by definition (17) of d_h and (27) we have

$$\int_h^T \frac{1}{2h^2} d_h^2(\chi_h(t), \chi_h(t-h))\, dt = \int_{(h,T)\times[0,1)^d} \frac{1}{h\sqrt{h}} |G_{\frac{h}{2}} * (\chi_h - \chi_h(\cdot - h))|^2 \, dxdt,$$

which motivates to introduce the non-negative (bounded) "dissipation measure" on $(0,T)\times[0,1)^d$ (after possibly passing to a subsequence)

$$\mu := \lim_{h\downarrow 0} \frac{1}{h\sqrt{h}} |G_{\frac{h}{2}} * (\chi_h - \chi_h(\cdot - h))|^2. \tag{48}$$

In fact, we shall establish (24) in the localized form of

$$c_0 V^2 |\nabla\chi|\, dt \leq \mu. \tag{49}$$

We now give an outline of the proof, which amounts to a better quantification of Step 2 in the proof of Lemma 3. In deriving this estimate on the distribution $\partial_t\chi$, we work on an intermediate time scale, which we fix to be an (eventually small) fraction of the characteristic spatial scale:

$$\tau := \alpha\sqrt{h} \quad \text{for some fixed } \alpha \in (0,\infty). \tag{50}$$

We consider the corresponding increments and their positive and negative parts

$$\delta\chi := \chi_h - \chi_h(\cdot - \tau) \quad \text{and} \quad \delta\chi_\pm := \max\{0, \pm\delta\chi\}. \tag{51}$$

Using $|\delta\chi| = \delta\chi_+ + \delta\chi_-$ we then write

$$\frac{1}{\sqrt{h}}|\delta\chi| = \frac{1}{\sqrt{h}}\big(\delta\chi_+ G_h * \delta\chi_+ + \delta\chi_+ G_h * (1-\delta\chi_+) + \delta\chi_- G_h * \delta\chi_- + \delta\chi_- G_h * (1-\delta\chi_-)\big). \tag{52}$$

In Step 1 we will show, in the sense of closeness between distributions,

$$\frac{1}{\sqrt{h}}|\delta\chi| \approx \frac{1}{\sqrt{h}}\big(\delta\chi G_h * \delta\chi + \delta\chi_+ G_h * (1-\delta\chi_+) + \delta\chi_- G_h * (1-\delta\chi_-)\big).$$

The main idea is to estimate the first r.h.s. by the dissipation measure μ (in Step 2) and to estimate the two last contributions by the surface measure $|\nabla\chi| dt$ (in Step 3). However, this just suffices to show that $\partial_t\chi$

is absolutely continuous w.r.t. $|\nabla\chi|dt$, as is carried out at the beginning of Step 4 on the level of $O(\alpha)$ as $\alpha\downarrow 0$. In order to retrieve (49), we need the finer estimate in Step 3 and look at level $O(\alpha^2)$.

STEP 1. We claim that the mixed term vanishes distributionally:

$$\lim_{h\downarrow 0}\frac{1}{\sqrt{h}}(\delta\chi_+ G_h*\delta\chi_- + \delta\chi_- G_h*\delta\chi_+) = 0. \tag{53}$$

Spelling out the z-integral, we want to show that the distributional limit of

$$\int G_h(z)\frac{1}{\sqrt{h}}\big(\delta\chi_+\delta\chi_-(\cdot - z) + \delta\chi_-\delta\chi_+(\cdot - z)\big)dz \tag{54}$$

vanishes. Fixing some unit vector ν_0, we split the expression into

$$\int_{\nu_0\cdot z\geq 0} G_h(z)\frac{1}{\sqrt{h}}\big(\delta\chi_+\delta\chi_-(\cdot - z) + \delta\chi_-\delta\chi_+(\cdot - z)\big)dz \tag{55}$$

and the analogous expression on $\{\nu_0\cdot z\leq 0\}$. We note that by definition of (51), we have $\delta\chi_+ = \chi_h(1-\chi_h)(\cdot-\tau)$ and $\delta\chi_- = \chi_h(\cdot-\tau)(1-\chi_h)$ and thus $\delta\chi_+\delta\chi_-(\cdot - z) = \chi_h(1-\chi_h)(\cdot-\tau)\chi_h(\cdot-\tau-z)(1-\chi_h)(\cdot-z)\leq (1-\chi_h)(\cdot-\tau)\chi_h(\cdot-\tau-z)$ and likewise $\delta\chi_-\delta\chi_+(\cdot-z) = \chi_h(\cdot-\tau)(1-\chi_h)\chi_h(\cdot-z)(1-\chi_h)(\cdot-\tau-z)\leq(1-\chi_h)\chi_h(\cdot-z)$. Here, with a slight abuse of notation, $\chi(\cdot-\tau-z)(t,x) = \chi(t-\tau,x-z)$. Hence the distributional limit of (55) is dominated by the one of

$$\begin{aligned}\int_{\nu_0\cdot z\geq 0} G_h(z)\frac{1}{\sqrt{h}}\big((1-\chi_h)&(\cdot-\tau)\chi_h(\cdot-\tau-z)\\ &+(1-\chi_h)\chi_h(\cdot-z)\big)dz.\end{aligned}$$

Since the first contribution differs from the second one just by a (vanishing) time shift, which we may put into the continuous test function, the distributional limit is equal to the weak limit of

$$2\int_{\nu_0\cdot z\geq 0} G_h(z)\frac{1}{\sqrt{h}}(1-\chi_h)\chi_h(\cdot-z)\,dz,$$

provided the latter exists. Appealing to the scaling (26), according to (22) in Lemma 3 which we test with the characteristic function of the closed set $\{\nu_0\cdot z\geq 0\}$, the weak limit of this term is dominated by the measure $2\int_{\nu_0\cdot z\geq 0}G_1(z)(\nu\cdot z)_-|\nabla\chi|dtdz$. Treating the contribution of $\{\nu_0\cdot z\leq 0\}$ in a similar way (exchanging the roles of χ and $1-\chi$), the weak limit of that contribution is dominated by $2\int_{\nu_0\cdot z\leq 0}G_1(z)(\nu\cdot z)_+|\nabla\chi|dtdz$.

Hence we have shown that the weak limit $\lambda \geq 0$ of (54) satisfies

$$\lambda \leq 2\left(\int_{\nu_0\cdot z\geq 0} G_1(\nu\cdot z)_- dz + \int_{\nu_0\cdot z\leq 0} G_1(\nu\cdot z)_+ dz\right)|\nabla\chi|\,dt. \tag{56}$$

In particular, we have $\lambda \leq 4c_0|\nabla\chi|dt$, cf. (28), so that there is a $\theta \in L^1(|\nabla\chi|dt)$ with $\lambda = \theta|\nabla\chi|dt$, which allows us the rewrite (56) as

$$\theta \leq 2\int_{\nu_0\cdot z\geq 0} G_1(\nu\cdot z)_- dz + 2\int_{\nu_0\cdot z\leq 0} G_1(\nu\cdot z)_+ dz$$
$$|\nabla\chi|\,dt - \text{a.e. and for all } \ \nu_0 \in \mathbb{R}^d.$$

An elementary separability argument allows to exchange the order between the $\forall$, so that we may choose $\nu_0 = \nu$, obtaining $\theta \leq 0$ $|\nabla\chi|dt$-a.e. and thus $\lambda \leq 0$, yielding (53).

STEP 2. We claim that in a distributional sense

$$\limsup_{h\downarrow 0} \frac{1}{\sqrt{h}}\delta\chi G_h * \delta\chi \leq \alpha^2\mu. \tag{57}$$

Indeed, by definition (48), we may split this into

$$\lim_{h\downarrow 0} \frac{1}{\sqrt{h}}\left(\delta\chi G_h * \delta\chi - |G_{\frac{h}{2}} * \delta\chi|^2\right) = 0 \quad \text{and} \tag{58}$$

$$\limsup_{h\downarrow 0}\left(\frac{1}{\sqrt{h}}\left|G_{\frac{h}{2}} * \delta\chi\right|^2 - \alpha^2\frac{1}{h\sqrt{h}}\left|G_{\frac{h}{2}} * (\chi_h - \chi_h(\cdot - h))\right|^2\right) \leq 0. \tag{59}$$

We start with (59) and assume w.l.o.g. that $\tau = Nh$ for some $N \in \mathbb{N}$ so that by telescoping and the Cauchy-Schwarz inequality in n

$$\frac{1}{\sqrt{h}}\left|G_{\frac{h}{2}} * (\chi_h - \chi_h(\cdot - \tau))\right|^2$$
$$\leq N\sum_{n=0}^{N-1}\frac{1}{\sqrt{h}}\left|G_{\frac{h}{2}} * \big(\chi_h(\cdot - nh) - \chi_h(\cdot - (n+1)h)\big)\right|^2.$$

Appealing to $N = \frac{\alpha}{\sqrt{h}}$, the r.h.s. may be rewritten as

$$\alpha^2\frac{1}{N}\sum_{n=0}^{N-1}\frac{1}{h\sqrt{h}}\left|G_{\frac{h}{2}} * \big(\chi_h(\cdot - nh) - \chi_h(\cdot - (n+1)h)\big)\right|^2.$$

Note that this is an average of time shifts of $\alpha^2\frac{1}{h\sqrt{h}}|G_{\frac{h}{2}}*(\chi_h-\chi_h(\cdot-h))|^2$; because of $Nh = O(\sqrt{h}) = o(1)$ all these time shifts are small, so that the (non-negative) expression has the same (bounded) weak limit as $\alpha^2 \frac{1}{h\sqrt{h}}|G_{\frac{h}{2}} * (\chi_h - \chi_h(\cdot - h))|^2$ itself. This yields (59).

We now turn to (58). By the semi-group property (27) and the symmetry of $G_{\frac{h}{2}}*$, we have for a smooth test function ζ

$$\int \zeta\big(\delta\chi G_h * \delta\chi - |G_{\frac{h}{2}} * \delta\chi|^2\big) = -\int [\zeta, G_{\frac{h}{2}} *]\delta\chi\, G_{\frac{h}{2}} * \delta\chi,$$

where $[\zeta, G_{\frac{h}{2}} *]$ denotes the commutator of multiplying with ζ and convolving with $G_{\frac{h}{2}}$. Hence by the boundedness of $\frac{1}{\sqrt{h}}|G_{\frac{h}{2}} * \delta\chi|^2$ as a sequence of measures, which follows from (59), it is enough to establish

$$\lim_{h\downarrow 0} \frac{1}{\sqrt{h}} \int \big|[\zeta, G_{\frac{h}{2}} *]\delta\chi\big|^2 = 0.$$

We spell out the integrand:

$$\big([\zeta, G_{\frac{h}{2}} *]\delta\chi\big)(t,x) = \int G_{\frac{h}{2}}(z)\big(\zeta(t,x) - \zeta(t,x-z)\big)\delta\chi(t,x-z)\,dz,$$

so that $|[\zeta, G_{\frac{h}{2}} *]\delta\chi|(t,x) \leq \sup|\nabla\zeta| \int G_{\frac{h}{2}}(z)|z||\delta\chi(t,x-z)|dz$ and thus

$$\frac{1}{\sqrt{h}}\int \big|[\zeta, G_{\frac{h}{2}} *]\delta\chi\big|^2 dx \leq \frac{1}{\sqrt{h}}\left(\sup|\nabla\zeta| \int G_{\frac{h}{2}}(z)|z|\,dz\right)^2 \int |\delta\chi|^2\,dxdt.$$

By the scaling (26), the r.h.s. is $O(\sqrt{h})$ and thus vanishing.

STEP 3. For given unit vector ν_0 and $V_0 \in (0,\infty)$ we claim that in a distributional sense

$$\begin{aligned}(60)\quad \limsup_{h\downarrow 0} \frac{1}{\sqrt{h}}\bigg(&\delta\chi_+ G_h * (1-\delta\chi_+) + \delta\chi_- G_h * (1-\delta\chi_-)\\ &- 2\int_{z\cdot\nu_0 > \alpha V_0} G_1(z)\,dz|\delta\chi|\bigg) \leq 2\int_{0\leq z\cdot\nu_0\leq\alpha V_0} G_1(z)|z\cdot\nu||\nabla\chi|\,dt.\end{aligned}$$

We split this into three steps: 1) The l.h.s. may be substituted according to

$$\begin{aligned}(61)\qquad \lim_{h\downarrow 0}\frac{1}{\sqrt{h}}\bigg(&\delta\chi_\pm G_h * (1-\delta\chi_\pm)\\ &- \int_{z\cdot\nu_0\geq 0} G_h(z)|\delta\chi_\pm - \delta\chi_\pm(\cdot - z)|\,dz\bigg) = 0.\end{aligned}$$

2) The integrand, which is a second-order difference, satisfies the two inequalities

$$
(62)\qquad \begin{aligned}
&|\delta\chi_+ - \delta\chi_+(\cdot - z)| + |\delta\chi_- - \delta\chi_-(\cdot - z)| \\
&\quad \le \left\{ \begin{array}{c} |\chi_h - \chi_h(\cdot - z)| + |\chi_h(\cdot - \tau) - \chi_h(\cdot - \tau - z)| \\ |\delta\chi| + |\delta\chi(\cdot - z)| \end{array} \right\},
\end{aligned}
$$

where the first inequality is "space-like" and we use it on the set $\{0 \le z \cdot \nu_0 \le \tau V_0\}$, while the second one is "time-like" and we use it on the complement $\{z \cdot \nu_0 > \tau V_0\}$. 3) We finally argue that

$$
(63)\quad \begin{aligned}
&\limsup_{h\downarrow 0} \frac{1}{\sqrt{h}} \int_{0\le z\cdot\nu_0\le\tau V_0} G_h(z)\big(|\chi_h - \chi_h(\cdot - z)| + |\chi_h(\cdot - \tau) \\
&\qquad - \chi_h(\cdot - z - \tau)|\big)\, dz \le 2 \int_{0\le z\cdot\nu_0\le\alpha V_0} G_1(z)|z\cdot\nu||\nabla\chi|\, dt dz,
\end{aligned}
$$

$$
(64)\quad \begin{aligned}
&\lim_{h\downarrow 0} \frac{1}{\sqrt{h}} \bigg(\int_{z\cdot\nu_0>\tau V} G_h(z)\big(|\delta\chi| + |\delta\chi(\cdot - z)|\big) dz \\
&\qquad - 2\frac{1}{\sqrt{h}} \int_{z\cdot\nu_0>\alpha V} G_1(z)\, dz |\delta\chi| \bigg) = 0.
\end{aligned}
$$

We start with (62); the second inequality is obvious by the triangle inequality in form of $|\delta\chi_\pm - \delta\chi_\pm(\cdot - z)| \le \delta\chi_\pm + \delta\chi_\pm(\cdot - z)$ and by $\delta\chi_+ + \delta\chi_- = |\delta\chi|$. In view of the definition (51), the first inequality in (62) follows from the elementary inequality

$$
(65)\qquad \begin{aligned}
&|(a - a')_+ - (b - b')_+| + |(a - a')_- - (b - b')_-| \\
&\qquad \le |a - b| + |a' - b'|,
\end{aligned}
$$

which can be seen by distinguishing two cases: Case 1): $(a - a')(b - b') \ge 0$, w.l.o.g. by symmetry $(a, a', b, b') \rightsquigarrow (-a, -a', -b, -b')$ we may assume $a - a', b - b' \ge 0$, in which case (65) turns into the obvious $|(a - a') - (b - b')| \le |a - b| + |a' - b'|$. Case 2): $(a - a')(b - b') \le 0$, by the same symmetry we may assume $a - a' \ge 0 \ge b - b'$, in which case (65) turns into the obvious $(a - a') + (b' - b) \le |a - b| + |a' - b'|$.

We now argue for (63) and (64). Putting the vanishing shift τ in the time variable on the continuous test function, (63) follows once we argue that

$$
(66)\qquad \begin{aligned}
&\limsup_{h\downarrow 0} \frac{1}{\sqrt{h}} \int_{0\le z\cdot\nu_0\le\tau V_0} G_h(z)|\chi_h - \chi_h(\cdot - z)|\, dz \\
&\qquad \le \int_{0\le z\cdot\nu_0\le\alpha V_0} G_1(z)|z\cdot\nu||\nabla\chi|\, dt dz,
\end{aligned}
$$

which because of non-negativity implies that the l.h.s. admits a limit as a measure on $(0,T)\times[0,1)^d$. Appealing to scaling (26) in form of $G_h(z)dz = G_1(\frac{z}{\sqrt{h}})d\frac{z}{\sqrt{h}}$ and noting that $z\cdot\nu\le\tau V_0$ is equivalent to $\frac{z}{\sqrt{h}}\cdot\nu\le\alpha V_0$, cf. (50), (66) follows from taking the sum of (21) and (22), appealing to (40), testing with the characteristic function of the closed set $\{0\le z\cdot\nu_0\le\alpha V_0\}$, and integrating out z.

We now turn to (64) and note that it follows from

$$\lim_{h\downarrow 0}\frac{1}{\sqrt{h}}\int_{z\cdot\nu_0>\tau V} G_h(z)(|\delta\chi(\cdot-z)|-|\delta\chi|)\,dz=0,$$

which testing with $\zeta\in C_0^\infty((0,T)\times[0,1)^d)$ assumes the form

$$\lim_{h\downarrow 0}\frac{1}{\sqrt{h}}\int_{z\cdot\nu_0>\tau V} G_h(z)(\zeta(\cdot+z)-\zeta)|\delta\chi|\,dxdtdz=0.$$

This holds, since the integral can be estimated by

$$\text{(67)}\qquad \begin{aligned}&\sup|\nabla\zeta|\int\frac{|z|}{\sqrt{h}}G_h(z)|\delta\chi|\,dxdtdz\\ &\overset{(26),(51)}{=}\sup|\nabla\zeta|\int|z|G_1(z)\,dz\int|\chi_h-\chi_h(\cdot-\tau)|\,dxdt,\end{aligned}$$

and the last contribution vanishes in the limit since τ does, and since $\{\chi_h\}_{h\downarrow 0}$ is compact in $L^1((0,T)\times[0,1)^d)$, cf. part i) of Lemma 3.

We finally address (61) and focus on the +-part. We first argue that

$$\text{(68)}\qquad \lim_{h\downarrow 0}\frac{1}{\sqrt{h}}\big(\delta\chi_+G_h*(1-\delta\chi_+)-(1-\delta\chi_+)G_h*\delta\chi_+\big)=0.$$

Spelling out the z-integral, and using that G_h is even, this follows from

$$\text{(69)}\qquad \begin{aligned}&\lim_{h\downarrow 0}\frac{1}{\sqrt{h}}\int G_h(z)\delta\chi_+(1-\delta\chi_+)(\cdot-z)\,dz\\ &=\lim_{h\downarrow 0}\frac{1}{\sqrt{h}}\int G_h(z)\delta\chi_+(\cdot+z)(1-\delta\chi_+)\,dz.\end{aligned}$$

Since the second function differs from the first just by a spatial shift of z, the limits coincide provided the l.h.s. limit is finite, which by the non-negativity of the functions follows if their integral remains bounded. The l.h.s. integral indeed remains bounded since $\delta\chi_+(1-\delta\chi_+)(\cdot-z)\le\chi_h(1-\chi_h)(\cdot-z)+(1-\chi_h)(\cdot-\tau)\chi_h(\cdot-\tau-z)$, which follows from $\delta\chi_+=\chi_h(1-\chi_h)(\cdot-\tau)$, and since the integral of the first summand remains bounded by (21), whereas the integral of the second summand is oblivious to the (vanishing) time shift and thus remains bounded by (22).

Equipped with (68), we now may substitute $\frac{1}{\sqrt{h}}\delta\chi_+ G_h * (1-\delta\chi_+)$ by $\frac{1}{2}\frac{1}{\sqrt{h}}\big(\delta\chi_+ G_h * (1-\delta\chi_+) + (1-\delta\chi_+) G_h * \delta\chi_+\big)$, which we may write as

$$\frac{1}{2\sqrt{h}}\int G_h(z)|\delta\chi_+ - \delta\chi_+(\cdot - z)|\,dz,$$

where we used for any $a, b \in \{0,1\}$ that $a(1-b) + (1-a)b = |a-b|$. Hence in order to obtain (61), we need the two function sequences

$$\frac{1}{\sqrt{h}}\int_{\pm z\cdot\nu_0\geq 0} G_h(z)|\delta\chi_+ - \delta\chi_+(\cdot - z)|\,dz$$

to have the same limit. Again by the evenness of G_h, these two functions only differ by a spatial shift z. The same argument as for (69) shows that the limits agree.

Step 4. Conclusion. We start from the identity (52) in form of

$$\begin{aligned}\frac{1}{\sqrt{h}}|\delta\chi| = \frac{1}{\sqrt{h}}\big(&\delta\chi G_h * \delta\chi \\ &+ \delta\chi_+ G_h * \delta\chi_- + \delta\chi_- G_h * \delta\chi_+ \\ &+ \delta\chi_+ G_h * (1-\delta\chi_+) + \delta\chi_- G_h * (1-\delta\chi_-)\big),\end{aligned}$$

or rather, using $2\int_{z\cdot\nu_0\geq 0} G_1 dz = 1$,

$$\begin{aligned}&2\int_{0\leq z\cdot\nu_0\leq\alpha V_0} G_1 dz \frac{1}{\sqrt{h}}|\delta\chi| = \frac{1}{\sqrt{h}}\Big(\delta\chi G_h * \delta\chi \\ &+ \delta\chi_+ G_h * \delta\chi_- + \delta\chi_- G_h * \delta\chi_+ \\ &+ \delta\chi_+ G_h * (1-\delta\chi_+) + \delta\chi_- G_h * (1-\delta\chi_-) - 2\int_{z\cdot\nu_0>\alpha V_0} G_1 dz|\delta\chi|\Big).\end{aligned}$$

We note that by an elementary lower-semi-continuity argument based on the definitions (50) and (51), we have the distributional inequality $\alpha|\partial_t\chi| \leq \liminf_{h\downarrow 0}\frac{1}{\sqrt{h}}|\delta\chi|$, provided the r.h.s. is a finite measure. Hence we obtain from (53), (57), and (60) the distributional inequality

$$\begin{aligned}2\alpha&\int_{0\leq z\cdot\nu_0\leq\alpha V_0} G_1\,dz|\partial_t\chi| \\ &\leq \alpha^2\mu + 2\int_{0\leq z\cdot\nu_0\leq\alpha V_0} G_1|z\cdot\nu|\,dz\;|\nabla\chi|\,dt,\end{aligned} \tag{70}$$

which in particular shows that $\partial_t\chi$ is a measure. Letting $V_0 \uparrow \infty$ and appealing to (28), this yields in particular $\alpha|\partial_t\chi| \le \alpha^2\mu + 4c_0|\nabla\chi|dt$ which we divide by α:

$$|\partial_t\chi| \le \alpha\mu + \frac{4}{\alpha}c_0|\nabla\chi|\,dt.$$

Letting $\alpha \downarrow 0$, we learn from the latter estimate that null sets of $|\nabla\chi|dt$ are null sets of $\partial_t\chi$, so that there exists $V \in L^1(|\nabla\chi|dt)$ such that (23) holds.

Since $|\partial_t\chi| = |V||\nabla\chi|dt$ is absolutely continuous w.r.t. to $|\nabla\chi|dt$, (70) even holds with μ replaced by its absolutely continuous part μ' w.r.t. $|\nabla\chi|dt$. Writing $\mu' = \theta|\nabla\chi|dt$ with $\theta \in L^1(|\nabla\chi|dt)$, (70) assumes the form

$$2\alpha\int_{0\le z\cdot\nu_0\le\alpha V_0} G_1\,dz|V| \le \alpha^2\theta + 2\int_{0\le z\cdot\nu_0\le\alpha V_0} G_1|z\cdot\nu|\,dz \quad |\nabla\chi|\,dt\text{-a.e.}$$

As in Step 1, a separability argument now allows to choose $\nu_0 = \nu$, so that by radial symmetry of G_1, the above assumes the form

$$2\alpha\int_{0\le z_1\le\alpha V_0} G_1\,dz|V| \le \alpha^2\theta + 2\int_{0\le z_1\le\alpha V_0} G_1 z_1\,dz \quad |\nabla\chi|\,dt\text{-a.e.}$$

Dividing by α^2 and momentarily writing $\alpha' := \alpha V_0$, this turns into

$$2\frac{V_0}{\alpha'}\int_{0\le z_1\le\alpha'} G_1\,dz|V| \le \theta + 2\frac{V_0^2}{\alpha'^2}\int_{0\le z_1\le\alpha'} G_1 z_1\,dz \quad |\nabla\chi|\,dt\text{-a.e.}$$

We now appeal to the limiting relations (which follow from factorizing G_1 into the $(d-1)$-dimensional standard Gaussian and $G_1^{d=1}$)

$$\lim_{\alpha'\downarrow 0}\frac{1}{\alpha'}\int_{0\le z_1\le\alpha'} G_1\,dz = G_1^{d=1}(0) = c_0,$$
$$\lim_{\alpha'\downarrow 0}\frac{1}{\alpha'^2}\int_{0\le z_1\le\alpha'} G_1 z_1\,dz = \frac{1}{2}G_1^{d=1}(0) = \frac{c_0}{2},$$

to see that the above turns into

$$2c_0V_0|V| \le \theta + c_0V_0^2 \quad |\nabla\chi|dt\text{-a.e.}$$

Again, by a separability argument for V_0, we may assume $V_0 = |V|$ so that the above yields (49) in form of $c_0V^2 \le \theta$.

Proof of Proposition 2.

Step 1. Metric slope and functional derivative. We claim the following relation between the metric slope $|\partial E(u)|$ of a functional E on $\mathcal{M}$, cf. (14), at a configuration u, and its first variation $\delta E(u).\xi$ in direction of a smooth vector field ξ:

$$\frac{1}{2}|\partial E(u)|^2 \geq \delta E(u).\xi - \frac{1}{2}\big(\delta d(u,\cdot)(u).\xi\big)^2. \tag{71}$$

As usual, first variation δ is defined by considering the curve $s \mapsto u_s$ of configurations characterized via the transport equation (to be interpreted distributionally or solved explicitly with help of the flow map Φ_s via $u_s \circ \Phi_s = u$)

$$\frac{\partial u_s}{\partial s} + \xi \cdot \nabla u_s = 0 \quad \text{and} \quad u_{s=0} = u, \tag{72}$$

and setting

$$\delta E(u).\xi := \frac{d}{ds}\Big|_{s=0} E(u_s) \quad \text{and} \quad \delta d(u,\cdot)(u).\xi := \frac{d}{ds}\Big|_{s=0} d(u,u_s), \tag{73}$$

with the understanding that both derivatives exist (and define linear functionals in ξ which is the case for $E = E_h$ and $d(u,\cdot) = d_h(u,\cdot)$, cf. Steps 2 and 3). Inequality (71) is then a direct consequence of the definition $|\partial E(u)|$, cf. (13), which yields

$$\begin{aligned}|\partial E(u)| &\geq \limsup_{s\downarrow 0} \frac{(E(u)-E(u_s))_+}{d(u,u_s)} \\ &\geq \frac{\lim_{s\downarrow 0}\frac{1}{s}(E(u)-E(u_s))}{\lim_{s\downarrow 0}\frac{1}{s}d(u,u_s)} = \frac{-\delta E(u).\xi}{\delta d(u,\cdot)(u).\xi},\end{aligned}$$

and Young's inequality.

STEP 2. Representation of $\delta E_h(u).\xi$; we claim:

$$\begin{aligned}\delta E_h(u).\xi = \frac{1}{\sqrt{h}}\int \Big(&\nabla\cdot\xi\big((1-u)G_h * u + uG_h*(1-u)\big) \\ &+ u[\xi,\nabla G_h *](1-u)\Big),\end{aligned} \tag{74}$$

where $[\xi,\nabla G_h *] = \sum_{i=1}^d [\xi_i, \partial_i G_h *]$ denotes the commutator of multiplying with ξ and convolving with ∇G_h. In checking this formula we may by approximation assume that u is smooth; by definition (72) and (73) of δ we obtain from the definition (15) of E_h

$$\delta E_h(u).\xi = \frac{1}{\sqrt{h}}\int \big((\xi\cdot\nabla u)G_h * u - (1-u)G_h * (\xi\cdot\nabla u)\big),$$

which by the symmetry of G_h* we rewrite as

$$\delta E_h(u).\xi = -\frac{1}{\sqrt{h}}\int \big(\xi\cdot\nabla(1-u)\,G_h*u + \xi\cdot\nabla u\,G_h*(1-u)\big).$$

We write $\xi\cdot\nabla u = \nabla\cdot(u\xi) - u\nabla\cdot\xi$ and $\xi\cdot\nabla(1-u) = \nabla\cdot((1-u)\xi) - (1-u)\nabla\cdot\xi$, so that (74) reduces to the identity

$$-\int \big(\nabla\cdot((1-u)\xi)\,G_h*u + \nabla\cdot(u\xi)\,G_h*(1-u)\big) \\ = \int u[\xi,\nabla G_h*](1-u),$$

which follows from integration by parts and the anti-symmetry of ∇G_h*.

STEP 3. Representation of $\delta d_h(u,\cdot)(u).\xi$:

(75) $$\frac{1}{2}\big(\delta d_h(u,\cdot)(u).\xi\big)^2 \\ = \sqrt{h}\int\Big(u\xi\cdot\nabla^2 G_h*((1-u)\xi) - u\xi\cdot\nabla G_h*((1-u)\nabla\cdot\xi) \\ + u\nabla\cdot\xi\,\nabla G_h*((1-u)\xi) - u\nabla\cdot\xi\,G_h*((1-u)\nabla\cdot\xi)\Big).$$

In the notation of Step 2, $\frac{1}{2}\big(\delta d_h(u,\cdot)(u).\xi\big)^2 = \frac{1}{2}\big(\frac{d}{ds}\big|_{s=0} d_h(u,u_s)\big)^2 = \frac{1}{4}\frac{d^2}{ds^2}\big|_{s=0} d_h^2(u,u_s)$, so that by definition (17) of d_h and by (72), we have $\frac{1}{2}\big(\delta d_h(u,\cdot)(u).\xi\big)^2 = \sqrt{h}\int \frac{\partial u_s}{\partial s}\big|_{s=0} G_h*\frac{\partial u_s}{\partial s}\big|_{s=0} = \sqrt{h}\int \xi\cdot\nabla u\,G_h*(\xi\cdot\nabla u)$. Rewriting the second factor $\xi\cdot\nabla u = \nabla\cdot(u\xi) - u\,\nabla\cdot\xi$ and using the symmetry of G_h*, an integration by parts, and $-\nabla u = \nabla(1-u)$, we obtain

$$\frac{1}{2}\big(\delta d_h(u,\cdot)(u).\xi\big)^2 \\ = \sqrt{h}\int\big(u\xi\cdot\nabla G_h*(\xi\cdot\nabla(1-u)) + u\nabla\cdot\xi\,G_h*(\xi\cdot\nabla(1-u))\big).$$

We write $\xi\cdot\nabla(1-u) = \nabla\cdot((1-u)\xi) - (1-u)\nabla\cdot\xi$ to obtain (75).

STEP 4. Passage to the limit in δE_h; we claim that

(76) $$\lim_{h\downarrow 0}\int_0^T \delta E_h(u_h).\xi\,dt = c_0\int(\nabla\cdot\xi - \nu\cdot\nabla\xi\nu)|\nabla\chi|\,dt.$$

According to (74), we may split into two statements. The first statement is

$$\lim_{h\downarrow 0}\frac{1}{\sqrt{h}}\int \nabla\cdot\xi\big((1-u_h)\,G_h*u_h+u_h\,G_h*(1-u_h)\big) = 2c_0\int \nabla\cdot\xi|\nabla\chi|\,dt,$$

which is an immediate consequence of testing (21) and (22) with $\nabla\cdot\xi$, appealing to the scaling (26) and to the formula (28). The second statement is

$$\lim_{h\downarrow 0}\frac{1}{\sqrt{h}}\int u_h[\xi,\nabla G_h*](1-u_h) = -c_0\int(\nu\cdot\nabla\xi\nu+\nabla\cdot\xi)|\nabla\chi|\,dt, \tag{77}$$

for which we now give the argument. Spelling out

$$\big([\xi,\nabla G_h*](1-u_h)\big)(t,x) = \int(\xi(t,x)-\xi(t,x-z))\cdot\nabla G_h(z)(1-u_h)(t,x-z)\,dz,$$

we see that

$$\Big|\big([\xi,\nabla G_h*](1-u_h)\big)(t,x) - \int\nabla G_h(z)\cdot\nabla\xi(t,x)z(1-u_h)(t,x-z)\,dz\Big| \le \frac{1}{2}\sup|\nabla^2\xi|\int|z|^2|\nabla G_h(z)|(1-u_h)(t,x-z)\,dz. \tag{78}$$

Appealing to (26) in form of $\nabla G_h(z)dz=\frac{1}{\sqrt{h}}\nabla G_1(\frac{z}{\sqrt{h}})d\frac{z}{\sqrt{h}}$, we learn that the limit of the contribution of the main term can be computed by testing (21) with $\frac{\nabla G_1(z)\cdot\nabla\xi(t,x)z}{G_1(z)}=-z\cdot\nabla\xi(t,x)z$ (which is of polynomial growth in z); it assumes the value

$$\int\nabla G_1\cdot\nabla\xi z(\nu\cdot z)_+|\nabla\chi|\,dtdz,$$

which yields (77) by formula (83) below. The contribution of the r.h.s. error term in (78) is vanishing of $O(\sqrt{h})$, as follows from appealing to (21) tested with $\frac{|\nabla G_1(z)||z|^2}{G_1(z)}=|z|^3$.

STEP 5. Passage to the limit in δd_h; we claim

$$\lim_{h\downarrow 0}\frac{1}{2}\int_0^T\left(\delta d_h(u_h,\cdot)(u_h).\xi\right)^2 dt = c_0\int(\xi\cdot\nu)^2|\nabla\chi|\,dt. \tag{79}$$

According to (75), this statement may be split into a leading-order statement

$$\lim_{h\downarrow 0}\sqrt{h}\int u_h\xi\cdot\nabla^2 G_h*((1-u_h)\xi)\,dxdt = c_0\int(\xi\cdot\nu)^2|\nabla\chi|\,dt, \tag{80}$$

and the higher-order statements

$$\begin{aligned}\int u_h\xi\cdot\nabla G_h*((1-u_h)\nabla\cdot\xi)\,dxdt &= O(1),\\ \int u_h\nabla\cdot\xi\,\nabla G_h*((1-u_h)\xi)\,dxdt &= O(1),\\ \frac{1}{\sqrt{h}}\int u_h\nabla\cdot\xi\,G_h*((1-u_h)\nabla\cdot\xi)\,dxdt &= O(1).\end{aligned} \tag{81}$$

The statement (80) itself splits into the main part

$$\lim_{h\downarrow 0}\sqrt{h}\int\xi\cdot\left(u_h\nabla^2 G_h*(1-u_h)\right)\xi\,dxdt = c_0\int(\xi\cdot\nu)^2|\nabla\chi|\,dt$$

and the higher-order commutator

$$\int u_h\xi\cdot[\xi,\nabla^2 G_h]*(1-u_h)\,dxdt = O(1). \tag{82}$$

The main part follows from appealing to (26) in form of $\sqrt{h}\nabla^2 G_h(z)dz = \frac{1}{\sqrt{h}}\nabla^2 G_1(\frac{z}{\sqrt{h}})d\frac{z}{\sqrt{h}}$, testing (21) with $\frac{\xi(t,x)\cdot\nabla^2 G_1(z)\xi(t,x)}{G_1(z)} = (\xi(t,x)\cdot z)^2 - |\xi(t,x)|^2$, and appealing to formula (84). The estimate of the error term (82) follows from estimating the integrand by $\sup|\xi|\sup|\nabla\xi|$ $\int|z||\nabla^2 G_h(z)|\,u_h(t,x)(1-u_h)(t,x-z)dz$, and then using the scaling (26) further by

$$\sup|\xi|\sup|\nabla\xi|\int|z||\nabla^2 G_1(z)|\,\frac{1}{\sqrt{h}}\,u_h(t,x)(1-u_h)(t,x-\sqrt{h}z)\,dz,$$

so that another application of (21) yields (82). Statement (81) and the other two higher-order estimates follow along the same lines: For instance, the integrand in (81) is $\le\sup|\xi|\sup|\nabla\cdot\xi|\int|\nabla G_h(z)|\;u_h(t,x)$ $(1-u_h)(t,x-z)dz$. By rescaling, $\int|\nabla G_h(z)|\,u_h(t,x)\,(1-u_h)(t,x-z)dz = \int|z|G_1(z)\,\frac{1}{\sqrt{h}}u_h(t,x)(1-u_h)(t,x-\sqrt{h}z)dz$, which is $O(1)$ by (21).

Step 6. Conclusion. By Riesz' representation theorem in $L^2(|\nabla\chi|dt)$ and an approximation argument in the (arbitrary) smooth vector field ξ, the statement of Proposition 2 is a consequence of

$$\liminf_{h\downarrow 0}\int_0^T \frac{1}{2}\,|\partial E_h(u_h)|^2\,dt \geq c_0\int\big(\nabla\cdot\xi-\nu\cdot\nabla\xi\nu-(\xi\cdot\nu)^2\big)|\nabla\chi|\,dt.$$

The latter follows starting from the inequality (71) for $(E,d,u)=(E_h,d_h,u_h(t))$, integrating in $t\in(0,T)$, and appealing to (76) and (79) to pass to the limit on the r.h.s..

Step 7. Two formulas: For any unit vector ν, any matrix A, and any vector ξ, we have

$$-\int\nabla G_1(z)\cdot Az\,(\nu\cdot z)_+dz=c_0(\nu\cdot A\nu+\mathrm{trA}), \tag{83}$$

$$\int\xi\cdot\nabla^2G_1(z)\xi\,(\nu\cdot z)_+dz=c_0(\xi\cdot\nu)^2. \tag{84}$$

Since $G_1(z)=-zG_1(z)$, for the first formula we may assume that A is symmetric; by linearity we may assume that $A=e\otimes e$ for some unit vector e; by radial symmetry of G_1, it thus remains to show

$$-\int\partial_1G_1z_1(\nu\cdot z)_+dz=c_0(\nu_1^2+1),$$

which by one integration by parts, taking into account (28), reduces to

$$\int_{\nu\cdot z>0}G_1z_1\,dz=c_0\nu_1,$$

which in view of $G_1z_1=-\partial_1G_1=-\nabla\cdot(G_1e_1)$ by the divergence theorem reduces to

$$\int_{\nu\cdot z=0}G_1=c_0, \tag{85}$$

which follows since $c_0=G_1^{d=1}(0)$. We now turn to (84). By radial symmetry of G_1 and homogeneity, it suffices to show

$$\int\partial_1^2G_1(\nu\cdot z)_+dz=c_0\nu_1^2, \tag{86}$$

which by two integration by parts reduces to (85).

References

[1] Fred Almgren, Jean E. Taylor and Lihe Wang, Curvature-driven flows: a variational approach, *SIAM Journal on Control and Optimization*, 31(2):387–438, 1993.

[2] Luigi Ambrosio, Nicola Gigli and Giuseppe Savaré, *Gradient flows in metric spaces and in the space of probability measures*, Birkhäuser, 2008.

[3] Selim Esedoğlu and Felix Otto, Threshold dynamics for networks with arbitrary surface tensions, *Communications on Pure and Applied Mathematics*, 68(5):808–864, 2015.

[4] Lawrence C. Evans, Convergence of an algorithm for mean curvature motion, *Indiana University Mathematics Journal*, 42(2):533–557, 1993.

[5] Tim Laux and Felix Otto, Convergence of the thresholding scheme for multiphase mean-curvature flow, *Calculus of Variations and Partial Differential Equations*, 55(5):1–74, 2016.

[6] Tim Laux and Felix Otto, Brakke's inequality for the thresholding scheme, *Calculus of Variations and Partial Differential Equations*, 59:39, 2020.

[7] Stephan Luckhaus and Thomas Sturzenhecker, Implicit time discretization for the mean curvature flow equation, *Calculus of Variations and Partial Differential Equations*, 3(2):253–271, 1995.

[8] Barry Merriman, James K. Bence and Stanley J. Osher, Diffusion generated motion by mean curvature, CAM Report 92-18, 1992, Department of Mathematics, University of California, Los Angeles.

[9] Peter W. Michor and David Mumford, Riemannian geometries on spaces of plane curves, *Journal of the European Mathematical Society*, 8(1):1–48, 2006.

[10] Tiago Salvador and Selim Esedoğlu, A simplified threshold dynamics algorithm for isotropic surface energies, *Journal of Scientific Computing*, 79, 648–669, 2019.

Hausdorff Center for Mathematics,
University of Bonn,
Villa Maria, Endenicher Allee 62,
D-53115 Bonn,
Germany
E-mail address: `tim.laux@hcm.uni-bonn.de`

Max Planck Institute for Mathematics in the Sciences,
Inselstrase 22,
D-04103 Leipzig,
Germany
E-mail address: `Felix.Otto@mis.mpg.de`

Advanced Studies in Pure Mathematics 85, 2020
The Role of Metrics in the Theory of Partial Differential Equations
pp. 95–104

Stochastic approaches to Lagrangian coherent structures

Sanjeeva Balasuriya

Abstract.

This Note discusses a connection between deterministic Lagrangian coherent structures (robust fluid parcels which move coherently in unsteady fluid flows according to a deterministic ordinary differential equation), and the incorporation of noise or stochasticity which leads to the Fokker–Planck equation (a partial differential equation governing a probability density function). The link between these is via a stochastic ordinary differential equation. It is argued that a closer investigation of the stochastic differential equation offers additional insights to both the other approaches, and in particular to uncertainty quantification in Lagrangian coherent structures.

§1. Lagrangian coherent structures

The 'Lagrangian coherent structure' problem typically takes the following form. For some geophysical (oceanic [9], atmospheric [42], extraterrestrial [22]) or laboratory-scale [12, 33] flow, fluid velocities u are obtained as observational/experimental data, usually on a spatiotemporal grid defined for positions $x \in \Omega \subset \mathbb{R}^n$ (typically, the dimension is $n = 2$ or 3) and times $t \in [0, T]$. The Lagrangian evolution of fluid particle positions x is then given by the *ordinary* differential equation

$$\frac{\mathrm{d}x}{\mathrm{d}t} = u\,(x, t)\ , \tag{1}$$

for which—given the fact that u is available as discrete data in a finite-time interval $[0, T]$—*numerical* methods are needed. Determining sets in Ω at time 0 which 'remain coherent' under some given criterion after

Received October 5, 2018.
Revised April 10, 2019.
2010 *Mathematics Subject Classification.* 34C60, 35J15, 37H99, 76R99.
Key words and phrases. Unsteady fluid flow, diffusion, uncertainty quantification, Perron–Frobenius transfer operator, Itô stochastic differential equations.

evolution by (1) to time T will be referred to as 'the standard Lagrangian coherent structure problem' [7]. The adjective 'Lagrangian' in this case refers to the fluid-dynamical nomenclature of 'following fluid particles' as opposed to the 'Eulerian' framework of specifying flow properties as functions of (x, t). Influential observations [27] pointed to the inadequacy of using Eulerian information in determining coherent fluid regions which had a dominant influence on how fluid transport occurred, if u were nonautonomous. However, defining exactly what is meant in saying that a time-varying structure is 'coherent' in a Lagrangian sense is less clear. Thus, there are many approaches to this, each based on a particular interpretation of 'coherence.' Both theoretical (variational formulations [23, 25] and stable/unstable manifold approaches [2, 3] to define coherent structure boundaries) and numerical (stretching of fluid elements [44], entangling of trajectories [1], complexity of motion [29]) methods have been developed. In general, these methods tend to be *deterministic* in that they rely on advecting Ω according to (1), and then extracting sets at time 0 based on some specific information gleaned from such advection. Review and comparison articles on the extensive field of Lagrangian coherent structure methods are available [7, 21, 24].

An issue that is gradually coming to the fore is that *the velocity u in* (1) *is not known with certainty.* This is because in typical applications, u is obtained from experimental/observational data. Uncertainties therefore include measurement error, known situations in which the velocity measurements have large errors (e.g., regions of cloud cover when taking satellite observations of sea-surface heights in order to infer oceanic velocities), modeling errors in converting observations to velocities (as in using the geostrophic approximation to relate sea-surface height to a streamfunction [45]), as well as errors arising from interpolating data only available on a spatiotemporally discrete grid. The latter situation arises even when using velocity field from numerical simulations, since any computational fluid dynamics simulation itself has a finite resolution. How can subgrid processes, or velocity uncertainties, be encapsulated in Lagrangian coherent structure methods?

This Note highlights emerging approaches to this issue which are connected to stochastic differential equations. Furthermore, the relationship to a probability density function—whose evolution is governed by a corresponding Fokker–Planck equation [11, 16] or alternatively a Perron–Frobenius (transfer) operator [14, 15, 42]—is discussed.

§2. Stochastic differential equation

A straightforward idea of modeling the velocity uncertainties described previously is to include them as perturbations in the velocity field, which are in some way *stochastic*. That is, the velocity field has some *dominant* smooth behavior (which is captured by the data), plus a small *stochastic* term which incorporates the range of uncertainties. This viewpoint implies the replacement of (1) with an appropriate *stochastic differential equation* [4, 5, 6, 8, 11, 16, 20]. A fairly general formulation for this could be [4, 6, 26]

$$\mathrm{d}x_t = u(x_t, t)\,\mathrm{d}t + \sigma(x_t, t)\,\mathrm{d}W_t\,, \tag{2}$$

in which W_t is n-dimensional Brownian noise in Ω, and σ is an $n \times n$ matrix, assumed small in some sense, which encodes the possibility that the stochasticity be modulated in a heterogenous anisotropic way. The potential for x-dependence in σ means that the noise is *multiplicative* [41], and here (2) will be thought of in an Itô (as opposed to Stratonovich) sense [39]. Thus, rather than deterministic trajectories $x(t)$ to (1), one can think of $t = 0$ conditions to (2) evolving as *random trajectories* x_t.

Stochastic ordinary differential equations of the form (2) are well-established in financial mathematics and stochastic integration theory [30, 36, 39, e.g.], but their usage in Lagrangian coherence is only emerging recently; hence, this Note.

How can one make sense of how individual random trajectories of (2) contribute collectively? Numerically, one can determine x_t using, say, an Euler–Maruyama algorithm [34]. By performing many simulations, the statistics of various scalar fields can then be numerically computed [4, 8, 20]. Such computational experiments have been performed for finite-time Lyapunov exponents [4, 6, 20] as well as so-called Lagrangian descriptors (taking a suitable average along trajectories) [8]. In these cases, σ is usually chosen to be proportional to the identity.

Another method to combine ensemble information is to seek theoretical means and variances resulting from (2), through the usage of Itô's isometry [31, 39]. With the assumption that σ is small, but that the velocity field need not be divergence-free, several such results have recently been developed in the context of two-dimensional unsteady flows operating over a finite time interval. The first is an expression for the'fattening' of a curve when advected according to (2) [4]. The second specializes to curves which have importance in Lagrangian coherent structures: unsteadily evolving flow barriers which demarcate distinct Lagrangian coherent structures [6]. The fuzziness of such flow barriers under stochasticity in the form (2) allows for a spatial characterisation of *mixing layers* between coherent structures. The evolution of the spatial structure of these layers with time, as well as a limiting behavior, is determined [6]. The direct connection of this to *mixing* is verified by numerically evolving density fields initially supported on the coherent structures on the two sides of the mixing layer, and showing that their evolution obeys the predicted spatial characterization of the mixing layer. A third addresses (2) in the context of rigorously quantifying how sensitive each and every Lagrangian trajectory's final position x_T is to noise [5]. In this, the anisotropic variance of a scaled displacement from the deterministic location is defined as a field on Ω, which can then be related to a measure of *robustness* of sets [5]. Implications of this theory towards realistic data is ongoing, and shows promise for evaluating the interdependence between velocity measurement uncertainties, spatial resolution, and Lagrangian structures that can be resolved subject to these.

Building on the various approaches described in this section—-utilizing Monte Carlo simulations, stochastic calculus methods, or new approaches on (2)—seems to be a promising way forward in understanding the impact of noise and uncertainty on Lagrangian structures.

§3. Diffusion and the Fokker–Planck equation

Stochasticity at the microscopic level corresponds to diffusion at the macroscopic level. Thus, an alternative way of understanding 'coherence' associated with trajectories of the deterministic ordinary differential equation (1) is to consider instead an evolving density $\rho(x,t)$ governed by the advection–diffusion partial differential equation

$$\frac{D\rho}{dt} := \frac{\partial \rho}{\partial t} + u \cdot \nabla \rho = \nabla^2 \rho \,, \tag{3}$$

in which the left-hand side is the convective derivative (derivative following the flow) of (2). 'Coherence' can now be attributed to the structure of the evolving scalar field $\rho(x,t)$. For example, if there is a collection of closed contours of $\rho(x,t)$ at some time t which evolve without developing significant filamentation [24] then these might be construed as demarcating an unsteady coherent structure. Alternatively, material surfaces across which there is minimum diffusive flux might be taken as defining a boundary of a coherent structure [26].

How is this connected to the stochastic differential equation (2)? The impact of evolving random trajectories of (2) generically results in a nonuniform distribution of trajectory densities $\rho(x,t)$ for $t \in [0,T]$. Standard stochastic techniques [35, 40, 43] enable the derivation that this evolves according to

$$\frac{\partial \rho}{\partial t} + \nabla \cdot (\rho u) = \frac{1}{2} \nabla \cdot \nabla \cdot \left(\rho \sigma \sigma^{\top} \right) , \tag{4}$$

which is the *Fokker–Planck equation* corresponding to the stochastic dynamics (2) [40]. In the simplest case where σ is proportional to the identity [11, 18, 32] and when the fluid is incompressible ($\nabla \cdot u = 0$), this collapses to the classical advection–diffusion equation. Preserving the (x, t)-dependence in σ is relevant, allowing for the possibility of modeling anisotropic diffusion which is further modulated by time and space.

While (2) and (4) may be considered equivalent, there are some practical differences in the *usage* of these two equations. Complicated σ causes considerable difficulties in analyzing, and in simulating, (4). (For example, lack of spatial periodicity will complicate using Fourier methods.) In contrast, such σ is easily incorporated into numerical simulations of (2) using, for example, the Euler-Maruyama algorithm [34]. Itô calculus [30], relevant to analyzing (2), forms an alternative methodology for addressing Lagrangian coherence information, contrasting with classical parabolic PDE methods required for (4). One interesting approach for the general Fokker–Planck equation (4) is by Haller et al [26], who develop a variational approach for determining surfaces across which there is the least diffusive flux (as a potential definition for demarcating a Lagrangian coherent structure). Another is by Denner et al [11], who (with σ being the identity) propose a method which can be numerically applied to (4) to investigate coherence.

Another direct comparison between (2) and (4) is that the probability density obtained from the collection of random trajectories of (2) all starting at the *same* initial position x_0, is equivalent to considering (4) with the initial density distribution $\rho(x, 0) = \delta(x - x_0)$, where $\delta(.)$ is the Dirac-delta distribution. Thus, statistical information from the trajectories of (2) can be used to reach conclusions on the Fokker–Planck equation's evolution. The linearity of (4) may allow for amalgamation of this information for ensembles of trajectories with different x_0-vaiues.

§4. Perron–Frobenius transfer operator

Given a map M, the *Perron–Frobenius* or *transfer* operator $\mathfrak{P}$ encodes how probability densities are pushed forward by M [10]. If imagining M to be the deterministic flow map of the ordinary differential equation (1) from time 0 to T, an approximation for $\mathfrak{P}$ is obtainable by the *Ulam matrix* P with elements P_{ij}, by partitioning Ω into boxes

B_i and by deterministic evolution, determining the transition probabilities into boxes C_j in $M(\Omega)$. The Perron–Frobenius operator can be made compact by convolving with a diffusive kernel prior to and after applying M [11, 14]; this enables the extraction of singular values λ_k and corresponding singular vectors v_k of the approximant P. Thus $P^\top P v_k = \lambda_k^2 v_k$, with the singular vector v_k (a constant on each of the B_is) forming an approximation for a field on Ω. Consequently, the transfer operator satisfies $\mathfrak{P}^\top \mathfrak{P} v_k \approx \lambda_k^2 v_k$. Singular vectors associated with the largest (near unity) λ_ks therefore identify 'almost coherent' sets [14, 15, 42] in Ω, with the level of coherence quantified by λ_k. This method, popularized and extended by Froyland and collaborators, is a well-established method for the detection of coherent structures associated with Lagrangian trajectories if (1) [14, 15, 37, 42]. It is often viewed as *deterministic* because M is, and nice theoretical results related to the singular limit of diffusion approaching zero are available [15].

In the above implementation, there is usually no explicit reference to a stochastic ordinary differential equation. However, the Perron–Frobenius operator is also often mentioned as the solution operator $\mathfrak{P}'$ of a Fokker–Planck equation [11, 16, 18]; specifically, $\mathfrak{P}'\rho(x,0) = \rho(x,T)$. Since the Fokker–Planck equation (4) is generated from the stochastic differential equation (2), this Perron–Frobenius operator $\mathfrak{P}'$ captures the *continuing* impact of uncertainties in the velocity field over the *entire* time duration $t \in [0,T]$. This is different from the understanding of the transfer operator $\mathfrak{P}$ in the most commonly known deterministic Lagrangian coherent structure approaches [14, 15, 37, 42]. However, it *is* a method for incorporating the uncertainty in the velocity field, and singular vectors of $\mathfrak{P}'$ give similar information on coherence [11] to those of $\mathfrak{P}$, albeit with a different interpretation of diffusion.

It should be pointed out that even with a deterministic u, if it were only known on a spatio-temporal grid, there is uncertainty in u at the subgrid level. When performing advection numerically, implicit or explicit interpolation usually occurs between gridpoints. These values, therefore, are in reality not known exactly. A possible modeling approach is to apply the stochastic differential equation (2) with a σ matrix which is zero at the gridpoints, but *unknown* off these gridpoints. Thus, (2) and its Fokker–Planck equation (4) may be apply even if the velocity is known with infinite accuracy, if only known on a discrete grid.

It is worth summarizing the two different Perron–Frobenius operator interpretations that have been discussed, within the context of the Fokker–Planck equation (4) and the corresponding stochastic differential equation (2). In both cases, consider an initial condition x_0 for (2).

(1) $\mathfrak{P}$: The impact of a diffusive kernel prior to advection is associated with applying (2) and (4) with $\sigma \equiv 0$. In other words, rather than (2), the understanding is that the deterministic advection (1) is in operation, while the Fokker–Planck equation (4) collapses to the pure advection/transport equation $\partial \rho / \partial t + \nabla \cdot (\rho u) = 0$. However, the initial condition $\rho(x, 0)$ is then not a Dirac mass at x_0, but is a diffused version of this.

(2) $\mathfrak{P}'$: The impact of velocity uncertainties can be modeled by applying (2) and (4) with $\sigma \neq 0$, but with the initial condition to the Fokker–Planck equation being $\rho(x, 0) = \delta(x - x_0)$. This applies if there is genuine uncertainty in u (measurement error, say), or if u is only known at gridpoints.

Additionally, to evaluate the impact across *all* initial conditions x_0 (in terms of a field on Ω), a broadening of the above understanding will be necessary.

The two transfer operators approaches above have a nice connection to how *uncertainties in data impact Lagrangian predictions*, which can have profound consequences (e.g., attributing uncertainties to the movement of coherent weather systems). Uncertainties in the prediction of any model can arise because of two main reasons: (i) uncertainty in initial position, and (ii) uncertainty in the model. These are quite naturally connected to $\mathfrak{P}$ and $\mathfrak{P}'$ respectively, because the former reflects diffusion imparted on the initial condition $\delta(x - x_0)$ but evolved with no noise, while the latter has an exact initial condition subject to a noisy evolution. Can these two effects be *combined* within the Fokker–Planck framework by having a diffused initial condition (rather than $\delta(x - x_0)$) at each point x_0, and also letting σ be nonzero (but chosen in a reasonable way to model system noise, e.g., having it zero at gridpoints but nonzero elsewhere to model resolution error)? Alternatively, is there an advantage to thinking of $\mathfrak{P}$ and $\mathfrak{P}'$ in terms of the stochastic differential equation (2), where the former might be associated with a weighting of initial conditions?

§5. Concluding remarks

This Note discusses the link between deterministic advection (which reflects the 'standard' Lagrangian coherent structure methodology) as given in (1), and the Fokker–Planck equation (4). The relationship is via the stochastic equation (2), which hitherto has only seen limited attention from this community. It is suggested that analyzing the stochastic ordinary differential equation (2) may offer considerable new insight into Lagrangian coherence, noise and diffusion.

Acknowledgements. The author acknowledges with gratitude support from the Japan Science and Technology Agency to attend the 'Second Joint Australia-Japan workshop on dynamical systems with applications in life science' in conjunction with the '11th Mathematical Society of Japan Seasonal Institute' in Hokkaido. Partial support from the Australian Research Council via grants FT130100484 and DP200101764 is also acknowledged. The author also thanks two anonymous referees, whose perceptive comments have improved this Note tremendously.

References

[1] M. Allshouse and J.-L. Thiffeault, Detecting coherent structures using braids, *Physica D*, **241** (2012), 95–105.

[2] S. Balasuriya, A tangential displacement theory for locating perturbed saddles and their manifolds, *SIAM J. Appl. Dyn. Sys.*, **10** (2011), 1100–1126.

[3] S. Balasuriya, *Barriers and transport in unsteady flows: a Melnikov approach*, volume 21 of *Mathematical Modeling and Computation*, Society for Industrial and Applied Mathematics (SIAM), Philadelphia, 2016.

[4] S. Balasuriya, Stochastic uncertainty of advected curves in finite-time unsteady flows, *Phys. Rev. E*, **95** (2017), 062201.

[5] S. Balasuriya, Stochastic sensitivity: A computable Lagrangian uncertainty measure for unsteady flows, *SIAM Review, in press*, (2020).

[6] S. Balasuriya and G. Gottwald, Estimating stable and unstable sets and their role as transport barriers in stochastic flows, *Phys. Rev. E*, **98** (2018), 013106.

[7] S. Balasuriya, N. Ouellette and I. Rypina, Generalized Lagrangian coherent structures, *Physica D*, **372** (2018), 31–51.

[8] F. Balibrea-Iniesta, C. Lopesino, S. Wiggins and A. Mancho, Lagrangian descriptors for stochastic differential equations: a tool for revealing the phase portrait of stochastic dynamical systems, *Intern. J. Bifurc. Chaos*, **26** (2016), 1630036.

[9] J. Bettencourt, V. Rossi, E. Hernández-Garcia, M. Marta-Almeida and C. López, Characterization of the structure and cross-shore transport porperties of a coastal upwelling filament using three-dimensional finite-size Lyapunov exponents, *J. Geophys. Res. Oceans*, **122** (2017), 7433–7448.

[10] M. Dellnitz and O. Junge, On the approximation of complicated dynamical behavior, *SIAM J. Numer. Anal.*, **36** (1999) 491–515.

[11] A. Denner, O. Junge and D. Matthes, Computing coherent sets using the Fokker–Planck equation, *J. Comp. Dyn.*, **3** (2016) 163–177.

[12] L. Fang, S. Balasuriya and N. Ouellette, Local linearity, coherent structures, and scale-to-scale coupling in two-dimensional flows, *Phys. Rev. Fluids*, **4** (2019) 014501.

[13] G. Froyland, Finite approximation of Sinai-Bowen-Ruelle measure for Anosov systems in two dimensions, *Random Comput. Dyn.*, **3** (1995), 251–264.

[14] G. Froyland, An analytical framework for identifying finite-time coherent sets in time-dependent dynamical systems, *Physica D*, **250** (2013), 1–19.

[15] G. Froyland, Dynamic isoperimetry and the geometry of Lagrangian coherent structures, *Nonlinearity*, **28** (2015), 3587–3622.

[16] G. Froyland and P. Koltai, Estimating long-term behavior of periodically driven flows without trajectory integration, *Nonlinearity*, **30** (2017), 1948–1986.

[17] G. Froyland and K. Padberg-Gehle, Finite-time entropy: A probabilistic approach for measuring nonlinear stretching, *Physica D*, **241** (2012), 1612–1628.

[18] G. Froyland and N. Santitissadeekorn, Optimal mixing enhancement, *SIAM J. Appl. Math.*, **77** (2017) 1444–1470.

[19] J. Guckenheimer and P. Holmes, *Nonlinear oscillations, dynamical systems, and bifurcations of vector fields*, Springer, New York, 1983.

[20] H. Guo, W. He, T. Peterka, H.-W. Shen, S. Collis and J. Helmus, Finite-time Lyapunov exponents and Lagrangian coherent structures in uncertain unsteady flows, *IEEE Trans. Visual. Comp. Graphics*, **22** (2016), 1672–1682.

[21] A. Hadjighasem, M. Farazmand, D. Blazevski, G. Froyland and G. Haller, A critical comparison of Lagrangian methods for coherent structure detection, *Chaos*, **27** (2017), 053104.

[22] A. Hadjighasem and G. Haller, Geodesic transport barriers in Jupiter's atmosphere: a video-based analysis, *SIAM Review*, **58** (2016), 69–89.

[23] G. Haller, A variational theory of hyperbolic Lagrangian Coherent Structures, *Physica D*, **240** (2011), 574–598.

[24] G. Haller, Lagrangian Coherent Structures, *Annu. Rev. Fluid Mech.*, **47** (2015), 137–162.

[25] G. Haller and F. Beron-Vera, Geodesic theory of transport barriers in two-dimensional flows, *Physica D*, **241** (2012), 1680–1702.

[26] G. Haller, D. Karrasch and F. Kogelbauer, Material barriers to diffusive and stochastic transport, *Proc. Nat. Acad. Sci.*, **115** (2018), 9074–9079.

[27] G. Haller and G. Yuan, Lagrangian coherent structures and mixing in two-dimensional turbulence, *Physica D*, **147** (2000), 352–370.

[28] H. Hwang, J. Jang and J. Jung, On the kinetic Fokker–Planck equation in a half-space with absorbing barriers, *Indiana Univ. Math. J.*, **64** (2015), 1767–1804.

[29] K. Ide, D. Small and S. Wiggins, Distinguished hyperbolic trajectories in time dependent fluid flows: Analytical and computational approach for velocity fields defined as data sets, *Nonlin. Proc. Geophys.*, **9** (2002), 237–263.

[30] K. Itô, Stochastic integral, *Proc. Imp. Acad.*, **20** (1944), 519–524.

[31] G. Kallianpur and P. Sundar, *Stochastic analysis and diffusion processes*, Oxford University Press, Oxford, 2014.

[32] D. Karrasch and J. Keller, A geometric heat-flow theory of Lagrangian coherent structures, *submitted*, (2018).

[33] D. Kelley and N. Ouellette, Onset of three-dimensionality in electromagnetically forced thin-layer flow, *Phys. Fluids*, **23** (2011), 045103.

[34] P. Kloeden, E. Platen and H. Schurz, *Numerical solution of SDE through computer experiments*, Universitext. Springer-Verlag, Berlin, 1994.

[35] B. Matkowsky, Singular perturbations, stochastic differential equations, and applications, In: *Singular Perturbations and Asymptotices*, (Eds. R. Meyer and S. Parter), pages 109–147, Academic Press, 1980.

[36] R. Merton, Lifetime portfolio selection under uncertainty: the continuous time case, *Rev. Econom. Statist.*, **51** (1969), 247–257.

[37] P. Miron, F. Beron-Vera, M. Olascoaga, J. Sheinbaum, P. Pérez-Brunius and G. Froyland, Lagrangian dynamic geography of the Gulf of Mexico, *Scientific Reports*, **7** (2017), 7021.

[38] F. Nier, Boundary conditions and subelliptic estimates for geometric Kramers–Fokker–Planck operators on manifolds with boundaries, *Mem. Amer. Math. Soc.*, **252** (2018), 1–145.

[39] B. Øksendahl, *Stochastic differential equations*, Springer, 4 edition, 1995.

[40] H. Risken, *The Fokker–Planck equation: Methods of solution and applications*, Springer, New York, 1984.

[41] J. Sancho, M. San Miguel, S. Katz and J. Gunton, Analytical and numerical studies of multiplicative noise, *Phys. Rev. A*, **26** (1982), 1589–1609.

[42] N. Santitissadeekorn, G. Froyland and A. Monahan, Optimally coherent sets in geophysical flows: a new approach to deliminating the stratospheric polar vortex, *Phys. Rev. E*, **82** (2010), 056311.

[43] Z. Schuss, Singular perturbation methods in stochastic differential equations of mathematical physics, *SIAM Review*, **22** (1980), 119–155.

[44] S. Shadden, F. Lekien and J. Marsden, Definitions and properties of Lagrangian coherent structures from Lyapunov exponents in two-dimensional aperiodic flows, *Physica D*, **212** (2005), 271–304.

[45] D. Stammer, Global characteristics of ocean variability estimated from regional TOPEX/POSEIDON altimeter measurements, *J. Phys. Oceanography*, **27** (1997), 1743–1769.

School of Mathematical Sciences,
University of Adelaide,
Adelaide SA 5005, Australia
E-mail address: `sanjeeva.balasuriya@adelaide.edu.au`

Advanced Studies in Pure Mathematics 85, 2020
The Role of Metrics in the Theory of Partial Differential Equations
pp. 105–114

Geometric nonlinear problems and GIT stability through moment maps

Akito Futaki and Hajime Ono

Abstract.

In this paper we re-visit the standard principle that if a geometric nonlinear PDE problem is set up in the moment map picture then the existence of the solution to the PDE should be related to GIT stability in algebraic geometry. We also take up Kähler metrics with constant Cahen–Gutt moment map and an extension of conformally Kähler, Einstein-Maxwell metrics.

§1. Introduction

In this paper we re-visit the standard principle that if a geometric nonlinear PDE problem is set up in the moment map picture then the existence of the solution to the PDE should be related to GIT stability in algebraic geometry. Here moment map picture of the PDE problem is meant by a picture that the relevant function space has an infinite dimensional symplectic manifold structure with Hamiltonian group action and the solutions of the PDE are zero set of the moment map.

The typical successful examples are
(i) Kobayashi-Hitchin correspondence of Hermitian-Einstein metrics on stable bundles in the sense of Mumford and Takemoto. [6], [29].
(ii) Kähler-Einstein metrics on K-stable Fano manifolds. [5], [28].

The examples we expect this principle can be applied are
(a) Kähler metrics of constant scalar curvature on K-stable polarized manifolds (M, L). Here $L \to M$ is a holomorphic line bundle with positive first Chern class $c_1(L)$ so that $c_1(L)$ can be considered as the space of Kähler forms ([7]).

Received October 30, 2018.
Revised May 31, 2019.
2010 *Mathematics Subject Classification.* Primary 53C55; Secondary 53C21, 55N91.
Key words and phrases. moment map, scalar curvature, star product, Einstein-Maxwell equation.

(b) Perturbed scalar curvature incorporating higher Chern classes ([10], [11]).
(c) Conformally Kähler, Einstein-Maxwell metrics ([2], [13], [14], [15], [17], [22]).
(d) Kähler metrics with constant Cahen-Gutt moment map, which is related to closedness of Fedosov star product ([4], [20], [21], [16]).
(e) Kähler-Yang-Mills-Higgs equations. ([1]).

In the examples (a), (b), (c), (d), (e), using the derivative of the norm squared moment map, one obtains two obstructions to the existence of the PDE. One is a Lie algebra character Fut : $\mathfrak{g} \to \mathbf{R}$ of the Lie algebra $\mathfrak{g}$ of the holomorphic vector fields on M, typically given in [9] for Kähler-Einstein metrics on Fano manifolds, and a structure theorem of the same Lie algebra $\mathfrak{g}$, [26]. The Lie algebra character Fut motivates to define "K-stability". The conjecture is the equivalence of the existence of the solution to the PDE and K-stability.

In sections from 2 to 6, we outline the above principle. In Section 7 and 8 we respectively take up the cases of Cahen-Gutt moment map and an extension of Conformally Kähler, Einstein-Maxwell metrics. The latter extension refines the one given in [17].

§2. GIT stablity and moment map

Geometric Invariant Theory (GIT) is a theory due to Mumford [27] to construct good moduli space when the equivalence classes are given by orbits of a group action. The idea is to discard "unstable orbits" and take the quotient of (semi)stable orbits. Then one will get a Hausdorff and compactifiable moduli space. There is a moment map interpretation due to Kempf and Ness [19] (see also [8]) of stable orbits. Let (N, Ω) be a compact Kähler manifold and $\pi : L \to (N, \Omega)$ a holomorphic line bundle with $c_1(L) = [\Omega]$. Suppose a reductive complex Lie group G is a complexification K^c of a compact Lie group K where K acts on N in the Hamiltonian way, i.e. for any $X \in \mathfrak{k} := \mathrm{Lie}(K)$ we have

$$i(X)\Omega = -d\mu_X$$

for some smooth function $\mu_X \in C^\infty(N)$. Then $\mu : N \to \mathfrak{k}^*$ is called the moment map for the action of K if μ is K-equivariant and

$$\langle \mu, X \rangle = \mu_X.$$

Suppose the action of K^c lifts to L. Let $p \in N$.

Definition 2.1. *The orbit $K^c \cdot p$ is polystable if, for $\widetilde{p} \in L^{-1}$ with $\pi(\widetilde{p}) = p$, $K^c \cdot \widetilde{p}$ is closed.*

Kempf-Ness theorem claims as follows. The orbit $K^c \cdot p$ is polystable if and only if $K^c \cdot p$ has a zero point of μ. That is,

$$K^c \cdot p \cap \mu^{-1}(0) \neq \emptyset.$$

§3. Donaldson-Fujiki picture (cscK case)

Let M be a compact Kähler manifold and ω a fixed symplectic form on M. In this section we take N to be the following infinite dimensional Kähler manifold:

$$N = \{J \in \Gamma(\mathrm{End}(TM)) \mid \omega\text{-compatible integrable complex structure}\}.$$

This means that $J^2 = -\mathrm{id}$ and that $g_J(\cdot, \cdot) := \omega(\cdot, J\cdot)$ is a Kähler metric. It is well-known that

$$T_J N \cong \mathrm{Sym}(T'_J M \otimes T'_J M).$$

The standard L^2-inner product gives N a Kähler structure, thus a symplectic structure. Let K be the group of all Hamiltonian symplectomorphisms. Its Lie algebra is given by

$$\mathfrak{k} = C_0^\infty(M) = \left\{ u \in C^\infty(M) \,\middle|\, \int_M u\, \omega^m = 0 \right\}.$$

Theorem 1 (Donaldson, Fujiki). *The map $\mu : N \to \mathfrak{k}^* \cong C^\infty(M)/\mathbf{R}$ given by $\mu(J) = (S(J), \cdot)_{L^2}$ is the moment map for the action of K where $S(J)$ is the scalar curvature of the Kähler metric g_J.*

Noteice that, because of the normalization $\int_M u\, \omega^m = 0$, the zeros of μ are J's for which $S(J)$ is constant. Thus, this is a moment map interpretation for constant scalar curvature Kähler (cscK) metrics. A K^c-orbit can be identified with a "Kähler class" since if X is a Hamiltonian vector field and u_X its Hamiltonian function,

$$L_{JX}\omega = i\partial\bar{\partial} u_X.$$

That is, $\mu^{-1}(0)$ is identified with the set of cscK metrics in a fixed Kähler class.

Kempf-Ness Theorem suggests that cscK problem should be a GIT stability issue.

§4. Hilbert-Mumford criterion

In this section N is a finite dimensional compact Kähler manifold again as in Section 2. Hilbert-Mumford criterion says that $p \in N$ is stable with respect to K^c-action if and only if $p \in N$ is stable with respect to every one parameter subgroup $\sigma : \mathbb{C}^* \to K^c$. If $\lim_{t\to 0}\sigma(t)p = p_0$ then p_0 is a fixed point of σ, and $\sigma(t)L_{p_0} = L_{p_0}$. $\sigma(t) : L_{p_0} \to L_{p_0}$ is a linear action. Let α be its weight so that $z \mapsto t^\alpha z$. Then $p \in N$ is stable w.r.t. σ if and only if $\alpha < 0$. We call the weight α the Mumford weight.

§5. K-stability

Now we return to the infinite dimensional Kähler manifold N as in Section 3, and try to apply Hilbert-Mumford criterion in Section 4. If X is a Hamiltonian vector field and u_X its Hamiltonian function then the moment map formula shows

$$\left.\frac{d}{dt}\right|_{t=0} \langle \mu(J_t), u_X\rangle = (L_X J, \dot{J})_{L^2}.$$

If $L_X J = 0$,

$$\mathrm{Fut}(X) = \langle \mu(J), u_X\rangle$$

is independent of $J \in N$.

A pair (M, L) of a compact complex manifold M and an ample line bundle L over M is said to be **K-stable** if it is stable when $-\mathrm{Fut}$ is chosen as the Mumford weight. But strictly speaking, we need to consider at p_0 as in Section 4. This limit point represents a limit of smoth complex manifolds, and may represents a singular variety. There is a formulation due to Donaldson [7] for such singular limits, and the refined invariant Fut is called the Donaldson-Futaki invariant. Yau-Tian-Donaldson conjecture is stated as follows: (M, L) should admit a cscK metric with the Kähler form in $c_1(L)$ if and only if it is K-stable.

Chen-Donaldson-Sun [5] and Tian [28] proved the following: When $L = K_M^{-1}$, the conjecture is true, that is, a Fano manifold admits a Kähler-Einstein metric if and only if it is K-stable.

§6. Other nonlinear problems

The cscK problem is a nonlinear PDE problem. The scalar curvature of $g_{i\bar{j}}$ is expressed as $-\Delta \log\det(g_{i\bar{j}})$. If we wish to find a cscK metric $\tilde{g}_{i\bar{j}} = g_{i\bar{j}} + \frac{\partial^2\varphi}{\partial z^i \partial \bar{z}^j}$ in the same Kähler class as g we need to solve the equation

$$-\tilde{\Delta} \log \det(\tilde{g}_{i\bar{j}}) = \text{constant}$$

for a smooth function φ, and this is a 4-th order elliptic PDE.

Recall that we saw in Section 3 that in the cscK problem case the symplectic structure on N (= the space of J's) was given by the Kähler structure coming form the standard L^2-inner product on $\mathrm{Sym}(T'_J M \otimes T'_J M)$. If we change the symplectic structure on N, we obtain other nonlinear problems which can be treated as a problem of GIT issue.

§7. Cahen–Gutt moment map

In this section we apply the ideas in previous sections for symplectic connections, see [3] for basics of symplectic connections.

Let (M, ω) be a symplectic manifold of dimension $2m$. A symplectic connection ∇ is a torsion free affine connection such that $\nabla\omega = 0$. There always exists a symplectic connection on any symplectic manifold. Unlike the Levi-Civita connection on a Riemannian manifold, a symplectic connection is not unique on a symplectic manifold. Given two symplectic connections ∇ and ∇', we write the difference by S:

$$\nabla_X Y - \nabla'_X Y = S(X, Y).$$

Then $\omega(S(X,Y), Z)$ is totally symmetric in X, Y and Z. Conversely, if ∇ is a symplectic connection and $\omega(S(X,Y), Z)$ is totally symmetric, then $\nabla' := \nabla + S$ is a symplectic connection. Thus, on a symplectic manifold (M, ω), the space of symplectic connections, denoted by $\mathcal{E}(M, \omega)$, is an affine space modeled on the set of all smooth sections $\Gamma(S^3(T^*M))$ of symmetric covariant 3-tensors. Thus we may identify $\mathcal{E}(M, \omega)$ as $\mathcal{E}(M, \omega) \cong \nabla + \Gamma(S^3(T^*M))$.

Now we assume M is a closed manifold. On $\mathcal{E}(M, \omega)$ there is a natural symplectic structure $\Omega^{\mathcal{E}}$ defined at ∇ given by

$$\Omega^{\mathcal{E}}_{\nabla}(\underline{A}, \underline{B}) = \int_M \omega^{i_1 j_1} \omega^{i_2 j_2} \omega^{i_3 j_3} \underline{A}_{i_1 i_2 i_3} \underline{B}_{j_1 j_2 j_3} \, \omega_m$$

for $\underline{A}$, $\underline{B} \in T_\nabla \mathcal{E}(M, \omega) \cong \Gamma(S^3(T^*M))$ where $\omega_m := \frac{\omega^m}{m!}$.

There is a natural action of the group of symplectomorphisms (i.e. symplectic diffeomorphisms) of (M, ω) on $\mathcal{E}(M, \omega)$, which is given for a symplectomorphism φ by

$$(\varphi(\nabla))_X Y = \varphi_*(\nabla_{\varphi_*^{-1} X} \varphi_*^{-1} Y)$$

for any $\nabla \in \mathcal{E}(M, \omega)$ and any smooth vector fields X and Y on M. This action preserves the symplectic structure $\Omega^{\mathcal{E}}$ on $\mathcal{E}(M, \omega)$. For $\nabla \in$

$\mathcal{E}(M,\omega)$ we put

$$\mu(\nabla) = \nabla_p\nabla_q \mathrm{Ric}(\nabla)^{pq} - \frac{1}{2}\mathrm{Ric}(\nabla)_{pq}\mathrm{Ric}(\nabla)^{pq} + \frac{1}{4}\mathrm{R}(\nabla,\omega)_{pqrs}\mathrm{R}(\nabla,\omega)^{pqrs}$$

where

$$\mathrm{R}(\nabla,\omega)(X,Y,Z,W) = \omega(R(X,Y)Z,W)$$

and

$$\mathrm{Ric}(X,Y) = -\mathrm{tr}(Z\mapsto R(X,Z)Y).$$

Theorem 2 (Cahen–Gutt). *The function* μ *on* $\mathcal{E}(M,\omega)$ *gives a moment map for the action of* $\mathrm{Ham}(M,\omega)$.

This follows from the formula

(1) $$\left.\frac{d}{dt}\right|_{t=0}\int_M \mu(\nabla+tA)\, f\,\omega_m = \Omega^{\mathcal{E}}(\underline{L_{X_f}\nabla, A}).$$

Now we assume that M is a compact Kähler manifold and that ω is a fixed symplectic form. We set as before

$$N = \{J \text{ integrable complex structure} \mid (M,\omega,J) \text{ is a Kähler manifold.}\}$$

La Fuente-Gravy considered the *Levi-Civita map* $lv : N \to \mathcal{E}(M,\omega)$ sending J to the Levi-Civita connection ∇^J of the Kähler manifold (M,ω,J).
Then $lv^*\Omega^{\mathcal{E}}$ gives a new symplectic structure on N, though the non-degeneracy is not obvious. We can ignore this point.

Proposition 7.1 ([16]). *For a real smooth function* f, $L_{X_f}\nabla^J = 0$ *if and only if* $L_{X_f}J = 0$. *In this case,* X_f *is a holomorphic Killing vector field.*

Let $\mathfrak{h}(M)$ denote the complex Lie algebra of all holomorphic vector fields on a compact Kähler manifold M. We set

$$\mathfrak{h}_{red}(M) = \{X\in\mathfrak{h}(M)\mid X \text{ has a zero}\}$$

and call it the reduced Lie algebra of holomorphic vector fields. It is a well-known result (see e.g. [18]) that for $X\in\mathfrak{h}_{red}(M)$ there exists uniquely up to constant functions a complex-valued smooth function u_X such that

(2) $$i(X)\omega = \sqrt{-1}\,\bar\partial u_X.$$

The reductive part of $\mathfrak{h}_{red}(M)$ can be expressed as a complexification of the real Lie subalgebra $\mathfrak{g}_{\mathbf{R}}$ consisting of Hamiltonian functions for holomorphic Killing vector fields. We call $\mathfrak{g}_{\mathbf{R}}$ the real reduced Lie algebra.

Using the previous proposition and (1) we can show

Corollary 7.1 ([21]). *Let (M, ω) be a compact Kähler manifold, and $\mathfrak{g}_{\mathbf{R}}$ be the real reduced Lie algebra of holomorphic vector fields. Then* $\mathrm{Fut} : \mathfrak{g}_{\mathbf{R}} \to \mathbb{R}$ *deifened by*

$$\mathrm{Fut}(\mathrm{grad}' f) := \int_M \mu(\nabla^J)\, f\ \omega_m$$

is independent of the choice of $J \in N$ where f is a Hamiltonian function for a holomorphic Killing vector field in $\mathfrak{g}_{\mathbf{R}}$ with normalization $\int_M f\omega_m = 0$.

Theorem 3 ([16]). *Let M be a compact Kähler manifold. If there exists a Kähler metric of non-negative Ricci curvature such that $\mu(\nabla)$ is constant for the Cahen–Gutt moment map μ and the Levi-Civita connection ∇ then the reduced Lie algebra $\mathfrak{g}$ of holomorphic vector fields is reductive.*

In [12], the invariant of K-stability condition for the existence of Cahen-Gutt momentum equation is formulated as a cohomological formula.

§8. An extension of Conformally Kähler, Einstein-Maxwell metrics

In this section we take up conformally Kähler, Einstein-Maxwell metrics. The relevance of Einstein-Maxwell equation to Kähler geometry was found by LeBrun [24], [25]. Apostolov and Maschler [2] started a more systematic study including the Donaldson-Fujiki picture.

Let (M, ω) be a compact Kähler manifold of complex dimension m. Let T be a maximal torus of the reduced automorphism group $\mathrm{Aut}_r(M)$. We take N to be the infinite dimensional manifold consisting of all J's in $\Gamma(\mathrm{End}(TM))$ such that J is a T-invariant ω-compatible integrable complex structure.

When we fix $a \in \mathbf{R}$ and a smooth positive function f such that the Hamiltonian vector field X_f is an element of the Lie algebra of T, N carries the following formal Kähler structure $(\boldsymbol{J}, \Omega^{f^a})$

$$\boldsymbol{J}_J(\dot{J}) := J\dot{J}$$

$$\Omega_J^{f^a}(\dot{J}_1, \dot{J}_2) := \frac{1}{2}\int_M \mathrm{Tr}(J\dot{J}_1\dot{J}_2) f^a \omega_m,$$

where a tangent vector of N at J is identified with a smooth T-invariant section of $\mathrm{End}(TM)$ satisfying

$$\dot{J}J + J\dot{J} = 0 \quad \text{and} \quad \omega(\dot{J}\cdot, \cdot) + \omega(\cdot, \dot{J}\cdot) = 0.$$

Let K be the group of all T-equivariant Hamiltonian symplectomorphisms. Its Lie algebra is given by

$$\mathfrak{k} = (C_0^\infty(M))^T = \{u \in C_0^\infty(M) \mid u \text{ is } T\text{-invariant}\}.$$

For each fixed $b \in \mathbf{R}$, $\mathfrak{k}$ has the following inner product

$$(\varphi, \psi)_{f^b} := \int_M \varphi\psi f^b \omega_m.$$

We define the (f, a, b)-scalar curvature $S_{f,a,b}(J)$ of g_J as

$$S_{f,a,b}(J) := f^{a-b-2}\left\{f^2 S(J) + 2af\Delta_{g_J} f - a(a-1)|df|^2_{g_J}\right\}.$$

When $S_{f,a,b}(J)$ is constant, we call g_J a constant (f, a, b)-scalar curvature Kähler metric.

Theorem 4 ([2], [13], [17], [22], [23]). *The map $\mu : N \to \mathfrak{k}^* \cong (C^\infty(M)^T)/\mathbf{R}$ given by $\mu(J) = \langle S_{f,a,b}(J), \cdot\rangle_{f^b}$ is the moment map for the action of K.*

As the argument in Section 3, $\mu^{-1}(0)$ is identified with the set of constant (f, a, b)-scalar curvature Kähler metrics in a fixed Kähler class.

Corollary 8.1 ([2]). *Let (M, ω) be a compact Kähler manifold, and $\mathfrak{t}$ be the Lie algebra of a maximal torus T in $\mathrm{Aut}_r(M)$. Then $\mathrm{Fut}_{f,a,b} : \mathfrak{t} \to \mathbf{R}$ defined by*

$$\mathrm{Fut}_{f,a,b}(X_h) := (S_{f,a,b}(J), h)_{f^b}$$

is independent of the choice of $J \in N$. If there exists J such that g_J is a constant (f, a, b)-scalar curvature Kähler metric then $\mathrm{Fut}_{f,a,b}$ vanishes identically.

Theorem 5 ([14], [22]). *Let g_J be a constant (f, a, b)-scalar curvature Kähler metric. Then the centralizer*

$$\mathrm{Aut}_r(M)^{X_f} := \{g \in \mathrm{Aut}_r(M) \mid Ad(g)X_f = X_f\}$$

of the Hamiltonian vector field X_f is the complexification of $\mathrm{Isom}(M, g_J) \cap \mathrm{Aut}_r(M)^{X_f}$. In particular $\mathrm{Aut}_r(M)^{X_f}$ is reductive.

For special a and b, (f, a, b)-scalar curvature represent the scalar curvature of some Riemannian metric. For example, when $a - 1 - 2m$ and $b = -1 - 2m$, (f, a, b)-scalar curvature of g_J is the scalar curvature of the Hermitian metric g_J/f^2. When $S_{f,1-2m,-1-2m}(J)$ is constant, the Hermitian metric g_J/f^2 is called conformally Kähler, Einstein-Maxwell metric [2].

References

[1] L. Álvarez-Cónsul, M. Garcia-Fernandez and O. Garcia-Prada, On the Kähler-Yang-Mills-Higgs equations, preprint, arXiv:1807.10691.

[2] V. Apostolov and G. Maschler, Conformally Kähler, Einstein-Maxwell geometry, J. Eur. Math. Soc. (JEMS) **21** (2019), no. 5, 1319–1360.

[3] P. Bieliavsky, M. Cahen, S. Gutt, J. Rawnsley and L. Schwachhöfer, Symplectic connections, Int. J. Geom. Methods Mod. Phys., **3** (2006), 375–420.

[4] M. Cahen and S. Gutt, Moment map for the space of symplectic connections, Liber Amicorum Delanghe, F. Brackx and H. De Schepper eds., Gent Academia Press, 27–36 (2005).

[5] X.-X. Chen, S.K. Donaldson and S. Sun, Kähler-Einstein metrics on Fano manifolds, III: limits as cone angle approaches 2π and completion of the main proof, J. AMS., **28** (2015), 235–278.

[6] S. K. Donaldson, Anti-self-dual Yang-Mills connections on complex algebraic surfaces and stable vector bundles, Proc. London Math. Soc. **50** (1985), 1–26.

[7] S. K. Donaldson, Scalar curvature and stability of toric varieties, J. Differential Geometry, **62** (2002), 289–349.

[8] S. K. Donaldson and P. B. Kronheimer, The geometry of four manifolds, Oxford Mathematical Monographs, Claren Press, Oxford, 1990.

[9] A. Futaki, An obstruction to the existence of Einstein Kähler metrics, Invent. Math., **73** (1983), no. 3, 437–443.

[10] A. Futaki, Harmonic total Chern forms and stability, Kodai Math. J., **29**, no. 3 (2006), 346–369.

[11] A. Futaki, Holomorphic vector fields and perturbed extremal Kähler metrics, J. Symplectic Geom., **6**, no. 2 (2008), 127–138.

[12] A. Futaki and L. La Fuente-Gravy, Deformation quantization and Kähler geometry with moment map, preprint, arXiv:1904.11749.

[13] A. Futaki and H. Ono, Volume minimization and Conformally Kähler, Einstein-Maxwell geometry, J. Math. Soc. Japan, **70**, no. 4 (2018) pp. 1493–1521.

[14] A. Futaki and H. Ono, Conformally Einstein-Maxwell Kähler metrics and the structure of the automorphisms, Math. Zeit., **292** (2019), 571–589.

[15] A. Futaki and H. Ono, Volume minimization and obstructions to solving some problems in Kähler geometry, ICCM Not., **6** (2018), no. 2, 51–60.

[16] A. Futaki and H. Ono, Cahen-Gutt moment map, closed Fedosov star product and structure of the automorphism group, J. Symplectic Geom., **18** (2020), no. 1, 123–145.
[17] A. Futaki and H. Ono, On the existence problem of Einstein-Maxwell Kähler metrics, Progress in Math., Vol. 333, 93–111, Birkhauser.
[18] P. Gauduchon, Calabi's extremal metrics, An elementary introduction, Lecture Notes.
[19] G. Kempf and L. Ness, On the lengths of vectors in representation spaces, Lecture Notes in Math., **732**, pp. 233–242, Springer-Verlag.
[20] L. La Fuente-Gravy, Infinite dimensional moment map geometry and closed Fedosov's star products, Annals of Global Analysis and Geometry, **49** (2016), 1–22.
[21] L. La Fuente-Gravy, Futaki invariant for Fedosov's star products, J. Symplectic Geom., **17** (2019), no. 5, 1317–1330.
[22] A. Lahdili, Automorphisms and deformations of conformally Kähler, Einstein-Maxwell metrics, J. Geom. Anal., **29** (2019), 542–568.
[23] A. Lahdili, Kähler metrics with constant weighted scalar curvature and weighted K-stability, Proc. London Math. Soc., **119** (2019), 1065–1114.
[24] C. LeBrun, The Einstein-Maxwell equations, Kähler metrics, and Hermitian geometry, J. Geom. Phys., **91** (2015), 163–171.
[25] C. LeBrun, The Einstein-Maxwell equations and conformally Kähler geometry, Commun. Math. Phys., **344** (2016), 621–653.
[26] Y. Matsushima, Sur la structure du groupe d'homéomorphismes analytiques d'une certaine variété kaehlérienne, Nagoya Math. J., **11** (1957), 145–150.
[27] D. Mumford, Stability of projective varieties, L'Enseignement Mathematiques, **23** (1977), 39–110.
[28] G. Tian, K-Stability and Kähler-Einstein Metrics, Comm. Pure Appl. Math., **68** (2015), 1085–1156, Corrigendum, 2082–2083.
[29] K. Uhlenbeck and S.-T.Yau, On the existence of Hermitian-Yang-Mills connections in stable vector bundles, Comm. Pure Appl. Math. **39** (1986), 257–293, Comm. Pure Appl. Math. **42** (1989), 703–707.

Yau Mathematical Sciences Center, Tsinghua University,
Haidian district, Beijing 100084, China
E-mail address: futaki@tsinghua.edu.cn

Department of Mathematics, Saitama University,
255 Shimo-Okubo, Sakura-Ku, Saitama 380-8570, Japan
Current Address:
Department of Mathematics, University of Tsukuba,
1-1-1 Tennodai, Tsukuba, Ibaraki 305-8571, Japan
E-mail address: h-ono@math.tsukuba.ac.jp

Advanced Studies in Pure Mathematics 85, 2020
The Role of Metrics in the Theory of Partial Differential Equations
pp. 115–126

On the inclusion of damping terms in the hyperbolic MBO algorithm

Elliott Ginder and Ayumu Katayama

Abstract.

The hyperbolic MBO is a threshold dynamic algorithm which approximates interfacial motion by hyperbolic mean curvature flow. We introduce a generalization of this algorithm for imparting damping terms onto the equation of motion. We also construct corresponding numerical methods, and perform numerical tests. We also use our results to show that the generalized hyperbolic MBO is able to approximate motion by the standard mean curvature flow.

§1. Introduction

We develop an approximation method for computing the damped motion of interfaces under hyperbolic mean curvature flow (HCMF):

$$\alpha \boldsymbol{x}_{tt}(t,s) + \beta \boldsymbol{x}_t(t,s) = -\gamma\kappa(t,s)\nu(t,s). \tag{1}$$

In the above, $\boldsymbol{x} : [0,T) \times I \to \mathbf{R}^2$ denotes a closed curve in $\mathbf{R}^2$ (parameterized over an interval I), $T > 0$ is a final time, κ denotes the curvature of the interface, and ν is the outward unit normal of the interface. The nonnegative parameters α, β, and γ, designate mass, damping, and surface tension coefficients, respectively. The subscripts signify differentiation with respect to their variables, so that $\boldsymbol{x}_{tt}$ refers to the normal acceleration of the interface, and $\boldsymbol{x}_t$ denotes the normal velocity. We remark that the presence of the inertial term signifies that the HMCF is an oscillatory interfacial motion.

The equation of motion (1) is accompanied by two initial conditions: one for the initial shape of the interface, and another prescribing the initial velocity field along the interface. It can be shown [4] that, when

---

Received April 5, 2019.
Revised June 13, 2019.
2010 *Mathematics Subject Classification.* Primary 35A35; Secondary 53C44.

the initial velocity field is normal to the interface, the velocity field of the interface remains normal for the remainder of the flow. Although tangental velocities can be used to impart features such as rotation into the interfacial dynamics, our study assumes the initial velocity field to act in the normal direction of the interface.

§2. A generalized HMBO algorithm

The original threshold dynamical (TD) algorithm (the so-called MBO algorithm, see [5]) is a method for approximating motion by mean curvature flow (MCF). Borrowing on such ideas, a TD algorithm for hyperbolic mean curvature flow was introduced in [3]. Whereas previous TD algorithms utilize properties of the diffusion equation to approximate MCF, properties of wave propagation (along with a particular choice of initial condition) were used to design an approximation method for HMCF. For a time step size $\tau > 0$, the error of the approximation was shown to be of the order $O(\tau)$. In this study, we will use properties of wave propagation, together with a suitable initial velocity field, to incorporate damping terms into the HMCF.

Let time be discretized with a step size $\tau > 0$, and n be a non-negative integer. For the sake of simplicity in the exposition, let V_n denote the normal velocity of the interface at the time step n, $\dot{V}_n$ be the normal acceleration, and κ_n be the corresponding curvature of the interface. For the time being, we take the mass, damping, and surface tension coefficients to be unity, and proceed to construct an approximation method for the following interfacial dynamics:

$$\dot{V}_n - V_n = -\kappa_n. \tag{2}$$

Our approach is to observe the propagation of interfaces under the wave equation:

$$\begin{cases} u_{tt} = c^2 \Delta u, & \text{in } (0,\tau) \times \Omega \\ u(0,\boldsymbol{x}) = u_0(\boldsymbol{x}), & \text{in } \Omega \\ u_t(0,\boldsymbol{x}) = -v_0(\boldsymbol{x}), & \text{in } \Omega \\ \partial_{\boldsymbol{n}} u = 0 & \text{on } (0,\tau) \times \partial\Omega, \end{cases}$$

where Ω is a given domain with smooth boundary, c^2 sets the wave speed, u_0 is an initial profile, v_0 designates the initial velocity, and τ is the time step. Although we have prescribed a Neumann boundary condition, $\partial_{\boldsymbol{n}} u = 0$, we will only focus on the motion of interfaces located away from the boundary of the domain. In particular, away from the boundary, the short-time solution of the wave equation can be expressed using the Poisson formula:

$$u(t,\boldsymbol{x}) = \frac{1}{2\pi ct}\int_{B(\boldsymbol{x},ct)} \frac{u_0(\boldsymbol{y}) + \nabla u_0(\boldsymbol{y})\cdot(\boldsymbol{y}-\boldsymbol{x}) - tv_0(\boldsymbol{y})}{\sqrt{c^2t^2 - |\boldsymbol{y}-\boldsymbol{x}|^2}}\,d\boldsymbol{y}, \tag{3}$$

where $B(\boldsymbol{x}, ct)$ denotes the ball centered at $\boldsymbol{x}$ with radius ct.

Let Γ^n be the closed curve at time step n, described as the boundary of a set S_n, and denote its signed distance function by

$$d_n(\boldsymbol{x}) = \begin{cases} \inf_{\boldsymbol{y}\in\Gamma^n} \|\boldsymbol{x}-\boldsymbol{y}\| & \boldsymbol{x}\in S_n \\ -\inf_{\boldsymbol{y}\in\Gamma^n} \|\boldsymbol{x}-\boldsymbol{y}\| & \text{otherwise.} \end{cases}$$

We remark that d_0 is constructed from the given initial configuration of the interface, and that d_{-1} can be constructed using the initial velocity field along the interface. This allows us to define $u_0(\boldsymbol{x})$ as follows, for any non-negative integer:

$$u_0(\boldsymbol{x}) = 2d_n(\boldsymbol{x}) - d_{n-1}(\boldsymbol{x}).$$

By taking $v_0(\boldsymbol{x}) = 0$ in (3), it can be shown (see [3]) that

$$\delta_n = \delta_{n-1} - \frac{c^2}{2}(2\kappa_n - \kappa_{n-1})\tau^2 + O(\tau^3), \tag{4}$$

where δ_n denotes the distance traveled in the normal direction at time step n (see Figure 1). Denoting the average velocity within the time interval $[(n-1)\tau, n\tau]$ by $\bar{V}_n$, one has

$$\delta_n = \bar{V}_n\tau, \qquad \delta_{n-1} = \bar{V}_{n-1}\tau,$$

and hence equation (4) can be written:

$$\bar{V}_n\tau = \bar{V}_{n-1}\tau - \frac{c^2}{2}(2\kappa_n - \kappa_{n-1})\tau^2 + O(\tau^3).$$

Formally assuming $|\kappa_n - \kappa_{n-1}| < C\tau$ for some non-negative C, one obtains

$$\bar{V}_n\tau = \bar{V}_{n-1}\tau - \frac{c^2}{2}\kappa_n\tau^2 + O(\tau^3),$$

and hence

$$\dot{V}_n = -\frac{c^2}{2}\kappa_n + O(\tau) \qquad \text{(as } \tau\to 0\text{)}.$$

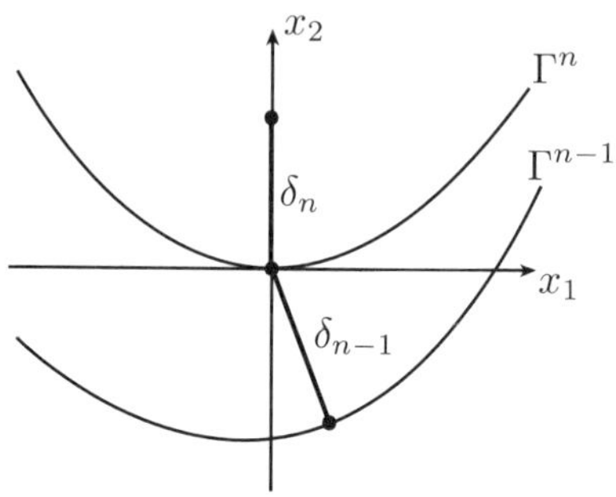

Fig. 1. Motion of a single point of the interface in the normal direction. The point moves a distance δ_n at step n. Without loss of generality, the direction of motion at the nth step is in the x_2 direction.

The damping term in equation (2) can be included by prescribing the initial velocity of the wave equation to be $v_0(\boldsymbol{x}) = d_n(\boldsymbol{x})$. This can be seen by expanding $d_n(\boldsymbol{x})$ in a Taylor series about $\boldsymbol{x} = \mathbf{0}$ (see [2]):

$$d_n(x_1, x_2) = x_2 + \frac{1}{2}\kappa_n x_1^2 + \frac{1}{6}(\kappa_n)_{x_1} x_1^3 - \frac{1}{2}\kappa_n^2 x_1^2 x_2 + O(|\boldsymbol{x}|^4),$$

and appealing to Poisson's formula (3):

$$u_v(t, \boldsymbol{x}) = \frac{1}{2\pi ct}\int_{B(\boldsymbol{x},ct)} \frac{-tv_0(\boldsymbol{y})}{\sqrt{c^2t^2 - |\boldsymbol{y}-\boldsymbol{x}|^2}}\, d\boldsymbol{y}.$$

The change of variables:

$$\boldsymbol{y} - \boldsymbol{x} = ct\boldsymbol{z},$$

transforms the domain of integration in Poisson's formula to that of a ball centered at the origin with radius one. In such a setting, we remark that

$$O(|\boldsymbol{y}|^4) = O(t^4) \qquad (\text{as } t \to 0).$$

We thus investigate the contribution of the first four terms in the Taylor expansion, u_v^1, u_v^2, u_v^3, and u_v^4. We begin with the lowest order term:

$$\begin{aligned} u_v^1(t, \boldsymbol{x}) &= \frac{1}{2\pi ct}\int_{B(\boldsymbol{x},ct)} \frac{-ty_2}{\sqrt{c^2t^2 - |\boldsymbol{y}-\boldsymbol{x}|^2}}\, d\boldsymbol{y} \\ &= \frac{-t}{2\pi ct}\int_{B(0,1)} \frac{ctz_2 + x_2}{ct\sqrt{1 - |\boldsymbol{z}|^2}}(ct)^2\, d\boldsymbol{z}. \end{aligned}$$

Appealing to function parity we have:

$$\int_{B(0,1)} \frac{z_2}{\sqrt{1-|\boldsymbol{z}|^2}}\, d\boldsymbol{z} = 0,$$

and it follows that

$$u_v^1(t,\boldsymbol{x}) = \frac{-t}{2\pi}\int_{B(0,1)} \frac{x_2}{\sqrt{1-|\boldsymbol{z}|^2}}\, d\boldsymbol{z}.$$

By making the change of variables:

$$z_1 = r\cos\theta, \qquad z_2 = r\sin\theta, \tag{5}$$

one arrives at

$$\begin{aligned} u_v^1(t,\boldsymbol{x}) &= \frac{-tx_2}{2\pi}\int_0^1\int_0^{2\pi} \frac{1}{\sqrt{1-r^2}}\, r d\theta dr \\ &= \frac{-tx_2}{2\pi}\int_0^1 \frac{2\pi r}{\sqrt{1-r^2}}\, dr \\ &= -tx_2\int_0^1 \frac{r}{\sqrt{1-r^2}}\, dr. \end{aligned}$$

Composite function integration yields

$$u_v^1(t,\boldsymbol{x}) = -tx_2\int_0^1 (1-r^2)^{-\frac{1}{2}}(-2r)\left(-\frac{1}{2}\right) dr = -tx_2. \tag{6}$$

We next consider the influence of the second term:

$$\begin{aligned} u_v^2(t,\boldsymbol{x}) &= \frac{1}{2\pi ct}\int_{B(\boldsymbol{x},ct)} \frac{-\frac{1}{2}t\kappa_n y_1^2}{\sqrt{c^2t^2-|\boldsymbol{y}-\boldsymbol{x}|^2}}\, d\boldsymbol{y} \\ &= \frac{-t\kappa_n}{4\pi ct}\int_{B(\boldsymbol{x},ct)} \frac{y_1^2}{\sqrt{c^2t^2-|\boldsymbol{y}-\boldsymbol{x}|^2}}\, d\boldsymbol{y} \\ &= \frac{-t\kappa_n}{4\pi}\int_{B(0,1)} \frac{c^2t^2z_1^2+2ctx_1z_1+x_1^2}{\sqrt{1-|\boldsymbol{z}|^2}}\, d\boldsymbol{z}. \end{aligned}$$

As before, function parity yields

$$\int_{B(0,1)} \frac{z_1}{\sqrt{1-|\boldsymbol{z}|^2}}\, d\boldsymbol{z} = 0,$$

and hence

$$u_v^2(t,\boldsymbol{x}) = \frac{-t\kappa_n}{4\pi}\int_{B(0,1)} \frac{c^2t^2z_1^2+x_1^2}{\sqrt{1-|\boldsymbol{z}|^2}}\, d\boldsymbol{z}.$$

Making the change of variables (5), we have

$$\begin{aligned}
u_v^2(t,\boldsymbol{x}) &= \frac{-t\kappa_n}{4\pi}\int_0^1\int_0^{2\pi}\frac{c^2t^2r^3\cos^2\theta + x_1^2 r}{\sqrt{1-r^2}}\,d\theta dr\\
&= \frac{-t\kappa_n}{4\pi}\int_0^1\int_0^{2\pi}\frac{c^2t^2r^3\dfrac{1+\cos 2\theta}{2} + x_1^2 r}{\sqrt{1-r^2}}\,d\theta dr\\
&= \frac{-t\kappa_n}{4\pi}\int_0^1\frac{\pi\left(c^2t^2r^3 + 2x_1^2 r\right)}{\sqrt{1-r^2}}\,dr.
\end{aligned}$$

Using another change of variables:

$$r = \cos\theta, \tag{7}$$

allows one to obtain:

$$\begin{aligned}
u_v^2(t,\boldsymbol{x}) &= \frac{-t\kappa_n}{4}\int_0^{\frac{\pi}{2}}\frac{c^2t^2\cos^3\theta + 2x_1^2\cos\theta}{\sqrt{1-\cos^2\theta}}\sin\theta\, d\theta\\
&= \frac{-t\kappa_n}{4}\int_0^{\frac{\pi}{2}}\left(c^2t^2\frac{\cos 3\theta + 3\cos\theta}{4} + 2x_1^2\cos\theta\right)d\theta\\
&= \frac{-t\kappa_n}{4}\left(c^2t^2\left(-\frac{1}{12}+\frac{3}{4}\right) + 2x_1^2\right)\\
&= -t\kappa_n\left(\frac{c^2t^2}{6}+\frac{x_1^2}{2}\right).
\end{aligned}\tag{8}$$

The third term is similar:

$$\begin{aligned}
u_v^3(t,\boldsymbol{x}) &= \frac{1}{2\pi ct}\int_{B(\boldsymbol{x},ct)}\frac{-\dfrac{1}{6}t(\kappa_n)_{x_1}y_1^3}{\sqrt{c^2t^2-|\boldsymbol{y}-\boldsymbol{x}|^2}}\,d\boldsymbol{y}\\
&= \frac{-t(\kappa_n)_{x_1}}{12\pi ct}\int_{B(\boldsymbol{x},ct)}\frac{y_1^3}{\sqrt{c^2t^2-|\boldsymbol{y}-\boldsymbol{x}|^2}}\,d\boldsymbol{y}\\
&= \frac{-t(\kappa_n)_{x_1}}{12\pi}\int_{B(0,1)}\frac{z_1^3 + 3c^2t^2x_1z_1^2 + 3ctx_1^2z_1 + x_1^3}{\sqrt{1-|\boldsymbol{z}|^2}}\,d\boldsymbol{z}.
\end{aligned}$$

Again appealing to function parity:

$$\int_{B(0,1)}\frac{z_1^3}{\sqrt{1-|\boldsymbol{z}|^2}}\,d\boldsymbol{z} = 0,$$

and therefore

$$u_v^3(t,\boldsymbol{x}) = \frac{-t(\kappa_n)_{x_1}}{12\pi}\int_{B(0,1)}\frac{3c^2t^2x_1z_1^2 + x_1^3}{\sqrt{1-|\boldsymbol{z}|^2}}\,d\boldsymbol{z}.$$

The change of variables (5) gives

$$\begin{aligned} u_v^3(t,\boldsymbol{x}) &= \frac{-t(\kappa_n)_{x_1}}{12\pi}\int_0^1\int_0^{2\pi}\frac{3c^2t^2x_1r^3\cos^2\theta + x_1^3r}{\sqrt{1-r^2}}\,d\theta dr \\ &= \frac{-t(\kappa_n)_{x_1}}{12\pi}\int_0^1\int_0^{2\pi}\frac{3c^2t^2x_1r^3\dfrac{1+\cos 2\theta}{2} + x_1^3r}{\sqrt{1-r^2}}\,d\theta dr \\ &= \frac{-t(\kappa_n)_{x_1}}{12\pi}\int_0^1\frac{\pi\left(3c^2t^2x_1r^3 + 2x_1^3r\right)}{\sqrt{1-r^2}}\,dr, \end{aligned}$$

while (7) allows one to express:

$$\begin{aligned} \text{(9)}\quad u_v^3(t,\boldsymbol{x}) &= \frac{-t(\kappa_n)_{x_1}}{12}\int_0^{\frac{\pi}{2}}\frac{3c^2t^2x_1\cos^3\theta + 2x_1^3\cos\theta}{\sqrt{1-\cos^2\theta}}\sin\theta d\theta \\ &= \frac{-t(\kappa_n)_{x_1}}{12}\int_0^{\frac{\pi}{2}}\left(3c^2t^2x_1\frac{\cos 3\theta + 3\cos\theta}{4} + 2x_1^3\cos\theta\right)d\theta \\ &= \frac{-t(\kappa_n)_{x_1}}{12}\left(3c^2t^2x_1\left(-\frac{1}{12}+\frac{3}{4}\right) + 2x_1^3\right) \\ &= \frac{-t(\kappa_n)_{x_1}}{6}\left(c^2t^2x_1 + x_1^3\right). \end{aligned}$$

The final term follows the same approach:

$$\begin{aligned} u_v^4(t,\boldsymbol{x}) &= \frac{1}{2\pi ct}\int_{B(\boldsymbol{x},ct)}\frac{\frac{1}{2}t\kappa_n^2y_1^2y_2}{\sqrt{c^2t^2-|\boldsymbol{y}-\boldsymbol{x}|^2}}\,d\boldsymbol{y} \\ &= \frac{t\kappa_n^2}{4\pi ct}\int_{B(\boldsymbol{x},ct)}\frac{y_1^2y_2}{\sqrt{c^2t^2-|\boldsymbol{y}-\boldsymbol{x}|^2}}\,d\boldsymbol{y} \\ &= \frac{t\kappa_n^2}{4\pi}\int_{B(0,1)}\frac{c^3t^3z_1^2z_2 + c^2t^2x_2z_1^2 + 2c^2t^2x_1z_1z_2}{\sqrt{1-|\boldsymbol{z}|^2}} \\ &\quad + \frac{2ctx_1x_2z_1 + ctx_1^2z_2 + x_1^2x_2}{\sqrt{1-|\boldsymbol{z}|^2}}\,d\boldsymbol{z}. \end{aligned}$$

Function parity tells us that

$$\int_{B(0,1)}\frac{z_1^2z_2}{\sqrt{1-|\boldsymbol{z}|^2}}\,d\boldsymbol{z} = 0,$$

and hence

$$u_v^4(t,\boldsymbol{x}) = \frac{t\kappa_n^2}{4\pi}\int_{B(0,1)}\frac{c^2t^2x_2z_1^2 + 2c^2t^2x_1z_1z_2 + x_1^2x_2}{\sqrt{1-|\boldsymbol{z}|^2}}\,d\boldsymbol{z}.$$

Applying the change of variables (5) and computing gives

$$\begin{aligned}
u_v^4(t,\boldsymbol{x}) &= \frac{t\kappa_n^2}{4\pi}\int_0^1\int_0^{2\pi}\frac{c^2t^2x_2r^3\cos^2\theta+2c^2t^2x_1r^3\cos\theta\sin\theta+x_1^2x_2r}{\sqrt{1-r^2}}\,d\theta dr\\
&= \frac{t\kappa_n^2}{4\pi}\int_0^1\int_0^{2\pi}\frac{c^2t^2x_2r^3\frac{1+\cos 2\theta}{2}+c^2t^2x_1r^3\sin 2\theta+x_1^2x_2r}{\sqrt{1-r^2}}\,d\theta dr\\
&= \frac{t\kappa_n^2}{4\pi}\int_0^1\frac{\pi\left(c^2t^2x_2r^3+2x_1^2x_2r\right)}{\sqrt{1-r^2}}\,dr.
\end{aligned}$$

Using the change of variables (7) leads us to the expession:

$$\begin{aligned}
&= \frac{t\kappa_n^2}{4}\int_0^{\frac{\pi}{2}}\frac{c^2t^2x_2\cos^3\theta+2x_1^2x_2\cos\theta}{\sqrt{1-\cos^2\theta}}\sin\theta d\theta \qquad (10)\\
&= \frac{t\kappa_n^2}{4}\int_0^{\frac{\pi}{2}}\left(c^2t^2x_2\frac{\cos 3\theta+3\cos\theta}{4}+2x_1^2x_2\cos\theta\right)d\theta\\
&= \frac{t\kappa_n^2}{4}\left(c^2t^2x_2\left(-\frac{1}{12}+\frac{3}{4}\right)+2x_1^2x_2\right)\\
&= t\kappa_n^2\left(\frac{c^2t^2x_2}{6}+\frac{x_1^2x_2}{2}\right).
\end{aligned}$$

Equations (6), (8), (9), and (10) together express

$$\begin{aligned}
u_v(t,\boldsymbol{x}) = -t\left(x_2+\kappa_n\left(\frac{c^2t^2}{6}+\frac{x_1^2}{2}\right)+\frac{(\kappa_n)_{x_1}}{6}\left(c^2t^2x_1+x_1^3\right)\right)\\
+t\kappa_n^2\left(\frac{c^2t^2x_2}{6}+\frac{x_1^2x_2}{2}\right).
\end{aligned}$$

Upon taking $t=\tau$ and $\boldsymbol{x}=(0,\delta_n)$, we arrive at

$$0=-\tau\left(\delta_n+\kappa_n\frac{c^2\tau^2}{6}-\kappa_n^2\frac{c^2\tau^2\delta_n}{6}\right)=-\delta_n\tau+O(\tau^3), \tag{11}$$

where the terms involving the curvatures are formally of order $O(\tau^3)$. Combining this equation with our previous results yields:

$$\delta_n-\delta_{n-1}-\delta_n\tau=-\frac{c^2}{2}(2\kappa_n-\kappa_{n-1})\tau^2+O(\tau^3). \tag{12}$$

Writing $\delta_n=\bar{V}_n\tau$ and $\delta_{n-1}=\bar{V}_{n-1}\tau$ in equation (12) expresses

$$\bar{V}_n\tau-\bar{V}_{n-1}\tau-\bar{V}_n\tau^2=-\frac{c^2}{2}(2\kappa_n-\kappa_{n-1})\tau^2+O(\tau^3).$$

Formally assuming $|\kappa_n - \kappa_{n-1}| < C\tau$, for some constant C, and dividing both sides by τ^2 gives

$$\frac{\bar{V}_n - \bar{V}_{n-1}}{\tau} - \bar{V}_n = -\frac{c^2}{2}\kappa_n + O(\tau).$$

It follows that the damping term enters the equation of motion:

$$\dot{V}_n - V_n = -\frac{c^2}{2}\kappa_n + O(\tau).$$

By linearity, taking $u_0(\boldsymbol{x}) = a(2d_n(\boldsymbol{x}) - d_{n-1}(\boldsymbol{x}))$ and $v_0(\boldsymbol{x}) = bd_n(\boldsymbol{x})$ one can obtain the interfacial motion:

$$a\dot{V}_n - bV_n = -\frac{ac^2}{2}\kappa_n + O(\tau),$$

where a and b are real parameters. Therefore, one can rewrite the parameters:

$$\alpha = a, \qquad \beta = -b, \qquad \gamma = \frac{ac^2}{2},$$

to approximate a prescribed interfacial motion:

$$\alpha\dot{V} + \beta V = -\gamma\kappa + O(\tau) \qquad \text{(as } \tau \to 0).$$

The previous results show that the wave equation's initial velocity can be used in the HMBO algorithm to impart damping terms. In the next section, by choosing parameters, we will make a numerical investigation into using the HMBO to approximate interfacial motion by the standard mean curvature flow.

§3. The HMBO approximation of mean curvature flow

An approximation method for mean curvature flow can be obtained by returning to equation (3) and choosing appropriate initial conditions. For a predetermined time step $\tau > 0$, we take $u_0(\boldsymbol{x}) = 0$, $v_0(\boldsymbol{x}) = d_n(\boldsymbol{x})$, and $c^2 = \lambda/\tau$. Then equation (11) gives

$$\begin{aligned} 0 &= -\tau\left(\delta_n + \kappa_n\frac{\lambda\tau}{6} - \kappa_n^2\frac{\lambda\delta_n\tau}{6}\right) + O(\tau^3), \\ &= \delta_n + \kappa_n\frac{\lambda\tau}{6} - \kappa_n^2\frac{\lambda\delta_n\tau}{6} + O(\tau^2). \end{aligned}$$

Proceeding as in the previous section, we obtain

$$\bar{V}_n = -\frac{\lambda}{6}\kappa_n + O(\tau).$$

Since λ is a free parameter, we find that the corresponding threshold dynamics can approximate curvature flow with a parameter γ:

$$V_n = -\gamma\kappa_n + O(\tau) \qquad (\text{as } \tau \to 0). \tag{13}$$

§4. Numerical investigation

We will now perform a numerical error analysis of the HMBO approximation of MCF. The numerical method's performance will be compared to the case of a circle evolving by MCF. In such a setting, the evolution of the circle's radius is governed by the solution of the following ordinary differential equation:

$$\begin{cases} \dot{r}(t) = -\dfrac{1}{r(t)} & t > 0, \\ r(0) = r_0, \end{cases} \tag{14}$$

where r_0 is the initial radius of the circle. We remark that the radius decreases until its extinction time $t_e = r_0^2/2$, and that $r(t) = \sqrt{r_0^2 - 2t}$.

The HMBO approximation method solves the following wave equation for a small time $\tau > 0$:

$$\begin{cases} u_{tt} = c^2\Delta u & \text{in } (0,\tau)\times\Omega \\ u(0,\boldsymbol{x}) = 0 & \text{in } \Omega \\ u_t(0,\boldsymbol{x}) = d_k(\boldsymbol{x}) & \text{in } \Omega \\ \partial_{\boldsymbol{n}} u = 0 & \text{on } \partial\Omega, \end{cases}$$

where $\Omega = (-2,2)\times(-2,2)$ and k denotes the kth step of the HMBO algorithm.

We choose the initial interface to be a circle with radius one, so that

$$d_0(\boldsymbol{x}) = \|\boldsymbol{x}\| - 1.$$

The initial velocity at the kth step is then defined as the signed distance function to the zero level set of the solution to the wave equation:

$$d_k(\boldsymbol{x}) = \begin{cases} \displaystyle\inf_{\boldsymbol{y}\in\partial\{u(\boldsymbol{x},\tau)>0\}} \|\boldsymbol{x}-\boldsymbol{y}\| & \boldsymbol{x}\in\{u(\boldsymbol{x},\tau)>0\} \\ \displaystyle -\inf_{\boldsymbol{y}\in\partial\{u(\boldsymbol{x},\tau)>0\}} \|\boldsymbol{x}-\boldsymbol{y}\| & \text{otherwise.} \end{cases}$$

Since the extinction time t_e depends on r_0, we set $\tau = t_e/N_\tau$. Here $r_0 = 1$ (hence $t_e = 0.5$), and we set $N_\tau = 150$ to ensure a level of precision. The time step is then $\tau = 3.33\times10^{-3}$. The target problem (14) corresponds to $\gamma = 1$ in equation (13), and we thus set $c^2 = 6/\tau$.

Finite differences are used to numerically solve the wave equation with a time step $\Delta t = 2.22 \times 10^{-6}$. The grid spacing in the x and y directions are equal to $\Delta x = 2/(N-1)$, where N is a natural number. We examine the numerical error when $N = 2^j$, for $j = 4, 5, 6, 7, 8$. The numerical results are shown in Figure 2, where the radius of the numerical solution is defined to be the average distance $\tilde{r}(t)$ of the level set's point cloud to the origin. The error is measured using the quantity:

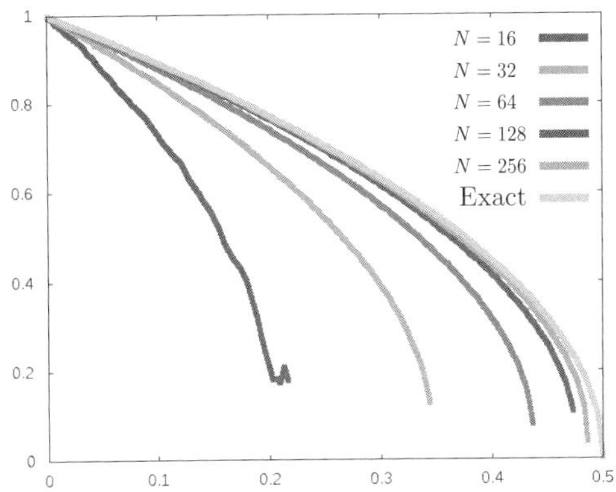

Fig. 2. Convergence of the approximation method as N is increased.

$$Err(t) = \int_0^T |r(t) - \tilde{r}(t)| dt.$$

Since the extinction time of the numerical solution differs from the exact solution, the actual error is computed as follows:

$$Err(t) \approx \sum_{i=0}^{N_s} |r(i\tau) - \tilde{r}(i\tau)|\tau, \tag{15}$$

where N_s denotes the number of time steps until the numerical solution's radius disappears (the corresponding time is $N_s\tau$). Our results are summarized in table (1), where we observe the convergence of our method to the exact solution.

§5. Acknowledgments

E. Ginder would like to acknowledge the support of JSPS Kakenhi Grant Number 17K14229, as well as that from the Presto Research Program of the Japan Science and Technology Agency. The authors also

Table 1. Error Table with respect to Δx.

N	$N_s\tau$	Err
16	0.223333	0.044613
32	0.343333	0.039463
64	0.436667	0.022746
128	0.473333	0.008509
256	0.486667	0.003907

express their gratitude to the referees and editors for their informative suggestions.

References

[1] Y. G. Chen, Y. Giga and S. Goto, Uniqueness and existence of viscosity solutions of generalized mean curvature flow equations, Journal of Differential Geometry, Vol. 33, Number 3 (1991), 749–786.

[2] S. Essedoglu, S. Ruuth and R. Tsai, Diffusion generated motion using signed distance functions, J. Comp. Phys., 229, 4 (2010), 1017–1042.

[3] E. Ginder, K. Svadlenka, Wave-type threshold dynamics and the hyperbolic mean curvature flow, Japan Journal of Industrial Applied Mathematics, doi 10.1007/s13160-016-0221-0, (2016).

[4] P. G. LeFloch and K. Smoczyk, The hyperbolic mean curvature flow, Journal de Mathématiques Pures et Appliquées, Vol. 90, Issue 6 (2008), 591–614.

[5] B. Merriman, J. Bence and S. Osher, Diffusion Generated Motion by Mean Curvature, UCLA CAM, (1992), 1–11.

[6] S. Osher and R. Fedkiw, Level Set Methods and Dynamic Implicit Surfaces, Applied Mathematical Science, (2003).

[7] R. C. Reilly, Mean Curvature, The Laplacian, and Soap Bubbles, The American Mathematical Monthly, Vol. 89, No. 3 (1982), 180–188.

[8] S. Shin and D. Juric, High Order Level Contour Reconstruction Method, Journal of Mechanical Science and Technology, Vol. 21, (2007), 311–326.

4-21-1 Nakano, R-1023
Nakano-ku, Tokyo 164-8525
E-mail address: eginder@meiji.ac.jp

Advanced Studies in Pure Mathematics 85, 2020
The Role of Metrics in the Theory of Partial Differential Equations
pp. 127–136

An analytic solution for a Gompertz-like reaction-diffusion model for tumour growth

Bronwyn H. Bradshaw-Hajek and Philip Broadbridge

Abstract.

Here, we construct spherically symmetric exact analytic solutions to a reaction-diffusion model for tumour growth with a Gompertz-like growth term. This is achieved by making use of a nonclassical symmetry that is valid whenever the nonlinear reaction term and the nonlinear diffusivity satisfy a particular relationship. Example solutions are given in two different cases. In the first, the tumour size and the cell density within the tumour change, while in the second, the cell density changes but the tumour size remains constant.

§1. Introduction

Nonlinear reaction-diffusion equations have been used to model many situations in biology, and modelling tumour growth is no exception. This type of modelling has provided powerful tools for predicting the growth of tumours and enhancing our understanding of the mechanisms underpinning tumour development [1].

Reaction-diffusion models relating to tumour growth are typically characterised by the equation

$$\frac{\partial u}{\partial t} = \nabla \cdot (D(u)\nabla u) + R(u) \tag{1}$$

where $u(r,t)$ is the density of tumour cells and $R(u)$ is the reaction term incorporating cell proliferation, cell death, and some tumour growth dynamics. Typically, the diffusion coefficient, $D(u)$ is taken as a constant.

A number of growth rate functions, $R(u)$ have been suggested. Here we are interested in so-called Gompertz growth [2], which accounts for

Received April 4, 2019.
2010 *Mathematics Subject Classification.* 5K57, 35B06, 35K55.
Key words and phrases. nonlinear reaction-diffusion, nonclassical symmetry analysis, Gompertz function, tumour growth.

the slow down of tumour growth with increasing size, as observed in the clinic and the laboratory [1, 3]. The Gompertz growth term is given by

$$R(u) = -su \ln u \tag{2}$$

where the cell density has already been scaled with the carrying capacity [3] (in some cases, the carrying capacity is taken to be time dependent in order to better account for control processes related to the tumour vasculature, but here we assume it to be constant).

In this paper, we construct analytic solutions to the generalised nonlinear reaction-nonlinear diffusion equation (1), with a Gompertz-like growth term in spherical coordinates with spherical symmetry, so that $u(\mathbf{x}, t) = u(r, t)$. To our knowledge, these are the first analytic solutions for Gompertz-like growth with diffusion.

Exact analytic solutions to equations of type (1) can be constructed under certain circumstances using nonclassical symmetry analysis. The result we use here was first discovered by Arrigo and Hill [4] in the case of one Cartesian dimension, and by Goard and Broadbridge [5] in the case of two Cartesian dimensions. More recently, it has been shown to be valid in any number of dimensions and in any coordinate system [6, 7], including in spherical coordinates [8]. A summary of the existing known exact analytic solutions of equations of type (1) with various reaction terms is given elsewhere (see for example [7, 9]).

§2. Nonclassical symmetry reduction of the nonlinear reaction-diffusion equation

Symmetry based methods are some of the most powerful techniques for constructing exact analytic solutions to nonlinear PDEs. These methods all stem from the classical Lie symmetry method [10] which provides a systematic technique to search for transformations that allow a reduction in the number of independent variables in a differential equation (there are many books devoted to the Lie symmetry method, see for example [11, 12, 13, 14]). Many extensions to the classical method have been developed, among them the method of nonclassical symmetries [15]. This type of conditional symmetry, which restricts the solution set to the surface of invariance of an unknown symmetry, can sometimes give rise to solutions not obtainable using the classical method.

As described in the Introduction, equation (1) admits a nonclassical symmetry that can be used in any number of dimensions and in any coordinate system. By making use of this nonclassical symmetry, equations of type (1) can be reduced to a linear second order differential equation with one fewer variables whenever the diffusion and reaction terms are related via

$$R(u) = \left(\frac{A}{D(u)} + \kappa\right) \int_{u^*}^{u} D(u')\, du' \tag{3}$$

where κ and A are constants arising in the solution of the nonclassical determining equations, and u^* is an appropriately chosen zero of the reaction term, that is $R(u^*) = 0$.

If relationship (3) is satisfied, the nonclassical symmetry generator

$$\Gamma = \frac{\partial}{\partial t} + Au\frac{\partial}{\partial u}$$

may be used to reduce equation (1) to

$$\nabla^2\Psi(\mathbf{x}) + \kappa\Psi(\mathbf{x}) = 0 \tag{4}$$

where $\Psi(\mathbf{x})$ is related to $u(\mathbf{x})$ using the Kirchhoff variable,

$$\mu = \int_{u^*}^{u} D(u')\, du' = \mathrm{e}^{At}\Psi(\mathbf{x}). \tag{5}$$

Equation (4) is the Helmholtz ($\kappa > 0$), modified Helmholtz ($\kappa < 0$) or Laplace equation ($\kappa = 0$), and solutions in spherical coordinates are well known; those with spherical symmetry can be written in terms of spherical Bessel functions and modified spherical Bessel functions.

§3. Analytic solution for a Gompertz-like model

In this paper, we will examine two different Gompertz-like models described below. In both cases, we assume the tumour is a spherical mass with spherical symmetry, so that the Laplacian in equation (4) is given by

$$\nabla^2 = \frac{1}{r^2}\frac{\partial}{\partial r}\left(r^2\frac{\partial}{\partial r}\right).$$

In the first case, we consider the Gompertz-like nonlinear reaction term given by

$$R(u) = Au\ln\left(\frac{su + \kappa D_0}{s + \kappa D_0}\right) \tag{6}$$

with $s > 0$, with the nonlinear diffusivity

$$D(u) = -\frac{AD_0}{su + \kappa D_0} \tag{7}$$

where we have chosen $u^* = 1$, and where $D_0 = -(A+s)/\kappa$ is the value of the diffusion coefficient at $u = u^*$. This combination of reaction and diffusion satisfies relationship (3). Since we require the diffusion to be positive, if $\kappa > 0$ then $s < -A$ and A must be negative. Alternatively, if $\kappa < 0$ then $s > -A$. In the example solution below, we choose $\kappa < 0$.

If $\kappa < 0$, the spherically symmetric solution to the modified Helmholtz equation (4) is

$$\Psi(r) = c_1 i_0(\sqrt{-\kappa}r) + c_2 k_0(\sqrt{-\kappa}r)$$

where $i_0(\sqrt{-\kappa}r)$ and $k_0(\sqrt{-\kappa}r)$ are modified spherical Bessel functions of the first and second kinds respectively, and c_1 and c_2 are constants of integration. In order to ensure that the solution is finite at the centre of the tumour, we choose $c_2 = 0$. The solution for the tumour cell concentration may be found using the Kirchhoff variable (5), and integrating and rearranging to obtain

$$u(r,t) = \frac{1}{s}\left[-A\exp\left(-\frac{s}{AD_0}\mu\right) - \kappa D_0\right] \tag{8}$$

where

$$\mu = c_1 \mathrm{e}^{At}\frac{\sinh\sqrt{-\kappa}r}{\sqrt{-\kappa}r}.$$

Solution (8) is an exact analytic solution to equation (1) with the nonlinear reaction and diffusion terms given by (6) and (7) respectively. The constant c_1 can be chosen by specifying the initial cell density at the centre of the tumour, that is, $u(0,0) = u_0 < 1$.

It is interesting to note that the time dependence of the solution in (8) takes the form $\sim \mathrm{e}^{-\alpha \mathrm{e}^{-\beta t}}$. This is precisely the form of the sigmoid Gompertz function [2].

The diffusion and reaction terms are plotted in Figure 1, and the time evolution of the tumour cell density is plotted in Figure 2. The nonlinear diffusivity is positive and an increasing function of cell density, while the nonlinear reaction term takes a similar form to the Gompertz growth term, or a logistic term (for example, a Fisher source term [16]). The solution for the cell density shows that the radius at which the cell density is zero decreases with time, indicating that the tumour is shrinking. At the same time, the cell density within the tumour is decreasing.

The second Gompertz-like model we examine that also satisfies relationship (3) is characterised by the nonlinear reaction term given by

$$R(u) = A(u-1)\ln\left(-\frac{1}{A}(su-A)\right) \tag{9}$$

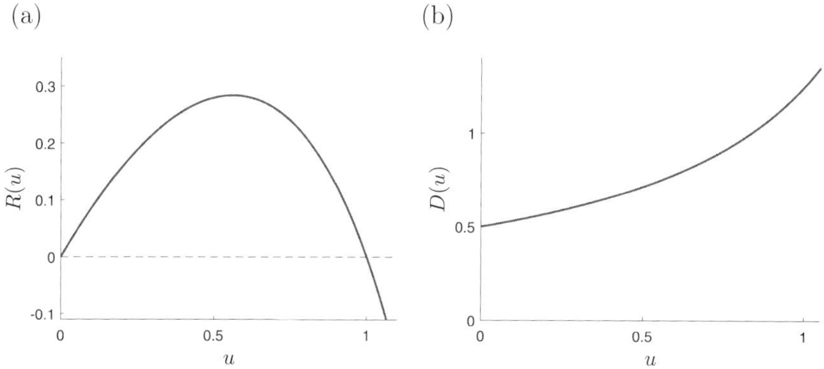

Fig. 1. (a) Nonlinear reaction and (b) diffusivity as given by (6) and (7), respectively. Here, we have chosen $A = 1$, $\kappa = -2$, $s = 1.5$, $u_0 = 0.8$.

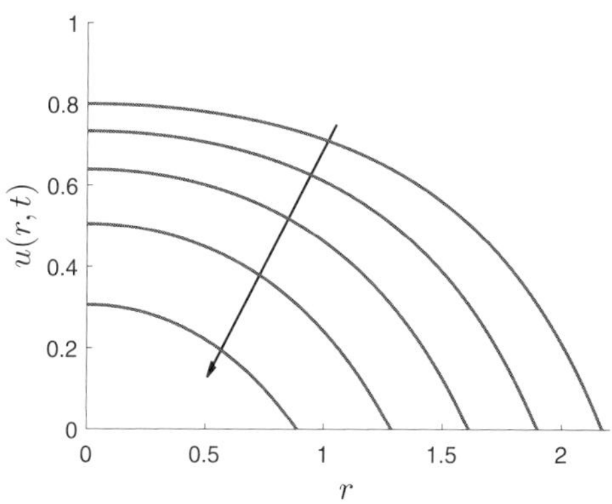

Fig. 2. Nondimensional cell density as a function of radius (8), the arrow shows increasing time ($t = 0, 0.25, \ldots, 1$). The size of the tumour and the cell density within the tumour decrease with time. Parameter values are given in Figure 1.

with $s > 0$, and the nonlinear diffusivity

$$D(u) = -\frac{AD_0}{su - A} \tag{10}$$

where $D_0 = (s - A)/\kappa$, and we have chosen $u^* = 0$. In order to demonstrate a different type of behaviour, we choose $\kappa > 0$ and, since we require $D_0 > 0$ and the diffusion coefficient to be positive, we restrict ourselves to $s > A$ and $A < 0$. For positive κ, the solution to the Helmholtz equation (4) is

$$\Psi(r) = c_1 j_0(\sqrt{\kappa}r) + c_2 y_0(\sqrt{\kappa}r)$$

where $j_0(\sqrt{\kappa}r)$ and $y_0(\sqrt{\kappa}r)$ are spherical Bessel functions of the first and second kinds respectively, and c_1 and c_2 are constants of integration. To ensure that the solution is finite at the centre of the tumour, we set $c_2 = 0$. Once again, the solution for the tumour cell concentration may be found using the Kirchhoff transformation (5), and integrating and rearranging for $u(r,t)$ to obtain

$$u(r,t) = \frac{1}{s}\left[-A\exp\left(-\frac{s}{AD_0}\mu\right) + A\right] \tag{11}$$

where

$$\mu = c_1 \mathrm{e}^{At}\frac{\sin\sqrt{\kappa}r}{\sqrt{\kappa}r}.$$

This is an exact analytic solution to equation (1) when the nonlinear reaction and diffusion terms are given by (9) and (10) respectively. Again, the constant c_1 can be chosen so that the initial cell density in the centre of the tumour is known, $u(0,0) = u_0$.

The nonlinear reaction and diffusion terms are shown in Figure 3, while the solution given by (11) is shown in Figure 4. The nonlinear reaction term again takes a similar form to the standard Gompertz growth term (2) or the logisitc Fisher source term [16], while the nonlinear diffusivity is a positive and slowly decreasing function of cell density. In this case, and in contrast to the first reacion-diffusion combination considered here, the solution shows that the tumour size remains constant, while the cell density within the tumour decreases with time.

§4. Discussion and final remarks

Reaction-diffusion modelling has long been used to provide insight into the growth of tumours. Here, we have constructed exact analytic solutions for two example reaction terms similar to the Gompertz growth term commonly used in tumour modelling. To the authors' knowledge, these are the first analytic solutions for Gompertz like growth with diffusion. We note that the method used here can be used for many nonlinear

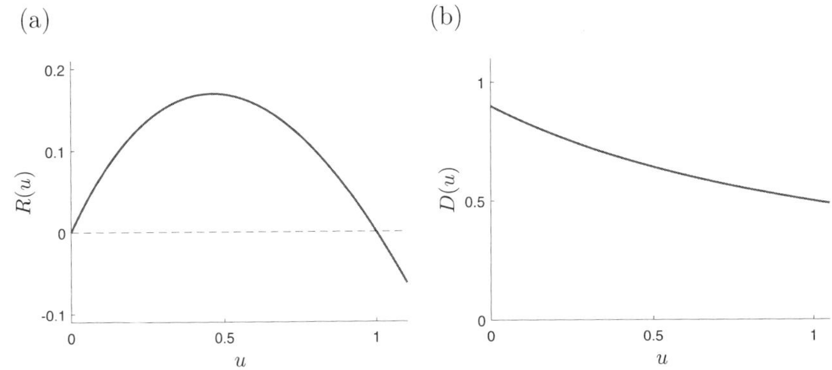

Fig. 3. (a) Nonlinear reaction and (b) diffusivity terms as given by (9) and (10), respectively. Here, we have chosen $A = -1$, $\kappa = 2$, $s = 0.8$, $u_0 = 1$.

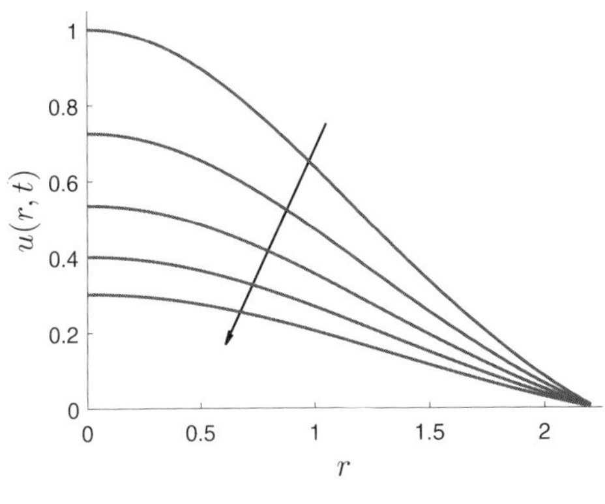

Fig. 4. Nondimensional cell density as a function of radius (11), the arrow shows increasing time ($t = 0, 0.25, \ldots, 1$). The cell density within the tumour decreases with time, but the tumour size remains the same. Parameter values are given in Figure 3.

reaction-diffusion equations, provided that the nonlinear diffusivity and nonlinear reaction term satisfy the relationship in equation (3).

The results for the two examples shown here are quite different. In both cases, the cell density within the tumor decreases with time, however in one case the tumour size decreases, while in the other the tumour remains the same size. The main difference between the two models is that the first has a diffusivity that increases with cell density, while the second decreases with cell density. In both cases, the decrease in cell density in the tumour is due to an outwards flux at the boundary, for example due to the removal of cancer cells due to radiation or directed chemotherapy.

One benefit of constructing an analytic solution is that it enables the analytic investigation of features that might otherwise be obscured. For example, using the formulation described here it is simple to find an analytic expression for the flux. Examining the Kirchhoff variable, we see that μ is the flux potential since $-\nabla\mu = -D(u)\nabla u$. Therefore, if we define the tumour to be of radius $r_0(t)$, where $u(r_0, t) = 0$, then the flux through the surface of the tumour in the second example is given by

$$-4\pi r_0^2 \nabla\mu = -\frac{4\pi^2 A D_0}{s\sqrt{\kappa}} \ln\left(-\frac{s u_0 - A}{A}\right) \mathrm{e}^{At}. \tag{12}$$

From this expression it is straightforward to determine how the flux depends on the parameters of the problem. In addition, we can calculate the total number of cells in the tumour

$$\begin{aligned} 4\pi \int_0^{r_0} r^2 u\, dr \\ = \frac{4\pi A}{s} \int_0^{r_0} r^2 \left(1 - \exp\left(-\frac{s c_1}{A D_0} e^{At} \frac{\sin\sqrt{\kappa} r}{\sqrt{\kappa} r}\right)\right) dr, \end{aligned} \tag{13}$$

and the rate of decrease in the total number of cells

$$\begin{aligned} -4\pi \frac{d}{dt} \int_0^{r_0} r^2 u\, dr \\ = -\frac{4\pi A c_1}{D_0} \mathrm{e}^{At} \int_0^{r_0} r^2 \frac{\sin\sqrt{\kappa} r}{\sqrt{\kappa} r} \exp\left(-\frac{s c_1}{A D_0} \mathrm{e}^{At} \frac{\sin\sqrt{\kappa} r}{\sqrt{\kappa} r}\right) dr. \end{aligned} \tag{14}$$

Figure 5 shows how these quantities change with time for the second example. Figure 5(a) plots the total number of cells in the tumour (13), while Figure 5(b) shows both the outward flux at the surface of the tumour (12), and the rate of decrease in the total number of cells (14). The outward flux at the boundary is greater than the total decrease in the number of cells, indicating that the cells within the tumour are still proliferating. This could correspond to therapy that targets miotic cells at the surface of the tumour.

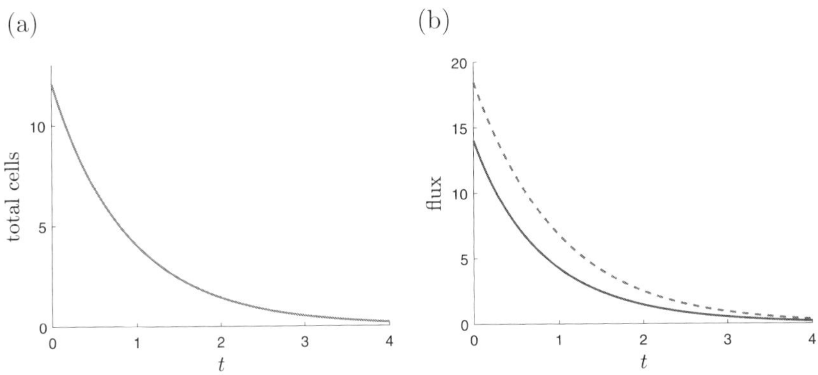

Fig. 5. (a) Total number of cells as a function of time for the second example in Section 3. (b) Outward flux at the tumour boundary (dashed), and rate of decrease of total number of cells (solid) for the second example in Section 3. The parameter values are the same as those given in Figure 3.

References

[1] H. Enderling and M. A. J. Chaplain, Mathematical modeling of tumor growth and treatment, Current Pharmaceutical Design, 20 (2014), 4934–4940.

[2] B. Gompertz, On the nature of the function expressive of the law of human mortality, and on a new mode of determining the value of life contingencies, Philos. Trans. Royal Soc. Lond., **115** (1825), 513–583.

[3] P. Hahnfeldt, D. Panigrahy, J. Folkman and L. Hlatky, Tumor development under angiogenic signaling: A dynamical theory of tumor growth, treatment response, and postvascular dormancy, Cancer Res., **59** (1999), 4770–4775.

[4] D. J. Arrigo and J. M. Hill, Nonclassical symmetries for nonlinear diffusion and absorption, Stud. Appl. Math., **94** (1995), 21–39.

[5] J. M. Goard and P. Broadbridge, Nonclassical symmetry analysis of nonlinear reaction-diffusion equations in two spatial dimensions, Nonlinear Anal.-Theor., **26** (1996), 735–754.

[6] P. Broadbridge, B. H. Bradshaw-Hajek and D. Triadis, Exact nonclassical symmetry solutions of Arrhenius reaction-diffusion, P. Roy. Soc. Lond. A Mat., **471** (2015), 20150580.

[7] P. Broadbridge and B. H. Bradshaw-Hajek, Exact solutions for logistic reaction-diffusion equations in biology, Zeitschrift für angewandte Mathematik und Physik, **67** (2016), p.93.

[8] B. H. Bradshaw-Hajek and P. Broadbridge, Analytic solutions for calcium ion fertilisation waves on the surface of eggs, Mathematical Medicine and Biology, (2019).
[9] R. Cherniha, M. Serov and O. Pliukhin, Nonlinear Reaction-Diffusion-Convection Equations: Lie and Conditional Symmetry, Exact Solutions and their Applications, CRC Press, USA, 2018.
[10] S. Lie, Theorie der Transformationsgruppen, Mathematishe Annalen, **16** (1880), 441–528.
[11] P. J. Olver, Applications of Lie Group to Differential Equations, AMS no. 107, Springer, New York, 1982.
[12] G. W. Bluman and S. Kumei, Symmetries and Differential Equations, AMS no. 81, Springer, New York, 1989.
[13] G. W. Bluman and S. C. Anco, Symmetry and Integration Methods for Differential Equations, Applied Mathematical Sciences, 2002.
[14] J. M. Hill, Differential Equations and Group Methods for Scientists and Engineers, CRC Press, Florida, 1992.
[15] G. W. Bluman and J. D. Cole, The general similarity solution of the heat equation, J. Math. Mech., **18** (1969), 1025–1042.
[16] R. A. Fisher, The wave of advance of advantageous genes, Annals of Eugenics, **7** (1937) 355–369.

School of Information Technology and Mathematical Sciences,
University of South Australia,
Mawson Lakes, SA 5095, Australia
E-mail address: bronwyn.hajek@unisa.edu.au

Department of Mathematics and Statistics,
La Trobe University,
Victoria 3086, Australia
E-mail address: p.broadbridge@latrobe.edu.au

Advanced Studies in Pure Mathematics 85, 2020
The Role of Metrics in the Theory of Partial Differential Equations
pp. 137–150

Pinned pulse solutions in a FitzHugh–Nagumo model with a bump-type heterogeneity

Peter van Heijster and Takashi Teramoto

Abstract.

In this article, we use geometric singular perturbation techniques combined with an action functional approach to study pinned pulse solutions in a three-component FitzHugh–Nagumo model with a bump-type heterogeneity. We show that the model in the original activator-2 inhibitor framework only supports symmetric pinned pulse solutions that are located in the middle of the bump region. In contrast, asymmetric pinning is possible for parameters not in the activator-2 inhibitor framework.

§1. Introduction

We study a nondimensionalised singularly perturbed three-component FitzHugh–Nagumo defect model

$$(1)\qquad \begin{cases} U_t = \varepsilon^2 U_{xx} + U - U^3 - \varepsilon(\alpha V + \beta W + \gamma(x)), \\ \tau V_t = V_{xx} + U - V, \\ \theta W_t = D^2 W_{xx} + U - W, \end{cases}$$

with a symmetric bump-type heterogeneity

$$(2)\qquad \gamma(x) = \begin{cases} \gamma_1, & \text{for } |x| \geq L, \\ \gamma_2, & \text{for } |x| < L. \end{cases}$$

Here, $0 < \varepsilon \ll 1; D > 1; L, \alpha, \beta, \tau, \theta > 0; (x,t) \in \mathbb{R} \times \mathbb{R}^+; \gamma_{1,2} \in \mathbb{R}$ and all parameters are assumed to be strictly $\mathcal{O}(1)$ with respect to ε.

Received November 30, 2018.
Revised May 30, 2019.
2010 *Mathematics Subject Classification.* 34A36, 35B25, 35B35, 35K57, 49J40.
Key words and phrases. Reaction-diffusion equations, defects, singular perturbations, pinning.

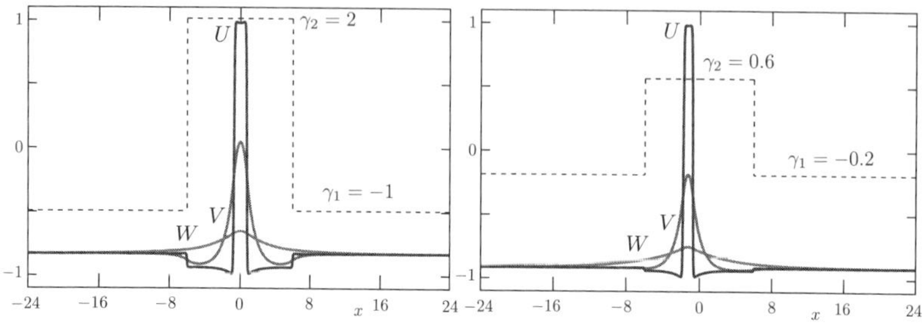

Fig. 1. Pinned pulse solutions located inside the bump region $[-6, 6]$ obtained by simulating the time independent version of (1) with (2). Left panel: for $\alpha\beta > 0$ the pulse is necessarily pinned around the middle of the bump region. The system parameters used are $(\alpha, \beta, D, \varepsilon) = (3, 2, 5, 0.05)$ and $(\gamma_1, \gamma_2) = (-1, 2)$. Right panel: for $\alpha\beta < 0$ the pulse is not necessarily pinned around the middle of the bump region. The system parameters used are $(\alpha, \beta, D, \varepsilon) = (4, -1, 5, 0.05)$ and $(\gamma_1, \gamma_2) = (-0.2, 0.6)$.

A dimensional homogeneous version of this model was introduced in the nineties to study gas-discharge experiments on a phenomenological level [6]. See also [5], and references therein, for a good review article related to the development of the model. Subsequently, homogeneous and heterogeneous versions of the above model have been intensively studied. See, for instance, [3, 7, 8, 9, 10] and references therein. In [10], the system (1) with a jump-type heterogeneity was studied (that is, $\gamma(x) = \gamma_1$ for $x < 0$ and $\gamma(x) = \gamma_2$ for $x \geq 0$) and numerical simulations of (1) with a bump-type heterogeneity (2) were also presented at the end. These simulations suggested that, under appropriate conditions on the bump-type heterogeneity, the model supports a stable pinned pulse solution pinned asymmetrically inside the bump region $[-L, L]$ and $\mathcal{O}(1)$-away from the boundaries of the bump region. That is, the pinned pulse solution did not appear to be centred around $x = 0$, see Figure 8 of [10].

In this article, we are going to derive the conditions for the existence and stability for these type of stationary pulse solutions. That is, we determine under what conditions on the system parameters and the heterogeneity the model supports pinned pulse solutions located inside the bump region and away from the bump edges. Most notably, we find that for $\alpha\beta > 0$[1] these pinned pulse solutions are necessarily centred around the middle of the bump region (see the left panel of Figure 1), while this is not necessarily the case for $\alpha\beta < 0$ (see the right panel of Figure 1). Hence, we argue that the pulse solution presented in Figure 8 of [10] is not fully pinned yet (since $\alpha\beta > 0$ in the simulation) and that it may move on (very slowly) towards the centre of the bump region.

To obtain these results, we adopt the methodology of [7, 8]. In [7], the homogeneous version of (1) was studied (that is, (1) with $\gamma(x) \equiv \gamma$), while (1) with a jump-type heterogeneity was studied in [8]. This methodology consists of two parts. First, we use geometric singular perturbation techniques (e.g., [4]) to construct the profile of a pinned pulse solution with undetermined width and location. Second, we compute an action functional [1, 2] associated to this profile. The width and location of the pinned pulse solution are then to be determined by the extrema of this action functional. Furthermore, only minimisers of the action functional are associated to stable pinned pulse solutions. See [7, 8] for more details.

As expected from the ε-term in (1) premultiplying the bump-type heterogeneity (2), the heterogeneity does not influence the width of a pinned pulse solution (when compared to the homogeneous case studied in [3, 7]). That is, for L large enough[2] the width of the pulse x_w is to leading order determined by the positive solutions of

$$(3) \qquad f(x_w) = \gamma_2, \qquad \text{with} \qquad f(x) := \alpha e^{-x} + \beta e^{-x/D}.$$

In addition, the first stability condition is not influenced by the bump-type heterogeneity either (when compared to the homogeneous case studied in [9]) and it is given by

$$(4) \quad g(x_w) > 0, \qquad \text{with} \qquad g(x) := -f'(x) = \alpha e^{-x} + \frac{\beta}{D} e^{-x/D}.$$

In contrast – and as for the jump-type heterogeneity [8] – the bump-type heterogeneity (2) determines the location x_ℓ of a pinned pulse solution (from a continuous family of stationary pulse solutions for the homogeneous case [3]), as well as the eigenvalue related to the translation invariance eigenvalue at the origin for the homogeneous case (since the problem (1) is no longer translation invariant). In particular, for $\alpha\beta > 0$,

[1]In the original activator-2 inhibitor framework of the homogeneous version of (1) both α and β were assumed to be positive such that $\alpha\beta > 0$.

[2]L needs to be larger than the width of the pulse to be able to support this type of local defect pulse solution.

pinned pulse solutions will be pinned around the centre of the bump region and they are stable if $\alpha > 0$ and $\gamma_2 > \gamma_1$.

This article is organised as follows, in §2 we derive the profile of the pinned pulse solution with undetermined width and location. In §3, we compute the associate action functional for this profile and we derive the existence and stability conditions. We end the article with a short summary and discussion related to future work.

§2. The profile of a pinned pulse solution

To study a pinned pulse solution $\bar{Z}(x) = (\bar{U}, \bar{V}, \bar{W})(x)$ located inside the bump region and $\mathcal{O}(1)$-away from its boundaries, we divide the spatial domain into five slow regions $I_s^{1,4,6,8,11}$ and six fast regions $I_f^{2,3,5,7,9,10}$:

$$
\begin{aligned}
&I_s^1 := (-\infty, -L-\sqrt{\varepsilon}\,],\ I_f^2 := (-L-\sqrt{\varepsilon}, -L],\\
&I_f^3 := (-L, -L+\sqrt{\varepsilon}\,),\ I_s^4 := [-L+\sqrt{\varepsilon}, x_1-\sqrt{\varepsilon}\,],\\
&I_f^5 := (x_1-\sqrt{\varepsilon}, x_1+\sqrt{\varepsilon}\,),\ I_s^6 := [x_1+\sqrt{\varepsilon}, x_2-\sqrt{\varepsilon}\,],\\
&I_f^7 := (x_2-\sqrt{\varepsilon}, x_2+\sqrt{\varepsilon}\,),\ I_s^8 := [x_2+\sqrt{\varepsilon}, L-\sqrt{\varepsilon}\,],\\
&I_f^9 := (L-\sqrt{\varepsilon}, L),\ I_f^{10} := [L, L+\sqrt{\varepsilon}\,),\ I_s^{11} := [L+\sqrt{\varepsilon}, \infty),
\end{aligned} \tag{5}
$$

where $-L < x_1 < x_2 < L$ are the locations of the interfaces of the pinned pulse solution, that is, $\bar{U}(x_{1,2}) = 0$, see also §2 of [8]. We define the width of a pulse as $x_w := x_2 - x_1$, and its location as $x_\ell := (x_2 + x_1)/2$. The profile of $\bar{Z}$ is to leading order the same as the homogeneous one derived in [3] since the bump-type heterogeneity (2) has no leading order influence on (1), see also [8]. Upon using a regular expansion in ε, $\bar{Z}(x) = \bar{Z}_0(x) + \varepsilon \bar{Z}_1(x) + \mathcal{O}(\varepsilon^2)$, we get

$$
\bar{U}_0(x) = \begin{cases}
-1, & x \in I_s^1 \cup I_f^2 \cup I_f^3 \cup I_s^4,\\
\tanh\left(\dfrac{x-x_1}{\sqrt{2}\varepsilon}\right), & x \in I_f^5,\\
1, & x \in I_s^6,\\
-\tanh\left(\dfrac{x-x_2}{\sqrt{2}\varepsilon}\right), & x \in I_f^7,\\
-1, & x \in I_s^8 \cup I_f^9 \cup I_f^{10} \cup I_s^{11},
\end{cases}
$$

$$
\bar{V}_0(x;x_1,x_2) = \begin{cases} -1+e^{x-x_1}-e^{x-x_2}, & x \in I_s^1 \cup I_f^2 \cup I_f^3 \cup I_s^4, \\ -e^{-x_w}, & x \in I_f^5, \\ 1-e^{x-x_2}-e^{-(x-x_1)}, & x \in I_s^6, \\ -e^{-w_w}, & x \in I_7^7, \\ -1+e^{-(x-x_2)}-e^{-(x-x_1)}, & x \in I_s^8 \cup I_f^9 \cup I_f^{10} \cup I_s^{11}, \end{cases}
$$

and $\bar{W}_0(x;x_1,x_2) = \bar{V}_0(x/D;x_1/D,x_2/D)$.

We follow [8] to determine the higher order correction terms. Substituting the regular expansion into the time-independent version of (1) gives that in the slow regions the first order correction terms are determined by

$$
\begin{cases} 0 = \bar{U}_1 - 3(\bar{U}_0)^2\bar{U}_1 - \big(\alpha\bar{V}_0 + \beta\bar{W}_0 + \gamma(x)\big), \\ 0 = (\bar{V}_1)_{xx} + \bar{U}_1 - \bar{V}_1, \\ 0 = D^2(\bar{W}_1)_{xx} + \bar{U}_1 - \bar{W}_1. \end{cases} \tag{6}
$$

Since $\bar{U}_0 = \pm 1$ in the slow regions, this yields

$$
\bar{U}_1 = -\frac{1}{2}\big(\alpha\bar{V}_0 + \beta\bar{W}_0 + \gamma(x)\big), \qquad x \in I_s^1 \cup I_s^4 \cup I_s^6 \cup I_s^8 \cup I_s^{11}.
$$

Substituting $\bar{U}_1$ given above, we can exactly solve the linear equations for $\bar{V}_1$ and $\bar{W}_1$ of (6), which gives the slow profiles in the slow regions (they include up to 16 integration constants). In the fast regions, $\bar{V}_1$ and $\bar{W}_1$, as well as their derivatives, are constant, and this determines the remaining 16 integration constants, see [8] for more detail. For brevity of the presentation the resulting correction terms $\bar{V}_1$ and $\bar{W}_1$ are given in Appendix A. Finally, in the fast regions, the first order correction term $\bar{U}_1$ solves

$$
0 = \left(\bar{U}_1\right)_{\xi\xi} + \bar{U}_1 - 3\left(\bar{U}_0\right)^2\bar{U}_1 - \big(\alpha\bar{V}_0 + \beta\bar{W}_0 + \gamma(\xi)\big), \tag{7}
$$

where $\xi = x/\varepsilon$. This gives $\bar{U}_1 = 0$[3] for the fast fields $I_f^{5,7}$, and for the remaining fast fields $I_f^{2,3,9,10}$ we get

$$\bar{U}_1(x) = \begin{cases} \frac{1}{4}(\gamma_1 - \gamma_2)e^{\sqrt{2}(x+L)/\varepsilon} - \frac{1}{2}\big(\alpha\bar{V}_0(-L) + \beta\bar{W}_0(-L) + \gamma_1\big), & x \in I_f^2, \\ \frac{1}{4}(\gamma_2 - \gamma_1)e^{-\sqrt{2}(x+L)/\varepsilon} - \frac{1}{2}\big(\alpha\bar{V}_0(-L) + \beta\bar{W}_0(-L) + \gamma_2\big), & x \in I_f^3, \\ \frac{1}{4}(\gamma_2 - \gamma_1)e^{\sqrt{2}(x-L)/\varepsilon} - \frac{1}{2}\big(\alpha\bar{V}_0(L) + \beta\bar{W}_0(L) + \gamma_2\big), & x \in I_f^9, \\ \frac{1}{4}(\gamma_1 - \gamma_2)e^{-\sqrt{2}(x-L)/\varepsilon} - \frac{1}{2}\big(\alpha\bar{V}_0(L) + \beta\bar{W}_0(L) + \gamma_1\big), & x \in I_f^{10}. \end{cases}$$

§3. The action functional

The action functional for a pinned pulse solution supported by (1) with (2) is similar to that for the homogeneous case [7] and the heterogeneous case with a jump-heterogeneity [8]. In particular, the action functional for a pinned pulse solution $\bar{Z} = (\bar{U}, \bar{V}, \bar{W})$, whose profile with unknown width $x_w := x_2 - x_1$ and location $x_\ell := (x_2 + x_1)/2$ has been computed in the previous section, is given by

$$(8) \quad J(\bar{U}) = \int_{-\infty}^{\infty} \bigg(\frac{1}{2}\varepsilon^2 \bar{U}_x^2 + F(\bar{U}) - F(U_b) + \frac{1}{2}\varepsilon\alpha(\bar{U}\mathcal{L}_1\bar{U} - U_b^2) + \frac{1}{2}\varepsilon\beta(\bar{U}\mathcal{L}_2\bar{U} - U_b^2) + \varepsilon\gamma(x)(\bar{U} - U_b) \bigg)\, dx,$$

with $\mathcal{L}_1 := \big(-(d^2/dx^2) + 1\big)^{-1}$, $\mathcal{L}_2 := \big(-D^2(d^2/dx^2) + 1\big)^{-1}$, $F(U) := U^4/4 - U^2/2$, and with U_b being the zero of $U^3 - U + \varepsilon((\alpha+\beta)U + \gamma_1)$ near $U \approx -1$. That is, (U_b, U_b, U_b) is the asymptotic end state of the pinned pulse solutions under consideration. The existence and the stability conditions now follow from the critical points of J, and its minimisers, respectively [7, 8].

Because the bump-type heterogeneity (2) does not influence the leading order profiles of a pinned pulse solution $\bar{Z}$, the leading order analysis of the action functional J (8) is similar to analysis of that associated to the homogeneous pulse solution given in [7][4]. The computation

[3] Here, we have already used the relation $\alpha\bar{V}_0(x_{1,2}) + \beta\bar{W}_0(x_{1,2}) + \gamma_2 = 0$ in $I_f^{5,7}$, see the upcoming leading order computation of the action functional in §3.

[4] See also [8] for a discussion regarding the influences of the heterogeneities in the fast fields $I_f^{2,3,9,10}$.

of the next order term, however, is a bit more involved. In particular, after some algebra and upon splitting the indefinite integral in (8) into the five slow and six fast regions (5), we get $J(\bar{U}) = \varepsilon J_1(\bar{U}) + \varepsilon^2 J_2(\bar{U}) + \cdots$ [5], with, in terms of x_w and x_ℓ,

$$J_1(\bar{U}) = 2\left(\alpha\big(e^{-x_w} - 1\big) + \beta D\big(e^{-x_w/D} - 1\big) + \gamma_2 x_w + \frac{2\sqrt{2}}{3}\right), \tag{9}$$

and

(10)

$$\begin{aligned}
J_2(\bar{U}) = \alpha\bigg(& e^{-x_w} - 1 - \gamma_2 x_w + (\gamma_1 - \gamma_2)e^{-L}\Big(\sinh\left(x_\ell - \frac{x_w}{2}\right) \\
& - \sinh\left(x_\ell + \frac{x_w}{2}\right)\Big)\bigg) + \frac{1}{2}\alpha^2\left(3 + e^{-x_w}(-3 - x_w)\right) \\
+ \beta\bigg(& \frac{1}{D}\left(e^{-x_w/D} - 1\right) - \gamma_2 x_w \\
& + D(\gamma_1 - \gamma_2)e^{-L/D}\Big(\sinh\left(\frac{x_\ell}{D} - \frac{x_w}{2D}\right) - \sinh\left(\frac{x_\ell}{D} + \frac{x_w}{2D}\right)\Big)\bigg) \\
+ & \frac{1}{2}\beta^2\left(3D + e^{-x_w/D}(-3D - x_w)\right) \\
+ & \frac{2\alpha\beta\left(e^{-x_w} - 1 - D^3\left(e^{-x_w/D} - 1\right)\right)}{D^2 - 1} - \frac{1}{2}L(\gamma_1 - \gamma_2)^2.
\end{aligned}$$

Thus, as expected, the leading order term of the action functional only depends on the width x_w of the pinned pulse $\bar{Z}$ and its location x_ℓ only appears in the expression at the next order. Consequently, the critical points of $J(\bar{U})$ are determined by the solutions of $dJ_1(\bar{U})/dx_w = 0$ and $dJ_2(\bar{U})/dx_\ell = 0$, and these critical points are minimisers if and only if both $d^2J_1(\bar{U})/dx_w^2 > 0$ and $d^2J_2(\bar{U})/dx_\ell^2 > 0$[6]. Thus, we differentiate (9) once and twice with respect to x_w to get that the first existence and stability conditions of $\bar{Z}$. In particular, we find that x_w needs to solve $f(x_w) = \gamma_2$ in (3) for $\bar{Z}$ to exist. For this pinned pulse solution to be stable, we also need $g(x_w) > 0$ as shown in (4).

---

[5]The $\mathcal{O}(\varepsilon\sqrt{\varepsilon})$-term of the action functional $J(\bar{U})$, resulting from the artificial boundaries of the slow and fast regions (5), cancels out, see [8] for more details.

[6]The second derivative test is simplified, since $J_1(\bar{U})$ is independent of x_ℓ, see also [8].

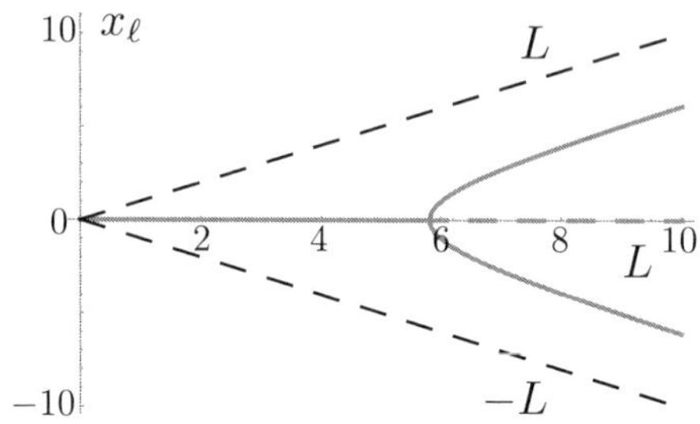

Fig. 2. Bifurcation diagram of x_ℓ versus L for $(\alpha, \beta, D, \gamma_1, \gamma_2) = (4, -1, 5, -0.2, 0.6)$ (as in Figure 1) obtained by solving the existence conditions (13) of Theorem 1. The dashed black lines indicate the boundaries of the bump region. It follows from the stability conditions (14) of Theorem 1 that the pinned pulse solutions centred in the middle of the bump region (i.e. $x_\ell = 0$) are unstable, while the asymmetrically pinned pulse solutions (i.e. $x_\ell \neq 0$) bifurcating from the symmetric ones are stable.

Differentiating (10) with respect to x_ℓ and equating it to zero gives

$$\begin{aligned} (11) \qquad & 2\alpha(\gamma_2 - \gamma_1)e^{-L}\Big(\sinh(x_\ell)\sinh\left(\frac{x_w}{2}\right)\Big) \\ & + 2\beta(\gamma_2 - \gamma_1)e^{-L/D}\left(\sinh\left(\frac{x_\ell}{D}\right)\sinh\left(\frac{x_w}{2D}\right)\right) = 0. \end{aligned}$$

For $\alpha\beta > 0$ this necessarily yields that $x_\ell = 0$. Consequently, the pulse is pinned in the middle of the bump region (and $x_{1,2} = \pm x_w/2$). For $\alpha\beta < 0$, the location $x_\ell \neq 0$ solving

$$|\alpha|e^{-L}\Big(\sinh(x_\ell)\sinh\left(\frac{x_w}{2}\right)\Big) = |\beta|e^{-L/D}\left(\sinh\left(\frac{x_\ell}{D}\right)\sinh\left(\frac{x_w}{2D}\right)\right),$$

where x_w solves (3), with $L - |x_\ell \pm x_w/2|$ being positive and strictly $\mathcal{O}(1)$, would result in a pinned pulse solution located asymmetrically within the bump region. See, for instance, Figure 2.

Differentiating (10) twice with respect to x_ℓ gives

$$\begin{aligned} (12) \qquad \left.\frac{d^2 J_2(\bar{U})}{dx_\ell^2}\right|_{x_\ell=0} &= 2\alpha(\gamma_2 - \gamma_1)e^{-L}\Big(\cosh(x_\ell)\sinh\left(\frac{x_w}{2}\right)\Big) \\ & + 2\frac{\beta}{D}(\gamma_2 - \gamma_1)e^{-L/D}\left(\cosh\left(\frac{x_\ell}{D}\right)\sinh\left(\frac{x_w}{2D}\right)\right), \end{aligned}$$

and for the pinned pulse solution to be stable, this formula needs to be positive.

By combining the above, we arrive at the main result of this article.

Theorem 1. *Let $\varepsilon > 0$ be small enough and let the system parameters of* (1) *with* (2) *be such that there exist $x_w > 0$ and x_ℓ solving*

(13)
$$\begin{cases} \alpha e^{-x_w} + \beta e^{-x_w/D} = \gamma_2, \\ \alpha e^{-L}\Big(\sinh(x_\ell)\sinh\left(\frac{x_w}{2}\right)\Big) + \beta e^{-L/D}\left(\sinh\left(\frac{x_\ell}{D}\right)\sinh\left(\frac{x_w}{2D}\right)\right) = 0, \end{cases}$$

such that $L - |x_\ell \pm x_w/2|$ is positive and strictly $\mathcal{O}(1)$. Then, (1) *with* (2) *supports a pinned pulse solution that asymptotes to $(-1,-1,-1) + \mathcal{O}(\varepsilon)$ as $x \to \pm\infty$ and is pinned in the bump region centred around x_ℓ and has leading order width x_w. Furthermore, the pinned pulse solution is stable if, and only if,*[7]

$$\text{(14)} \qquad \begin{cases} \alpha e^{-x_w} + \dfrac{\beta}{D} e^{-x_w/D} > 0, \\ \alpha(\gamma_2-\gamma_1)e^{-L}\Big(\cosh(x_\ell)\sinh\left(\frac{x_w}{2}\right)\Big) \\ \quad + \dfrac{\beta}{D}(\gamma_2-\gamma_1)e^{-L/D}\left(\cosh\left(\frac{x_\ell}{D}\right)\sinh\left(\frac{x_w}{2D}\right)\right) > 0. \end{cases}$$

For given system parameters $(\alpha, \beta, D, \gamma_1, \gamma_2)$, the first existence condition of (13) has at most two solutions $x_w^{1,2}$ [3]. In turn, for such a given solution x_w^i, the second existence condition of (13) has at most three solutions: $x_\ell = 0$ and $x_\ell^\pm = \pm\tilde{x}_\ell$ (which follows from the particulars of the hyperbolic sinus), see also Figure 2. Observe that the numerically observed widths and locations of the simulations presented in Figure 1 coincide (to leading order) with those predicted by Theorem 1.

Corollary 1. *Pinned pulse solutions are always centred in the middle of the bump region, that is, $x_\ell = 0$, in the original activator-2 inhibitor framework where $\alpha, \beta > 0$ (and as long as $\alpha + \beta > \gamma_2 > 0$). Such a solution is stable if $\gamma_2 > \gamma_1$, since the second stability condition in Theorem* 1 *reduces to*

$$(\gamma_2 - \gamma_1)\left(\alpha e^{-L}\sinh\left(\frac{x_w}{2}\right) + \frac{\beta}{D}e^{-L/D}\sinh\left(\frac{x_w}{2D}\right)\right) > 0.$$

---

[7]The stability conditions (14) of Theorem 1 are independent of the time constants τ, θ. This is as expected since we assumed that these time constants are strictly $\mathcal{O}(1)$ with respect to the small parameter ε. In contrast, we expect that large time constants ($\tau, \theta = \mathcal{O}(\varepsilon^{-2})$) play a crucial role for the stability of pinned pulse solutions. See, for instance, [7, 8, 9, 10].

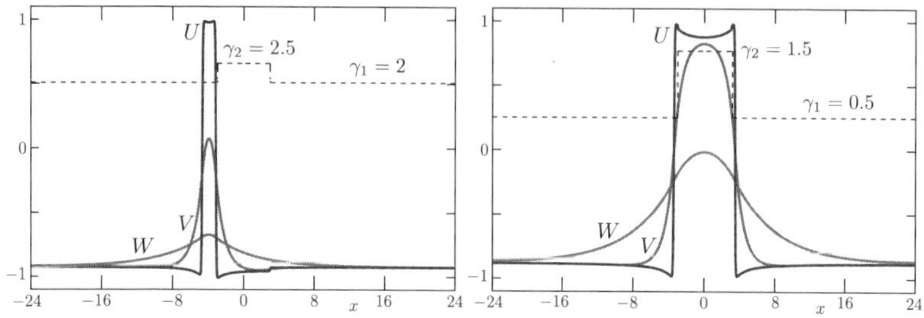

Fig. 3. Two different types of pinned pulse solutions supported by (1). These are obtained by simulating the time independent version of (1) with $(\alpha, \beta, D, \varepsilon) = (3, 2, 5, 0.02)$ and $(\gamma_1, \gamma_2) = (2, 2.5)$, respectively $(\gamma_1, \gamma_2) = (0.5, 1.5)$. In both cases the bump region is given by $[-3, 3]$.

§4. Concluding remarks

In this article, we used geometric singular perturbation techniques and an action functional to show that a singularly perturbed three-component FitzHugh–Nagumo defect model (1) with a bump-type heterogeneity (2) supports both stable and unstable pinned pulse solutions located within the bump region (and $\mathcal{O}(1)$-away from the bump edges). This pinning always occurs in the middle of the bump region for $\alpha\beta > 0$, while asymmetric pinning is also possible for $\alpha\beta < 0$.

The singularly perturbed three-component FitzHugh–Nagumo defect model (1) with (2) also potentially supports other type of pinned pulse solutions, for instance pinned fully either to the left or to the right of the bump region, or with one (or both) of its interfaces located at the boundary of the bump region. See, for instance, Figure 3. These types of pinned solutions can be analysed with the same techniques as used in this article, or by the techniques in [10]. This is left as further research.

Acknowledgements

TT is partially supported by KAKENHI Grant-in-Aid for Scientific Research 17K05355. The authors would also like to thank the 2nd Joint Australia-Japan workshop on dynamical systems with applications in life science (AJwsDSALS2, Biei, Japan, July 15–17, 2018) for the opportunity to work on this project.

References

[1] C. N. Chen and Y. S. Choi, Standing pulse solutions to FitzHugh–Nagumo equations, Arch. Ration. Mech. Anal., **206** (2012), 741–777.

[2] C. N. Chen and X. Hu, Stability analysis for standing pulse solutions to FitzHugh–Nagumo equations, Calc. Var. Partial Differ. Equ., **49** (2014), 827–845.

[3] A. Doelman, P. van Heijster and T. J. Kaper, Pulse dynamics in a three-component system: existence analysis, J. Dyn. Differ. Equ., **21** (2009) 73–115.

[4] T. J. Kaper, An introduction to geometric methods and dynamical systems theory for singular perturbation problems, In: Analyzing Multiscale Phenomena Using Singular Perturbation Methods, Baltimore, MD, 1998, (eds. R.E. O'Malley and J. Cronin), Proc. Sym. Ap., **56**, American Math. Soc., 1999, pp. 85–131.

[5] H. G. Purwins and L. Stollenwerk, Synergetic aspects of gas-discharge: lateral patterns in dc systems with a high ohmic barrier, Plasma Phys. Contr. F., **56** (2015), 123001.

[6] C. P. Schenk, M. Or-Guil, M. Bode adn H. G. Purwins, Interacting pulses in three-component reaction-diffusion systems on two-dimensional domains, Phys. Rev. Lett., **78** (1997), 3781–3784.

[7] P. van Heijster, C.-N. Chen, Y. Nishiura and T. Teramoto, Localized patterns in a three-component FitzHugh–Nagumo model revisited via an action functional, J. Dyn. Differ. Equ., **30** (2018), 521–555.

[8] P. van Heijster, C.-N. Chen, Y. Nishiura and T. Teramoto, Pinned solutions in a heterogeneous three-component FitzHugh–Nagumo model, J. Dyn. Differ. Equ., **31** (2019), 153–203.

[9] P. van Heijster, A. Doelman adn T. J. Kaper, Pulse dynamics in a three-component system: stability and bifurcations, Physica D, **237** (2008), 3335–3368.

[10] P. van Heijster, A. Doelman, T. J. Kaper, Y. Nishiura and K.-I. Ueda, Pinned fronts in heterogeneous media of jump type, Nonlinearity, **24** (2011), 127–157.

§Appendix A. Higher order correction terms of the slow components

The correction term $\bar{V}_1$ is given by

$$
\bar{V}_1(x;x_1,x_2) =
\begin{cases}
\frac{\alpha}{4}\left(e^{x-x_1}(x-x_1-2)-e^{x-x_2}(x-x_2-2)\right) \\
+\frac{\beta}{2(D^2-1)}\left(e^{x-x_1}-D^2e^{(x-x_1)/D}-e^{x-x_2}+D^2e^{(x-x_2)/D}\right) \\
+\frac{1}{2}(\alpha+\beta-\gamma_1)+\frac{1}{2}\sinh(L)e^{x}(\gamma_1-\gamma_2), \ x\in I_s^1, \\
\frac{\alpha}{4}\left(e^{x-x_1}(x-x_1-2)-e^{x-x_2}(x-x_2-2)\right) \\
+\frac{\beta}{2(D^2-1)}\left(e^{x-x_1}-D^2e^{(x-x_1)/D}-e^{x-x_2}+D^2e^{(x-x_2)/D}\right) \\
+\frac{1}{2}(\alpha+\beta-\gamma_2)-\frac{1}{2}e^{-L}\cosh(x)(\gamma_1-\gamma_2), \ x\in I_s^4, \\
\frac{\alpha}{4}\left(e^{-(x-x_1)}(x-x_1+2)-e^{x-x_2}(x-x_2-2)\right) \\
+\frac{\beta}{2(D^2-1)}\left(-e^{-(x-x_1)}+D^2e^{-(x-x_1)/D}-e^{x-x_2}\right. \\
\left.+D^2e^{(x-x_2)/D}\right)-\frac{1}{2}(\alpha+\beta+\gamma_2)-\frac{1}{2}e^{-L}\cosh(x)(\gamma_1-\gamma_2), \ x\in I_s^6, \\
\frac{\alpha}{4}\left(e^{-(x-x_1)}(x-x_1+2)-e^{-(x-x_2)}(x-x_2+2)\right) \\
+\frac{\beta}{2(D^2-1)}\left(-e^{-(x-x_1)}+D^2e^{-(x-x_1)/D}+e^{-(x-x_2)}\right. \\
\left.-D^2e^{-(x-x_2)/D}\right)+\frac{1}{2}(\alpha+\beta-\gamma_2)-\frac{1}{2}e^{-L}\cosh(x)(\gamma_1-\gamma_2), \ x\in I_s^8, \\
\frac{\alpha}{4}\left(e^{-(x-x_1)}(x-x_1+2)-e^{-(x-x_2)}(x-x_2+2)\right) \\
+\frac{\beta}{2(D^2-1)}\left(-e^{-(x-x_1)}+D^2e^{-(x-x_1)/D}+e^{-(x-x_2)}\right. \\
\left.-D^2e^{-(x-x_2)/D}\right)+\frac{1}{2}(\alpha+\beta-\gamma_1)+\frac{1}{2}\sinh(L)e^{-x}(\gamma_1-\gamma_2), \ x\in I_s^{11},
\end{cases}
$$

and $\bar{W}_1$ is given by

$$
\bar{W}_1(x; x_1, x_2) =
\begin{cases}
\frac{\beta}{4D}\Big(e^{(x-x_1)/D}(x - x_1 - 2D) - e^{(x-x_2)/D}(x - x_2 - 2D)\Big) \\
\quad + \frac{\alpha}{2(D^2-1)}\Big(e^{x-x_1} - D^2 e^{(x-x_1)/D} - e^{x-x_2} + D^2 e^{(x-x_2)/D}\Big) \\
\qquad + \frac{1}{2}(\alpha + \beta - \gamma_1) + \frac{1}{2}\sinh(L/D) e^{x/D}(\gamma_1 - \gamma_2), \; x \in I_s^1, \\
\frac{\beta}{4D}\Big(e^{(x-x_1)/D}(x - x_1 - 2D) - e^{(x-x_2)/D}(x - x_2 - 2D)\Big) \\
\quad + \frac{\alpha}{2(D^2-1)}\Big(e^{x-x_1} - D^2 e^{(x-x_1)/D} - e^{x-x_2} + D^2 e^{(x-x_2)/D}\Big) \\
\qquad + \frac{1}{2}(\alpha + \beta - \gamma_2) - \frac{1}{2} e^{-L/D}\cosh(x/D)(\gamma_1 - \gamma_2), \; x \in I_s^4, \\
\frac{\beta}{4D}\Big(e^{-(x-x_1)/D}(x - x_1 + 2D) - e^{(x-x_2)/D}(x - x_2 - 2D)\Big) \\
\quad + \frac{\alpha}{2(D^2-1)}\Big(-e^{-(x-x_1)} + D^2 e^{-(x-x_1)/D} - e^{x-x_2} \\
\qquad + D^2 e^{(x-x_2)/D}\Big) - \frac{1}{2}(\alpha + \beta + \gamma_2) \\
\qquad - \frac{1}{2} e^{-L/D}\cosh(x/D)(\gamma_1 - \gamma_2), \; x \in I_s^6, \\
\frac{\beta}{4D}\Big(e^{-(x-x_1)/D}(x - x_1 + 2D) - e^{-(x-x_2)/D}(x - x_2 + 2D)\Big) \\
\quad + \frac{\alpha}{2(D^2-1)}\Big(-e^{-(x-x_1)} + D^2 e^{-(x-x_1)/D} + e^{-(x-x_2)} \\
\qquad - D^2 e^{-(x-x_2)/D}\Big) + \frac{1}{2}(\alpha + \beta - \gamma_2) \\
\qquad - \frac{1}{2} e^{-L/D}\cosh(x/D)(\gamma_1 - \gamma_2), \; x \in I_s^8, \\
\frac{\beta}{4D}\Big(e^{-(x-x_1)/D}(x - x_1 + 2D) - e^{-(x-x_2)/D}(x - x_2 + 2D)\Big) \\
\quad + \frac{\alpha}{2(D^2-1)}\Big(-e^{-(x-x_1)} + D^2 e^{-(x-x_1)/D} + e^{-(x-x_2)} \\
\qquad - D^2 e^{-(x-x_2)/D}\Big) + \frac{1}{2}(\alpha + \beta - \gamma_1) \\
\qquad + \frac{1}{2}\sinh(L/D) e^{-x/D}(\gamma_1 - \gamma_2), \; x \in I_s^{11}.
\end{cases}
$$

School of Mathematical Sciences,
Queensland University of Technology,
Brisbane, QLD 4001, Australia

and

Biometris
Wageningen University and Research
6700 AA Wageningen, Netherlands
E-mail address: peter.vanheijster@wur.nl

School of Medicine,
Asahikawa Medical University,
Asahikawa, 078-8510, Japan
E-mail address: teramoto@asahikawa-med.ac.jp

Advanced Studies in Pure Mathematics 85, 2020
The Role of Metrics in the Theory of Partial Differential Equations
pp. 151–161

Rigidity for positive Ricci curvature via metric measure geometry

Shouhei Honda

Abstract.

In this survey we introduce a main result in [AH17], which gives an almost rigidity related to positive Ricci curvature via RCD theory. This is new even for cloased Riemannian manifolds.

§1. Introduction

Let (M^n, g) be an n-dimensional complete Riemannian manifold with $\partial M^n = \emptyset$ and Ricci curvature bounded below by $n-1$;

$$\text{Ric} \geq n-1. \tag{1}$$

Then Myers' theorem [My41] states that the diameter is at most π;

$$\text{diam}\,(M^n, g) \leq \pi = \text{diam}\,(\mathbb{S}^n, g_{\mathbb{S}^n}). \tag{2}$$

In particular M^n is compact. Moreover the rigidity holds, that is, the equality in (2) holds if and only if (M^n, g) is isometric to the standard unit n-sphere $(\mathbb{S}^n, g_{\mathbb{S}^n})$, which is known as Cheng's maximal diameter theorem [Ch75].

On the other hand Lichnerowicz's theorem [L58] states that the first positive eigenvalue $\lambda_1(M^n, g)$ of the (minus) Laplacian $-\Delta$ is at least n;

$$\lambda_1(M^n, g) \geq n = \lambda_1(\mathbb{S}^n, g_{\mathbb{S}^n}). \tag{3}$$

The rigidity in (3) also holds which is known as Obata's theorem [O62].

Finally let us recall that Lévy-Gromov inequality [Gr80] implies

$$h(M^n, g) \geq \frac{\mathcal{H}^{n-1}(\mathbb{S}^{n-1})}{\mathcal{H}^n(\mathbb{S}^n_+)} = h(\mathbb{S}^n, g_{\mathbb{S}^n}), \tag{4}$$

Received March 8, 2019.
Revised June 14, 2019.
2010 *Mathematics Subject Classification.* 53C21, 53C23.
Key words and phrases. Ricci curvature, Sobolev space and Laplacian.

where $\mathbb{S}^n_+$ is the hemisphere and $h(M^n, g)$ denotes the Cheeger constant defined by

$$h(M^n, g) := \inf_{\Omega \subset M^n} \frac{\mathcal{H}^{n-1}(\partial\Omega)}{\mathcal{H}^n(\Omega)}, \tag{5}$$

where the infimum runs over all open subset Ω of M^n with the smooth boundary $\partial\Omega$ and $\mathcal{H}^n(\Omega) \le \mathcal{H}^n(M^n)/2$. Moreover the rigidiy in (4) holds.

These *almost* rigidities also hold, that is, roughly speaking, the following three conditions are equivalent;

(a) the diameter is close to π;
(b) the first positive eigenvalue is close to n;
(c) the Cheeger constant is close to that of $(\mathbb{S}^n, g_{\mathbb{S}^n})$.

Moreover these conditions are also equivalent to the following condition;

(d) (M^n, g) is Gromov-Hausdorff close to the spherical suspension $\mathbb{S}^0 * X$ of a compact metric space (X, d) with $\mathrm{diam}\,(X, d) \le \pi$.

Recall that the Gromov-Hausdorff distance defines a complete distance on the space of all isometry classes of compact metric spaces. See [B04, C96, Cr82] for the proofs of these almost rigidity results. We emphasize that the proofs were done by different and independent ways each other.

It is worth pointing out that for the condition (d), in general, we can not take (X, d) as $(X, d) = (\mathbb{S}^{n-1}, g_{\mathbb{S}^{n-1}})$. Thus in general, even if the conditions above are satisfied, (M^n, g) is *not* necessary Gromov-Hausdorff close to $(\mathbb{S}^n, g_{\mathbb{S}^n})$.

The aim of this survey is to introduce one of main results in [AH17], which gives a *unified* relationship between them, including their new unified proofs. This was conjectured in [H13]. For that, we use two key notions.

The first one is the first positive eigenvalue $\lambda_{1,p}(M^n, g)$ of the p-Laplacian for all $p \in (1, \infty)$;

$$-\mathrm{div}(|\nabla f|^{p-2}\nabla f) = \lambda |f|^{p-2} f \tag{6}$$

in weak sense, where a nonnegative constant λ is called an eigenvalue of the p-Laplacian if there exists a nontrivial Sobolev function $f \in H^{1,p}(M^n, g)$ such that (6) holds. Note that $\lambda_{1,2}(M^n, g) = \lambda_1(M^n, g)$. Then the corresponding inequality with the rigidity for $\lambda_{1,p}(M^n g)$ also hold by Matei [M00];

$$\lambda_{1,p}(M^n, g) \ge \lambda_{1,p}(\mathbb{S}^n, g_{\mathbb{S}^n}). \tag{7}$$

See [V12] for the almost rigidity.

The second one is the synthetic treatment of 'Ricci bounds from below for metric measure spaces'. In [LV09, S06] Lott-Villani and Sturm introduce such notion independently, so-called $\mathrm{CD}(K, N)$ metric measure spaces (Y, d, m). This means roughly that the Ricci curvature of (Y, d, m) is bounded below by $K \in \mathbb{R}$ and the dimension of (Y, d, m) is bounded above by $N \in [1, \infty]$;

$$\mathrm{Ric}_{(Y,d,m)} \geq K, \ \dim_{(Y,d,m)} \leq N. \tag{8}$$

For example the metric measure space $([0, \pi], d_{\mathbb{R}}, \sin^{N-1} t dt)$ satisfies $\mathrm{Ric} \geq N-1$ and $\dim \leq N$, that is, it is a $\mathrm{CD}(N-1, N)$ space.

Later, Ambrosio-Gigli-Savaré [AGS14b], Gigli [G13, G15], Erbar-Kuwada-Sturm [EKS15] and Ambrosio-Mondino-Savaré [AMS15] introduce the notion of $\mathrm{RCD}(K, N)$ metric measure spaces by adding a condition that the Sobolev space $H^{1,2}$ is a Hilbert space, to $\mathrm{CD}(K, N)$ condition. The metric measure space $([0, \pi], d_{\mathbb{R}}, \sin^{N-1} t dt)$ is a typical example (more strongly, it is a model space. See Theorem 7) of $\mathrm{RCD}(N-1, N)$ spaces. Further development on the axiomatization of $\mathrm{RCD}(K, N)$ has been done actively (see e.g. [CavMil16]).

We should refer Cheeger-Colding theory which is closely related to this toplic as written below. In [CC96, CC97, CC00a, CC00b] they established the structure theory on measured Gromov-Hausdorff (mGH) limit spaces (so-called Ricci limit spaces) of Riemannian manifolds with solving several major conjectures. Such limit spaces are also typical examples of RCD spaces. It is worth pointing out that thanks to recent very quick developments on RCD spaces, most of parts on Cheeger-Colding's works are covered by RCD theory.

The following notation is useful to state the main result;

$$\lambda_{1,1}(M^n, g) := h(M^n, g), \ (\lambda_{1,\infty}(M^n, g))^{1/\infty} := \frac{2}{\mathrm{diam}\,(M^n, g)}. \tag{9}$$

Note that the inequalitis (2), (3) and (4) are written simply as

$$(\lambda_{1,p}(M^n, g))^{1/p} \geq (\lambda_{1,p}(\mathbb{S}^n, g_{\mathbb{S}^n}))^{1/p} \quad \forall p \in [1, \infty] \tag{10}$$

and that the rigidities are stated as; if the equality in (10) holds for some $p \in [1, \infty]$, then this also holds for all $p \in [1, \infty]$.

The main result is the almost rigidity which is stated roughly as follows (see Theorem 8 for the precise statement);

Theorem 1. *The following two conditions are equivalent;*

(e) $(\lambda_{1,p}(M^n, g))^{1/p}$ *is close to* $(\lambda_{1,p}(\mathbb{S}^n, g_{\mathbb{S}^n}))^{1/p}$ *for some* $p \in [1, \infty]$;

(f) $(\lambda_{1,p}(M^n, g))^{1/p}$ *is close to* $(\lambda_{1,p}(\mathbb{S}^n, g_{\mathbb{S}^n}))^{1/p}$ *for all* $p \in [1, \infty]$.
Moreover the above conditions are equivalent to that (M^n, g) *is Gromov-Hausdorff close to the spherical suspension of a compact metric space with the diameter at most* π.

It is worth pointing out that the compactness of a moduli space consists of RCD spaces plays a key role in the proof, which is an important point for using RCD theory.

Because of the restriction of the number of pages, we will forcus to explain only main ideas to prove Theorem 1, that is, most of statements have no detailed proofs. See references for that.

Acknowledgement. The author would like to express his appreciation to the referees for his/her careful reading and useful suggestions, in particular, on the history of RCD theory. He acknowledges supports of the Grantin-Aid for Young Scientists (B) 16K17585 and of Grant-in-Aid for Scientific Research (B) of 18H01118.

§2. Metric measure spaces with Ricci bounds from below; RCD(K, N) spaces

2.1. Compactness of moduli space

We say that a triple (X, d, m) is a metric measure space if (X, d) is a complete separable metric space and m is a Borel measure on X with full support. For $K \in \mathbb{R}$ and $N \in [1, \infty]$, (X, d, m) is an RCD(K, N) space if the following four conditions are satisfied;

(g) $\exists x \in X, \exists C_i \geq 1$ s.t. $m(B_r(x)) \leq C_1 r^{C_2 r^2}, \forall r > 0$;
(h) the Sobolev space $H^{1,2}(X, d, m)$ is a Hilbert space;
(i) the Bochner inequality;

$$\frac{1}{2}\Delta|\nabla f|^2 \geq \frac{(\Delta f)^2}{N} + \langle \nabla \Delta f, \nabla f \rangle + K|\nabla f|^2 \tag{11}$$

holds in weak sense for all test functions $f : X \to \mathbb{R}$;
(j) the Sobolev-to-Lipschitz property holds.

See [AGS14b, AMS15, EKS15, CavMil16] for the precise definition (including the equivalent formulations).

The following compactness is due to [AGS14b, GMS13, LV09, S06];

Theorem 2 (Compactness). *Let* $K \in \mathbb{R}$, *let* $N \in [1, \infty)$, *let* $d > 0$ *and let us denote by* $\mathcal{M}(K, N, d)$ *the set of all isometry classes of* RCD(K, N) *spaces* (X, d, m) *with* diam $(X, d) \leq d$ *and* $m(X) = 1$. *Then* $\mathcal{M}(K, N, d)$ *is compact with respect to measured Gromov-Hausdorff convergence (which determines a metrizable topology).*

This compactness theorem fits nicely several compactness results on functional analysis. We will discuss them in section 3.

2.2. The first positive eigenvalue of p-Laplacian and the Cheeger constant

Let (X, d, m) be a metric measure space with $m(X) < \infty$. Then let us define the first positive eigenvalue $\lambda_{1,p}(X, d, m)$ of p-Laplacian for $p \in (1, \infty)$ by

$$\lambda_{1,p}(X,d,m) := \inf_f \frac{\|\nabla f\|_{L^p}^p}{\inf_{c\in\mathbb{R}} \|f-c\|_{L^p}^p}, \tag{12}$$

where the infimum runs over all nonconstant Lipschitz functions $f : X \to \mathbb{R}$. Moreover the Cheeger constant $h(X, d, m)$ is defined by;

$$h(X,d,m) := \inf_{A\subset X} \frac{\mathrm{Per}(A)}{m(A)}, \tag{13}$$

where the infimum runs over all Borel subset A of X with $m(A) \le m(X)/2$ and $\mathrm{Per}(A)$ denotes the perimeter of A defined by;

$$\mathrm{Per}(A) := \liminf_{\epsilon\to+0} \frac{m(B_\epsilon(A)) - m(A)}{\epsilon}. \tag{14}$$

Note that these coincide with the usual one defined in Section 1 if (X, d, m) is a closed Riemanifold with the standard probability measure.

§3. Convergence of functions

Throughout this section we fix a measured Gromov-Hausdorff convergent sequence of compact $\mathrm{RCD}(K, N)$ spaces $(N < \infty)$;

$$(X_i, d_i, m_i) \overset{mGH}{\to} (X, d, m) \tag{15}$$

and fix a convergent sequence $p_i \to p$ in $[1, \infty)$. Let us recall the definition of L^{p_i}-weak/strong convergence of functions $f_i \in L^{p_i}(X_i, m_i)$ to $f \in L^p(X, m)$. In order to simplify our argument we write them only in the case when $p > 1$. See [AH17, AST16, H13, H15] for the details in this section.

Definition 3.1 (L^{p_i}-weak/strong convergence of functions). *Assume $p > 1$. We say that $f_i \in L^{p_i}(X_i, m_i)$ L^{p_i}-weakly converge to $f \in L^p(X, m)$ if $\sup_i \|f_i\|_{L^{p_i}} < \infty$ and*

$$\int_{X_i} \phi_i f_i dm_i \to \int_X \phi f dm \tag{16}$$

are satisfied for all uniform convergent sequence $\phi_i \in C^0(X_i)$ *to* $\phi \in C^0(X)$ *(that is,* $\{\phi_i\}_i$ *is equi-continuous and* $\phi_i(x_i) \to \phi(x)$ *whenever* $x_i \in X_i \to x \in X$*, where they are well-defined). Moreover we say that it is* L^{p_i}*-strong convergence if it holds that*

$$\limsup_{i\to\infty} \|f_i\|_{L^{p_i}} \le \|f\|_{L^p}. \tag{17}$$

Let us introduce fundamental results on L^{p_i} convergence. It is worth pointing out that the Sobolev space $H^{1,p}(X,d,m)$ (for all $p \in [1,\infty)$) and the space of functions of bounded variations $BV(X,d,m)$ are also well-defined (see [ADM14, Ch99, Sh00]. See also [AH17]). Let us use the following notation;

$$\hat{H}^{1,p}(X,d,m) := \begin{cases} H^{1,p}(X,d,m) & \text{if } p \in (1,\infty), \\ BV(X,d,m) & \text{if } p = 1. \end{cases} \tag{18}$$

The following play key roles in section 4.

Theorem 3 (L^{p_i}-weak compactness). *Let* $f_i \in L^{p_i}(X_i, m_i)$ *be an* L^{p_i}*-bounded sequence, that is,* $\sup_i \|f_i\|_{L^{p_i}} < \infty$*. Then there exist a subsequence* $i(j)$ *and* $f \in L^p(X,m)$ *such that* f *is the* $L^{p_{i(j)}}$*-weak limit function of* $f_{i(j)}$*.*

Theorem 4 (The existence of approximating sequence). *Let* $f \in \hat{H}^{1,p}(X,d,m)$*. Then there exists a sequence* $f_i \in \hat{H}^{1,p_i}(X_i,d_i,m_i)$ *such that* f_i L^{p_i}*-strongly converge to* f *and that* $\lim_{i\to\infty} \|f_i\|_{\hat{H}^{1,p_i}} = \|f\|_{\hat{H}^{1,p}}$*.*

Theorem 5 (Stability). *Let* $f_i \in \hat{H}^{1,p_i}(X_i,d_i,m_i)$ *and let* $f \in L^p(X,m)$ *be the* L^{p_i}*-weak limit function of* f_i*. If* $\sup_i \|f_i\|_{\hat{H}^{p_i}} < \infty$*, then* f *is the* L^{p_i}*-strong limit function of* f_i *with* $f \in \hat{H}^{1,p}(X,d,m)$ *and*

$$\liminf_{i\to\infty} \|f_i\|_{\hat{H}^{1,p_i}} \ge \|f\|_{\hat{H}^{1,p}}. \tag{19}$$

Note that these results are well-known in the case when the spaces and the exponents are fixed, that is, $(X_i,d_i,m_i) \equiv (X,d,m)$ and $p_i \equiv p$. However the measured Gromov-Hausdorff convergence version as above is nontrivial. These in the case when $p_i \equiv 2$ come from [GMS13] which are equivalent to the spectral convergence of Laplacian. The spectral convergence (in the case when the spaces are Ricci limit spaces) was conjectured in [F87] first, and was proved in [CC97, CC00a, CC00b]. Roughly speaking, Theorems 4 and 5 mean that if two RCD(K,N) spaces $(Y_i,d_i,m_i)(i=1,2)$ are measured Gromov-Hausdorff close, and if two exponents p_i are close, then the two functional spaces $\hat{H}^{1,p_i}(Y_i,d_i,m_i)$ are also close in some sense.

§4. Continuity of $(\lambda_{1,p})^{1/p}$

Let us recall our useful notation;

(20) $\lambda_{1,1}(X,d,m) := h(X,d,m), \ (\lambda_{1,\infty}(X,d,m))^{1/\infty} := \dfrac{2}{\operatorname{diam}(X,d)}.$

Theorem 6 (Continuity of $(\lambda_{1,p})^{1/p}$). *Under the same setting as in* (15), *we see that for all* $q_i \to q$ *in* $[1,\infty]$,

(21) $$\lim_{i\to\infty} (\lambda_{1,q_i}(X_i,d_i,m_i))^{q_i} = (\lambda_{1,q}(X,d,m))^{1/q}.$$

Proof. The upper semicontinuity comes from Theorem 4. Moreover we can check the lower semicontinuity by Theorems 3 and 5. See [AH17] for the detail. Q.E.D.

Note that Grosjean proved in [Gros05] the continuity $(\lambda_{1,p}(M^n,g))^{1/p} \to 2/\operatorname{diam}(M^n,g)$ as $p \to \infty$ for a fixed closed Riemannian manifold (M^n,g).

§5. Proof of Theorem 1

Let

(22) $$(\mathbb{S}^N_1, d_N, m_N) := ([0,\pi], d_{\mathbb{R}}, \sin^{N-1} t dt) \quad \forall N \in (1,\infty)$$

and let us introduce the following results which generalize results introduced in section 1;

Theorem 7 (Inequalities and rigidities for positive Ricci curvature in RCD setting). *We see that*

(k) $(\mathbb{S}^N_1, d_N, m_N)$ *is an* RCD$(N-1,N)$ *space;*

(l) *it holds that*

(23) $$(\lambda_{1,p}(X,d,m))^{1/p} \ge (\lambda_{1,p}(\mathbb{S}^N_1, d_N, m_N))^{1/p} \quad \forall p \in [1,\infty]$$

for all RCD$(N-1,N)$ *space* (X,d,m);

(m) *for an* RCD$(N-1,N)$ *space* (X,d,m), *the equality in* (23) *is satisfied for some* p *if and only if it also holds for all* p. *Moreover it is equivalent to that* (X,d,m) *is a (measured) spherical suspension of* $(\mathbb{S}^N_1, d_N, m_N)$.

(n) *if* $N \in \mathbb{N}$, *then*

(24) $$(\lambda_{1,p}(\mathbb{S}^N_1, d_N, m_N))^{1/p} = (\lambda_{1,p}(\mathbb{S}^N, g_{\mathbb{S}^N}))^{1/p} \quad \forall p \in [1,\infty].$$

See [CavMon17a, CavMon17b, K15a, K15b] for the details (see also [AH17]).

Theorem 8 (Precise formulation of Theorem 1). *For any $N \in (1, \infty)$ and any $\epsilon > 0$ there exists $\delta := \delta(N, \epsilon) > 0$ depending only on N, ϵ such that if an* RCD$(N-1, N)$ *space (X, d, m) satisfies*

$$(\lambda_{1,p}(X, d, m))^{1/p} \le (\lambda_{1,p}(\mathbb{S}^N_1, d_N, m_N))^{1/p} + \delta \tag{25}$$

for some $p \in [1, \infty]$, then

$$(\lambda_{1,q}(X, d, m))^{1/q} \le (\lambda_{1,q}(\mathbb{S}^N_1, d_N, m_N))^{1/q} + \delta \tag{26}$$

for all $q \in [1, \infty]$.

Proof. The proof is done by a contradiction. If the assertion is not satisfied, then there exist $\epsilon_0 > 0$, sequences $p_i, q_i \in [1, \infty]$ and a sequence of RCD$(N-1, N)$ spaces (X_i, d_i, m_i) with $m_i(X_i) = 1$ such that $(\lambda_{1,p_i}(X_i, d_i, m_i))^{1/p_i} \le (\lambda_{1,p_i}(\mathbb{S}^N_1, d_N, m_N))^{1/p_i} + i^{-1}$ holds and that $(\lambda_{1,q_i}(X_i, d_i, m_i))^{1/q_i} \ge (\lambda_{1,q_i}(\mathbb{S}^N_1, d_N, m_N))^{1/q_i} + \epsilon_0$. Then by Theorem 2 with no loss of generality we can assume that there exist an RCD$(N-1, N)$ space (X, d, m) and $p, q \in [1, \infty]$ such that $(X_i, d_i, m_i) \stackrel{mGH}{\to} (X, d, m)$ and that $p_i, q_i \to p, q$ respectively. Then Theorem 6 yields that $(\lambda_{1,p}(X, d, m))^{1/p} = (\lambda_{1,p}(\mathbb{S}^N_1, d_N, m_N))^{1/p}$ and that $(\lambda_{1,q}(X, d, m))^{1/q} \neq (\lambda_{1,q}(\mathbb{S}^N_1, d_N, m_N))^{1/q}$, which contradicts Theorem 7. Q.E.D.

§6. Questions

It is natural to ask what is a higher order analogue of Theorem 6, that is, justify the continuity of $(\lambda_{k,p})^{1/p}$, where $\lambda_{k,p}$ denotes the k-th positive eigenvalue of p-Laplacian. The interesting parts are in taking limits as $p \to 1$ and $p \to \infty$. In order to catch the higher order eigenvalues it might be useful to use the Krasnoselskii spectrum (see also [AHP18]). Moreover the asymptotic behavior of $(\lambda_{k,p})^{1/p}$ as $k \to \infty$ is also interesting, which should be related to Weyl's law (see also [AHT18, ZZ17]).

References

[ADM14] L. Ambrosio and S. DiMarino, *Equivalent definitions of BV space and of total variation on metric measure spaces*, Journal of Functional Analysis, **266** (2014), 4150–4188.

[AGS14a] L. Ambrosio, N. Gigli and G. Savaré, *Calculus and heat flow in metric measure spaces and applications to spaces with Ricci bounds from below*, Invent. Math., **195** (2014), 289–391.

[AGS14b] L. Ambrosio, N. Gigli and G. Savaré, *Metric measure spaces with Riemannian Ricci curvature bounded from below*, Duke Math. J., **163** (2014), 1405–1490.

[AH17] L. Ambrosio and S. Honda, *New stability results for sequences of metric measure spaces with uniform Ricci bounds from below*, in Measure Theory in Non-Smooth Spaces, 1–51, De Gruyter Open, Warsaw, 2017.

[AHP18] L. Ambrosio, S. Honda and J. W. Portegies, *Continuity of nonlinear eigenvalues in* $\mathrm{CD}(K,\infty)$ *spaces with respect to measured Gromov-Hausdorff convergence*, Calc. Var. Partial Differential Equations, **57** (2018), no. 2, Art. 34, 23 pp.

[AHT18] L. Ambrosio, S. Honda and D. Tewodrose, *Short-time behavior of the heat kernel and Weyl's law on* $\mathrm{RCD}^*(K,N)$*-spaces*, Ann. Glob. Anal. Geom., **53**(1) (2018), 97–119.

[AMS15] L. Ambrosio, A. Mondino and G. Savaré, *Nonlinear diffusion equations and curvature conditions in metric measure spaces*, to appear in Mem. Am. Math. Soc., arXiv preprint: 1509.07273 (2015).

[AST16] L. Ambrosio, F. Stra and D. Trevisan, *Weak and strong convergence of derivations and stability of flows with respect to MGH convergence*, J. Funct. Anal., **272** (2017), 1182–1229.

[B04] V. Bayle, *A differential inequality for the isoperimetric profile*, Int. Math. Res. Not., **7** (2004), 311–342.

[CavMil16] F. Cavalletti and E. Milman, *The Globalization Theorem for the Curvature Dimension Condition*, arXiv:1612.07623.

[CavMon17a] F. Cavalletti and A. Mondino, *Sharp and rigid isoperimetric inequalities in metric-measure spaces with lower Ricci curvature bounds*, Invent. Math., **208** (2017), no. 3, 803–849.

[CavMon17b] F. Cavalletti and A. Mondino, *Sharp geometric and functional inequalities in metric measure spaces with lower Ricci curvature bounds*, Geom. Topol., **21** (2017), no. 1, 603–645.

[Ch99] J. Cheeger, *Differentiability of Lipschitz functions on metric measure spaces*, Geom. Funct. Anal., **9** (1999), 428–517.

[CC96] J. Cheeger and T. H. Colding, *Lower bounds on Ricci curvature and the almost rigidity of warped products*, Annals of Math., Second Series, **144**(1) (1996), 189–237.

[CC97, CC00a, CC00b] J. Cheeger and T. H. Colding, *On the structure of spaces with Ricci curvature bounded below.* I, II, III, J. Differential Geom., **46**, **54**, **54** (1997, 2000, 2000), 406–480, 13–35, 37–74.

[Ch75] S. Y. Cheng, *Eigenvalue comparison theorems and its geometric applications*, Math. Z., **143** (1975) 289–297.

[C96] T. H. Colding, *Shape of manifolds with positive Ricci curvature*, Invent. Math. **124** (1996), 175–191.

[Cr82] C. B. Croke, *An eigenvalue pinching problem*, Invent. Math., **68** (1982), 253–256.

[EKS15] M. Erbar, K. Kuwada and K.-T. Sturm, *On the equivalence of the entropic curvature-dimension condition and Bochner's inequality on metric measure spaces*, Invent. Math., **201** (2015), 993–1071.

[F87] K. Fukaya, *Collapsing of Riemannian manifolds and eigenvalues of Laplace operator*, Invent. Math., **87** (1987), 517–547.

[G13] N. Gigli, *The splitting theorem in non-smooth context*, arxiv:1302.5555.

[G15] N. Gigli, *On the differential structure of metric measure spaces and applications*, Mem. Amer. Math. Soc., **236** (2015), no. 1113.

[GMS13] N. Gigli, A. Mondino and G. Savaré, *Convergence of pointed non-compact metric measure spaces and stability of Ricci curvature bounds and heat flows*, Proceedings of the London Mathematical Society, **111** (2015), 1071–1129.

[Gr80] M. Gromov, *Paul Lévy's isoperimetric inequality*, Preprint IHES.

[Gros05] J.-F. Grosjean, *p-Laplace operator and diameter of manifolds*, Ann. Global Anal. Geom., **28** (2005), 257–270.

[H13] S. Honda, *Cheeger constant, p-Laplacian and Gromov-Hausdorff convergence*, arxiv:1310.0304v3.

[H15] S. Honda, *Ricci curvature and L^p-convergence*, J. Reine Angew Math., **705** (2015), 85–154.

[K15a] C. Ketterer, *Cones over metric measure spaces and the maximal diameter theorem,* J. Math. Pures Appl., (9) **103** (2015), 1228–1275.

[K15b] C. Ketterer, *Obata's rigidity theorem for metric measure spaces,* Anal. Geom. Metr. Spaces, **3** (2015), 278–295.

[L58] A. Lichnerowicz, *Géométrie des Groupes de Transformations*, Travaux et Recherches Mathematiques, III, Dunod, Paris (1958).

[LV09] J. Lott and C. Villani, *Ricci curvature for metric-measure spaces via optimal transport*, Ann. of Math., **169** (2009), 903–991.

[M00] A.-M. Matei, *First eigenvalue for the p-Laplace operator*, Nonlinear Anal., **39** (2000), 1051–1068.

[My41] S. B. Myers, *Riemannian manifolds with positive mean curvature*, Duke Math. J. **8** (2) (1941), 401–404.

[O62] M. Obata, *Certain conditions for a Riemannian manifold to be iosometric with a sphere*, J. Math. Soc. Japan, **14** no. 3 (1962), 333–340.

[Sh00] N. Shanmugalingam, *Newtonian spaces: an extension of Sobolev spaces to metric measure spaces*, Rev. Mat. Iberoamericana, **16** (2000), 243–279.

[S06] K.-T. Sturm, *On the geometry of metric measure spaces,* I *and* II, Acta Math., **196** (2006), 65–131 and 133–177.

[V12] D. Valtorta, *On the p-Laplace operator on Riemannian manifolds*, PhD Thesis, available at arXiv:1212.3422.

[ZZ17] H.-C. Zhang and X.-P. Zhu, *Weyl's law on* $\mathrm{RCD}^*(K,N)$ *metric measure spaces*, arXiv:1701.01967v4, to appear in Comm. Anal. Geom.

Mathematical Institute,
Tohoku University,
Sendai 980-8578, Japan
E-mail address: shouhei.honda.e4@tohoku.ac.jp

Advanced Studies in Pure Mathematics 85, 2020
The Role of Metrics in the Theory of Partial Differential Equations
pp. 163–170

Blow-up of solutions to semilinear wave equations with a scaling invariant critical damping

Masahiro Ikeda

Abstract.

In this paper, we survey our recent results [10, 11] with Motohiro Sobajima, which give blow-up results and upper estimates of lifespan for certain small data to semilinear wave equations with the scaling invariant space-dependent or time-dependent damping:

$$(\mathrm{DW}_{V_0})\qquad \begin{cases} \partial_t^2 u - \Delta u + \dfrac{V_0}{|x|}\partial_t u = |u|^{p_{V_0}}, & (t,x) \in (0,T)\times\mathbb{R}^N, \\ (u,\partial_t u)|_{t=0} = \varepsilon(u_0,u_1), & x\in\mathbb{R}^N, \end{cases}$$

$$(\mathrm{DW}_{\mu})\qquad \begin{cases} \partial_t^2 v - \Delta v + \dfrac{\mu}{1+t}\partial_t v = |v|^{p_{\mu}}, & (t,x) \in (0,T)\times\mathbb{R}^N, \\ (v,\partial_t v)|_{t=0} = \varepsilon(v_0,v_1), & x\in\mathbb{R}^N, \end{cases}$$

where $N\in\mathbb{N}$, $V_0 \in [0,(N-1)^2/(N+1))$, $p_{V_0}\in(N/(N-1), p_S(N+V_0)]$, $\mu\in[0,(N^2+N+2)/(N+2))$, $p_\mu \in (p_F(N), p_S(N+\mu)]$. Here p_S is the Strauss exponent and p_F is the Fujita exponent. And (u_0,u_1) and (v_0,v_1) are pairs of given functions, whose support is compact, and $\varepsilon>0$ is a small parameter.

§1. Introduction

In this paper, we review our recent results[1] [10, 11], which give blow-up results and upper estimates of lifespan for some non-negative small data to semilinear wave equations with a scaling invariant critical

Received November 28, 2018.
Revised February 16, 2019.
2010 *Mathematics Subject Classification.* 35L70.
Key words and phrases. Wave equation with a scaling invariant space or time-dependent damping, Small data blowup, Strauss exponent, Critical and subcritical case, The Gauss hypergeometric functions.

[1]These are joint works with Motohiro Sobajima in Tokyo University of Science.

damping. Here "critical" means the borderline case between heat-like and wave-like phenomena. For the undamped case ($V_0 = \mu = 0$), see [6, 17], for the classical damping case, see [19, 21], for another type of space-dependent damping case, see [15, 16], and for other types of time-dependent damping case, see [1, 2, 18, 20], for example, in which heat-like phenomenon occurs.

The Cauchy problem to a semilinear wave equation with a space-dependent critical damping is

$$(\mathrm{DW}_{V_0}) \qquad \begin{cases} \partial_t^2 u - \Delta u + \dfrac{V_0}{|x|}\partial_t u = |u|^p, & (t,x) \in (0,T) \times \mathbb{R}^N, \\ (u, \partial_t u)|_{t=0} = \varepsilon(u_0, u_1), & x \in \mathbb{R}^N. \end{cases}$$

Here $N \geq 3$, $(u_0, u_1) \in H^2 \times H^1$ is a given function satisfying supp $(u_0, u_1) \subset \{|x| \leq R_0\}$ for someR_0, which denotes the shape of the data, $\varepsilon > 0$ is a (small) parameter, which denotes the size of the data, $V_0 \geq 0$ and $p > 1$ are constants, $u \in C([0,T) : H^2) \cap C^1([0,T) : H^1)$ is an unknown function,

$$T = T(\varepsilon) := \sup\{T \in (0, \infty] : \exists \text{ sol } u \text{ to } (\mathrm{DW}_{V_0}) \text{ on } [0,T)\}$$

denotes the maximal existence time of the function u. We note that if $u = u(t,x)$ is a solution to the wave equation (DW_{V_0}) on $[0,T)$, then $u_\lambda(t,x) := \lambda^{\frac{2}{p-1}} u(\lambda t, \lambda x)$ with $\lambda > 0$ is also a solution to the equation on $[0, T/\lambda)$, which means "scaling invariant damping".

To the author's knowledge, there were no results about local well-posedness to (DW_{V_0}), because of existence of the singular potential $\frac{V_0}{|x|}\partial_t u$. In [10], it is firstly proved that local well-posedness in $H^2 \times H^1$ to (DW_{V_0}) for arbitrary data hold via a contraction mapping principle with the Hardy inequality $\||x|^{-1} f\|_{L^2} \lesssim \|\nabla f\|_{L^2}$ due to $N \geq 3$ and the Sobolev inequality, in the case $V_0 \geq 0$ and $p < \frac{N-2}{N-4}$ ($p < \infty$, when $N = 3, 4$).

Remark 1.1. 1. *We do not know whether* (DW_{V_0}) *is locally well-posed or not in two dimensional case* $N = 2$.
2. *It is also interesting whether Strichartz type estimates to the linear problem for* (DW_{V_0}) *can be obtained or not. However, we do not know such estimates.*
3. *In the almost similar manner as in Section* 2 *in* [10], *we can prove local well-posedness in the lower-order space* $H^1 \times L^2$ *if* $p < N/(N-2)$.

When $V_0 = 0$, the following Strauss conjecture and the sharp estimate of lifespan for small data are verified by several mathematicians:

$$\begin{cases} p > p_S(N) \Longrightarrow \forall (u_0, u_1), \exists \varepsilon_0 > 0 \text{ s.t } \forall \varepsilon \in [0, \varepsilon_0],\ T = \infty, \\ p \leq p_S(N) \Longrightarrow \exists (u_0, u_1), \exists \varepsilon_0 > 0 \text{ s.t } \forall \varepsilon \in (0, \varepsilon_0],\ T < \infty, \end{cases}$$

$$T \sim \begin{cases} C\varepsilon^{-\frac{2p(p-1)}{\gamma(N;p)}}, & \text{if } p < p_S(N), \\ \exp(C\varepsilon^{-p(p-1)}), & \text{if } p = p_S(N), \end{cases}$$

where $\gamma(n;p) := 2 + (n+1)p - (n-1)p^2$,
$p_S(n) :=$ (the positive root of $\gamma(n;p) = 0$), for $n \geq 2$, $p_S(1) := \infty$ and C is independent of ε.

The following result is the main one in [10] and extends the above blow-up results to $V_0 \geq 0$:

Theorem 1.1 ([10]). *Let $V_0 < \frac{(N-1)^2}{N+1}$, $\frac{N}{N-1} < p \leq p_S(N+V_0)$, $u_0, u_1 \geq 0$ with $u_0 + u_1 \neq 0$. Then there exists $\varepsilon_0 > 0$ such that for any $\varepsilon \in (0, \varepsilon_0]$, $T < \infty$. And the following estimate holds:*

$$T \leq \begin{cases} \exp(C\varepsilon^{-p(p-1)}) & \text{if } p = p_S(N+V_0), \\ C_\delta \varepsilon^{-\frac{2p(p-1)}{\gamma(N+V_0;p)}-\delta} & \text{if } \exists p_0(N, V_0) \leq p < p_S(N+V_0), \end{cases}$$

where C, C_δ are independent of ε and $\delta > 0$ can be taken arbitrarily small and $p_0(N, V_0) \in (N/(N-1), p_S(N+V_0))$.

Remark 1.2. 1. *The proof of the theorem is based on a test-function method with properties of Gauss' hypergeometric functions initiated by Zhou-Han* [25] *($V_0 = 0$).*
2. *Zhou-Han* [25] *uses a time decay estimate of solutions from below to the free wave equation proved by Yordanov-Zhang '06* [24], *though our argument does not need such estimate. Instead we use Lemma* 1.2 *below, which is the key lemma and enables us to treat the wave equations with the scaling invariant damping.*
3. *The value $V_* := \frac{(N-1)^2}{N+1}$ is reasonable, since the identity*

$$p_S(N+V_*) = p_F(N-1)$$

holds, where for $n \in \mathbb{N}$, $p_F(n) := 1+2/n$, which is called Fujita exponent and is known as the critical exponent for the semilinear heat equation and the semilinear damped wave equation.
4. *In Proposition* 3.8 *in* [12], *a blow-up result and an upper estimate of lifespan for some small data belonging to $L^1(\mathbb{R}^N)$ or a smaller space than $L^1(\mathbb{R}^N)$ are proved in the case $N/(N-1) < p \leq p_F(N-1)$, $V_0 > 0$. We note that V_0 might be larger than V_*. The same result holds for the corresponding semilinear heat equation $-\Delta u + \frac{V_0}{|x|}\partial_t u = |u|^p$ under the same assumption. This means that this situation is the heat-like case.*

5. *It is conjectured that small data global existence holds in the opposite case* $p > p_S(N+V_0)$*. However, it is still open (see also* [7] *for a related result).*
6. *Ikehata-Todorova-Yordanov* [15] *studied the Cauchy problem of a wave equation with a different type space-dependent damping term such as* $(1+|x|^2)^{-\frac{\alpha}{2}}\partial_t u$ *with* $\alpha \in [0,1)$ *and the same nonlinearity* $|u|^p$*. They proved that the critical exponent is given by* $p_F(N-\alpha)$ *for this problem.*

In order to state the key lemma, we assume that $R_0 = 1$ for simplicity and introduce a test function $\Phi_\beta(t,x)$ with $\beta > 0$ given by

$$\Phi_\beta(t,x) := (2+t+|x|)^{-\beta} F\left(\beta, \frac{N-1+V_0}{2}, N-1; \frac{2|x|}{2+t+|x|}\right),$$

where $F(a,b,c;z)$ is the Gauss hypergeometric function.

Lemma 1.2 (Lemma 4.5 in [10]). *Let* $p > N/(N-1)$ *and* $V_0 \in [0, V_*)$*, let* $q > \max(p, 2/(N-1-V_0))$*, set* $\beta := (N-1-V_0)/2 - 1/q$ *and let* u *be a solution to* (DW_{V_0}) *on* $[0,T)$*. Then there exists a constant* $C > 0$ *such that the estimate*

$$\begin{aligned}\varepsilon \int_{\mathbb{R}^N} f(x)\Phi_\beta(0,x) + \varepsilon E_{\beta,1} t + \int_0^t (t-s) G_\beta(s) ds \\ \le C\left\{\|u(t)\|_{L^p}(2+t)^{\frac{N}{p'}-\beta} + \int_0^t \|u(s)\|_{L^p}(2+s)^{\frac{N}{p'}-\frac{N-1-V_0}{2}-\frac{1}{p'}} ds\right\}\end{aligned}$$

holds for any $t \in [0,T)$*. Here*

$$\begin{aligned} E_{\beta,1} &:= \int_{\mathbb{R}^N} g(x)\Phi_\beta(0,x) + \beta \int_{\mathbb{R}^N} f(x)\Phi_{\beta+1}(0,x)dx \\ &\quad + V_0 \int_{\mathbb{R}^N} \frac{1}{|x|} f(x)\Phi_\beta(x)dx \\ G_\beta(t) &:= \int_{\mathbb{R}^N} |u(t,x)|^p \Phi_\beta(t,x)dx. \end{aligned}$$

Our method for proving the theorem is applicable to the case of the scaling invariant time-dependent damping:

$$(\mathrm{DW}_\mu) \qquad \begin{cases} \partial_t^2 v - \Delta v + \dfrac{\mu}{1+t}\partial_t v = |v|^p, & (t,x) \in (0,T)\times\mathbb{R}^N, \\ (v, \partial_t v)|_{t=0} = \varepsilon(v_0, v_1), & x \in \mathbb{R}^N. \end{cases}$$

Here $N \ge 1$, $(v_0, v_1) \in H^1 \times L^2$ is a given function satisfying supp $(v_0, v_1) \subset \{|x| \le r_0\}$ for some $r_0 < 1$, $\mu \ge 0$ is constant, $v \in C([0,T) : H^1) \cap C^1([0,T) : L^2)$ is an unknown function. We also note that if $v = v(t,x)$

is a solution to (DW_μ) on $[0,T)$, then $v_\lambda(t,x)=\lambda^{\frac{2}{p-1}}v(\lambda(1+t)-1,\lambda x)$ with $\lambda>0$ is also a solution to (DW_μ) on $[0,(T+1)/\lambda-1)$, which means "scaling invariant damping".

There are many results about well-posedness to (DW_μ) (see [4, 8, 11, 23] and the references therein), compared to the space-dependent damping case (DW_{V_0}). Also, not only linear problems but also nonlinear problems to (DW_μ) with other types of time-dependent coefficients like effective, non-effective or overdamping case, were studied by several authors (see [5, 9, 13, 14] and their references).

By the similar argument as in Theorem 1.1, we get the following blow-up result to (DW_μ):

Theorem 1.3 ([11]). *Let $N\in\mathbb{N}$, $0<\mu<\frac{N^2+N+2}{N+2}=:\mu_*$, $p_F(N)<p\le p_S(N+\mu)$ and $v_0,v_1\ge 0$ with $v_0+v_1\ne 0$. Then there exists $\varepsilon_0>0$ such that for any $\varepsilon\in(0,\varepsilon_0)$, $T<\infty$. And the following estimate holds:*

$$T\le\begin{cases}\exp(C\varepsilon^{-p(p-1)}) & \text{if } p=p_S(N+\mu),\\ C_\delta\varepsilon^{-\frac{2p(p-1)}{\gamma(N+\mu;p)}-\delta} & \text{if } \exists p_0(N,\mu)\le p<p_S(N+\mu).\end{cases}$$

Here $C,C_\delta>0$ are independent of ε, and $\delta>0$ can be taken arbitrarily small.

Remark 1.3. 1. *Theorem* 1.3 *extends the results obtained in* [8, 23]. *Small data global existence to* (DW_μ) *is proved in some* μ,p *in* [3, 4]. *The relation between the previous results and ours is as follows:*

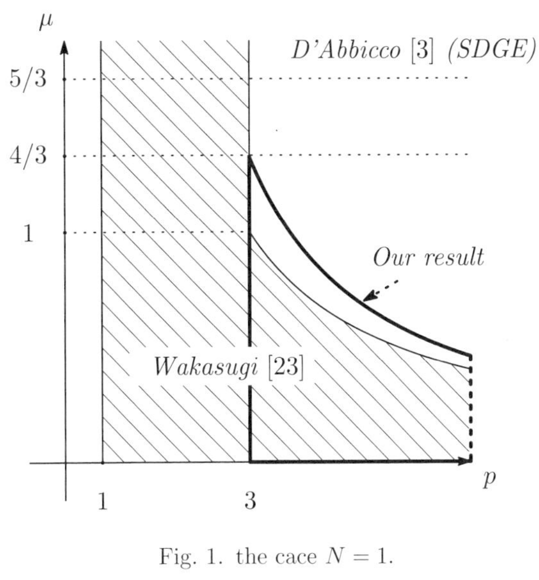

Fig. 1. the cace $N=1$.

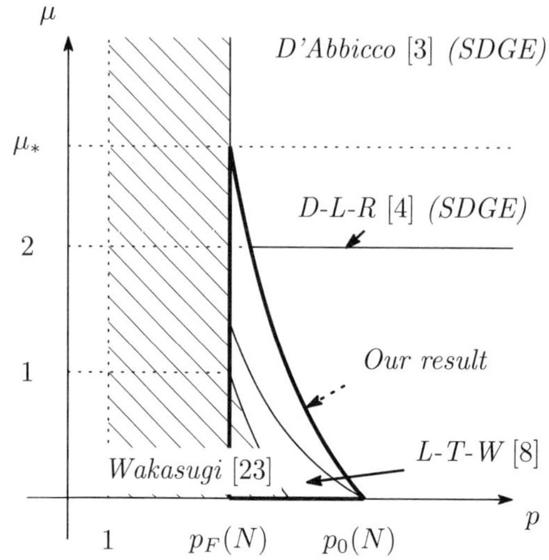

Fig. 2. the case $N \geq 3$.

2. *Recently, in [22], an improvement of the result in Theorem* 1.2 *in the sub-critical case* $p < p_S(N+\mu)$ *is given.*
3. *It is also conjectured that small data global existence holds in the opposite case* $p > p_S(N+\mu)$. *However, it is still open (except for the special case* $\mu = 2$).

References

[1] M. D'Abbicco and S. Lucente, *A modified test function method for damped wave equations*, Adv. Nonlinear Studies, **13** (2013), 867–892.

[2] M. D'Abbicco, S. Lucente and M. Reissig, *Semilinear wave equations with effective damping*, Chinese Ann. Math., **34B** (2013), 3, 345–380.

[3] M. D'Abbicco, *The threshold of effective damping for semilinear wave equations*, Math. Methods, Appl. Sci., **38** (2015), 1032–1045.

[4] M. D'Abbicco, S. Lucente and M. Reissig, *A shift in the Strauss exponent for semilinear wave equations with a not effective damping*, J. Diff. Equs., **259** (2015), 5040–5073.

[5] K. Fujiwara, M. Ikeda and Y. Wakasugi, *Estimates of lifespan and blow-up rates for the wave equation with a time-dependent damping and a power-type nonlinearity*, to appear in Funkcialaj Ekvacioj, arXiv:1609.01035.

[6] V. Georgiev, H. Lindblad and C. D. Sogge, *Weighted Strichartz estimates and global existence for semilinear wave equations*, Amer. J. Math., **119** (1997), 1291–1319.

[7] V. Georgiev, H. Kubo and K. Wakasa, *Critical exponent for nonlinear damped wave equations with non-negative potential in 3D*, arXiv:1810.05956.

[8] N. A. Lai, H. Takamura and K. Wakasa, *Blow-up for semilinear wave equations with the scaling invariant damping and super-Fujita exponent*, J. Diff. Equs., **263** (2017), 5377–5394.

[9] M. Ikeda and T. Inui, *The sharp estimate of the lifespan for the semilinear wave equation with time-dependent damping*, Diff. Int. Equs., **32** (2019), 1–36.

[10] M. Ikeda and M. Sobajima, *Life-span of blowup solutions to semilinear wave equation with space-dependent critical damping*, to appear in Funkcialaj Ekvacioj, arXiv:1709.04401.

[11] M. Ikeda and M. Sobajima, *Life-span of solutions to semilinear wave equation with time-dependent critical damping for specially localized initial data*, Math. Ann., **372** (2018), 1017–1040.

[12] M. Ikeda and M. Sobajima, *Sharp upper bound for lifespan of solutions to some critical semilinear parabolic, dispersive and hyperbolic equations via a test function method*, Nonlinear anal., **182** (2019), 57–74.

[13] M. Ikeda, M. Sobajima and Y. Wakasugi, *Sharp lifespan estimates of blowup solutions to semilinear wave equations with time-dependent effective damping*, submitted, arXiv:1808.06189.

[14] M. Ikeda and Y. Wakasugi, *Global well-posedness for semilinear wave equation with time dependent damping in the overdamping case*, to appear in Proc. Amer. Math. Soc., arXiv:1708.08044.

[15] R. Ikehata, G. Todorova and B. Yordanov, *Critical exponent for semilinear wave equations with a subcritical potential*, Funkcial. Ekvac., **52** (2009), 411–435.

[16] R. Ikehata, G. Todorova and B. Yordanov, *Optimal decay rate of the energy for wave equations with critical potential*, J. Math. Soc. Japan, **65** (2013), 183–236.

[17] F. John, *Blow-up of solutions of nonlinear wave equations in three space dimensions*, Manuscripta Math., **28** (1979), 235–268.

[18] J. Lin, K. Nishihara and J. Zhai, *Critical exponent for the semilinear wave equation with time-dependent damping*, Discrete and Continuous Dynamical Systems, **32** (2012), 4307–4320.

[19] A. Matsumura, *On the asymptotic behavior of solutions of semi-linear wave equations*, Publ. RIMS., **12** (1976), 169–189.

[20] K. Nishihara, *Asymptotic behavior of solutions to the semilinear wave equation with time-dependent damping*, Tokyo J. of Math., **34** (2011), 327–343.

[21] G. Todorova and B. Yordanov, *Critical exponent for a nonlinear wave equation with damping*, J. Diff. Equs., **174** (2001), 464–489.

[22] Z. Tu and J. Lin, *A note on the blowup of scale invariant damping wave equation with sub-Strauss exponent*, preprint, arXiv:1709.00866.

[23] Y. Wakasugi, *Critical exponent for the semilinear wave equation with scale invariant dampint*, Fourier analysis, 375-390, Trends Math., Birkhauser/Springer, Cham, 2014.

[24] B. Yordanov and Qi S. Zhang, *Finite time blow up for critical wave equations in high dimensions*, J. Funct. Anal., **231** (2006), 361–374.

[25] Y. Zhou and W. Han, *Life-span of solutions to critical semilinear wave equations*, Commun. P.D.E., **39** (2014), 439–451.

Center for Advanced Intelligence Project, RIKEN,
Nihonbashi 1-chome Mitsui Building, 15th floor 1-4-1 Nihonbashi,
Chuo-ku, Tokyo 103-0027,
Japan/Department of Mathematics,
Faculty of Science and Technology,
Keio University,
3-14-1 Hiyoshi, Kohoku-ku, Yokohama, 223-8522, Japan
E-mail address: masahiro.ikeda@riken.jp/masahiro.ikeda@keio.jp

Advanced Studies in Pure Mathematics 85, 2020
The Role of Metrics in the Theory of Partial Differential Equations
pp. 171–180

Asymptotic behavior of the energy critical nonlinear wave equation with a special scale invariant damping

Takahisa Inui

Abstract.

We are interested in the asymptotic behavior of the solution to the energy critical nonlinear wave equation with a special scale invariant damping. In the present paper, we show that a global solution to this equation goes to the solution to the linear wave equation with the scale invariant damping as time tends to infinity. We also discuss its time order.

§1. Introduction

We consider the following energy critical nonlinear wave equation with a scale invariant damping:

$$(1.1) \qquad \begin{cases} \partial_t^2 u - \Delta u + \dfrac{2}{1+t}\partial_t u = |u|^{\frac{d+2}{d-2}}, & (t,x) \in [0,\infty) \times \mathbb{R}^d, \\ (u(0), \partial_t u(0)) = (u_0, u_1), & x \in \mathbb{R}^d, \end{cases}$$

where $d \geq 3$, (u_0, u_1) is given, and u is an unknown function. The nonlinear equation

$$(1.2) \qquad \partial_t^2 u - \Delta u + \frac{\mu}{1+t}\partial_t u = |u|^p, \quad (t,x) \in [0,\infty) \times \mathbb{R}^d,$$

where $\mu > 0$ and $p > 1$, is well studied from the viewpoint of the critical exponent p_c. See [2, 3, 6, 10, 11] and references therein. In the present paper, we are interested in the global behavior of the solution to (1.1), which is (1.2) as $\mu = 2$ and $p = (d+2)/(d-2)$ (i.e. $p+1 = 2d/(d-2)$)

Received November 27, 2018.
Revised June 5, 2019.
2010 *Mathematics Subject Classification.* 35L70, 35P25.
Key words and phrases. Scattering, Damped wave equation, Scale invariant damping, Additional time decay.

is the critical Sobolev exponent). Setting $w = (1+t)u$ for (1.1), we get the equation

$$(1.3) \qquad \begin{cases} \partial_t^2 w - \Delta w = \dfrac{|w|^{\frac{d+2}{d-2}}}{(1+t)^{\frac{4}{d-2}}} & (t,x) \in [0,\infty) \times \mathbb{R}^d, \\ (w(0), \partial_t w(0)) = (u_0, u_0 + u_1), & x \in \mathbb{R}^d. \end{cases}$$

Therefore, we can use the Strichartz estimates for the wave equation to investigate the global behavior of the solution to (1.3).

First, we give a definition of the solution to (1.3). Let

$$\mathcal{W}(t) = \frac{\sin(t|\nabla|)}{|\nabla|},$$

which is related to the solution map for the wave equation.

Definition 1.1. Let $T \in (0,\infty]$. We say that w is a solution to (1.3) on $[0,T)$ if w satisfies $(w, \partial_t w) \in C([0,T) : \dot{H}^1(\mathbb{R}^d) \times L^2(\mathbb{R}^d))$, $|\nabla|^{1/2} w \in L_{t,x}^{2(d+1)/(d-1)}(I)$ and $w \in L_{t,x}^{2(d+1)/(d-2)}(I)$ for any compact interval $I \subset [0,T)$, $(w(0), \partial_t w(0)) = (u_0, u_0+u_1)$, and the Duhamel's formula

$$w(t,x) = \mathcal{W}(t)(u_0+u_1) + \partial_t \mathcal{W}(t) u_0 + \int_0^t \mathcal{W}(t-s) \frac{|w(s)|^{\frac{d+2}{d-2}}}{(1+s)^{\frac{4}{d-2}}} ds$$

for all $t \in [0,T)$. We say that w is global if $T = \infty$. Moreover, we have the solution u to (1.1) by the transformation $u = (1+t)^{-1} w$.

We show the local well-posedness result of (1.3).

Theorem 1.1 (local well-posedness). *Let $d \geq 3$ and $T \in (0,\infty]$. Let $(u_0,u_1) \in H^1(\mathbb{R}^d) \times L^2(\mathbb{R}^d)$ satisfy $\|(u_0,u_1)\|_{H^1 \times L^2} \leq A$. Then, there exists $\delta = \delta(A) > 0$ such that if*

$$(1.4) \qquad \|\mathcal{W}(t)(u_0+u_1) + \partial_t \mathcal{W}(t) u_0\|_{L_{t,x}^{\frac{2(d+1)}{d-2}}([0,T))} \leq \delta,$$

then there exists a solution $(w, \partial_t w) \in C([0,T) : \dot{H}^1(\mathbb{R}^d) \times L^2(\mathbb{R}^d))$ to (1.3) with $\|w\|_{L_{t,x}^{2(d+1)/(d-2)}([0,T))} \leq 2\delta$.

Remark 1.1. From Theorem 1.1, small data global existence can be shown, that is, if the initial data (u_0,u_1) is sufficiently small, then the solution is global and satisfies $\|w\|_{L_{t,x}^{2(d+1)/(d-2)}([0,\infty))} < \infty$.

Next, we obtain the global behavior of the solutions to (1.3).

Theorem 1.2. *Let $d \geq 3$. If w is a global solution of* (1.3) *which satisfies $\|w\|_{L^{2(d+1)/(d-2)}_{t,x}([0,\infty))} < \infty$, then there exists $(w_0^+, w_1^+) \in \dot{H}^1(\mathbb{R}^d) \times L^2(\mathbb{R}^d)$ such that*

$$\lim_{t\to\infty}(1+t)^{\frac{4}{d-2}}\left\|(w(t),\partial_t w(t)) - \mathcal{A}(t)(w_0^+, w_1^+)\right\|_{\dot{H}^1\times L^2} = 0,$$

where

$$\mathcal{A}(t) = \begin{pmatrix} \partial_t \mathcal{W}(t) & \mathcal{W}(t) \\ \partial_t^2 \mathcal{W}(t) & \partial_t \mathcal{W}(t) \end{pmatrix}.$$

This additional time decay comes from the decay $(1+t)^{-4/(d-2)}$ in the nonlinearity of (1.3). Such phenomena can be seen in the damped nonlinear Schrödinger equation [7].

The following theorem follows from the translation $w = (1+t)u$. We note that the assumption $\|(1+t)u\|_{L^{2(d+1)/(d-2)}_{t,x}([0,\infty))} < \infty$ in the following theorem is satisfied if the initial data (u_0, u_1) is sufficiently small. See Remark 1.1.

Theorem 1.3. *Let $d \geq 3$. Assume that u is a global solution of* (1.1) *and satisfies $\|(1+t)u\|_{L^{2(d+1)/(d-2)}_{t,x}([0,\infty))} < \infty$. Then there exists $(w_0^+, w_1^+) \in \dot{H}^1(\mathbb{R}^d) \times L^2(\mathbb{R}^d)$ such that*

$$\lim_{t\to\infty}(1+t)^{\frac{d+2}{d-2}}\left\|u(t) - (1+t)^{-1}\{\partial_t \mathcal{W}(t) w_0^+ + \mathcal{W}(t) w_1^+\}\right\|_{\dot{H}^1} = 0.$$

We note that $\phi(t) := (1+t)^{-1}\{\partial_t \mathcal{W}(t) w_0^+ + \mathcal{W}(t) w_1^+\}$ is the solution with the initial data $(\phi(0), \partial_t\phi(0)) = (w_0^+, w_1^+ - w_0^+)$ of the linear wave equation with the scale invariant damping

$$\partial_t^2 \phi - \Delta\phi + \frac{2}{1+t}\partial_t \phi = 0. \tag{1.5}$$

Thus, this means that the nonlinear solution (1.1) goes to the linear solution (1.5) as $t \to \infty$ with additional decay order $(1+t)^{-(d+2)/(d-2)}$.

§2. Proof

We say that (q, r) is a wave admissible pair if (q, r) satisfies

$$\frac{d-1}{2}\left(\frac{1}{2} - \frac{1}{r}\right) \geq \frac{1}{q}, \quad q, r, d \geq 2, \quad r \neq \infty.$$

Let p' denote the Hölder exponent of p. We have the following Strichartz estimates for the wave equation.

Lemma 2.1 ([5, 9]). *Let $d \geq 2$ and (q, r) be a wave admissible pair. Set $\gamma = d(1/2 - 1/r) - 1/q$. Then, we have*

$$\|\mathcal{W}(t)f\|_{L_t^q(I:L_x^r(\mathbb{R}^d))} \lesssim \left\||\nabla|^{\gamma-1}f\right\|_{L^2},$$
$$\|\partial_t\mathcal{W}(t)f\|_{L_t^q(I:L_x^r(\mathbb{R}^d))} \lesssim \||\nabla|^{\gamma}f\|_{L^2},$$

for f such that the right hand side is finite. Moreover, if $(\tilde{q}, \tilde{r})$ is a wave admissible pair, then it holds that

$$\left\|\int_0^t \mathcal{W}(t-s)F(s)ds\right\|_{L_t^q(I:L_x^r(\mathbb{R}^d))} \lesssim \left\||\nabla|^{\gamma+\tilde{\gamma}-1}F\right\|_{L_t^{\tilde{q}'}(I:L_x^{\tilde{r}'}(\mathbb{R}^d))},$$

where $\tilde{\gamma} = d(1/2 - 1/\tilde{r}) - 1/\tilde{q}$, for F such that the right hand side is finite.

Proof of Theorem 1.1. We only give a sketch of the proof of the local well-posedness for (1.1) when $d = 3, 4, 5$. See [4, 8, 12, 13] for details. We define the complete metric space

$$X(T, L, M) := \left\{v \text{ on } [0, T) \times \mathbb{R}^d \right.$$
$$\left. : \left\||\nabla|^{\frac{1}{2}}v\right\|_{L_{t,x}^{\frac{2(d+1)}{d-1}}([0,T))} \leq L, \|v\|_{L_{t,x}^{\frac{2(d+1)}{d-2}}([0,T))} \leq M\right\},$$
$$\|v\|_{X(T)} := \left\||\nabla|^{\frac{1}{2}}v\right\|_{L_{t,x}^{\frac{2(d+1)}{d-1}}([0,T))} + \|v\|_{L_{t,x}^{\frac{2(d+1)}{d-2}}([0,T))}.$$

We define

$$\Phi[w](t) := \mathcal{W}(t)(u_0 + u_1) + \partial_t\mathcal{W}(t)u_0 + \int_0^t \mathcal{W}(t-s)\frac{|w(s)|^{\frac{d+2}{d-2}}}{(1+s)^{\frac{4}{d-2}}}ds.$$

Since $(u_0, u_1) \in H^1(\mathbb{R}^d) \times L^2(\mathbb{R}^d)$, by the Strichartz estimates in Lemma 2.1, we obtain

$$\|\mathcal{W}(t)(u_0 + u_1) + \partial_t\mathcal{W}(t)u_0\|_{X(T)} \lesssim \|u_0\|_{H^1} + \|u_1\|_{L^2} < A. \tag{2.1}$$

We estimate the nonlinear term as follows. By the Strichartz estimates in Lemma 2.1 we get

$$\left\||\nabla|^{\frac{1}{2}} \int_0^t \mathcal{W}(t-s)\frac{|w(s)|^{\frac{d+2}{d-2}}}{(1+s)^{\frac{4}{d-2}}}ds\right\|_{L_{t,x}^{\frac{2(d+1)}{d-1}}([0,T))}$$
$$\lesssim \left\||\nabla|^{\frac{1}{2}} \frac{|w|^{\frac{d+2}{d-2}}}{(1+t)^{\frac{4}{d-2}}}\right\|_{L_{t,x}^{\frac{2(d+1)}{d+3}}([0,T))}.$$

It follows from the fractional Leibnitz rule (see Lemma 2.5 in [8] and references therein) that

$$\left\| |\nabla|^{\frac{1}{2}} |w|^{\frac{d+2}{d-2}} \right\|_{L_x^{\frac{2(d+1)}{d+3}}} \lesssim \|w\|_{L_x^{\frac{2(d+1)}{d-2}}}^{\frac{4}{d-2}} \left\| |\nabla|^{\frac{1}{2}} w \right\|_{L_x^{\frac{2(d+1)}{d-1}}} .$$

Combining these estimates and using Hölder inequality for time vaiables, we obtain

$$\begin{aligned} &\left\| |\nabla|^{\frac{1}{2}} \int_0^t \mathcal{W}(t-s) \frac{|w(s)|^{\frac{d+2}{d-2}}}{(1+s)^{\frac{4}{d-2}}} ds \right\|_{L_{t,x}^{\frac{2(d+1)}{d-1}}([0,T))} \\ &\lesssim \|w\|_{L_{t,x}^{\frac{2(d+1)}{d-2}}([0,T))}^{\frac{4}{d-2}} \left\| \frac{|\nabla|^{\frac{1}{2}} w}{(1+t)^{\frac{4}{d-2}}} \right\|_{L_{t,x}^{\frac{2(d+1)}{d-1}}([0,T))} . \end{aligned} \tag{2.2}$$

In the same way, we also get

$$\begin{aligned} &\left\| \int_0^t \mathcal{W}(t-s) \frac{|w(s)|^{\frac{d+2}{d-2}}}{(1+s)^{\frac{4}{d-2}}} ds \right\|_{L_{t,x}^{\frac{2(d+1)}{d-2}}([0,T))} \\ &\lesssim \left\| |\nabla|^{\frac{1}{2}} \frac{|w|^{\frac{d+2}{d-2}}}{(1+t)^{\frac{4}{d-2}}} \right\|_{L_{t,x}^{\frac{2(d+1)}{d+3}}([0,T))} \\ &\lesssim \|w\|_{L_{t,x}^{\frac{2(d+1)}{d-2}}([0,T))}^{\frac{4}{d-2}} \left\| \frac{|\nabla|^{\frac{1}{2}} w}{(1+t)^{\frac{4}{d-2}}} \right\|_{L_{t,x}^{\frac{2(d+1)}{d-1}}([0,T))} . \end{aligned} \tag{2.3}$$

Combining (2.1) and (2.2), we obtain

$$\begin{aligned} &\left\| |\nabla|^{\frac{1}{2}} \Phi[w] \right\|_{L_{t,x}^{\frac{2(d+1)}{d-1}}([0,T))} \\ &\leq \left\| |\nabla|^{\frac{1}{2}} \mathcal{W}(t)(u_0+u_1) + \partial_t \mathcal{W}(t) u_0 \right\|_{L_{t,x}^{\frac{2(d+1)}{d-1}}([0,T))} \\ &\quad + \left\| |\nabla|^{\frac{1}{2}} \int_0^t \mathcal{W}(t-s) \frac{|w(s)|^{\frac{d+2}{d-2}}}{(1+s)^{\frac{4}{d-2}}} ds \right\|_{L_{t,x}^{\frac{2(d+1)}{d-1}}([0,T))} \\ &\leq CA + CLM^{\frac{4}{d-2}} \\ &\leq L \end{aligned}$$

if we choose $L = 2CA$ and M such that $CM^{4/(d-2)} \leq 1/2$. By (2.1) and (2.3), recalling (1.4), we get

$$\begin{aligned}\|\Phi[w]\|_{L_{t,x}^{\frac{2(d+1)}{d-2}}([0,T))} &\leq \|\mathcal{W}(t)(u_0+u_1)+\partial_t\mathcal{W}(t)u_0\|_{L_{t,x}^{\frac{2(d+1)}{d-2}}([0,T))} \\ &\quad + \left\| \int_0^t \mathcal{W}(t-s)\frac{|w(s)|^{\frac{d+2}{d-2}}}{(1+s)^{\frac{4}{d-2}}}ds \right\|_{L_{t,x}^{\frac{2(d+1)}{d-2}}([0,T))} \\ &\leq \delta + CLM^{\frac{4}{d-2}} \\ &\leq M\end{aligned}$$

if we choose $\delta = M/2$ and $L \leq (2C)^{-1}M^{(d-6)/(d-2)}$ (which is possible if $d \leq 5$). Thus, Φ is a mapping on $X(T, L, M)$.

$$\begin{aligned}&\left\| |\nabla|^{\frac{1}{2}}(\Phi[w]-\Phi[v]) \right\|_{L_{t,x}^{\frac{2(d+1)}{d-1}}([0,T))} + \|\Phi[w]-\Phi[v]\|_{L_{t,x}^{\frac{2(d+1)}{d-2}}([0,T))} \\ &\lesssim \left\| |\nabla|^{\frac{1}{2}}\left(|w|^{\frac{d+2}{d-2}} - |v|^{\frac{d+2}{d-2}}\right) \right\|_{L_{t,x}^{\frac{2(d+1)}{d+3}}([0,T))} \\ &\leq CM^{\frac{4}{d-2}} \left\| |\nabla|^{\frac{1}{2}}(w-v) \right\|_{L_{t,x}^{\frac{2(d+1)}{d-1}}([0,T))} + CM^{\frac{6-d}{d-2}}L\|w-v\|_{L_{t,x}^{\frac{2(d+1)}{d-2}}([0,T))}.\end{aligned}$$

Taking M sufficiently small, Φ is a contraction mapping on $X(T, L, M)$. By the Banach fixed point theorem, we obtain the solution such that $w = \Phi[w]$. Then, $(w, \partial_t w)$ belongs to $C([0,T); \dot{H}^1(\mathbb{R}^d) \times L^2(\mathbb{R}^d))$ because of the Strichartz estimates (Lemma 2.1) and the nonlinear estimates. Q.E.D.

Remark 2.1. It is known that the restriction $d \in \{3, 4, 5\}$ is not essential. In the higher dimensional case, the nonlinearity is not smooth and thus the estimate of the difference is difficult. However, we can overcome this difficulty by the exotic Strichartz estimates. We omit the proof since it is same as that in [1].

Lemma 2.2. *Assume that w is a global solution of* (1.1) *and satisfies* $\|w\|_{L_{t,x}^{2(d+1)/(d-2)}([0,\infty))} < \infty$. *Then w satisfies*

$$\left\| |\nabla|^{\frac{1}{2}} w \right\|_{L_{t,x}^{\frac{2(d+1)}{d-1}}([0,\infty))} < \infty$$

Proof. Take $0 < \tau < T < \infty$ arbitrary. We note that the global solution belongs to $L_t^{\frac{2(d+1)}{d-1}}(K : \dot{W}_x^{\frac{1}{2},\frac{2(d+1)}{d-1}})$ for any compact interval $K \subset [0, \infty)$. It follows from the Duhamel's formula and the Strichartz estimates that

$$\left\| |\nabla|^{\frac{1}{2}} w \right\|_{L_{t,x}^{\frac{2(d+1)}{d-1}}((\tau,T))} \lesssim \|(w(\tau), \partial_t w(\tau))\|_{H^1 \times L^2} + \|w\|^{\frac{4}{d-2}}_{L_{t,x}^{\frac{2(d+1)}{d-2}}((\tau,T))} \left\| |\nabla|^{\frac{1}{2}} w \right\|_{L_{t,x}^{\frac{2(d+1)}{d-1}}((\tau,T))}.$$

Since $\|w\|_{L_{t,x}^{\frac{2(d+1)}{d-2}}((\tau,T))} \ll 1$ for large τ, we obtain

$$\left\| |\nabla|^{\frac{1}{2}} w \right\|_{L_{t,x}^{\frac{2(d+1)}{d-1}}((\tau,T))} \lesssim \|(w(\tau), \partial_t w(\tau))\|_{H^1 \times L^2}$$

for large $\tau > 0$. Fix such τ. Since T is arbitrary, we get

$$\left\| |\nabla|^{\frac{1}{2}} w \right\|_{L_{t,x}^{2(d+1)/(d-1)}((\tau,\infty))} \lesssim \|(w(\tau), \partial_t w(\tau))\|_{H^1 \times L^2}.$$

Therefore, we obtain $\left\| |\nabla|^{\frac{1}{2}} w \right\|_{L_{t,x}^{2(d+1)/(d-1)}([0,\infty))} < \infty$. Q.E.D.

Proof of Theorem 1.2. We have

$$\begin{pmatrix} w \\ \partial_t w \end{pmatrix} = \mathcal{A}(t) \begin{pmatrix} u_0 \\ u_0 + u_1 \end{pmatrix} + \int_0^t \mathcal{A}(t-s) \begin{pmatrix} 0 \\ \frac{|w(s)|^{\frac{d+2}{d-2}}}{(1+s)^{\frac{4}{d-2}}} \end{pmatrix} ds.$$

By (2.2), (2.3) and Lemma 2.2, it holds that

$$\begin{aligned} &\|\mathcal{A}(-t)(w(t), \partial_t w(t)) - \mathcal{A}(-\tau)(w(\tau), \partial_t w(\tau))\|_{\dot{H}^1 \times L^2} \\ &\lesssim \left\| \int_\tau^t \mathcal{W}(-s) \frac{|w(s)|^{\frac{d+2}{d-2}}}{(1+s)^{\frac{4}{d-2}}} ds \right\|_{\dot{H}^1} + \left\| \int_\tau^t \partial_t \mathcal{W}(-s) \frac{|w(s)|^{\frac{d+2}{d-2}}}{(1+s)^{\frac{4}{d-2}}} ds \right\|_{L^2} \\ &\lesssim \|w\|^{\frac{4}{d-2}}_{L_{t,x}^{\frac{2(d+1)}{d-2}}((\tau,t))} \left\| |\nabla|^{\frac{1}{2}} w \right\|_{L_{t,x}^{\frac{2(d+1)}{d-1}}((\tau,t))} \\ &\to 0, \end{aligned}$$

as $\tau, t \to \infty$. This means that there exists $(w_0^+, w_1^+) \in \dot{H}^1(\mathbb{R}^d) \times L^2(\mathbb{R}^d)$ such that $\mathcal{A}(-t)(w(t), \partial_t w(t)) \to (w_0^+, w_1^+)$ in $\dot{H}^1(\mathbb{R}^d) \times L^2(\mathbb{R}^d)$. Thus, we also have

$$
\begin{aligned}
&(1+t)^{\frac{4}{d-2}} \left\| \mathcal{A}(-t)(w(t), \partial_t w(t)) - (w_0^+, w_1^+) \right\|_{\dot{H}^1 \times L^2} \\
&\lesssim (1+t)^{\frac{4}{d-2}} \left\| \int_t^\infty \mathcal{W}(-s) \frac{|w(s)|^{\frac{d+2}{d-2}}}{(1+s)^{\frac{4}{d-2}}} ds \right\|_{\dot{H}^1} \\
&\quad + (1+t)^{\frac{4}{d-2}} \left\| \int_t^\infty \partial_t \mathcal{W}(-s) \frac{|w(s)|^{\frac{d+2}{d-2}}}{(1+s)^{\frac{4}{d-2}}} ds \right\|_{L^2} \\
&\lesssim (1+t)^{\frac{4}{d-2}} \|w\|^{\frac{4}{d-2}}_{L^{\frac{2(d+1)}{d-2}}_{t,x}((t,\infty))} \left\| \frac{|\nabla|^{\frac{1}{2}} w}{(1+\cdot)^{\frac{4}{d-2}}} \right\|_{L^{\frac{2(d+1)}{d-1}}_{t,x}([t,\infty))} \\
&\lesssim \|w\|^{\frac{4}{d-2}}_{L^{\frac{2(d+1)}{d-2}}_{t,x}((t,\infty))} \left\| |\nabla|^{\frac{1}{2}} w \right\|_{L^{\frac{2(d+1)}{d-1}}_{t,x}([t,\infty))} \\
&\to 0,
\end{aligned}
$$

as $t \to \infty$. Since $\mathcal{A}$ is unitary in $\dot{H}^1 \times L^2$, this implies the desired statement. Q.E.D.

Proof of Theorem 1.3. Let u be a global solution to (1.1) and we assumes that u satisfies $\|(1+t)u\|_{L^{2(d+1)/(d-2)}_{t,x}([0,\infty))} < \infty$. By the transformation $w = (1+t)u$, the function w is a global solution to (1.3) satisfying $\|w\|_{L^{2(d+1)/(d-2)}_{t,x}([0,\infty))} < \infty$. From Theorem 1.2, we get

$$
\lim_{t\to\infty} (1+t)^{\frac{4}{d-2}} \left\| (w(t), \partial_t w(t)) - \mathcal{A}(t)(w_0^+, w_1^+) \right\|_{\dot{H}^1 \times L^2} = 0.
$$

Since it follows that

$$
\begin{aligned}
&(1+t)^{\frac{4}{d-2}} \left\| w(t) - \{\partial_t \mathcal{W}(t) w_0^+ + \mathcal{W}(t) w_1^+\} \right\|_{\dot{H}^1} \\
&\quad = (1+t)^{\frac{d+2}{d-2}} \left\| u(t) - (1+t)^{-1} \{\partial_t \mathcal{W}(t) w_0^+ + \mathcal{W}(t) w_1^+\} \right\|_{\dot{H}^1},
\end{aligned}
$$

we obtain the desired statement. Q.E.D.

Open Problem. At last, we give an open problem. In the present paper, we only treat $\partial_t^2 u - \Delta u + \frac{2}{1+t}\partial_t u = |u|^{\frac{d+2}{d-2}}$. In general, we should treat $\partial_t^2 u - \Delta u + \frac{\mu}{1+t}\partial_t u = |u|^p$, where $\mu > 0$. By the so-called Liouville transform $w := (1+t)^{\mu/2} u$, this equation can be rewritten as

$$
\partial_t^2 w - \Delta w + \frac{\mu(2-\mu)}{4(1+t)^2} w = \frac{|w|^p}{(1+t)^{\frac{\mu(p-1)}{2}}}, \quad (t,x) \in [0,T) \times \mathbb{R}^d. \tag{2.4}
$$

Can we show the following for general μ and p when the initial data is small?

$$\lim_{t\to\infty}(1+t)^{\frac{\mu(p-1)}{2}}\|(w(t),\partial_t w(t))-(w_l(t),\partial_t w_l(t))\|_{\dot{H}^1\times L^2}=0,$$

where $w_l(t)$ is a solution of $\partial_t^2 w_l - \Delta w_l + \frac{\mu(2-\mu)}{4(1+t)^2}w_l = 0$.

We remark that p need to be relatively large because the small data blow-up holds for $p < p_c$. See [6] and references therein.

Acknowledgement. The author was supported by JSPS Grant-in-Aid for Early-Career Scientists JP18K13444. The author would like to thank the organizers of the 11th MSJ-SI and Professor Masahito Ohta for giving me the chance of talk.

References

[1] Aynur Bulut, Magdalena Czubak, Dong Li, Nataša Pavlović and Xiaoyi Zhang, *Stability and unconditional uniqueness of solutions for energy critical wave equations in high dimensions*, Comm. Partial Differential Equations, **38** (2013), no. 4, 575–607.

[2] Marcello D'Abbicco, Sandra Lucente and Michael Reissig, *A shift in the Strauss exponent for semilinear wave equations with a not effective damping*, J. Differential Equations, **259** (2015), 5040–5073.

[3] Marcello D'Abbicco and Sandra Lucente, *NLWE with a special scale invariant damping in odd space dimension*, Dynamical Systems, Differential Equations and Applications AIMS Proceedings, (2015), 312–319.

[4] Jean Ginibre, Avraham Soffer and Giorgio Velo, *The global Cauchy problem for the critical nonlinear wave equation*, J. Funct. Anal., **110** (1992), no. 1, 96–130.

[5] Jean Ginibre, Giorgio Velo, *Generalized Strichartz inequalities for the wave equation*, J. Funct. Anal., **133** (1995), no. 1, 50–68.

[6] Masahiro Ikeda, Motohiro Sobajima and Yuta Wakasugi, *Sharp lifespan estimates of blowup solutions to semilinear wave equations with time-dependent effective damping*, J. Hyperbolic Differ. Equ., **16** (2019), no. 3, 495–517.

[7] Takahisa Inui, *Asymptotic behavior of the nonlinear damped Schrödinger equation*, Proc. Amer. Math. Soc., **147** (2019), no. 2, 763–773.

[8] Carlos E. Kenig and Frank Merle, *Global well-posedness, scattering and blow-up for the energy-critical focusing non-linear wave equation*, Acta Math., **201** (2008), no. 2, 147–212.

[9] Markus Keel and Terence Tao, *Endpoint Strichartz estimates*, Amer. J. Math., **120** (1998), no. 5, 955–980.

[10] Ning-An Lai, Hiroyuki Takamura, Kyouhei Wakasa, *Blow-up for semilinear wave equations with the scale invariant damping and super-Fujita exponent*, J. Differential Equations, **263** (2017), no. 9, 5377–5394.

[11] Alessandro Palmieri and Michael Reissig, *A competition between Fujita and Strauss type exponents for blow-up of semi-linear wave equations with scale-invariant damping and mass*, J. Differential Equations, **266** (2019), no. 2–3, 1176–1220.

[12] Hartmut Pecher, *Nonlinear small data scattering for the wave and Klein-Gordon equation* Math. Z., **185** (1984), no. 2, 261–270.

[13] Jalal Shatah, Michael Struwe, *Regularity results for nonlinear wave equations*, Ann. of Math. (2), **138** (1993), no. 3, 503–518.

Department of Mathematics,
Graduate School of Science,
Osaka University,
Toyonaka, Osaka 560-0043, Japan
E-mail address: inui@math.sci.osaka-u.ac.jp

Advanced Studies in Pure Mathematics 85, 2020
The Role of Metrics in the Theory of Partial Differential Equations
pp. 181–191

Convergence of a threshold-type algorithm for curvature-dependent motions of hypersurfaces

Katsuyuki Ishii

Abstract.

We consider a threshold-type algorithm for curvature-dependent motions of hypersurfaces. This algorithm was numerically studied by Kimura–Notsu [11], Esedoḡlu–Ruuth–Tsai [4] and Mohammad–Švadlenka [13], where they used the signed distance function as the level set function for curvature-dependent motions. In this artcle we present the convergence and its optimal rate of this algorithm, according to Ishii–Kimura [9] and Ishii–Izumi [10].

§1. Introduction

Let $\{\Gamma(t)\}_{0\le t<T}$ be a family of compact hypersurfaces in $\mathbb{R}^N$. We say this family is a curvature-dependent motion (CDM for short) if $\Gamma(t)$ moves by the following equation:

$$V = \kappa + \langle \mathbf{b}, \mathbf{n}\rangle + g \quad \text{on } \Gamma(t), t \in (0,T). \tag{1}$$

Here $T > 0$, $\mathbf{n} = \mathbf{n}(t,x)$ is the inner unit normal vector field on $\Gamma(t)$, $V = V(t,x)$ is the velocity of $\Gamma(t)$ in the direction of $\mathbf{n}$, $\kappa = \kappa(t,x)$ $(= -\operatorname{div} \mathbf{n}(t,x))$ is the $((N-1)$-times) mean curvature of $\Gamma(t)$, $\mathbf{b} = \mathbf{b}(t,x)$ denotes a given vector field in $\mathbb{R}^N$, $g = g(t,x)$ is a forcing term and $\langle \cdot,\cdot\rangle$ denotes the inner product in $\mathbb{R}^N$. As is well known, in the case of $\mathbf{b}\equiv 0$ and $g \equiv 0$, $\{\Gamma(t)\}_{0\le t<T}$ is called a mean curvature flow (MCF for short). The CDM and MCF are mathematical models of the motion of an interface by its surface tension, a transport term and an external force and they arise in various fields such as two-phase flow problems, phase transitions, image processing and so on.

Received November 30, 2018.
Revised April 8, 2019.
2010 *Mathematics Subject Classification.* 35K15, 35K55, 65M15.
Key words and phrases. curvature-dependent motions, threshold-type algorithm, convergence.

From the viewpoints of these applications, many people have studied numerical methods for CDM. Among them, we treat the following algorithm: Let C_0 be a compact set in $\mathbb{R}^N$ and fix a time step $h > 0$. For $k = 0, 1, 2, \dots$, set $\mathbf{b}_k(t, x) := b(t + kh, x)$ and $g_k(t, x) := g(t + kh, x)$. Let $w_0 = w_0(t, x)$ be a solution of the initial value problem for the linear parabolic equation with $k = 0$:

$$w_t - \Delta w + \langle \mathbf{b}_k, Dw \rangle + g_k = 0 \quad \text{in } (0, h) \times \mathbb{R}^N, \tag{2}$$

$$w(0, x) = d(x, C_k) \quad \text{for } x \in \mathbb{R}^N, \tag{3}$$

where $d(x, E)$ is the signed distance function to ∂E defined by

$$d(x, E) := \begin{cases} \operatorname{dist}(x, \partial E) & \text{for } x \in E, \\ -\operatorname{dist}(x, \partial E) & \text{for } x \notin E, \end{cases} \tag{4}$$

for each closed subset $E (\neq \emptyset)$ of $\mathbb{R}^N$. We then set

$$C_1 := \{w_0(h, \cdot) \geq 0\} (:= \{x \in \mathbb{R}^N \mid w_0(h, x) \geq 0\}). \tag{5}$$

Let w_1 be a solution of (2)–(3) with $k = 1$. Again we define C_2 as the set in (5) with w_1 replacing w_0. Repeating this process, we have a sequence $\{C_k\}_{k=0}^{[T/h]}$ of compact subsets of $\mathbb{R}^N$. Here $[r]$ is the Gauss symbol for $r \geq 0$. Taking a limit of $\{C_k\}_{k=0}^{[T/h]}$ as $h \to 0$ and $kh \to t$, we formally obtain a flow $\{C(t)\}_{0 \leq t < T}$ of compact sets in $\mathbb{R}^N$ and observe that $\partial C(t)$ moves by (1) starting from $\partial C(0) = \partial C_0$. In the following of this article we call the above algorithm (ALGO).

The (ALGO) was numerically studied by Kimura–Notsu [11]. Esedoḡlu–Ruuth–Tsai [4] and Mohammand–Švadlenka [13] numerically considered various geometric motions. On the other hand, the (ALGO) is regarded as a variant of the Bence–Merriman–Osher (BMO for short) algorithm (cf. [3]). The convergence and generalizations of the BMO algorithm have been studied by many people. See, e.g., [1], [5], [12], [14] etc.

The main difference between the BMO algorithm and (ALGO) is the choice of the initial data. In the (generalized) BMO algorithm they choose the initial data

$$w(0, x) = \begin{cases} 1 & \text{for } x \in C_k, \\ -1 & \text{for } x \notin C_k, \end{cases}$$

instead of (3).

The main purpose of this article is to discuss the convergence of (ALGO) to the generalized CDM and the optimal rate of convergence to the smooth and compact CDM. The strategy in proving the convergence is based on the asymptotic analysis of the Allen–Cahn equation. See Barles–Souganidis [2] and Leoni [12]. The point in deriving the rate of convergence is to construct sub- and super-solutions of (2) which approximate the solution of (2)–(3). In the proof of the optimality, we treat the circle in the plane evolving by curvature and use an explicit solution of the level-set equation to MCF.

In Section 2, we state our assumptions, the solution of (2)–(3) and the smooth and compact CDM. We briefly present the convergence of (ALGO) to the generalized CDM in Section 3 and the rate of convergence and its optimality in Section 4. See Ishii–Kimura [9] and Ishii–Izumi [10] for details.

§2. Preliminaries

We make the following assumptions:

(A0) Γ_0 is a compact hypersurface in $\mathbb{R}^N$ and is of class C^4.

(A1) $\mathbf{b}$, $g \in C^{1,2}([0,+\infty)\times\mathbb{R}^N)\cap W^{1,2,\infty}((0,+\infty)\times\mathbb{R}^N)$.

It follows from [7] that under (A1) and $\|Dd(\cdot, C_k)\|_{L^\infty(\mathbb{R}^N)} = 1$, there is a unique classical solution w_k of (2)–(3) satisfying

$$\sup_{\substack{(t,x)\in[kh(k+1)h)\times\mathbb{R}^N \\ k=0,1,2,\dots,[T/h]-1,h>0}} \frac{|w_k(t,x)|}{1+|x|} < +\infty$$

for $h > 0$ and $k = 0, 1, \dots, [T/h] - 1$. By (A0) and (A1) we see that for some $T_0 > 0$, there uniquely exists a smooth and compact CDM $\{\Gamma(t)\}_{0\le t<T}$ with the initial data $\Gamma(0) = \partial C_0$ (cf. [6], [9]).

For any fixed $m \in \mathbb{N}$ and $y \in \mathbb{R}^N$, we define $v_m = v_m(t,x;y) = \sqrt{|x-y|^2 + 2mt}$ for $(t,x) \in [0,+\infty)\times\mathbb{R}^N$. Then

$$v_{m,t} - \Delta v_m \begin{cases} \le 0 & \text{if } m \le N-1, \\ \ge 0 & \text{if } m \ge N-1, \end{cases} \quad \text{in } (0,+\infty)\times\mathbb{R}^N. \tag{6}$$

The v_N is used in section 3 to construct sub- and super-solutions of (2). Note that for each $R > 0$, the set $\{v_{N-1}(t,\cdot;0) = R\}$ is a sphere evolving by mean curvature and shrinks to a point at $t = R^2/2(N-1)$.

§3. Convergence to the generalized CDM

In this section we breifly discuss the convergence of (ALGO) to the genralized CDM.

Let $u \in C([0,T) \times \mathbb{R}^N)$ be a unique viscosity solution of the initial value problem of the level-set equation to (1)

$$
(7) \quad \begin{cases} u_t - \Delta u + \dfrac{\langle D^2 u Du, Du\rangle}{|Du|^2} + \langle \mathbf{b}(t,x), Du\rangle + g(t,x)|Du| = 0 \\ \hfill \text{in } (0,T) \times \mathbb{R}^N, \\ u(0,x) = d(x, C_0) \quad \text{for } x \in \mathbb{R}^N. \end{cases}
$$

See Giga [8] for the well-posedness of this problem in the sense of viscosity solutions. Define $C(t) := \{u(t,\cdot) \geq 0\}$ for $t \in [0,T)$ and $C^h(t) = C_k$ for $t \in [kh, (k+1)h)$, $k = 0, 1, 2, \ldots, [T/h]$ and $h > 0$. Here $\{C_k\}_{k=0}^{[T/h]}$ is a sequence of compact sets constructed by (ALGO).

The convergence of (ALGO) is stated as follows.

Theorem 1 (Ishii–Kimura [9]). *Assume that* (A1) *holds and that* C_0 *is a compact set in* $\mathbb{R}^N$. *Let* $u \in C([0,T) \times \mathbb{R}^N)$ *be a unique viscosity solution of* (7) *and assume that*

$$
(8) \qquad \{u(t,\cdot) \geq 0\} = \overline{\{u(t,\cdot) > 0\}} \quad \text{for each } t \in (0,T).
$$

Then for any $\varepsilon > 0$,

$$
\lim_{h \to 0} \max_{t \in [0; T-\varepsilon]} d_H(\partial C(t), \partial C^h(t)) = 0.
$$

Here d_H *is the Hausdorff distance for sets* A, $B \in \mathbb{R}^N$, *defined by*

$$
d_H(A; B) := \max\left\{ \sup_{x \in A} \operatorname{dist}(x; B), \sup_{x \in B} \operatorname{dist}(x, A) \right\}.
$$

As mentioned in Introduction, the proof of this theorem is done by some modifications of the asymptotic analysis of the Allen–Cahn equation (cf. [12]). In the following of this section we always assume (A1).

We consider some semi-continuous limits. Set

$$
(9) \qquad d^h(t,x) := d(x, C^h(t)), \ w^h(t,x) := w_k(t - kh, x),
$$

for $(t,x) \in [kh, (k+1)h) \times \mathbb{R}^N$, $k = 0, 1, 2, \ldots, [T/h]$ and $h > 0$. Here $d(x, C^h(t))$ is given by (4) with $E = C^h(t)$ and w_k is the solution of (2)–(3). For $f^h = d^h$, w^h, define

$$
\overline{f}(t;x) := \limsup_{(h,s,y) \to (0,t,x)} f^h(s,y), \ \underline{f}(t;x) := \liminf_{(h,s,y) \to (0,t,x)} f^h(s,y)
$$

and set $\Omega^+(t) := \{\underline{d}(t,\cdot) > 0\}$, $\Omega^-(t) := \{\underline{d}(t;\cdot) < 0\}$. We have some properties of $\overline{d}$, $\underline{d}$, $\overline{w}$ and $\underline{w}$.

Proposition 1 (Ishii–Kimura [9]). (i) $\overline{d} = \overline{w}$ *and* $\underline{d} = \underline{w}$ *in* $[0,T) \times \mathbb{R}^N$.

(ii) *If* $\partial\Omega^+(t) = \partial\Omega^-(t)$ *for each* $t \in [0,T)$, *then*

$$\overline{w} = \overline{d} = \underline{w} = \underline{d} = 0 \quad \text{on} \quad \bigcup_{0 \le t < T} [\{t\} \times \partial\Omega^+(t)].$$

(iii) $\overline{w}(0,\cdot) = \underline{w}(0,\cdot) = \overline{d}(0,\cdot) = \underline{d}(0,\cdot) = d(\cdot, C_0)$ *in* $\mathbb{R}^N$.

Theorem 2 (Ishii–Kimura [9]). *The* $\overline{d}$ *and* $\underline{d}$ *satisfy, respectively,*

$$|D\overline{d}| - 1 \le 0 \quad \text{in } \{\overline{d} > 0\}, \ -|D\overline{d}| + 1 \le 0 \quad \text{in } \{\overline{d} < 0\},$$
$$|D\underline{d}| - 1 \ge 0 \quad \text{in } \{\underline{d} > 0\}, \ -|D\underline{d}| + 1 \ge 0 \quad \text{in } \{\underline{d} < 0\},$$

in the viscosity sense. Here $\{f > 0\} := \{(t,x) \in [0,T) \times \mathbb{R}^N \mid f(t,x) > 0\}$ *etc. for every function* f *defined in* $[0,T) \times \mathbb{R}^N$.

The sets $\Omega^+(t)$ and $\Omega^-(t)$ satisfy the following:

Theorem 3 (Ishii–Kimura [9]). *Assume* (8). *The* $\{\Omega^+(t)\}_{0 \le t < T}$ *and* $\{\mathbb{R}^N \backslash \overline{\Omega^-(t)}\}_{0 \le t < T}$ *are, respectively, a generalized superflow and a generalized subflow in the sense of Barles–Souganidis* [2]. *Moreover, the* $\{\Omega^+(t)\}_{0 \le t < T}$ *and* $\{\Omega^-(t)\}_{0 \le t < T}$ *satisfy*

$$\operatorname{int} C(t) \subset \Omega^+(t) \subset C(t), \ \mathbb{R}^N \backslash C(t) \subset \Omega^-(t) \subset \mathbb{R}^N \backslash \operatorname{int} C(t),$$

for any $t \in [0,T)$.

Outline of the proof of Theorem 1. It follows from (8) and Theorem 3 that $\operatorname{int} C(t) = \Omega^+(t) = \{\underline{d}(t,\cdot) > 0\}$, $\mathbb{R}^N \backslash C(t) = \Omega^-(t) = \{\overline{d}(t,\cdot) < 0\}$ and $\partial C(t) = \partial\Omega^+(t) = \partial\Omega^-(t)$. Then we apply Proposition 1 and Theorem 2 to obtain

$$\overline{d}(t,x) = \underline{d}(t,x) = d(t,x) := \begin{cases} \operatorname{dist}(x, \partial C(t)) & (x \in C(t)), \\ -\operatorname{dist}(x, \partial C(t)) & (x \notin C(t)), \end{cases}$$

and $d \in C([0,T) \times \mathbb{R}^N)$. Hence we get $\overline{w} = \underline{w} = d$ and

$$d^h, \ w^h \longrightarrow d \quad (h \to 0) \text{ locally uniformly in } [0,T) \times \mathbb{R}^N.$$

Since we observe that $C^h(t) = C_k = \{w^h(kh,\cdot) \ge 0\}$ and $C(t) = \{d(t,\cdot) \ge 0\}$, we obtain the result. Q.E.D.

§4. Rate of convergence

This section is devoted to the rate of convergence of (ALGO) to the smooth and compact CDM and its optimality.

4.1. Rate of convergence

In this subsection we always assume (A0) and (A1). Let $\{\Gamma(t)\}_{0\le t<T_0}$ be a smooth and compact CDM starting from Γ_0 and C_0 a compact set in $\mathbb{R}^N$ such that $\partial C_0 = \Gamma_0$. Let $D(t) \subset \mathbb{R}^N$ be a compact set enclosed by $\Gamma(t)$ and let $d = d(t,x)$ be defined by (4) with $E = D(t)$. In addition, we slightly modify $C^h(t)$ given in Section 3. Define

$$C^h(t) := \{w^h(t,\cdot) \ge 0\} \quad \text{for } t \in [0,T) \text{ and } h > 0,$$

where w^h is given by (9).

Theorem 4 (Ishii–Kimura [9], Ishii–Izumi [10]). *Assume* (A0) *and* (A1). *For any $\varepsilon > 0$, there exist L_0, $h_0 > 0$ such that*

$$\max_{t\in[0;T_0-\varepsilon]} d_H(\Gamma(t); \partial Ch(t)) \le L_0 h \quad \textit{for all } h \in (0,h_0). \tag{10}$$

The point of the proof of this theorem is to construct sub- and super-solutions of (2) which approximate the solution w_k of (2)–(3) near $\Gamma(t)$. Set $\mathbf{b} \equiv 0$ and $g \equiv 0$ to simplify the arguments. Thus $\{\Gamma(t)\}_{0\le t<T_0}$ is the smooth and compact MCF starting from Γ_0 and w_k is the solution of the heat equation satisfying (3).

Fix any $\varepsilon \in (0, T_0/3)$. We observe by the smoothness of $\{\Gamma(t)\}_{0\le t<T_0}$ that there exists a $\delta > 0$ such that $d \in C^{2,4}(\mathcal{N}_{\varepsilon,30\delta})$, where $\mathcal{N}_{\varepsilon,30\delta} := \cup_{t\in[0;T_0-\varepsilon]}[\{t\} \times \{|d(t,\cdot)| \le 30\delta\}]$. Besides it is seen that

$$|d_t - \Delta d - \kappa^s d| \le K_1 d^2 \quad \text{on } \mathcal{N}_{\varepsilon,30\delta}, \tag{11}$$

where κ^s is the sum of squares of all the principal curvatures at of $\Gamma(t)$ and $K_1 > 0$ is a constant (cf. [10]).

For any $\varepsilon \in (0, T_0/3)$, $h \in (0,1)$ and $k = 0, 1, \dots, [(T_0 - \varepsilon)/h] - 1$, we set

$$\begin{aligned}
&f_k = f_k(t,x) = f(t + kh, x) \quad \text{for } f = d \text{ and } \kappa^s,\\
&\underline{d}_k := (1 - t\kappa^s_k)d_k - K_1 t d_k^2 - K_2 t^2 - \alpha_k,\\
&\overline{d}_k := (1 - t\kappa^s_k)d_k + K_1 t d_k^2 + K_2 t^2 + \alpha_k,\\
&\mathcal{M}_k := \{(t,x) \mid 0 \le t < h,\ |d_k(0,x)| \le 15\delta\}.
\end{aligned}$$

Here $K_2 > 0$ and $\alpha_k \ge 0$ are selected later. Since $|d(kh,x) - d(t + kh, x)| \le \|d_t\|_{L^\infty(\mathcal{N}_{\varepsilon,30\delta})} t$ for $(t,x) \in \mathcal{M}_k$ and $k = 0, 1, 2, \dots, [(T_0 - \varepsilon)/h] - 1$, we are able to find a $t_1 > 0$ such that $\mathcal{M}_k \subset \mathcal{N}_{\varepsilon,30\delta}$ and $d_k \in C^{2,4}(\mathcal{M}_k)$ for $k = 0, 1, \dots, [(T_0 - \varepsilon)/h] - 1$ and $h \in (0, t_1)$. Since d satisfies (11), we are able to obtain the following lemma.

Lemma 1 (Ishii–Izumi [10]). *For large $K_2 > 0$, $\underline{d}_k$ and $\overline{d}_k$ are, respectively, a classical subsolution and a classical supersolution of the heat equation in $\mathcal{M}_k$ for all $k = 0, 1, \ldots, [(T_0 - \varepsilon)/h] - 1$ and $h \in (0, t_1)$. In addition, if $d_H(C^h(kh), C(kh)) \leq \alpha_k$ for some $\alpha_k \in [0, \delta)$, then $\underline{d}_k(0, \cdot) \leq d(\cdot, C^h(kh)) \leq \overline{d}_k(0, \cdot)$ in $\mathbb{R}^N$.*

Applying (6) with $m = N$ and Perron's method for viscosity solutions, we extend $\underline{d}_k$ (resp., $\overline{d}_k$) as a viscosity subsolution $\underline{w}_k$ (resp., a viscosity supersolution $\overline{w}_k$) of the head equation in $(kh, (k+1)h) \times \mathbb{R}^N$ satisfying $\underline{w}_k(0, \cdot) \leq d(\cdot, C(kh)) \leq \overline{w}_k(0, \cdot)$ in $\mathbb{R}^N$. The following is the outline of the extension of $\underline{d}_k$. Let $\underline{\rho}_k = \underline{\rho}_k(t, x)$ and $\underline{w}_k^{\pm} = \underline{w}_k^{\pm}(t, x; y)$ be functions defined by

$$\begin{aligned}
\underline{\rho}_k(t, x) &:= \underline{d}_k(t; x) - C_1(t, x) - \alpha_k, \\
\underline{w}_k^{+}(t, x; y) &:= 10\delta - v_N(t, x; y) - C_2(t, x) - \alpha_k, \\
\underline{w}_k^{-}(t, x; y) &:= -6\delta - v_N(t, x; y) - C_3(t, x) - \alpha_k.
\end{aligned}$$

Here $C_i(t, x)$ $(i = 1, 2, 3)$ is a correction term such that $\underline{\rho}_k$ and $\underline{w}_k^{+}$ are classical subsolutions of the heat equation. Define

$$\underline{w}_k = \underline{w}_k(t, x) : \begin{cases}
\sup_{y \in \{d_k(0,\cdot) \geq 10\delta\}} \underline{w}_k^{+}(t, x; y) \\
\qquad \text{for } (t, x) \in [0, h] \times \{d_k(0, \cdot) \geq 9\delta\}, \\
\max\left\{\underline{\rho}_k(t, x), \sup_{y \in \{d_k(0,\cdot) = 10\delta\}} \underline{w}_k^{+}(t, x; y)\right\} \\
\qquad \text{for } (t, x) \in [0, h] \times \{3\delta \leq d_k(0, \cdot) < 9\delta\}, \\
\underline{\rho}_k(t, x) \quad \text{for } (t, x) \in [0; h] \times \{|d_k(0, \cdot)| < 3\delta\}, \\
\max\left\{\underline{\rho}_k(t, x), \sup_{y \in \{d_k(0,\cdot) = -10\delta\}} \underline{w}_k^{-}(t, x; y)\right\} \\
\qquad \text{for } (t, x) \in [0, h] \times \{-12\delta \leq d_k(0, \cdot) < -3\delta\}, \\
\sup_{y \in \{d_k(0,\cdot) = 10\delta\}} \underline{w}_k^{-}(t, x; y) \\
\qquad \text{for } (t, x) \in [0, h] \times \{d_k(0, \cdot) < -12\delta\}.
\end{cases}$$

Then $\underline{w}_k$ is a viscosity subsolution of the heat equation in $(0, h) \times \mathbb{R}^N$. The extension of $\overline{d}_k$ is similar. See [10] for details.

By the comparison principle we have $\underline{w}_k \leq w_k \leq \overline{w}_k$ in $[0, h) \times \mathbb{R}^N$ for all $k = 0, 1, 2, \ldots, [(T_0 - \varepsilon)/h] - 1$ and $h > 0$. We can show the following estimate by use of the explicit form $\underline{w}_k$ and $\overline{w}_k$ near $\Gamma(t)$ and this inequality.

Proposition 2 (Ishii–Kimura [9]). *For any $\varepsilon > 0$, there exist $K_3 > 0$ and $t_2 \in (0, t_1)$ such that for any $h \in (0, t_2)$ and $k = 0, 1, \ldots, [(T_0 -$*

$\varepsilon)/h] - 1$, *if* $d_H(C^h(kh), C(kh)) \le \alpha_k$ *for some* $\alpha_k \in [0, \delta)$, *then*

$$d_H(C^h(t+kh), C(t+kh)) \le \frac{K_3 t^2 + \alpha_k}{1 - K_3 t} \quad \textit{for } \ t \in [0, h].$$

Outline of the proof of Theorem 4. Let $L_0 := (e^{K_3 T_0} - 1)$ and $h_0 := \min\{1/K_3, \delta, \delta/L, t_2\}/2$. We apply Proposition 2 with $k = 0$ and $\alpha_0 = 0$ to have

$$d_H(C^h(t), C(t)) \le \frac{K_3 t^2}{1 - K_3 t} (< \delta) \quad \text{for all } t \in [0, h] \text{ and } h \in (0, h_0).$$

For $k = 1, 2, \ldots, [(T_0 - \varepsilon)/h] - 1$, we inductively use Proposition 2 to obtain

$$d_H(C^h(t+kh), C(t+kh)) \le \frac{K_3 t^2 + \alpha_k}{1 - K_3 t} \quad \text{for all } t \in [0, h] \text{ and } h \in (0, h_0),$$

where $\alpha_k := (K_3 h^2 + \alpha_{k-1})/(1 - K_3 h) (< \delta)$. Since we calculate that

$$\begin{aligned}\frac{K_3 t^2 + \alpha_k}{1 - K_3 t} &\le \sum_{l=1}^{k+1} \frac{K_3 h^2}{(1 - K_3 h)^l} \le \left\{ \left(\frac{1}{1 - K_3 h} \right)^{[T_0/h]} - 1 \right\} h \\ &\le (e^{K_3 T_0} - 1) h \le L_0 h (< \delta)\end{aligned}$$

for all $t \in [0, h]$, $k = 0, 1, \ldots, [(T_0 - \varepsilon)/h] - 1$ and $h \in (0, h_0)$, we get the desired result. Q.E.D.

4.2. Optimality

In this subsection we discuss the optimality of (10) with respect to the order h. For this purpose we consider the planar and radial case. For simplicity, we set $R(t) := \sqrt{1 - 2t}$, $T_0 := 1/2$, $C(t) := \{x \in \mathbb{R}^2 \mid |x| \le R(t)\}$, $\mathbf{b} \equiv 0$ and $g \equiv 0$. Note that $\partial C(t)$ moves by curvature and shrinks to a point at T_0. Since it suffices to consider the radial solution, we treat the following problem:

$$w_{k,t} - w_{k,rr} - \frac{w_{k,r}}{r} = 0 \quad \text{in } (0, h] \times (0, +\infty), \tag{12}$$

$$w_{k,r}(t, 0) = 0 \quad \text{for } \ t > 0, \tag{13}$$

$$w_k(0, r) = R_k - r \quad \text{for } \ r \in [0, +\infty), \tag{14}$$

$$\begin{aligned}&C^h(t) := \{w_k(t - kh, |\cdot|) \ge 0\} \\ &\qquad \text{for } \ t \in [kh, (k+1)h), \ k = 0, 1, 2, \ldots, [T_0/h] - 1, \\ &C^h(0) = C_0 := \overline{B(0, 1)}, \\ &R^h(t) := \text{radius of } \ C^h(t), \ R_k := R^h(kh), \ R^h(0) = R_0 := 1.\end{aligned}$$

It follows from the maximum principle that $C^h(t) \subset C(t)$ for all $t \in [0, T_0)$ and $h > 0$. Thus $d_H(\partial C(t), \partial C^h(t)) = R(t) - R^h(t)$.

Theorem 5 (Ishii–Izumi [10]). *For any small $\varepsilon > 0$, there exist $L_1 > 0$, $h_1 > 0$ such that for all $h \in (0, h_1)$,*

$$|R^h(t) - \{R(t) - t^2\}| \le L_1 t^{5/2} \quad \textit{for } t \in [0, h],$$
$$\left| R^h(t) - \left\{ R(t) + \frac{h \log R(t)}{R(t)} \right\} \right| \le \frac{L_1 h^{3/2}}{R(t)} \quad \textit{for } t \in [h, T_0 - \varepsilon].$$

Remark 1. In Ishii–Kimura [9] we have already obtained $R(t) - R^h(t) \ge th/4$ for all $t \in [1/4, 1/2 - \varepsilon]$ and small $h > 0$. This estimate also assures the optimality of (10). The above theorem is a refinement of this lower estimate.

The strategy of the proof of Theorem 5 is similar to that of Theorem 4. The difference is to employ an explicit solution of the level-set equation to MCF.

Set $\rho(t, r) = 1 - \sqrt{r^2 + 2t}$ and $\widetilde{\rho}(t, x) := \rho(t, |x|)$. Then $C(t) = \{\widetilde{\rho}(t, \cdot) = 0\}$ and these functions satisfy

$$\widetilde{\rho}_t - \Delta\widetilde{\rho} + \frac{\langle D^2\widetilde{\rho} D\widetilde{\rho}, D\widetilde{\rho}\rangle}{|D\widetilde{\rho}|^2} = 0 \quad \text{in } (0, \infty) \times \mathbb{R}^N,$$
$$\rho_t - \rho_{rr} - \frac{1}{r}\rho_r \ge 0 \quad \text{in } (0, +\infty) \times (0, +\infty),$$
$$\rho_r(t, 0) = 0 \quad \text{for } t \in (0, +\infty).$$

For $t \in [0, h)$ and $r \in [0, +\infty)$, set

$$\underline{\rho}_k(t, r) := R^h(kh) - \sqrt{r^2 + 2t} - \frac{t^2}{(r^2 + 2t)^{3/2}} \quad (R^h(0) = 1).$$

Modifying $\underline{\rho}_k$, we have subsolutions and supersolutions of (12).

Lemma 2 (Ishii–Izumi [10]). *Take any small $\varepsilon > 0$. Then there exist t_3, $K_4 > 0$ such that for any $h \in (0, t_3)$, $\underline{w}_k := \underline{\rho}_k - K_4 t^{5/2}$ (resp., $\overline{w}_k := \underline{\rho}_k + K_4 t^{5/2}$) is a classical subsolution (resp, a classial supersolution) of (12) in $(0, h) \times (\varepsilon/2, +\infty)$.*

We modify $\underline{w}_k$ and $\overline{w}_k$ in $[0, h) \times [0, \varepsilon/2)$ with using the theory of viscosity solutions and still denote by wk and wk respectively. Then $\underline{w}_k$ (resp., $\overline{w}_k$) is a viscosity subsolution (resp., a viscosity supersolution) of (12) satisfying (13) in the classical sense and (14). Therefore by the comparison principle for viscosity solutions we have, in $[0, h) \times (\varepsilon, +\infty)$

$$\left| w_k(t,r) - \left\{ \rho_k(t,r) - \frac{t2}{(r^2+2t)^{3/2}} \right\} \right| \le K_4 t^{5/2}. \tag{15}$$

Outline of the proof of Theorem 5. We observe from (10) that for any small $\varepsilon > 0$, $R(t) \ge R^h(t) \ge R(t) - L_0 h \ge \sqrt{\varepsilon}(> \varepsilon)$ for all $t \in [0, T_0 - \varepsilon]$ and small $h > 0$. Thus we can use (15). Lengthy and careful calculations with using (15) yield that

$$|R^h(t) - \{R(t) - t^2\}| \le K_{12} t^{5/2} \quad \text{for } t \in [0,h],$$

$$\left| R(t) - R^h(t) - \frac{h}{R(t)} \int_0^t \frac{1}{(R(s))^2} ds \right| \le \frac{L_1 h^{3/2}}{R(t)} \quad \text{for } t \in [h, T_0 - \varepsilon].$$

Therefore we have the desired result. Q.E.D.

References

[1] G. Barles and C. Georgelin, A simple proof of convergence for an approximation scheme for computing motion by mean curvature, SIAM J. Numer. Anal., **32** (1995), 484–500,

[2] G. Barles and P. E. Souganidis, A new approach to front propagation problems: theory and applications, Arch. Rational Mech. Anal., **141** (1998), 237–296.

[3] J. Bence, B. Merriman and S. Osher, Diffusion generated motion by mean curvature, in "Computational Crystal Growers Workshop", J. Taylor ed. Selected Lectures in Math., Amer. Math. Soc., Province, 1992.

[4] S. Esedoḡlu, S. J. Ruuth and R. Tsai, Threshold dynamics for high order geometric motions, Interfaces Free Bound, **10** (2008), 263–282.

[5] L. C. Evans, Convergence of an algorithm for mean curvature motion, Indiana Univ. Math. J., **42** (1993), 533–557.

[6] L. C. Evans and J. Spruck, Motion of level sets by mean curvature II, Trans. Amer. Math. Soc., **330** (1992), 321–332.

[7] A. Friedman, Partial Differential Equations of Parabolic Type, Printice-Hall, New Jersey, 1964.

[8] Y. Giga, Surface Evolution Equations, Birkhäuser Basel/Boston/Berlin, 2006.

[9] K. Ishii and M. Kimura, Convergence of a threshold-type algorithm using the signed distance function, Interfaces Free Bound., **18** (2016), 479–522.

[10] K. Ishii and T. Izumi, Remarks on the convergence of an algorithm for curvature-dependent motions of hypersurfaces, Discrete Conti. Dyn. Syst., Ser. A, **38** (2018), 1103–1125.

[11] M. Kimura and H. Notsu, A level set method using the signed distance function, Japan J. Indust. Appl. Math., **19** (2002), 415–446.

[12] F. Leoni, Convergence of an approximation scheme for curvature-dependent motion of sets, SIAM J. Numer. Anal., **39** (2001), 1115–1131.

[13] R. Z. Mohammad and K. Švadlenka, Multiphase volume-preserving interface motion via localized signed distance vector scheme, Discrete Conti. Dyn. Syst., Ser. S., **8** (2015), 969–988.

[14] L. Vivier, Convergence of an approximation scheme for computing motions with curvature dependent velocities, Differential Integral Equations, **13** (2000), 1263–1288.

Graduate School of Maritime Sciences, Kobe University
Higashinada, Kobe 658-0022, Japan
E-mail address: ishii@maritime.kobe-u.ac.jp

Advanced Studies in Pure Mathematics 85, 2020
The Role of Metrics in the Theory of Partial Differential Equations
pp. 193–203

Spectral theory on manifolds

Kenichi Ito and Erik Skibsted

§1. Basic settings

This is a short and slightly simplified report on our article [IS2], in which we studied spectral theory for the Schrödinger operator

$$H = H_0 + V \quad \text{on } M; \quad H_0 = -\frac{1}{2}\Delta = \frac{1}{2}p_i^* g^{ij} p_j, \; p_i = -\mathrm{i}\partial_i.$$

Here (M, g) is a connected Riemannian manifold with Euclidean and/or hyperbolic ends, and V is a real-valued and bounded potential. Self-adjointness of H and H_0 on $\mathcal{H} = L^2(M)$ is realized by the Friedrichs extension. Our main results are Rellich's theorem, the LAP (Limiting Absorption Principle), radiation condition bounds and the Sommerfeld uniqueness result. Proofs depend intensively on commutator arguments with different conjugate operator from the Mourre theory, but we will omit them in this article. Based on the results here, stationary scattering theory and a characterization of asymptotics of generalized eigenfunction are developed in [IS3].

1.1. End structure

Now let us more precisely introduce our basic settings. We formulate and control end structure in a somewhat implicit manner. For examples of manifolds satisfying conditions below we refer to Section 3.

Condition 1. Let (M, g) be a connected Riemannian manifold of dimension $d \geq 1$. There exist $r \in C^\infty(M)$ and $r_0 \geq 2$ such that:

(1) The image of r is $r(M) = [1, \infty)$;

Received November 25, 2018.
Revised November 28, 2018.
2010 *Mathematics Subject Classification.* Primary 81U05; Secondary 47A40, 58J50.
Key words and phrases. spectral theory, Schrödinger operator, Riemannian manifold.

(2) The gradient vector field $\omega = \operatorname{grad} r \in \mathfrak{X}(M)$ is forward complete in the sense that its integral curves are defined for any initial point $x \in M$ and any non-negative time parameter $t \geq 0$;
(3) The identity $|\nabla r| = 1$ holds on $\{x \in M \,|\, r(x) > r_0/2\}$.

Each component of $E = \{x \in M \,|\, r(x) > r_0\}$ is called an *end* of M, and r models a distance function there. In fact, we can canonically construct the *spherical coordinates* on E with r being the radius function. Note that here Condition 1 (3) is simplified compared with [IS2].

Let us impose more assumptions on geometry of E. Let us set

$$\ell = g - \mathrm{d}r \otimes \mathrm{d}r,$$

which is exactly the "spherical part" of g. Let us also recall a local expression of the Levi–Civita connection ∇: Denote the Christoffel symbol by $\Gamma^k_{ij} = \frac{1}{2} g^{kl}(\partial_i g_{lj} + \partial_j g_{li} - \partial_l g_{ij})$, and then for any smooth function f

$$(\nabla f)_i = (\nabla_i f) = \partial_i f, \quad (\nabla^2 f)_{ij} = \partial_i \partial_j f - \Gamma^k_{ij} \partial_k f.$$

Note that $\nabla^2 f$ is the geometric Hessian of f.

Condition 2. There exist $\sigma, \tau, C > 0$ such that globally on M

$$r\nabla^2 r \geq \frac{1}{2}\sigma\ell - Cr^{-\tau} g \tag{1}$$

as quadratic forms on fibers of the tangent bundle of M, and

$$\Delta r \leq C, \quad \left|\ell^{\bullet i}\nabla_i \Delta r\right| \leq Cr^{-1-\tau/2}.$$

Condition 2 implicitly controls the volume growth rate of E as $r \to \infty$, admitting both asymptotically Euclidean and asymptotically hyperbolic ends. In fact, $\nabla^2 r$ can be identified with the second fundamental form of an r-sphere, and hence in particular Δr with its mean curvature.

Finally we impose a long-range type condition on V. With volume growth of E taken into account, it is more useful to consider an *effective potential*

$$q = V + \frac{1}{8}\big[(\Delta r)^2 + 2\nabla^r \Delta r\big]; \quad \nabla^r = g^{ij}(\nabla_i r)\nabla_j.$$

Such a quantity naturally shows up, if we rewrite H as

$$H = \frac{1}{2}A^2 + \frac{1}{2}L + q.$$

Here we have introduced "radial" and "spherical" differential operators

$$(2)\qquad A = \operatorname{Re} p^r = \frac{1}{2}\big(p^r + (p^r)^*\big), \quad L = p_i^* \ell^{ij} p_j; \quad p^r = -\mathrm{i}\nabla^r.$$

Note that L can be identified with the spherical part of $-\Delta$.

Condition 3. The potential $V \in L^\infty(M)$ is real-valued. Moreover, there exists a splitting by real-valued functions:

$$q = q_1 + q_2; \quad q_1 \in C^1(M) \cap L^\infty(M),\ q_2 \in L^\infty(M),$$

such that for some $\rho', C > 0$ the following bounds hold globally on M:

$$\nabla^r q_1 \le C r^{-1-\rho'}, \quad |q_2| \le C r^{-1-\rho'}.$$

1.2. Function spaces

Here we introduce several function spaces needed to state our main theorems. Define the *Agmon–Hörmander spaces* as

$$\mathcal{B} = \Big\{\psi \in L^2_{\mathrm{loc}}(M) \,\Big|\, \|\psi\|_{\mathcal{B}} := \sum_{\nu=0}^{\infty} 2^{\nu/2} \|F_\nu \psi\|_{\mathcal{H}} < \infty \Big\},$$

$$\mathcal{B}^* = \Big\{\psi \in L^2_{\mathrm{loc}}(M) \,\Big|\, \|\psi\|_{\mathcal{B}^*} := \sup_{\nu \ge 0} 2^{-\nu/2} \|F_\nu \psi\|_{\mathcal{H}} < \infty \Big\},$$

$$\mathcal{B}^*_0 = \Big\{\psi \in \mathcal{B}^* \,\Big|\, \lim_{\nu\to\infty} 2^{-\nu/2} \|F_\nu \psi\|_{\mathcal{H}} < \infty \Big\},$$

where F_ν denotes a characteristic function

$$F_\nu = F\big(\{x \in M \,|\, 2^\nu \le r(x) < 2^{\nu+1}\}\big) \ \text{ for } \ \nu = 0, 1, 2, \ldots.$$

Here $F(\Omega)$ is characteristic function of a subset $\Omega \subseteq M$. We remark that for any $s > 1/2$ we have the proper inclusions

$$\mathcal{H}_s \subsetneq \mathcal{B} \subsetneq \mathcal{H}_{1/2} \subsetneq \mathcal{H} \subsetneq \mathcal{H}_{-1/2} \subsetneq \mathcal{B}^*_0 \subsetneq \mathcal{B}^* \subsetneq \mathcal{H}_{-s}; \quad \mathcal{H}_s = r^{-s}\mathcal{H}.$$

Choose $\chi \in C^\infty(\mathbb{R})$ such that

$$\chi(t) = \begin{cases} 1 & \text{for } t \le 1, \\ 0 & \text{for } t \ge 2, \end{cases} \quad \chi \ge 0, \quad \chi' \le 0,$$

and define smooth cut-off functions $\chi_n, \bar\chi_n, \chi_{m,n} \in C^\infty(M)$ for $n > m \ge 0$ as

$$\chi_n = \chi(r/2^n), \quad \bar\chi_n = 1 - \chi_n, \quad \chi_{m,n} = \bar\chi_m \chi_n.$$

Then let us introduce an auxiliary space

$$\mathcal{N} = \{\psi \in \mathcal{H}_{\mathrm{loc}} \,|\, \chi_n \psi \in \mathcal{H}^1 \text{ for all } n \geq 0\}; \quad \mathcal{H}^s = (H_0 + 1)^{-s/2}\mathcal{H}.$$

We note that $\mathcal{N}$ is a space of functions that satisfy the Dirichlet boundary condition, possibly with infinite $\mathcal{H}^1$-norm on M.

§2. Results

2.1. Rellich's theorem

Our first theorem is Rellich's theorem, absence of generalized eigenfunctions in $\mathcal{B}_0^*$ with eigenvalues above *critical energy*

$$\lambda_H = \limsup_{r\to\infty} q_1 = \lim_{R\to\infty} \bigl(\sup\{q_1(x) \,|\, r(x) \geq R\}\bigr) < \infty.$$

In many examples we have $\sigma_{\mathrm{ess}}(H) = [\lambda_H, \infty)$, see [K1].

Theorem 4. *Suppose Conditions* 1–3, *and let* $\lambda > \lambda_H$. *If* $\phi \in L^2_{\mathrm{loc}}(M)$ *satisfies*

(1) $(H - \lambda)\phi = 0$ *in the distributional sense,*

(2) $\bar{\chi}_m \phi \in \mathcal{B}_0^* \cap \mathcal{N}$ *for some* $m \geq 0$,

then $\phi = 0$ *on* M.

Corollary 5. *One has* $\sigma_{\mathrm{pp}}(H) \cap (\lambda_H, \infty) = \emptyset$.

We refer to [FHH2O, FH, IS1] for related technique of proof, and for [L1, L2, H, K2, K3] related results.

2.2. LAP bounds

Next we discuss the LAP bounds, or a locally uniform bound as a map $\mathcal{B} \to \mathcal{B}^*$, for the resolvent

$$R(z) = (H - z)^{-1}.$$

Let us impose an additional condition.

Condition 6. In addition to Conditions 1–3, there exists an open subset $\mathcal{I} \subseteq (\lambda_H, \infty)$ such that for any $n \geq 0$ and compact interval $I \subseteq \mathcal{I}$ the mapping

$$\chi_n P_H(I) \colon \mathcal{H} \to \mathcal{H}$$

is compact, where $P_H(I)$ denotes the spectral projection onto I for H.

Due to Rellich's compact embedding theorem, 'boundedness' of the r-balls is a criterion for Condition 6, and in the affirmative case we have $\mathcal{I} = (\lambda_H, \infty)$. Note that so far the r-balls may not be bounded. Condition 6 in fact includes more general situations, see Section 3.

For notational simplicity we set for a large $C > 0$

$$h = \nabla^2 r + Cr^{-1-\tau} g \geq \frac{1}{2}\sigma r^{-1}|\mathrm{d}r|^2 \ell \geq 0,$$

cf. (1). For any subset $I \subseteq \mathcal{I}$ let us denote

$$I_\pm = \{z = \lambda \pm \mathrm{i}\Gamma \in \mathbb{C} \,|\, \lambda \in I,\ \Gamma \in (0,1)\},$$

respectively. We also use the notation $\langle T \rangle_\psi = \langle \psi, T\psi \rangle$.

Theorem 7. *Suppose Condition 6 and let $I \subseteq \mathcal{I}$ be a compact interval. Then there exists $C > 0$ such that for any $\phi = R(z)\psi$ with $z \in I_\pm$ and $\psi \in \mathcal{B}$*

$$\|\phi\|_{\mathcal{B}^*} + \|p^r \phi\|_{\mathcal{B}^*} + \langle p_i^* h^{ij} p_j \rangle_\phi^{1/2} + \|H_0 \phi\|_{\mathcal{B}^*} \leq C\|\psi\|_{\mathcal{B}}.$$

Corollary 8. *One has $\sigma_{\mathrm{sc}}(H) \cap \mathcal{I} = \emptyset$.*

2.3. Radiation condition

Theorem 7 does not immediately imply the LAP. To show that we establish radiation condition bounds under an additional but minor regularity and decay condition.

Condition 9. In addition to Condition 6, there exist $\rho, C > 0$ and splittings $q_1 = q_{11} + q_{12}$ and $q_2 = q_{21} + q_{22}$ by real-valued functions

$$q_{11} \in C^2(M) \cap L^\infty(M), \quad q_{12}, q_{21} \in C^1(M) \cap L^\infty(M), \quad q_{22} \in L^\infty(M)$$

such that for $k = 0, 1$

$$\begin{aligned}
|\nabla^r q_{11}| &\leq Cr^{-(1+\rho/2)/2}, & |\ell^{\bullet i}\nabla_i q_{11}| &\leq Cr^{-1-\rho/2},\\
|\nabla\nabla^r q_{11}| &\leq Cr^{-1-\rho/2}, & |\nabla q_{12}| &\leq Cr^{-1-\rho/2},\\
|(\nabla^r)^k q_{21}| &\leq Cr^{-k-\rho}, & q_{21}\nabla^r q_{11} &\leq Cr^{-1-\rho},\\
|q_{22}| &\leq Cr^{-1-\rho/2}.
\end{aligned}$$

Our radiation condition bounds are stated in terms of the radial derivative A from (2) and a complex phase a defined by the formula

$$a = a_z = \eta_\lambda \Big[\sqrt{2(z - q_1)} \pm \frac{1}{4}(p^r q_{11})/(z - q_1)\Big] \tag{3}$$

for $z = \lambda \pm \mathrm{i}\Gamma \in \mathcal{I} \cup \mathcal{I}_\pm$, respectively. Here branch of the square root is chosen such that $\mathrm{Re}\sqrt{w} > 0$ for $w \in \mathbb{C} \setminus (-\infty, 0]$, and smooth cut-off

$\eta_\lambda = 1 - \chi(2r/r_\lambda)$ is introduced to make $a = a_z$ smooth in $z \in \mathcal{I}_\pm$. To be precise, we picked a smooth decreasing function $r_\lambda \geq r_0$ of $\lambda > \lambda_H$ such that

$$\lambda + \lambda_H - 2q_1 \geq 0 \quad \text{for } r \geq r_\lambda/2,$$

and that $r_\lambda = r_0$ for all λ large enough. Note that a is an approximate solution to the radial Riccati equation

$$\pm p^r a + a^2 - 2|\mathrm{d}r|^2(z - q_1) = 0,$$

and that the second term in the brackets of (3) is added to have a better approximate solution.

Now let

$$\beta_c = \frac{1}{2}\min\{\sigma, \tau, \rho\} > 0.$$

Theorem 10. *Suppose Condition 9 and let $I \subseteq \mathcal{I}$ be a compact interval. Then for all $\beta \in [0, \beta_c)$ there exists $C > 0$ such that for any $\phi = R(z)\psi$ with $\psi \in r^{-\beta}\mathcal{B}$ and $z \in I_\pm$*

$$\|r^\beta(A \mp a)\phi\|_{\mathcal{B}^*} + \langle p_i^* r^{2\beta} h^{ij} p_j\rangle_\phi^{1/2} \leq C\|r^\beta\psi\|_{\mathcal{B}},$$

respectively.

2.4. Applications

Finally we provide several applications. We begin with the LAP.

Corollary 11. *Suppose Condition 9, and let $I \subseteq \mathcal{I}$ be a compact interval. For any $s > 1/2$ and $\epsilon \in (0, \min\{s - 1/2, \beta_c, (2+\rho)/4\})$ there exists $C > 0$ such that for $k = 0, 1$ and any $z, z' \in I_+$ or $z, z' \in I_-$*

$$\|p^k R(z) - p^k R(z')\|_{\mathcal{B}(\mathcal{H}_s, \mathcal{H}_{-s})} \leq C|z - z'|^{\min\{\epsilon, 1\}}.$$

In particular, $p^k R(z)$, $k = 0, 1$, attain uniform limits as $I_\pm \ni z \to \lambda \in I$ in the norm topology of $\mathcal{B}(\mathcal{H}_s, \mathcal{H}_{-s})$:

$$p^k R(\lambda \pm \mathrm{i}0) = \lim_{I_\pm \ni z \to \lambda} p^k R(z), \quad \lambda \in I,$$

respectively. These limits $p^k R(\lambda \pm \mathrm{i}0)$ belong to $\mathcal{B}(\mathcal{B}, \mathcal{B}^*)$, and, in addition, $R(\lambda \pm \mathrm{i}0)\mathcal{B} \subseteq \mathcal{B}^* \cap \mathcal{N}$.

Now we have the limiting resolvents, and Theorem 10 extends to them, cf. [Is, S, HS].

Corollary 12. *Suppose Condition 9 and let $I \subseteq \mathcal{I}$ be a compact interval. Then for all $\beta \in [0, \beta_c)$ there exists $C > 0$ such that for any $\phi = R(\lambda \pm \mathrm{i}0)\psi$ with $\psi \in r^{-\beta}\mathcal{B}$ and $\lambda \in I$*

$$\|r^{\beta}(A \mp a)\phi\|_{\mathcal{B}^*} + \langle p_i^* r^{2\beta} h^{ij} p_j \rangle_{\phi}^{1/2} \le C\|r^{\beta}\psi\|_{\mathcal{B}},$$

respectively.

As our final application, we have the *Sommerfeld uniqueness result*, or a characterization of $R(\lambda \pm \mathrm{i}0)$, cf. [Is].

Corollary 13. *Suppose Condition 9, and let $\lambda \in \mathcal{I}$, $\phi \in \mathcal{H}_{\mathrm{loc}}$ and $\psi \in r^{-\beta}\mathcal{B}$ with $\beta \in [0, \beta_c)$. Then $\phi = R(\lambda \pm \mathrm{i}0)\psi$ holds if and only if both of the following conditions hold:*

(i) *$(H - \lambda)\phi = \psi$ in the distributional sense.*
(ii) *$\phi \in (r^{\beta}\mathcal{B}^*) \cap \mathcal{N}$ and $(A \mp a)\phi \in r^{-\beta}\mathcal{B}_0^*$.*

§3. Examples

In this section we provide several examples satisfying our conditions.

3.1. Unbounded obstacles in the Euclidean space

For simplicity we let $V \equiv 0$, and consider $H = H_0$.

Example 14. Let $\Omega \subseteq \mathbb{R}^d$ be a connected open subset. Assume that for some $r_0 \ge 2$ and $x_0 \in \mathbb{R}^d$ restriction of the vector field

$$\nabla|x - x_0| = |x - x_0|^{-1}(x - x_0)$$

to $\{x \in \Omega \,|\, |x - x_0| > r_0\}$ is forward complete. Then Condition 9 is satisfied for $\sigma = 2$, any $\tau > 0$, $\rho' = 2$ and $\rho = 6$, and the critical energy is $\lambda_H = 0$. Construction of the function r is almost obvious; It suffices to smoothly and appropriately modify $|x - x_0|$ in a bounded region.

We refer to [C, Il1, Il2, Il3, Mi] for related works.

3.2. Ends of warped-product type

Here also we let $V \equiv 0$, and consider $H = H_0$.

Example 15. Let (M, g) be a connected and complete Riemannian manifold. Assume that there exist an open subset $B \subseteq M$ and a $(d-1)$-dimensional closed manifold S such that

$$M \setminus B \cong [2, \infty) \times S.$$

In addition, assume that there exist $f \in C^\infty((1,\infty))$ and a Riemannian metric h on S such that in local coordinates $(r,\sigma) \in [2,\infty) \times S$

$$g(r,\sigma) = \mathrm{d}r \otimes \mathrm{d}r + f(r)h(\sigma); \quad h(\sigma) = h_{\alpha\beta}(\sigma)\, \mathrm{d}\sigma^\alpha \otimes \mathrm{d}\sigma^\beta.$$

Here the Greek indices run over $2,\dots,d$. Then by appropriately modifying and extending r to B Condition 1 is fulfilled. We remark that here B can be unbounded.

In the sequel we examine other conditions by specifying f explicitly. Note that the Christoffel symbols are computed on $M \setminus B$:

$$\Gamma^r_{rr} = 0, \qquad \Gamma^r_{r\alpha} = \Gamma^r_{\alpha r} = 0, \qquad \Gamma^\alpha_{rr} = 0,$$

$$\Gamma^r_{\alpha\beta} = -\frac{1}{2} f' h_{\alpha\beta}, \quad \Gamma^\alpha_{r\beta} = \Gamma^\alpha_{\beta r} = \frac{1}{2}(f'/f)\delta^{\alpha\beta}, \quad \Gamma^\alpha_{\beta\gamma} = (\Gamma_S)^\alpha_{\beta\gamma},$$

where $\delta^{\alpha\beta}$ denotes Kronecker's δ and Γ_S the Christoffel symbol for (S,h). Hence we have

$$|\nabla r|^2 = 1, \quad \nabla^2 r = \frac{1}{2} f' h, \quad \Delta r = \frac{d-1}{2} f'/f.$$

Example 16. In the setting of Example 15 assume B is bounded.

(1) Let

$$f(r) = r^\theta; \quad \theta > 0.$$

Condition 9 is satisfied for $\sigma = \theta$, any $\tau > 0$, $\rho' = 2$ and $\rho = 6$, and the critical energy is $\lambda_H = 0$. The Euclidean space corresponds to $f(r) = r^2$ and S being the standard unit sphere.

(2) Let

$$f(r) = \exp(\delta r^\theta); \quad 0 < \delta,\ 0 < \theta < 1.$$

Condition 9 is satisfied for any $\sigma > 0$, any $\tau > 0$, $\rho' = 2 - 2\theta$ and $\rho = 6 - 4\theta$, and the critical energy is $\lambda_H = 0$.

(3) Let

$$f(r) = C \exp(\kappa r + \delta_\theta(r)); \quad C, \kappa > 0,\ \theta < 1.$$

Here θ is an order parameter in the sense that the derivatives

$$\delta_\theta^{(k)}(r) = O(r^{\theta - k}); \ k = 0, 1, 2, \dots.$$

Condition 9 is satisfied for any $\sigma > 0$, any $\tau > 0$, $\rho' = 1 - \theta$ and $\rho = 4 - 2\theta$, and the critical energy is $\lambda_H = (d-1)^2\kappa^2/32$. The hyperbolic space corresponds to $f(r) = (\sinh r)^2$ and S being the standard unit sphere, for which $\theta < 1$ may be arbitrary.

We can furthermore add certain unbounded obstacles to the models of Examples 16 as in Example 14. We refer to [D, Me, RT, K1, K2, K3, K4] for related setting on manifolds.

3.3. Multi-ends with different critical energies

Here we discuss Condition 6 in the simplest case: Let $M = \mathbb{R}$, which is exactly a Riemannian manifold with two ends. There we consider

$$H = -\frac{1}{2}\frac{\mathrm{d}^2}{\mathrm{d}x^2} + V \quad \text{on } \mathcal{H} = L^2(\mathbb{R}).$$

We assume that $V \in C^\infty(\mathbb{R})$ is equal to different constants $\lambda_0 < \lambda_1$ on the two ends:

$$V(x) = \begin{cases} \lambda_0 & \text{for } x \geq 1, \\ \lambda_1 & \text{for } x \leq -1. \end{cases}$$

If we choose $r \in C^\infty(\mathbb{R})$ such that

$$r = \begin{cases} x & \text{for } x \geq 2, \\ 1 & \text{for } x \leq 1, \end{cases}$$

then clearly Conditions 1–3 are satisfied with critical energy $\lambda_H = \lambda_0$. In this case, although the r-balls are unbounded, Condition 6 (and hence Condition 9) is certainly satisfied as follows:

Lemma 17. *Under the above setting Condition* 6 *holds with*

$$\mathcal{I} = (\lambda_0, \lambda_1).$$

Now by Lemma 17 all the results of Section 2 hold for $\mathcal{I} = (\lambda_0, \lambda_1)$. Furthermore, the same results hold also for $\mathcal{I} = (\lambda_1, \infty)$. In fact, we can retake $r \in C^\infty(\mathbb{R})$ such that

$$r = |x| \quad \text{for } |x| \geq 2.$$

Then Conditions 1–3 is satisfied with critical energy $\lambda_H = \lambda_1$, and also Condition 6 (and hence Condition 9) for $\mathcal{I} = (\lambda_1, \infty)$. Here we use Rellich's compact embedding theorem rather than Lemma 17, since now the r-balls are bounded. We remark that the only exceptional energy λ_1 is a 'threshold'.

These arguments easily generalize to a manifold with several ends possibly with different critical energies. For example, a manifold has different critical eneregies in Euclidean and hyperbolic ends, if $V \equiv 0$. We refer to [DS, K4] for related results.

References

[C] P. Constantin, *Scattering for Schrödinger operators in a class of domains with non-compact boundaries*, J. Funct. Anal., **44** (1981), 87–119.

[DS] E. B. Davies and B. Simon, *Scattering Theory for Systems with Different Spatial Asymptotics on the Left and Right*, Comm. Math. Phys., **63** (1978), 277–301.

[D] H. Donnelly, *Spectrum of the Laplacian on asymptotically Euclidean spaces*, Michigan Math. J., **46** (1999), 101–111.

[FH] R. Froese and I. Herbst, *Exponential bounds and absence of positive eigenvalues for N-body Schrödinger operators*, Comm. Math. Phys., **87**, no. 3 (1982/83), 429–447.

[FHH2O] R. Froese, I. Herbst, M. Hoffmann-Ostenhof and T. Hoffmann-Ostenhof, *On the absence of positive eigenvalues for one-body Schrödinger operators*, J. Analyse Math., **41** (1982), 272–284.

[HS] I. Herbst, E. Skibsted, *Time-dependent approach to radiation conditions*, Duke Math. J., **64**, no. 1 (1991), 119–147.

[H] L. Hörmander, *The analysis of linear partial differential operators.* II–IV, Berlin, Springer, 1983–85.

[Il1] E. M. Il'in, *The principle of limit absorption and scattering by noncompact obstacles.* I, Izv. Vyssh. Uchebn. Zaved. Mat., **1** (1984), 46–55.

[Il2] E. M. Il'in, *The principle of limit absorption and scattering by noncompact obstacles.* II, Izv. Vyssh. Uchebn. Zaved. Mat., **2** (1984), 27–34.

[Il3] E. M. Il'in, *Scattering by unbounded obstacles for second-order elliptic operators*, Proc. Steklov Inst. Math., **2** (1989), 85–107.

[Is] H. Isozaki, *Eikonal equations and spectral representations for long-range Schrödinger Hamiltonians*, J. Math. Kyoto Univ., **20** (1980), 243–261.

[IS1] K. Ito and E. Skibsted, *Absence of embedded eigenvalues for Riemannian Laplacians*, Adv. Math., **258** (2013), 945–962.

[IS2] K. Ito and E. Skibsted, *Radiation condition bounds on manifolds with ends*, J. Funct. Anal., **278** (2020), no. 9, 108449.

[IS3] K. Ito and E. Skibsted, *Stationary scattering theory on manifolds*, to appear in Ann. Inst. Fourier (Grenoble).

[K1] H. Kumura, *On the essential spectrum of the Laplacian on complete manifolds*, J. Math. Soc. Japan, **49**, no. 1 (1997), 1–14.

[K2] H. Kumura, *The radial curvature of an end that makes eigenvalues vanish in the essential spectrum* I, Math. Ann., **346**, no. 4 (2010), 795–828.

[K3] H. Kumura, *The radial curvature of an end that makes eigenvalues vanish in the essential spectrum* II, Bull. London Math. Soc., (2011) **43** (5), 985–1003.

[K4] H. Kumura, *Limiting absorption principle on manifolds having ends with various measure growth rate limits*, Proc. Lond. Math. Soc. (3), **107** (2013), no. 3, 517–548.

[L1] W. Littman, *Decay at infinity of solutions to partial differential equations with constant coefficients*, Trans. AMS., **123**, (1966), 449–459.

[L2] W. Littman, *Decay at infinity of solutions of higher order partial differential equations: removal of the curvature assumption*, Israel J. Math., **8**, (1970), 403–407.

[Me] R. Melrose, *Spectral and scattering theory for the Laplacian on asymptotically Euclidian spaces*, Spectral and scattering theory (Sanda, 1992) (M. Ikawa, ed.), Marcel Dekker (1994), 85–130.

[Mi] V. Minskii, *Sommerfeld radiation condition for second-order differential operator in a domain with infinite border*, J. Diff. Eq., **48** (1983), 157–176.

[RT] I. Rodnianski and T. Tao *Effective Limiting Absorption Principles, and Applications*, Commun. Math. Phys., **333** (2015), 1–95.

[S] Y. Saitō, *Spectral representations for Schrödinger operators with a long-range potentials*, Lecture Notes in Mathematics, **727**, Berlin, Springer, 1979.

Graduate School of Mathematical Sciences,
The University of Tokyo,
3-8-1 Komaba, Meguro-ku, Tokyo 153-8914, Japan
E-mail address: ito@ms.u-tokyo.ac.jp

Institut for Matematiske Fag,
Aarhus Universitet,
Ny Munkegade 8000 Aarhus C, Denmark
E-mail address: skibsted@math.au.dk

Advanced Studies in Pure Mathematics 85, 2020
The Role of Metrics in the Theory of Partial Differential Equations
pp. 205–214

A convergence of generalized Lagrangian mean curvature flow in Kähler manifold with positive weighted Ricci form

Toru Kajigaya and Keita Kunikawa

Abstract.

In this paper, we briefly summarize recent results on a generalization of the Lagrangian mean curvature flow in Kähler-Einstein manifolds. In particular, we focus on the geometry of generalized LMCF in a Kähler manifold with positive weighted Ricci-form and exhibit the extended results according to [6].

§1. Introduction

The *mean curvature flow (MCF)* is the gradient flow of the volume functional of a submanifold in a Riemannian manifold (M, g) and actively studied in geometric analysis. When M admits a Kähler-Einstein structure (ω, J) with the compatible Kähler metric g and the initial submanifold L is a *Lagrangian submanifold*, it is known that the MCF of initial L preserves the Lagrangian condition, and we call the flow *Lagrangian mean curvature flow (LMCF)* for this particular case. The LMCF is extensively studied especially when M is a Calabi-Yau manifold (Ricci flat case). For example, Thomas-Yau [17] introduced a notion of "stability" for Hamiltonian isotopy class $[L]$ of zero-Maslov Lagrangian L, and conjectured that if the initial Lagrangian L lies in $[L]$ and satisfies extra conditions, the LMCF of initial L exists for all time and converges to a special Lagrangian submanifold, that is, a submanifold with volume minimizing property in its homology class. This conjecture is recently

Received November 30, 2018.
Revised May 7, 2019.
2010 *Mathematics Subject Classification.* Primary 53D12; Secondary 53C44.
Key words and phrases. Lagrangian submanifolds, Hamiltonian stability, Fano manifolds, generalized Lagrangian mean curvature flow.

improved by Joyce [5], and he proposed a conjecture about the long-time existence of LMCF *with surgeries.*

On the other hand, as pointed out in [5], LMCF is likely to behave differently when M admits a non-Ricci flat Kähler metric, and hence, it is natural to ask whether we can establish a suitable analogue of Calabi-Yau for more general Kähler manifold. In the present paper, we consider LMCF especially when M admits a positive (modified) Ricci curvature, and summarize extended results according to [6]. To be precise, we mainly concern a Kähler manifold M such that the Ricci form ρ of M admits a potential function $f \in C^\infty(M)$, i.e., $\rho = C\omega + ndd^c f$ for $C > 0$. In particular, we suppose M is a Fano manifold equipped with a Kähler form $\omega \in 2\pi c_1(M)$ for example. We consider a modified MCF, that is, the gradient flow of the weighted volume functional Vol_f defined by f. It turns out that the modified MCF preserves the Lagrangian condition (although the usual MCF may not), and we call the flow *generalized Lagrangian mean curvature flow* (*GLMCF* for short). See Section 2 for a general frame work and properties of LMCF. In Section 3, we argue an issue about the volume minimizing property of Lagrangian submanifold. In particular, we introduce the notion of Hamiltonian-stability for the weighted volume, which is a generalization of Oh [12], in order to make a natural geometric assumption for initial data of the GLMCF. We give a simple criterion for the Hamiltonian-stability and show that examples of H-stable Lagrangian tori naturally appear in a toric Kähler manifold. In the last section, we show some results on the long-time existence and convergence of GLMCF generalizing the results of Haozhao Li [7].

§2. A general framework of Lagrangian mean curvature flow

In this section, we briefly summarize a general framework and properties of Lagrangian mean curvature flow in a symplectic manifold. We remark that some of results described in below can be extended to more general setting. See [8] and [14] for general descriptions.

Let (M, ω) be a real $2n$-dimensional symplectic manifold equipped with a symplectic form ω, and $\phi : L \to M$ a Lagrangian immersion of an n-dimensional compact connected manifold L, that is, an immersion of L satisfying $\phi^*\omega = 0$. We take a compatible almost complex structure J so that (M, ω, J) becomes an almost Kähler manifold, and we denote the compatible almost Kähler metric by g. The induced metric defines the volume form $d\mu_L$ on L, and $d\mu_L$ naturally determines (at least locally) a unit length section Ω_L of $\phi^*K(M)$, where $K(M) := \Lambda^{(n,0)}T^*M$ is the canonical line bundle, which is locally expressed by $\Omega_L = (e_1^* + J\sqrt{-1}e_1^*) \wedge \cdots \wedge (e_n^* + J\sqrt{-1}e_n^*)$ for orthonormal basis $\{e_i^*\}_{i=1}^n$ of T_p^*L.

Consider a Hermitian connection $\widetilde{\nabla}$ on TM (with torsion in general), namely, $\widetilde{\nabla}$ is an affine connection with $\widetilde{\nabla}g = \widetilde{\nabla}J = 0$. We use the same symbol for the induced connection on $K(M)$. Then, the connection 1-form for $\widetilde{\nabla}$ on $\phi^*K(M)$ is always imaginary, namely, there exists a real 1-form $\xi^{\widetilde{\nabla}}$ on L such that $\widetilde{\nabla}\Omega_L = \sqrt{-1}\xi^{\widetilde{\nabla}} \otimes \Omega_L$. According to [8], we call the 1-form $\xi^{\widetilde{\nabla}}$ *Maslov form* of L for the Hermitian connection $\widetilde{\nabla}$. We give another definition of the Maslov form which is more reasonable for computation: Define a $(0,3)$-tensor field $\widetilde{S}$ on L by

$$\widetilde{S}(X,Y,Z) := g(J(\widetilde{\nabla}_X Y)^\perp, Z).$$

for $X, Y, Z \in \Gamma(TL)$. Then $\widetilde{S}$ is symmetric with respect to the latter two components since $\widetilde{\nabla}$ is complex, and the Maslov 1-form is given by the trace of the symmetric components:

$$\xi^{\widetilde{\nabla}} = \sum_{i=1}^{n} \widetilde{S}(\cdot, e_i, e_i).$$

One notices that the Maslov 1-form coincides with the *mean curvature form* α_H of L when (M, ω, J) is Kähler and $\widetilde{\nabla}$ is taken as the Levi-Civita connection $\overline{\nabla}$. Thus, the Maslov form is regarded as a generalization of the mean curvature form, and therefore, the Maslov form is also referred as *generalized mean curvature form* (cf. [14], [15]). Moreover, we define a normal vector field K on L by $\xi^{\widetilde{\nabla}} = \phi^*(i_K\omega)$ or equivalently

$$K = -J(\xi^{\widetilde{\nabla}})^\sharp,$$

and K will be called the *generalized mean curvature vector* for Lagrangian immersion (see also Definition 2 and Lemma 3 in [14]).

Denote the curvature tensor of $\widetilde{\nabla}$ by $\widetilde{R}$, and consider a 2-form on M defined by

$$\widetilde{P}(X,Y) := \frac{1}{2}\mathrm{tr}\{J\widetilde{R}(X,Y)\}.$$

Then, it turns out that $\widetilde{P}$ is a closed form and represents the first Chern class $2\pi c_1(M)$ of (M, J). When (M, ω, J) is a Kähler manifold, the 2-form $\widetilde{P}$ coincides with the usual Ricci form, and hence, we call $\widetilde{P}$ the *Ricci form* for the Hermitian connection $\widetilde{\nabla}$. The Ricci form and the Maslov form are closely related by the following equation, which is a generalization of *Dazord's formula*:

$$d\xi^{\widetilde{\nabla}} = \phi^*\widetilde{P}.$$

A remarkable consequence of Dazord's formula is the Maslov form of a Lagrangian immersion ϕ is always closed if the Ricci form satisfies the *Einstein* condition, i.e.,

$$\widetilde{P} = C\omega \tag{1}$$

for some constant $C \in \mathbb{R}$. We call such a Hermitian connection *Einstein connection* [14]. By Cartan's formula, the closedness of the Maslov form and ω shows $\phi^* L_K \omega = \phi^* i_K d\omega + \phi^* d(i_K \omega) = d\xi^{\widetilde{\nabla}} = 0$, and hence, the infinitesimal deformation for the normal direction K of L preserves the Lagrangian condition. In fact, much more is true: Consider a family of Lagrangian immersions $F : L \times [0, T) \to M$ satisfying

$$\frac{\partial F}{\partial t} = K, \quad F_0 = \phi. \tag{2}$$

Then, by the results of Smoczyk-Wang (Theorem 1 and 2 in [14]), we see that the flow (2) exists for short time if $\widetilde{\nabla}$ is Einstein. According to Behrndt and Smoczyk-Wang, we call the flow (2) *generalized Lagrangian mean curvature flow* (*GLMCF* for short) as long as we take an Einstein connection. Obviously, the GLMCF coincides with the usual mean curvature flow when (M, ω, J) is Kähler-Einstein and $\widetilde{\nabla}$ is the Levi-Civita connection. There are other interesting classes of almost Kähler manifold admitting an Einstein connection. For example, any cotangent bundle T^*N equipped with the standard symplectic form over a Riemannian manifold N admits a natural Einstein connection so that $\widetilde{P} = 0$ (see [15]).

Another class of examples are obtained by a Kähler manifold such that the Ricci form ρ for the Levi-Civita connection satisfies

$$\rho = C\omega + ndd^c f \tag{3}$$

for some constant $C \in \mathbb{R}$ and $f \in C^\infty(M)$. In fact, we can define the Einstein connection $\widetilde{\nabla}$ by $\widetilde{\nabla} := \overline{\nabla} + d^c f \otimes J$ so that $\widetilde{P}$ coincides with the weighted Ricci form $\rho_f := \rho - ndd^c f$ and $\widetilde{P} = \rho_f = C\omega$.

Example 1. A complex n-dimensional Kähler manifold (M, ω, J) is called *almost Calabi-Yau* if it admits a non-vanishing holomorphic $(n, 0)$-form Ω ([4] §8.3). Define a weight function f by

$$e^{2nf} \frac{\omega^n}{n!} = (-1)^{\frac{n(n-1)}{2}} \left(\frac{\sqrt{-1}}{2}\right)^n \Omega \wedge \overline{\Omega}.$$

Then, one verifies that $\rho_f = 0$, i.e., (3) is satisfied with $C = 0$ (see [2]).

Example 2. Another typical example satisfying (3) is given by Fano manifold, i.e., a compact complex manifold (M, J) of positive first Chern class $c_1(M)$. Taking a Kähler form $\omega \in 2\pi c_1(M)$, we find a real function f satisfying $\rho = \omega + ndd^c f$ since ρ represents $2\pi c_1(M)$. Note that, for fixed J and the Kähler class $[\rho] = 2\pi c_1(M)$, the compatible symplectic forms in $[\rho]$ consist of an open and convex set, and it is well-known that there exists an obstruction for the existence of Kähler-Einstein metric on a Fano manifold. Also, we remark that a symplectic manifold equipped with a compatible almost complex structure J of $c_1(M) > 0$ is called *monotone* in symplectic geometry. We shall deal with this case in the following sections.

Suppose M is a Kähler manifold satisfying (3). Then, by the definition of Maslov form, it is easy to see that the generalized mean curvature vector for this case is given by $K = H - n(\overline{\nabla} f)^{\perp}$ a modification of the usual mean curvature vector H. In particular, the GLMCF has a variational meaning, namely, the GLMCF coincides with a negative gradient flow for the *weighted* volume functional $\mathrm{Vol}_f(\phi) := \int_L e^{nf} d\mu_L$, where $d\mu_L$ is the volume form induced by the Kähler metric. Moreover, $K \equiv 0$ if and only if ϕ is a critical point of the weighted volume functional Vol_f (cf. [2]). In this sense, we call the critical point of Vol_f (i.e., $K \equiv 0$) *f-minimal Lagrangian submanifold.*

For the GLMCF in a Kähler manifold satisfying (3), it is natural to ask the following question; *when does the GLMCF exist for all time and converge to an f-minimal Lagrangian submanifold?* In order to make a natural geometric assumption for the initial data, we mention another property of GLMCF.

Recall that the Maslov form $\xi^{\tilde{\nabla}}$ for an Einstein connection is always a closed form, and hence, defines a cohomology class $[\xi^{\tilde{\nabla}}] \in H^1(L, \mathbb{R})$. Then, under the assumption (3) (or (1)), it turns out that the class $[\xi^{\tilde{\nabla}}]$ is invariant under any *Hamiltonian* deformation of ϕ (see Proposition 2.14 in [6]). We note that $[\xi^{\tilde{\nabla}}] = 0$ is equivalent to the *monotone* Lagrangian in symplectic geometry when $C > 0$ (see [6]). Moreover, *the GLMCF of initial ϕ preserves the exactness of Maslov form* (see [13] or Proposition 4.3 in [6]). Since any f-minimal Lagrangian is $[\xi^{\tilde{\nabla}}] = 0$, it makes sense to assume the Maslov form of the initial Lagrangian is exact. Moreover, the GLMCF of L with $[\xi^{\tilde{\nabla}}] = 0$ generates a Hamiltonian deformation if the flow exists smoothly. This property will be used to build another geometric assumption for the initial data.

§3. Hamiltonian stability

When M is an almost Calabi-Yau manifold (Example 1), it turns out that any oriented and connected compact f-minimal Lagrangian in M is a calibrated submanifold with respect to the weighted metric (see Proposition 5.3 [2]), and hence it is volume minimizing in its homology class. More generally, if the modified Ricci form $\rho_f := \rho - ndd^c f$ for the Kähler manifold satisfies $\rho_f \leq 0$, any f-minimal Lagrangian L is automatically *stable*, i.e., $\frac{d^2}{dt^2}|_{t=0}\mathrm{Vol}_f(L_t) \geq 0$ for any infinitesimal deformation ϕ_t of $\phi = \phi_0$ (see Corollary 2.5 in [6]). Thus, any f-minimal Lagrangian as a stationary point of GLMCF is at least locally volume minimizing with respect to the weighted volume if $\rho_f \leq 0$.

However, the situation is different when $\rho_f > 0$. For example, it is known that there exists no minimal and stable Lagrangian in the complex projective space $\mathbb{C}P^n$ by the result of Lawson-Simons [9]. Note that such a non-existence result does not hold in a higher rank Hermitian symmetric space, e.g., some real forms in a Hermitian symmetric space of compact type are stable, however, the stable real form must have a simple topology, namely, the real form is stable if and only if it is simply-connected [16]. On the other hand, another stability criterion for Lagrangian submanifold was introduced by Y-G. Oh in [12], that is, the stability under *Hamiltonian deformation*. He pointed out that H-stable Lagrangian often appears more than stable ones. For example, the totally geodesic real projective space $\mathbb{R}P^n$ in $\mathbb{C}P^n$, or more generally, an *Einstein* real form in a Hermitian symmetric space of compact type is H-stable ([12], see also [10]). Besides the real forms, many examples of H-stable minimal Lagrangian have been found by several authors (e.g. [10, 11, 12] and references therein).

We mention details of the stability criterion and show that some H-stable Lagrangians for the weighted volume naturally appear. Recall that, for a Lagrangian immersion $\phi : L \to M$, a deformation $\{\phi_t\}_t$ of $\phi = \phi_0$ is called *Hamiltonian* if the variational vector $V = \frac{d}{dt}|_{t=0}\phi_t$ satisfies $\phi^*(i_V\omega) = dh$ for some $h \in C^\infty(L)$. We call an f-minimal Lagrangian *H-stable* with respect to Vol_f if $\frac{d^2}{dt^2}|_{t=0}\mathrm{Vol}_f(\phi_t) \geq 0$ for any Hamiltonian deformation. The following simple criterion is a generalization of the result of Chen-Leung-Nagano [3] and Oh [12].

Theorem 1 ([6]). *Suppose (M, ω, J) is a Kähler manifold satisfying $\rho_f = C\omega$. An f-minimal Lagrangian in M is H-stable with respect to Vol_f if and only if the first eigenvalue $\lambda_1(\Delta_f)$ of the weighted Laplacian $\Delta_f := \Delta + ng(\nabla f, \nabla \cdot)$ acting on $C^\infty(L)$ satisfies*

$$\lambda_1(\Delta_f) \geq C.$$

In order to find examples of H-stable Lagrangian, we consider the weighted projective space. Fix $\mathbf{a} := (a_1, \dots, a_{n+1}) \in \mathbb{N}^{n+1}$, and define a weighted S^1-action on $\mathbb{C}^{n+1}$ by $e^{\sqrt{-1}\theta} z := (e^{\sqrt{-1}a_1\theta} z^1, \dots, e^{\sqrt{-1}a_{n+1}\theta} z^{n+1})$. Then, the action is Hamiltonian and we denote the moment map by $\mu_{\mathbf{a}} : \mathbb{C}^{n+1} \to \mathbb{R}$. For a regular value $c \in \mathbb{R}$, the Kähler quotient space $\mathbb{C}P^n_{\mathbf{a}} := \mu^{-1}(c)/S^1$ is so called *weighted projective space*. In general, $\mathbb{C}P^n_{\mathbf{a}}$ is a toric Kähler orbifold, however, if any subset of $\{a_j\}_{j=1}^n$ are relatively prime for example, then it has a smooth structure (see [6]). The Ricci form ρ of the canonical Kähler structure (ω, J) at a smooth point is computed as follows:

$$\rho = C\omega + ndd^c f_{\mathbf{a}}, \quad f_{\mathbf{a}} := \frac{1}{n}\{-C\check{F} + \log vol(\mathcal{O})_x\},$$

where C is a positive constant depending on $\mathbf{a}$ and c, $\check{F}$ is a function induced from a S^1-invariant function $F := \frac{1}{4}|z|^2$, and $vol(\mathcal{O})_x$ is the volume of the fiber over the point $x \in \mathbb{C}P^n_{\mathbf{a}}$.

Let T^{n+1} be the *Clifford torus* in $\mathbb{C}^{n+1}$, that is, the product torus of the same radius; $T^{n+1} := S^1(r) \times \cdots \times S^1(r)$ for $r > 0$. Since T^{n+1} is a S^1-invariant Lagrangian, there exists a regular value c so that $T^{n+1} \subset \mu^{-1}(c)$. Thus, we obtain a reduced torus $T^n_{\mathbf{a}} := T^{n+1}/S^1 \subset \mathbb{C}P^n_{\mathbf{a}}$, and it turns out that $T^n_{\mathbf{a}}$ is a smooth Lagrangian submanifold. For this torus, we see the following result:

Theorem 2 ([6]). *The torus $T^n_{\mathbf{a}}$ is $f_{\mathbf{a}}$-minimal and H-stable in $\mathbb{C}P^n_{\mathbf{a}}$. In particular, the first eigenvalue of the weighted Laplacian $\Delta_{f_{\mathbf{a}}}$ satisfies $\lambda_1(\Delta_{f_{\mathbf{a}}}) \geq C$, and the equality holds if and only if at least two components of $\mathbf{a} = (a_1, \dots, a_{n+1})$ are equal.*

If $\mathbf{a} = (1, \dots, 1)$, then $\mathbb{C}P^n_{\mathbf{a}}$ is nothing but the complex projective space, and $T^n_{\mathbf{a}}$ is so called the *Clifford torus* in $\mathbb{C}P^n$, and the H-stability of the Clifford torus in $\mathbb{C}P^n$ was first proved by Oh [12]. We remark that $T^n_{\mathbf{a}}$ is a unique monotone torus orbit for the canonical torus action on $\mathbb{C}P^n_{\mathbf{a}}$ and is Hamiltonian non-displaceable as well as the Clifford torus in $\mathbb{C}P^n$ (see [1], [6]). It is conjectured by Oh [12] that the Clifford torus in $\mathbb{C}P^n$ is *Hamiltonian volume minimizing*, i.e., it has the least volume in its Hamiltonian isotopy class. Improving this, we conjecture that the torus $T^n_{\mathbf{a}}$ is Hamiltonian volume minimizing with respect to the weighted volume in $\mathbb{C}P^n_{\mathbf{a}}$.

§4. Convergence results

Since GLMCF in a Kähler manifold satisfying (3) is a gradient flow of the weighted volume functional and the flow generates a Hamiltonian deformation of L if the Maslov form of the initial L is exact (see Section 2), one may expect that the GLMCF of such an initial L converges to an *H-stable f-minimal* Lagrangian submanifold if the flow exists for all time. In this direction, we refer to Behrndt [2] for almost Calabi-Yau case, Joyce [5] for recent progress and conjectures when M is a Calabi-Yau manifold, and Li [7] for the case of Kähler-Einstein manifold of negative Ricci curvature (Recall that any f-minimal Lagrangian is stable in the classical sense for these cases). On the other hand, we have the following results when M has positive (modified) Ricci curvature, which are generalization of the results due to Li [7] when M is Kähler-Einstein:

Theorem 3 ([6], [7]). *Let (M, ω, J) be a compact Kähler manifold satisfying $\rho_f = C\omega$ with $C > 0$ and $f \in C^\infty(M)$. For any $V_0, \Lambda_0, \delta_0 > 0$, there exists an $\epsilon_0 = \epsilon_0(n, V_0, \Lambda_0, \delta_0, C, M) > 0$ such that if a Lagrangian immersion $\phi : L \to M$ of a compact manifold L with exact Maslov form satisfies*

$$(4) \quad \mathrm{Vol}_f(L) \le V_0, \quad |B| \le \Lambda_0, \quad \lambda_1(\Delta_f) \ge C + \delta_0, \quad \int_L |K|^2 d\mu_f \le \epsilon_0,$$

(where B is the second fundamental form of L), then the GLMCF of initial ϕ exists for all time and converges exponentially fast to an H-stable f-minimal Lagrangian immersion $\psi : L \to M$ smoothly.

The assumptions (4) are satisfied, for example, when we take a sufficiently small deformation of f-minimal Lagrangian tori satisfying $\lambda_1(\Delta_f) > C$ given in Theorem 2. Notice that Theorem 3 requires that the initial L satisfies $\lambda_1(\Delta_f) > C$ due to some technical issues, and hence, the limiting Lagrangian in Theorem 3 does not attain $\lambda_1(\Delta_f) = C$ since the inequality $\lambda_1(\Delta_f) > C$ is preserved under the flow. However, assuming the existence of f-minimal Lagrangian satisfying $\lambda_1(\Delta_f) = C$, we have another convergence result:

Theorem 4 ([6], [7]). *Let (M, ω, J) be a compact Kähler manifold satisfying $\rho_f = C\omega$ with $C > 0$. Suppose $\phi : L \to M$ an f-minimal Lagrangian immersion into M satisfying $\lambda_1(\Delta_f) = C$, and take a Hamiltonian deformation $\{\phi_s\}_s$ of $\phi_0 = \phi$ with variational vector field $V := d/ds|_{s=0}\phi_s$ such that $d^2/ds^2|_{s=0}\mathrm{Vol}_f(\phi_s) > 0$. Then, there exists an $\epsilon = \epsilon(V, L, M) > 0$ such that if ϕ_s satisfies*

$$\|\phi_s - \phi_0\|_{C^3} \le \epsilon,$$

then the GLMCF of initial ϕ_s converges exponentially fast to an H-stable f-minimal Lagrangian immersion $\psi : L \to M$.

In this theorem, it is a natural and interesting problem to study whether the limiting immersion ψ coincides with the immersion ϕ or not; we leave this problem as a future work.

Our proof of the convergence results is a generalization of argument developed by Li [7] when M is Kähler-Einstein. Key tools in our proof are the maximum principle and a volume ratio of submanifold so called *κ-noncollapsing on the scale r* (see [7]). In particular, the volume ratio condition is used to improve several estimates. In fact, the logic of our proof of long-time existence for Theorem 4 is as follows (See [6] and [7] for details): By assumption (4), the maximum principle and the volume ratio condition, we may assume the flow L_t is $\frac{1}{3}\kappa$-noncollapsed on the scale r and satisfies

$$(5) \qquad |B|(t) \leq 6\Lambda, \quad \lambda_1(t) \geq C + \frac{1}{3}\delta \quad |K|(t) \leq 2\epsilon^{\frac{1}{n+2}},$$

for $t \in [0, T)$, where Λ, δ and ϵ are determined by initial data. Notice that a C^0-estimate of $|K|(t)$ is obtained from (4). We consider a class $\mathcal{A}_1$ consisting of L_t such that (5) and the volume ratio condition holds. Suppose the maximal time interval $[0, T_*]$ such that $L_t \in \mathcal{A}_1$ is finite. Then, by using the maximum principle and the volume ratio condition, we improve the estimates (5), and show L_t is actually in the class $\mathcal{A}_2 = \{L_t; \frac{2}{3}\kappa\text{-noncollapsed on the scale } r, |B|(t) \leq 3\Lambda, \quad \lambda_1(t) \geq C + \frac{2}{3}\delta \quad |K|(t) \leq \epsilon^{\frac{1}{n+2}}\}$. Since the second fundamental form is uniformly bounded, we see the flow L_t exists in $[0, T_* + T_+]$ for some $T_+ > 0$. However, using the estimates for the class $\mathcal{A}_2$ and the key tools again, it turns out that $L_t \in \mathcal{A}_1$ for $[0, T_* + T_+]$. This contradicts to the assumption that T_* is the maximal time for the class $\mathcal{A}_1$.

Acknowlegement. The authors would like to thank all organizers of the workshop MSJ-SI 2018 for giving us an opportunity to talk, and referees for useful comments and suggestions. The first author was supported by JSPS KAKENHI Grant Number JP18K13420.

References

[1] M. Abreu and L. Macarini, *Remarks on Lagrangian intersections in toric manifolds*, Trans. Amer. Math. Soc., **365** (2013), no. 7, 3851–3875.

[2] T. Behrndt, *Generalized Lagrangian mean curvature flow in Kähler manifolds that are almost Einstein*, Complex and Differential Geometry, Springer Proceedings in Mathematics, **8** (2011), Springer-Verlag, Heidelberg, New York, 65–80.

[3] B.-Y. Chen, P.-F. Leung and T. Nagano, *Totally geodesic submanifolds of symmetric spaces*, III, unpublished (1980), arXiv:1307.7325v2 (2013).
[4] D. Joyce, *Riemannian holonomy groups and calibrated geometry*, Oxford Graduate Texts in Mathematics, 12, Oxford University Press, Oxford, 2007.
[5] D. Joyce, *Conjectures on Bridgeland stability for Fukaya categories of Calabi-Yau manifolds, special Lagrangians, and Lagrangian mean curvature flow*, EMS Surv. Math. Sci., **2** (2015), no. 1, 1–62.
[6] T. Kajigaya and K. Kunikawa, *Hamiltonian stability for weighted measure and generalized Lagrangian mean curvature flow*, J. Geom. Phys., **128** (2018), 140–168.
[7] H. Li, *Convergence of Lagrangian mean curvature flow in Kähler-Einstein manifolds*, Math. Z., **271** (2012), no. 1, 313–342.
[8] J. D. Lotay and T. Pacini, *From Lagrangian to totally real geometry: coupled flows and calibrations*, to appear in Comm. Anal. Geom., **28** (2020).
[9] H. B. Lawson Jr. and J. Simons, *On stable currents and their application to global problems in real and complex geometry*, Ann. of Math. (2), **98** (1973), 427–450.
[10] H. Ma and Y. Ohnita, *On Lagrangian submanifolds in complex hyperquadrics and isoparametric hypersurfaces in spheres*, Math. Z., **261** (2009), 749–785.
[11] H. Ma and Y. Ohnita, *Hamiltonian stability of the Gauss images of homogeneous isoparametric hypersurfaces.* I, J. Differential Geom., **97** (2014), no. 2, 275–348.
[12] Y.-G. Oh, *Second variation and stabilities of minimal lagrangian submanifolds in Kähler manifolds*, Invent. Math., **101** (1990), 501–519.
[13] K. Smoczyk, *Harnack inequality for the Lagrangian mean curvature flow*, Calc. Var. Partial Differential Equations, **8** (1999), 247–258.
[14] K. Smoczyk and M.-T. Wang, *Generalized Lagrangian mean curvature flows in symplectic manifolds*, Asian J. Math., **15** (2011), no. 1, 129–140.
[15] K. Smoczyk, M.-P. Tsui and M.-T. Wang, *Generalized Lagrangian mean curvature flows: The cotangent bundle case*, to appear in J. Reine. Angew. Math.
[16] M. Takeuchi, *Stability of certain minimal submanifolds of compact Hermitian symmetric spaces*, Tohoku Math. J. II. Ser., **36** (1984), 293–314.
[17] R. P. Thomas ans S. T. Yau, *Special Lagrangians, stable bundles and mean curvature flow*, Comm. Anal. Geom., **10** (2002), no. 5, 1075–1113.

Department of mathematics, School of Engineering, Tokyo Denki University, 5 Senju Asahi-cho, Adachi-ku, Tokyo 120-8551, Japan/
National Institute of Advanced Industrial Science and Technology (AIST), MathAM-OIL.
E-mail address: kajigaya@mail.dendai.ac.jp

Advanced Institute for Materials Research (AIMR), Tohoku University, 2-1-1 Katahira, Aoba-ku, Sendai, 980-8577, Japan
E-mail address: keita.kunikawa.e2@tohoku.ac.jp

Advanced Studies in Pure Mathematics 85, 2020
The Role of Metrics in the Theory of Partial Differential Equations
pp. 215–225

Global existence and the asymptotic behavior for systems of nonlinear wave equations violating the null condition

Soichiro Katayama

Abstract.

We consider the Cauchy problem for systems of quasi-linear wave equations in two and three space dimensions. We investigate the small data global existence and the asymptotic behavior of solutions under a condition in between the standard and weak null conditions.

§1. Introduction

Let $n = 2$ or $n = 3$. We consider the Cauchy problem for systems of nonlinear wave equations

$$\Box u = F(u, \partial u, \partial^2 u), \qquad (t, x) \in (0, \infty) \times \mathbb{R}^n, \tag{1}$$

$$u(0, x) = \varepsilon f(x), \ (\partial_t u)(0, x) = \varepsilon g(x), \quad x \in \mathbb{R}^n, \tag{2}$$

where $u = u(t, x) = \big(u_1(t, x), \ldots, u_N(t, x)\big)$ is an $\mathbb{R}^N$-valued unknown function, $\Box = \partial_t^2 - \Delta$, $\partial u = (\partial_a u)_{0 \le a \le n}$, and $\partial^2 u = (\partial_a \partial_b u)_{0 \le a, b \le n}$ with $\partial_0 = \partial_t$ and $\partial_k = \partial_{x_k}$ for $1 \le k \le n$. We suppose that $f = (f_1, \ldots, f_N), g = (g_1, \ldots, g_N) \in C_0^\infty(\mathbb{R}^n; \mathbb{R}^N)$ and that ε is a positive and small parameter. We assume that $F = (F_1, \ldots, F_N)$ is quasi-linear and diagonal in $\partial^2 u$, that is to say that each F_j has the form

$$F_j(u, \partial u, \partial^2 u) = \sum_{a,b=0}^{n} G_j^{ab}(u, \partial u) \partial_a \partial_b u_j + H_j(u, \partial u). \tag{3}$$

Received December 27, 2018.
Revised May 13, 2019.
2010 *Mathematics Subject Classification.* 35L72.
Key words and phrases. wave equation, null condition, weak null condition.

We also assume that $G_j^{ab}(u, \partial u)$ and $H_j(u, \partial u)$ are homogeneous polynomials in $(u, \partial u)$ of degree $p - 1$ and of degree p, respectively, with

$$p = \begin{cases} 2 & (n = 3) \\ 3 & (n = 2) \end{cases} \left(\Leftrightarrow \frac{n-1}{2}(p-1) = 1 \right). \tag{4}$$

We say that the small data global existence (of classical solutions) holds for (1)–(2) if, for any $f, g \in C_0^\infty(\mathbb{R}^n; \mathbb{R}^N)$, there is a positive constant ε_0 such that the Cauchy problem (1)–(2) admits a unique global solution $u \in C^\infty([0, \infty) \times \mathbb{R}^n; \mathbb{R}^N)$, provided that $0 < \varepsilon \le \varepsilon_0$. The power p in (4) is the critical one for the small data global existence for (1)–(2); indeed, the small data global existence fails for some nonlinearity of power p, while it always holds when the power is greater than p. Therefore, we need some additional condition to obtain the small data global existence in the present setting.

Given a nonlinearity $F = F(u, \partial u, \partial^2 u)$, we define its *reduced nonlinearity* $F^{\text{red}} = F^{\text{red}}(\omega, X, Y, Z)$ for $\omega = (\omega_1, \ldots, \omega_n) \in \mathbb{S}^{n-1}$ and $X, Y, Z \in \mathbb{R}^N$ by

$$F^{\text{red}}(\omega, X, Y, Z) = F\big(X, (\omega_a Y), (\omega_a \omega_b Z)\big), \tag{5}$$

where the right-hand side above means that X, $\omega_a Y$, and $\omega_a \omega_b Z$ are substituted in place of u, $\partial_a u$, and $\partial_a \partial_b u$ of $F(u, \partial u, \partial^2 u)$, respectively, with $\omega_0 = -1$. We say that the *null condition* is satisfied if

$$F^{\text{red}}(\omega, X, Y, Z) \equiv 0, \quad \omega \in \mathbb{S}^{n-1},\ X, Y, Z \in \mathbb{R}^N. \tag{6}$$

If the null condition is satisfied, then we have the small data global existence (see Klainerman [12] and Christodoulou [3] for $n = 3$; see the author [6] for $n = 2$; see also Alinhac [1], the author [8], and Zha [15] for the 2D case with quadratic nonlinearity satisfying the null condition). This global solution under the null condition is known to be *asymptotically free*, that is to say, there is *asymptotic data* $(f^+, g^+) \in \dot{H}^1(\mathbb{R}^n) \times L^2(\mathbb{R}^n)$, depending on f, g, and ε as well as the nonlinearity F, such that we have

$$\|(u - u^+)(t)\|_E := \big\|\partial(u - u^+)(t)\big\|_{L^2(\mathbb{R}^n)} \to 0, \quad t \to \infty$$

for the *free solution* u^+ to the *free wave equation* $\Box u^+ = 0$ with initial data $(u^+, \partial_t u^+) = (f^+, g^+)$ at $t = 0$, where $\dot{H}^1(\mathbb{R}^n)$ is the completion of $C_0^\infty(\mathbb{R}^n)$ with respect to the norm $\|\varphi\|_{\dot{H}^1} = \|\nabla\varphi\|_{L^2}$. Moreover,

$$\lim_{\varepsilon \to +0} \big\|\varepsilon^{-1}(f^+, g^+) - (f, g)\big\|_{\dot{H}^1 \times L^2} = 0. \tag{7}$$

Hence, quite roughly speaking, the asymptotic data (f^+, g^+) is close to the original initial data $(\varepsilon f, \varepsilon g)$ when ε is small enough.

We can easily see that the null condition is not necessary for the small data global existence if we consider a simple example $\Box(u_1, u_2) = \big((\partial_t u_2)^{p-1}(\partial_t u_1), 0\big)$ for which the null condition is violated, but a global solution apparently exists for any data, as we can first solve u_2, and then u_1. Therefore we would like to find a weaker sufficient condition for the small data global existence.

For this purpose, we briefly introduce the *reduced system*. We write $r = |x|$, $\omega = (\omega_1, \ldots, \omega_n) = x/|x|$ and $\omega_0 = -1$ in what follows. We also use the notation $\langle z \rangle = \sqrt{1 + |z|^2}$. For a free solution u^0 to $\Box u^0 = 0$ with C_0^∞-data, we know that the solution decays like $\langle t+r \rangle^{-(n-1)/2}$ near the light cone $t = r$, and decays faster away from the light cone (and thus, in a region where the main part of u^0 lies, we may consider $r = t$). We also know that $\partial_+ u^0 = (\partial_t + \partial_r)u^0$ decays faster than $\partial_a u^0$ for $0 \le a \le n$ (and consequently, for large t, we may expect $\partial_t \sim -\partial_r = \omega_0 \partial_r$, and $\partial_k \sim \omega_k \partial_r$ for $1 \le k \le n$ through the polar coordinates). We assume that these properties also hold for the solution u to the nonlinear system with small ε. Then, writing $\widetilde{u}(t,x) = r^{(n-1)/2} u(t,x)$, and recalling that $(p-1)(n-1)/2 = 1$, we expect

$$r^{(n-1)/2} \Box u \sim (\partial_t^2 - \partial_r^2)\widetilde{u} = \partial_+(-2\partial_r + \partial_+)\widetilde{u} \sim -2\partial_+(\partial_r \widetilde{u}),$$

$$r^{(n-1)/2} F(u, \partial u, \partial^2 u) \sim \frac{1}{r} F(\widetilde{u}, \partial\widetilde{u}, \partial^2 \widetilde{u}) \sim \frac{1}{t} F^{\mathrm{red}}(\omega, \widetilde{u}, \partial_r \widetilde{u}, \partial_r^2 \widetilde{u}).$$

Now, introducing $\phi(t, \sigma, \omega) = \widetilde{u}(t, r\omega)\big|_{r=t+\sigma}$, we can rewrite (1) as

$$\partial_t \partial_\sigma \phi = -\frac{1}{2t} F^{\mathrm{red}}(\omega, \phi, \partial_\sigma \phi, \partial_\sigma^2 \phi) + (\text{“remainder” terms}). \tag{8}$$

Neglecting the “remainder” terms (though we do not know whether they really are negligible at this point), and replacing ϕ with $U = U(t, \sigma, \omega)$, we reach the *reduced system*

$$\partial_t \partial_\sigma U = -\frac{1}{2t} F^{\mathrm{red}}(\omega, U, \partial_\sigma U, \partial_\sigma^2 U), \quad (t, \sigma, \omega) \in (1, \infty) \times \mathbb{R} \times \mathbb{S}^{n-1}. \tag{9}$$

We look for a solution U to (9) vanishing as $\sigma \to \infty$ and satisfying

$$U(1, \sigma, \omega) = \varepsilon U^0(\sigma, \omega), \tag{10}$$

where U^0 is a given function vanishing as $\sigma \to \infty$, and ε is small.

Let us assume that we have $F(u, 0, 0) = 0$ in addition, so that the energy method is easily applicable. Then it is believed that the small data global existence for the original system (1)–(2) is closely related to that for the reduced system (9)–(10). In fact, if the null condition is satisfied, then (9)–(10) trivially has the global solution $U(t,\sigma,\omega)=\varepsilon U^0(\sigma,\omega)$. In their study for the Einstein equations, Lindblad–Rodnianski [14] introduced the notion of the *weak null condition*. We say that the weak null condition is satisfied if we have the global existence of solutions for (9)–(10) with small $\varepsilon>0$ for all U^0 decaying sufficiently fast as $|\sigma|\to\infty$. One may conjecture that the weak null condition implies the small data global existence for the original Cauchy problem (1)–(2) with $F(u,0,0)=0$, but this is still an open problem.

Our aim is to find sufficient conditions for the small data global existence, which lie in between the null condition and the weak null condition. Under such conditions, we would also like to write the asymptotic behavior of the solution u to (1)–(2) through the solution U to the reduced system (9)–(10): If the "remainder" term in (8) can be really regarded as a remainder, one may choose a good profile U^0 such that $\partial_\sigma^k U(t,\sigma,\omega)\to\partial_\sigma^k\phi(t,\sigma,\omega)$ as $t\to\infty$ for $k=0,1$, and this gives the asymptotic behavior for u and $\partial_a u$, as $r^{(n-1)/2}u(t,x)=\phi(t,r-t,\omega)$ and $r^{(n-1)/2}\partial_a u(t,x)\sim\omega_a(\partial_\sigma\phi)(t,r-t,\omega)$. To be more precise, we want to find $\rho>0$, $\kappa\geq 0$, and a non-negative weight function $\mathcal{W}(t,r)$, satisfying

$$\mathcal{W}(t,r)\geq C_1\langle t-r\rangle \text{ for } |t-r|\geq C_0\varepsilon\log t \tag{11}$$

with some positive constants C_0 and C_1, such that we have the following:

(AP) For any $f,g\in C_0^\infty(\mathbb{R}^n;\mathbb{R}^N)$ and $0<\varepsilon\ll 1$, there is an *asymptotic profile* $U^0(\sigma,\omega)$, depending on f,g,ε, such that we have

$$r^{(n-1)/2}\partial_a u(t,x)=\omega_a(\partial_\sigma U)(t,r-t,\omega)+O\big(\varepsilon\langle t+r\rangle^{-\rho}\mathcal{W}(t,r)^{-\kappa}\big), \tag{12}$$

as well as (12) with $r^{(n-1)/2}\partial_a u(t,x)$ being replaced by $t^{(n-1)/2}\partial_a u(t,x)$, for $0\leq a\leq n$ and $(t,x)\in[0,\infty)\times\mathbb{R}^n$, where u and U are (global) solutions to (1)–(2) and (9)–(10), respectively. Moreover, we have

$$\lim_{t\to\infty}\sum_{0\leq a\leq n}\big\|\partial_a u(t,x)-\omega_a r^{-(n-1)/2}(\partial_\sigma U)(t,r-t,\omega)\big\|_{L^2(\mathbb{R}^n_x)}=0. \tag{13}$$

Remark 1. (i) From (12), we can show

$$r^{(n-1)/2}u(t,x)=\omega_a U(t,r-t,\omega)+O\big(\varepsilon\langle t+r\rangle^{-\rho}\mathcal{W}(t,r)^{-\kappa}\langle t-r\rangle\big).$$

(ii) If the error term in (12) is good enough, (13) follows immediately from (12).

(iii) It is known that u is asymptotically free if and only if there is $\Phi^+ = \Phi^+(\sigma, \omega) \in L^2(\mathbb{R} \times \mathbb{S}^{n-1})$ such that

$$\lim_{t\to\infty} \sum_{0\le a\le n} \left\|\partial_a u(t,x) - \omega_a r^{-(n-1)/2}\Phi^+(r-t,\omega)\right\|_{L^2(\mathbb{R}^n_x)} = 0$$

(see the author [8]). Moreover the correspondence between the asymptotic data $(f^+, g^+) \in \dot{H}^1(\mathbb{R}^n) \times L^2(\mathbb{R}^n)$ and $\Phi^+ \in L^2(\mathbb{R} \times \mathbb{S}^{n-1})$ above is one-to-one. If we have (13), then it is easy to see that u is asymptotically free if and only if there is $\Phi^+ \in L^2(\mathbb{R}\times\mathbb{S}^{n-1})$ such that $\partial_\sigma U(t,\cdot,\cdot) \to \Phi^+$ in $L^2(\mathbb{R} \times \mathbb{S}^{n-1})$ as $t \to \infty$. Therefore, by investigating the behavior of the solution to the reduced system, we can determine whether the solution u is asymptotically free or not.

Let us briefly mention the known results for the semilinear case in this direction. In what follows, an $\mathbb{R}^d$-vector is identified with a column vector for any $d \in \mathbb{N}$. Two different conditions are known to imply the small data global existence and (AP) with $\mathcal{W}(t,r) = \langle t-r\rangle$ and appropriate ρ, κ. Here we only give essential and simplified version of the conditions; see the references below for the full description.

The first condition is essentially due to Alinhac:

(A) The nonlinearity has the form $F = F(u, \partial u)$, and the reduced system can be written as

$$\begin{cases} \partial_t\partial_\sigma V = -\dfrac{1}{2t}\big(A(\omega, W, \partial_\sigma W)\partial_\sigma V + B(\omega, W, \partial_\sigma W)\partial_\sigma W\big), \\ \partial_t\partial_\sigma W = 0 \end{cases}$$

with $U = (V, W) \in \mathbb{R}^{N_1} \times \mathbb{R}^{N_2}$ for some N_1, N_2 satisfying $N_1 + N_2 = N$ and $1 \le N_2 < N$, where A and B are $N_1 \times N_1$ and $N_1 \times N_2$-matrix valued functions, respectively, whose components are homogeneous polynomials in $(W, \partial_\sigma W)$ of degree $p-1$ with smooth coefficients depending on ω.

Condition (A) implies the small data global existence and (AP) (see Alinhac [2], and the author [7, 8]).

Another kind of condition is due to Katayama, Matoba, Matsumura, and Sunagawa, and is an extension of the condition imposed by Agemi[1]:

[1]Agemi's condition is equivalent to (KMS) with $n = 2$ and $N = 1$. In some of his talks, he conjectured that his condition implies the small data global existence (and now the conjecture turns out to be true). As far as the author knows, there is no reference by himself on his condition or conjecture.

(KMS) The nonlinearity has the form $F = F(\partial u)$, and its reduced nonlinearity $F^{\mathrm{red}} = F^{\mathrm{red}}(\omega, Y)$ satisfies

$$Y \cdot F^{\mathrm{red}}(\omega, Y) \ge 0, \quad \omega \in \mathbb{S}^{n-1},\, Y \in \mathbb{R}^N.$$

Condition (KMS) implies the small data global existence for $n = 2, 3$, and (AP) for $n = 3$; however, we need some additional (and probably technical) assumption to obtain (AP) for $n = 2$; for example, if we assume $N = 1$, or if we assume a stronger assumption

$$Y \cdot F^{\mathrm{red}}(\omega, Y) = 0, \quad \omega \in \mathbb{S}^1,\, Y \in \mathbb{R}^N \tag{14}$$

in addition, then we have (AP) also for $n = 2$ (see the author *et al.* [8, 9, 10, 11]; see also Hoshiga [5] and Kubo [13] for $n = 2$ and $N = 1$).

It is trivial that the null condition leads to (A) and (KMS). (A) implies the weak null condition as one can first solve W, and then the equation for V is nothing but a linear one with variable coefficients. If (KMS) is satisfied, then the weak null condition is satisfied, because the solution U to (9)–(10) satisfies

$$\partial_t |\partial_\sigma U|^2 = 2\partial_\sigma U \cdot \partial_t \partial_\sigma U = -t^{-1}(\partial_\sigma U) \cdot F^{\mathrm{red}}(\omega, \partial_\sigma U) \le 0,$$

which leads to $|\partial_\sigma U(t)|^2 \le \varepsilon^2 |\partial_\sigma U^0|^2$ for $t \ge 1$, and U exists globally.

A typical example satisfying (A) is

$$\begin{cases} \Box v = \alpha(\partial_t w)^{p-1}\partial_t v + \beta(\partial_t w)^p + H_1(u, \partial u), \\ \Box w = H_2(u, \partial u) \end{cases}$$

with $u = (u_1, u_2) = (v, w)$, where $H = (H_1, H_2)$ satisfies the null condition. The null condition is violated when $(\alpha, \beta) \ne (0, 0)$. If $\alpha, \beta \ge 0$ and $(\alpha, \beta) \ne (0, 0)$, then, making use of (AP), we can show that there is some $(f, g) \in C_0^\infty(\mathbb{R}^n; \mathbb{R}^2)$ such that the uniform decay rate for v becomes really worse than that for the free solution, and $\|v(t)\|_E \to \infty$ as $t \to \infty$ (thus v cannot be asymptotically free), while w is asymptotically free.

An important example satisfying (KMS) (the stronger assumption (14) in fact) and violating the null condition is:

$$\begin{cases} \Box u_1 = (\partial_t u_1)^{p-1}\partial_t u_2 + H_1(u, \partial u), \\ \Box u_2 = -(\partial_t u_1)^p + H_2(u, \partial u), \end{cases}$$

where, as before, $H = (H_1, H_2)$ satisfies the null condition. In this case, for *any* $f, g \in C_0^\infty(\mathbb{R}^n; \mathbb{R}^2)$ and sufficiently small ε, by using (AP) we find that $\|u_1(t)\|_E \to 0$ as $t \to \infty$, while u_2 tends to a free solution which is non-zero in general. As the 0-function can be regarded as a free solution with initial data 0, we may say that the global solution $u = (u_1, u_2)$ with small data is always asymptotically free as in the case of the null condition; however, as the asymptotic data for u_1 is always 0, the asymptotic data (f^+, g^+) satisfies

$$\lim_{t\to\infty} \left\|\varepsilon^{-1}(f^+, g^+) - (f, g)\right\|_{\dot{H}^1\times L^2} \neq 0, \tag{15}$$

unless $(f_1, g_1) \equiv (0, 0)$. This makes a sharp contrast to (7) under the null condition.

In this paper, we would like to consider the quasi-linear version of the condition (A). This paper is intended to be an announcement of the main results in the next section, and the full proof will be given in a forthcoming paper.

§2. Main results

We introduce the quasi-linear version of the condition (A):

(QLA) The reduced system can be written as

$$\begin{cases} \partial_t\partial_\sigma V = -\dfrac{1}{2t}\big(G\partial_\sigma^2 V + A\partial_\sigma V + B\partial_\sigma W\big), \\ \partial_t\partial_\sigma W = 0 \end{cases}$$

with $U = (V, W) \in \mathbb{R}^{N_1} \times \mathbb{R}^{N_2}$ for some N_1, N_2 satisfying $N_1 + N_2 = N$ and $1 \le N_2 < N$, where $G = G(\omega, W, \partial_\sigma W)$ and $A = A(\omega, W, \partial_\sigma W)$ (resp. $B = B(\omega, W, \partial_\sigma W)$) are $N_1 \times N_1$ (resp. $N_1 \times N_2$)-matrix valued functions, whose components are homogeneous polynomials in $(W, \partial_\sigma W)$ of degree $p - 1$ with smooth coefficients depending on ω.

For $f, g \in C_0^\infty(\mathbb{R}^n; \mathbb{R}^N)$, we define the *Friedlander radiation field*

$$\mathcal{F}_0[f, g](\sigma, \omega) = \lim_{t\to\infty} t^{(n-1)/2} u^0\big(t, (t + \sigma)\omega\big),$$

where u^0 is a solution to $\Box u^0 = 0$ with $(u^0, \partial_t u^0) = (f, g)$ at $t = 0$.

The following is our main result:

Theorem 1. *Let $n = 2, 3$. If* (QLA) *is satisfied, then we have the small data global existence of classical solutions for* (1)–(2). *Moreover, if* (QLA) *holds with $N_1 = 1$, then we have* (AP) *with appropriate choice of $\rho > 0$, $\kappa \ge 0$, and $\mathcal{W}(t, r)$ satisfying* (11). *The asymptotic profile U^0 in* (AP) *satisfies $\partial_\sigma U^0 \to \partial_\sigma \mathcal{F}_0[f, g]$ uniformly, and also in $L^2(\mathbb{R} \times \mathbb{S}^{n-1})$, as $\varepsilon \to +0$.*

Remark 2. When $n = 3$ and $F = F(\partial u, \partial^2 u)$, the small data global existence under (QLA) was already established by Hidano–Yokoyama [4].

The proof will be given in a forthcoming paper, but we make some comments on the technical assumption related to the proof. Similarly to the proof for the semilinear case in [7], the existence part of Theorem 1 can be proved by the (standard) vector field method, with the help of the so-called ghost weight energy (see, for example, [1, 2]). In *a priori* estimates for $u = (v, w) \in \mathbb{R}^{N_1} \times \mathbb{R}^{N_2}$, we allow slow growth of the generalized energy for v, while we require the boundedness of that for w. To accomplish these estimates, we need to evaluate the generalized energy of the highest order separately for v and w, and this is the point where we need the diagonality of the nonlinearity.

To show (AP), we follow the same strategy as that for the semilinear case in [7]. The main difficulty here is that the reduced equations for V are PDEs for $\partial_\sigma V$ in the quasi-linear case, while they look like ODEs for $\partial_\sigma V$ in the semilinear case. To overcome this difficulty, we use the method of characteristics, and the assumption $N_1 = 1$ is required here.

The null condition is closely related to the asymptotic freedom of solutions. In fact, we can prove the following:

Theorem 2. *Let $n = 3$, and we assume that* (QLA) *is satisfied with $N_1 = 1$. Then the following two are equivalent to each other*:

(i) *For any $f, g \in C_0^\infty(\mathbb{R}^3; \mathbb{R}^N)$ and small ε, the global solution u to* (1)–(2) *is asymptotically free, and the asymptotic data satisfy* (7).

(ii) *The null condition is satisfied.*

Note that (i) without (7) does not imply (ii) in the above. We will give a counterexample in the next section. We believe that the same result holds also for $n = 2$, but it is very difficult in 2D to find (f, g) such that $\mathcal{F}_0[f, g]$, a good approximation of U^0 for small ε, has the desired property required in the proof.

§3. The asymptotic behavior for simple examples

To illustrate the application of (AP) in Theorem 1, we consider the following toy model and investigate what kind of asymptotic behavior can happen for quasi-linear systems:

$$
\begin{cases} \Box v(t,x) = \displaystyle\sum_{a,b=0}^{n} g^{ab}(w, \partial w)\partial_a \partial_b v, \\ \Box w(t,x) = 0, \end{cases} \tag{16}
$$

where $u = (v, w)$ is $\mathbb{R}^2$-valued, and each g^{ab} is a homogeneous polynomial of degree $p-1$ in $(w, \partial w)$. Let $U^0 = (V^0, W^0)$ be the asymptotic profile, and $U = (V, W)$ be a global solution to the reduced system with initial data $U(1) = \varepsilon U^0$. Then we get $W(t, \sigma, \omega) = \varepsilon W^0(\sigma, \omega)$, and V satisfies

$$\partial_t \partial_\sigma V = -(2t)^{-1} \varepsilon^{p-1} G(\omega, W^0, \partial_\sigma W^0) \partial_\sigma^2 V. \tag{17}$$

Observe that $G^0 := G(\omega, W^0, \partial_\sigma W^0)$ is independent of t. We assume that the null condition is violated. This assumption is equivalent to $G(\omega, X, Y) \not\equiv 0$.

We can easily check that $W^0 = \mathcal{F}_0[f_2, g_2]$. Hence W^0 and G^0 are independent of ε. Let $f_2(x) = g_2(x) = 0$ for $|x| \geq R$. When $n = 3$, the strong Huygens principle implies $W^0(\sigma, \omega) = 0$ for $|\sigma| \geq R$. In contrast, when $n = 2$, we only have $W^0(\sigma, \omega) = 0$ for $\sigma \geq R$ (by the finite speed propagation). In fact, we can even show the existence of f_2, g_2 such that $W^0(\sigma, \omega) > 0$ for all $\sigma < R$.

By introducing a variable $\tau = \log t$, we can eliminate the factor t^{-1} from (17), so it is essential to investigate the following type of equation:

$$\begin{cases} \partial_\tau \psi + a(\sigma) \partial_\sigma \psi = 0 \\ \psi(0, \sigma) = \psi^0(\sigma), \end{cases} \tag{18}$$

where a is a bounded C^1-function with a bounded derivative, and ψ^0 is a bounded continuous function with $\psi^0(\sigma) \to 0$ as $|\sigma| \to \infty$. Let $\sigma_* = \sigma_*(\tau; s)$ be a solution to $\dot{\sigma_*} = a(\sigma_*)$ with $\sigma_*(0; s) = s$. Then we have $\psi\big(\tau, \sigma_*(\tau; s)\big) = \psi^0(s)$. By the implicit function theorem, there is $s_* = s_*(\tau, \sigma)$ such that $\sigma_*(\tau; s_*) = \sigma$, and we get $\psi(\tau, \sigma) = \psi^0\big(s_*(\tau, \sigma)\big)$. This is the well-known method of characteristics.

As the equation for σ_* is autonomous, σ_* can converge only to a zero of a as $\tau \to \infty$. Moreover, $a\big(\sigma_*(\tau; s)\big) = 0$ for some $\tau > 0$ if and only if $a(s) = 0$. From these facts, we obtain the following crucial lemma.

Lemma 1. *We define*

$$s_\infty(\sigma) := \begin{cases} \sup\{s \in (-\infty, \sigma); a(s) = 0\}, & \text{if } a(\sigma) > 0, \\ \sigma, & \text{if } a(\sigma) = 0, \\ \inf\{s \in (\sigma, \infty); a(s) = 0\}, & \text{if } a(\sigma) < 0 \end{cases}$$

with the convention $\sup \emptyset = -\infty$ *and* $\inf \emptyset = \infty$. *Then* $\psi(\tau, \sigma) \to \psi^\infty(\sigma) := \psi^0\big(s_\infty(\sigma)\big)$ *as* $\tau \to \infty$ *for any* $\sigma \in \mathbb{R}$, *where* $\psi^0(\pm\infty) = 0$.

If $a(\sigma) = 0$ for $|\sigma| \geq R$ with some $R > 0$, then, by using the Lebesgue convergence theorem for $|\sigma| \leq R$, we get $\psi(\tau) \to \psi^\infty$ in $L^2(\mathbb{R})$ as $\tau \to \infty$, because $\psi(\tau, \sigma) = \psi^0(\sigma) = \psi^\infty(\sigma)$ for $|\sigma| \geq R$.

We apply this lemma with $\psi = \partial_\sigma V$ to see the asymptotic behavior for (16). Let $n = 3$. Then, as $G^0 = 0$ for $|\sigma| \geq R$ with some $R > 0$, there is $s_\infty = s_\infty(\sigma, \omega)$ such that $\partial_\sigma V(t)$ converges to $\varepsilon(\partial_\sigma V^0)\big(s_\infty(\sigma,\omega),\omega\big)$ in $L^2(\mathbb{R} \times \mathbb{S}^2)$ as $t \to \infty$. From this, we see that the solution $u = (v, w)$ is asymptotically free, and (16) violating the null condition is a counter example for Theorem 2 without (7). For this example, we can directly prove that (7) is violated for some (f, g) in the following way: As s_∞ is determined just by the sign of G^0, it is independent of ε. Hence we have $(\partial_\sigma V^0)\big(s_\infty(\sigma,\omega),\omega\big) \to \big(\partial_\sigma \mathcal{F}_0[f_1, g_1]\big)\big(s_\infty(\sigma,\omega),\omega\big)$ in $L^2(\mathbb{R}\times\mathbb{S}^2)$ as $\varepsilon \to +0$. Accordingly, we see that (7) is true, if and only if $\big(\partial_\sigma \mathcal{F}_0[f_1, g_1]\big)\big(s_\infty(\sigma,\omega),\omega\big) = \partial_\sigma \mathcal{F}_0[f_1, g_1](\sigma,\omega)$ for almost every (σ,ω). From this, we can easily construct (f, g) violating (7).

In two space dimensions, the situation is completely different. Let $\sum_{a,b} g^{ab}\partial_a\partial_b v = -w^2\Delta v$. We take f_2, g_2 such that $W^0 > 0$ for $\sigma < R$ and $W^0 = 0$ for $\sigma \geq R$. We can take f_1, g_1 such that $\partial_\sigma \mathcal{F}_0[f_1, g_1](R,\omega)$ is positive. Accordingly $\partial_\sigma V^0(R,\omega) > 0$ for sufficiently small ε. As $G^0 = -(W^0)^2 < 0$ for $\sigma < R$ and $G^0 = 0$ for $\sigma = R$, we see that $\partial_\sigma V(t,\sigma,\omega) \to \varepsilon\partial_\sigma V^0(R,\omega)(> 0)$ as $t \to \infty$ for all $\sigma \leq R$ and $\omega \in \mathbb{S}^1$, which implies $\|\partial_\sigma V(t)\|_{L^2} \to \infty$ as $t \to \infty$, and accordingly we have $\|v(t)\|_E \to \infty$ as $t \to \infty$. In other words, the growth of the energy occurs, and v (hence $u = (v, w)$) cannot be asymptotically free, differently from the 3D case.

Acknowledgments. The author would like to thank two anonymous referees for their valuable comments. This work is partially supported by JSPS KAKENHI Grant No. 18H01128.

References

[1] S. Alinhac, The null condition for quasilinear wave equations in two space dimensions I, Invent. Math., **145** (2001), 597–618.

[2] S. Alinhac, Semilinear hyperbolic systems with blowup at infinity, Indiana Univ. Math. J., **55** (2006), 1209–1232.

[3] D. Christodoulou, Global solutions of nonlinear hyperbolic equations for small initial data, Comm. Pure Appl. Math., **39** (1986), 267–282.

[4] K. Hidano and K. Yokoyama, Global existence for a system of quasi-linear wave equations in 3D satisfying the weak null condition, IMRN, Volume 2020 (2020), 39–70.

[5] A. Hoshiga, The existence of the global solutions to semilinear wave equations with a class of cubic nonlinearities in 2-dimensional space, Hokkaido Math. J., **37** (2008), 669–688.

[6] S. Katayama, Global existence for systems of nonlinear wave equations in two space dimensions, II, Publ. RIMS, Kyoto Univ., **31** (1995), 645–665.
[7] S. Katayama, Asymptotic pointwise behavior for systems of semilinear wave equations in three space dimensions, J. Hyperbolic Differ. Equ., **9** (2012), 263–323.
[8] S. Katayama, Global Solutions and the Asymptotic Behavior for Nonlinear Wave Equations with Small Initial Data, MSJ Memoirs, **36**, Math. Soc. Japan, Tokyo, 2017.
[9] S. Katayama, T. Matoba and H. Sunagawa, Semilinear hyperbolic systems violating the null condition, Math. Ann., **31** (2015), 275–312.
[10] S. Katayama, A. Matsumura and H. Sunagawa, Energy decay for systems of semilinear wave equations with dissipative structure in two space dimensions, NoDEA, **22** (2015), 601–628.
[11] S. Katayama, D. Murotani and H. Sunagawa, The energy decay and asymptotics for a class of semilinear wave equations in two space dimensions, J. Evol. Equ., **12** (2012), 891–916.
[12] S. Klainerman, The null condition and global existence to nonlinear wave equations, In: Nonlinear Systems of Partial Differential Equations in Applied Mathematics, Part 1, pp. 293–326, Lectures in Appl. Math., **23**, AMS, Providence, 1986.
[13] H. Kubo, Asymptotic behavior of solutions to semilinear wave equations with dissipative structure, Discrete Contin. Dynam. Systems, Supplement Volume 2007 (2007), 602–613.
[14] H. Lindblad and I. Rodnianski, The weak null condition for Einstein's equations, C. R. Math. Acad. Sci. Paris, **336** (2003), 901–906.
[15] D. Zha, Global and almost global existence for general quasilinear wave equations in two space dimensions, J. Math. Pures Appl., **123** (2019), 270–299.

Department of Mathematics, Graduate School of Science,
Osaka University,
1-1 Machikaneyama-cho, Toyonaka,
Osaka 560-0043, Japan
E-mail address: katayama@math.sci.osaka-u.ac.jp

Advanced Studies in Pure Mathematics 85, 2020
The Role of Metrics in the Theory of Partial Differential Equations
pp. 227–238

A note on traveling waves for area-preserving geometric flows

Takashi Kagaya and Yoshihito Kohsaka

Abstract.

The traveling waves for area-preserving geometric flows will be discussed. For an evolving plane curve, which is governed by the area-preserving curvature flow and has two endpoints moving freely on the x-axis with generating constant contact angles, we prove that there exists only a convex traveling wave. In the proof, a boundary value problem for an angle function associated with the Gauss map of the profile curve is analyzed by means of a shooting argument.

§1. Introduction

The traveling waves of plane curves governed by the are-preserving geometric evolution laws will be discussed. Set $\Pi := \{(x, y) \in \mathbb{R}^2 \mid y = 0\}$ and let $\Gamma(t) \subset \mathbb{R}^2$ be an evolving plane curve with two endpoints which are named $\partial\Gamma(t) := \{P_\pm(t)\}$. The motion of $\Gamma(t)$ is governed by the free boundary problem for the area-preserving curvature flow

$$(1) \qquad \begin{cases} V = \kappa - \kappa_{av} \quad \text{on} \quad \Gamma(t), \\ P_\pm(t) \in \Pi, \quad \sphericalangle(\Gamma(t), \Pi) = \psi_\pm \quad \text{at} \quad P_\pm(t), \end{cases}$$

or that for the surface diffusion

$$(2) \qquad \begin{cases} V = -\kappa_{ss} \quad \text{on} \quad \Gamma(t), \\ P_\pm(t) \in \Pi, \quad \sphericalangle(\Gamma(t), \Pi) = \psi_\pm, \quad \kappa_s = 0 \quad \text{at} \quad P_\pm(t). \end{cases}$$

Here V and κ are the normal velocity and the curvature of $\Gamma(t)$, respectively. In our sign convention, the curvature κ for a circle with an outer

---

Received December 5, 2018.
Revised April 17, 2019.
2010 *Mathematics Subject Classification.* Primary 35C07; Secondary 34B08, 53C44.
Key words and phrases. area-preserving curvature flow, traveling waves.

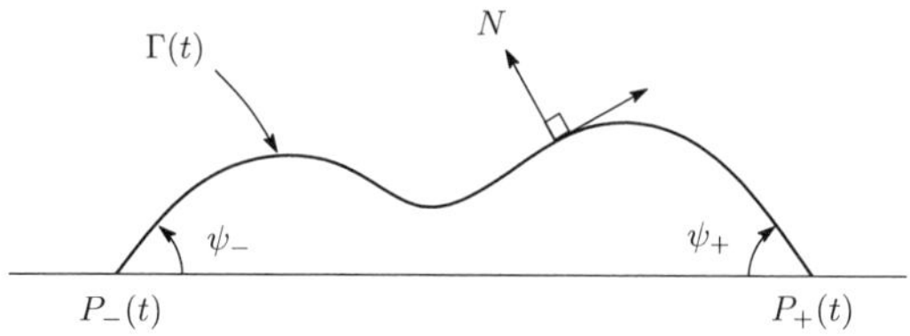

Fig. 1. Setting.

unit normal is negative. s is the arc-length parameter along $\Gamma(t)$ and κ_{av} is the mean value of κ, which is given by

$$\kappa_{av}(t) = \frac{1}{\text{Length}\,[\Gamma(t)]} \int_{\Gamma(t)} \kappa(s,t)\,ds.$$

From the boundary condition $P_\pm(t) \in \Pi$, the endpoints $P_\pm(t)$ are characterized by x-coordinate, that is, the endpoint with a smaller x-coordinate is $P_-(t)$ and the other is $P_+(t)$. The direction of the arc-length parameter s is taken from $P_-(t)$ to $P_+(t)$ and the unit normal N is obtained by rotating the unit tangent $\pi/2$ counterclockwise. Furthermore, $\sphericalangle(\Gamma(t), \Pi)$ means the interior contact angle between $\Gamma(t)$ and Π, and ψ_- and ψ_+ are measured counter-clockwise and clockwise from x-axis, respectively. It is well-known that both of these geometric flows for closed curves have a variational structure that the length of the curves is non-increasing in time under the area constraint. On the other hand, evolving plane curves by the area-preserving curvature flow preserve the convexity (cf. [2]), whereas ones by the surface diffusion can lose it (cf. [3]).

Our aim in the present paper is to obtain a traveling wave $\mathcal{W}(t)$ for (1) and to compare the property of it with that of traveling waves for (2), which are studied in [5]. Here, a traveling wave $\mathcal{W}(t)$ is a classical solution represented as

$$\mathcal{W}(t) = \mathcal{W}(0) + ct\vec{e}, \tag{3}$$

where $\vec{e} = (1,0)$, $\mathcal{W}(0)$ is a smooth plane curve and c is a constant. Note that $\mathcal{W}(0)$ represents the profile of $\mathcal{W}(t)$ and c is the wave speed. Consequently, in order to achieve our aim, we have only to find the pair $(\mathcal{W}(0), c)$, which is solution of (1), and to analyze the property of it.

Now, we introduce Shimojo and the first author [6] that has already shown the existence of a traveling wave for (1). But, in the proof of it, they needed the convexity of the profile curve $\mathcal{W}(0)$ in advance since they used the angle associated with the Gauss map of $\mathcal{W}(0)$ as the variable.

Our motivation for the present paper is to improve this point. Thus we use an approach different from [6] for the proof. As a result, we will prove the existence of a traveling wave for (1) without assuming the convexity of $\mathcal{W}(0)$ in advance and also show that there exists only a convex one for (1).

For the evolution of curves by other geometric flows with a similar boundary condition to ours, see [1, 4]. In particular, Guo, Matano, Shimojo, and Wu [4] studied the geometric flow $V = \kappa + A$ for a constant $A > 0$ and proved that any global solution to $V = \kappa + A$ with finite area becomes convex eventually under the restrictions that a initial curve is represented by a graph and embedded in the upper half space. We remark that traveling waves to our problem fulfill the geometric flow $V = \kappa + A$ for a suitable constant A. However, we do not assume the restrictions above, so that the eventual convexity by [4] can not be directly applied to ours. Hence, the convexity of traveling waves is unclear in the beginning of our proof.

This paper is organized as follows. In Section 2, we state the main theorem and make a comment on a difference between a traveling wave of (1) and that of (2). Furthermore, we derive a boundary value problem on a profile curve of a traveling wave for (1). Section 3 is devoted to find a solution to the boundary value problem obtained in Section 2 by means of a shooting argument.

§2. Main theorem and a profile equation of traveling wave

First, we observe that if $c = 0$ in (3), the pair $(\mathcal{W}(0), c)$ constructs a stationary solution. It is easy to see that stationary solutions to (1) exist only if $\psi_- = \psi_+$ and its shape is an arc. Also, if the pair $(\mathcal{W}(0), c)$ constructs the traveling wave $\mathcal{W}(t)$ for (1) defined as (3), the pair of

$$\widetilde{\mathcal{W}}(0) = \{(-x, y) \mid (x, y) \in \mathcal{W}(0)\}, \quad \tilde{c} = -c$$

constructs a reflected traveling wave. Therefore, through this paper, we consider only for the case $c > 0$ without loss of generality.

In addition, it follows that $(\lambda\mathcal{W}(0) + a\vec{e}_1, c/\lambda)$ constructs a traveling wave for any $a \in \mathbb{R}$ and $\lambda > 0$ if $(\mathcal{W}(0), c)$ constructs a traveling wave. Hence, it is sufficient to find a traveling wave $\mathcal{W}(t)$ defined as (3) such that

(W1) the length of $\mathcal{W}(0)$ is equal to 1,
(W2) the left endpoint of $\mathcal{W}(0)$ is $(0, 0)$ in the (x, y)-coordinate plane.

Furthermore, in this paper, we find a traveling wave $\mathcal{W}(t)$ under the restriction related to the winding number of $\mathcal{W}(t)$ that

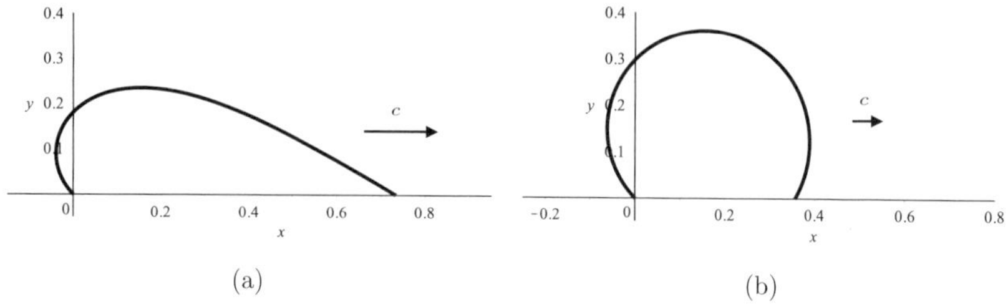

Fig. 2. The profile of a traveling wave $\mathcal{W}(t)$ for (1) with (a) $\psi_- = 3\pi/4$, $\psi_+ = \pi/6$, $c \approx 5.50$ and (b) $\psi_- = 3\pi/4$, $\psi_+ = 2\pi/3$, $c \approx 0.35$.

(W3) $\displaystyle\int_{\mathcal{W}(0)} \kappa_{\mathcal{W}}\, ds + \psi_- \in (-\pi, \pi)$,

where $\kappa_{\mathcal{W}}$ is the curvature of $\mathcal{W}(0)$. Now we state our main result.

Theorem 1. *Let $\psi_\pm$ be given constants and assume that $\psi_- \in (0, \pi)$. Then there exists a traveling wave $\mathcal{W}(t)$ for (1) defined as (3) such that $c > 0$ and a profile curve $\mathcal{W}(0)$ satisfies (W1)–(W3) if and only if $\psi_+ \in (0, \psi_-)$. Furthermore, the traveling wave $\mathcal{W}(t)$ is convex and the wave speed c fulfills*

$$\begin{cases} \psi_+ + \psi_- < c < \dfrac{\psi_+ + \psi_-}{\sin \psi_+} & \text{if } \psi_+ \in (0, \psi_*), \\ 0 < c \le \psi_+ + \psi_- & \text{if } \psi_+ \in [\psi_*, \psi_-), \end{cases} \tag{4}$$

where $\psi_* \in (0, \pi/2)$ is a unique solution of the equation

$$(\psi_- + \psi_*)\left(1 + \tan\frac{\psi_-}{2}\right)\left(1 - \tan\frac{\psi_*}{2}\right) - 2\left(\tan\frac{\psi_-}{2} + \tan\frac{\psi_*}{2}\right) = 0.$$

In addition, $\mathcal{W}(t)$ is unique up to the translation and the scaling $(\lambda\mathcal{W}(0) + a\vec{e}_1, c/\lambda)$ for any $a \in \mathbb{R}$ and $\lambda > 0$.

Remark 2. *The angle condition $\psi_+ \in (0, \psi_-)$ which guarantees the existence of a traveling wave is the same as that in [6]. Note that Theorem 1 also includes non-graph cases (cf. Fig. 2).*

On the other hand, in [6], the estimate $0 < c < (\psi_+ + \psi_-)/\sin\psi_+$ is derived directly from the convexity assumption of the profile curve $\mathcal{W}(0)$. But, in this paper, we obtain the estimate (4) from a different approach and, in addition, can prove that if $c \ge (\psi_+ + \psi_-)/\sin\psi_+$, there is no traveling wave defined as (3).

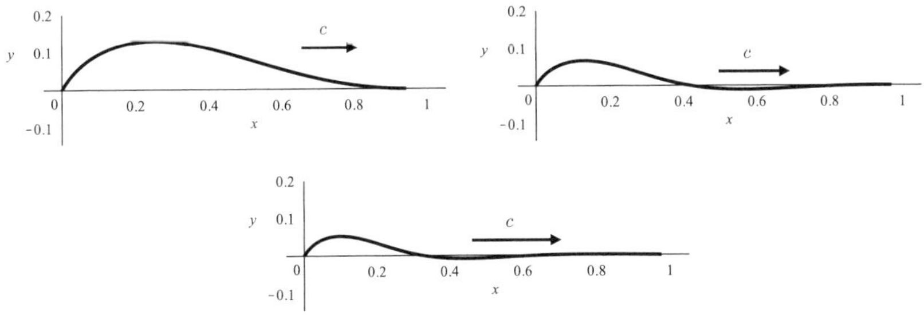

Fig. 3. The profile of traveling waves $\mathcal{W}(t)$ for (2) with $\psi_- = \pi/3$ and $\psi_+ \approx 0.0087$. At least, there are three configurations. The uniqueness and convexity are lost in this case.

Remark 3. *According to [5], we can find "oscillating" traveling waves for (2), which is a free boundary problem for surface diffusion, if one of the contact angles is sufficiently small. As a result, the uniqueness and convexity of traveling waves can be lost for (2) (see Fig. 2). For details of the proof, see [5]. For (1), this kind of situation can not occur since there exists only a convex traveling wave.*

In order to prove Theorem 1, we will derive a profile equation under (W1)–(W3). First, by virtue of (W1), the arc-length parameterization of $\mathcal{W}(0)$ can be given by

$$\mathcal{W}(0) = \{(x(s), y(s)) \mid 0 \le s \le 1\},$$

where s is the arc-length parameter along $\mathcal{W}(0)$. Let us introduce the smooth function $\Theta : [0,1] \to \mathbb{R}$ associated with the Gauss map defined as

$$x'(s) = \cos\Theta(s), \quad y'(s) = \sin\Theta(s) \text{ for } 0 \le s \le 1,$$

where $'$ means the differential with respect to s. By using this function Θ, the traveling wave $\mathcal{W}(t)$ is parametrized as

$$\mathcal{W}(t) = \left\{ \left(ct + \int_0^s \cos\Theta(\sigma)\, d\sigma, \int_0^s \sin\Theta(\sigma)\, d\sigma \right) \middle| \, 0 \le s \le 1 \right\} \tag{5}$$

since $\mathcal{W}(t)$ is defined as (3) and the profile curve $\mathcal{W}(0)$ satisfies (W2). Then the normal velocity $V_{\mathcal{W}}(s,t)$ and the curvature $\kappa_{\mathcal{W}}(s,t)$ of $\mathcal{W}(t)$ are represented as

$$V_{\mathcal{W}}(s,t) = -c\sin\Theta(s), \quad \kappa_{\mathcal{W}}(s,t) = \Theta'(s).$$

Therefore, if we choose the function Θ such that $\Theta(0) = \psi_-$,

$$\int_{\mathcal{W}(t)} \kappa_{\mathcal{W}}(s,t)\,ds = \Theta(1) - \Theta(0) = \Theta(1) - \psi_-,$$

and the restriction (W3) is equivalent to $\Theta(1) \in (-\pi, \pi)$. Furthermore, taking account of Length $[\mathcal{W}(t)] =$ Length $[\mathcal{W}(0)] = 1$, $\kappa_{\mathcal{W},av}(t)$ is given by $-\psi_+ - \psi_-$. If $\mathcal{W}(t)$ is a solution to (1), Θ satisfies

$$\begin{cases} \Theta' = -c\sin\Theta - (\psi_+ + \psi_-), & s \in (0,1), \\ \Theta(0) = \psi_-, \quad \Theta(1) = -\psi_+. \end{cases} \tag{6}$$

Therefore, our aim is reduced to proving the existence of the solution (Θ, c) to the boundary value problem (6). Note that we can construct the traveling wave $\mathcal{W}(t)$ from the solution (Θ, c) to (6) through the parametrization (5). In the next section, we will prove the existence of a solution to (6) with deriving the condition of the angle ψ_+ which guarantees the existence of the solution.

§3. The existence of a solution to the boundary value problem

Let us prove the existence of a solution (Θ, c) to the boundary value problem (6). Our strategy is based on the shooting method. In the first step, we will derive the representations of the solution to the initial value problem

$$\begin{cases} \Psi' = -c\sin\Psi + \beta - \psi_-, & s > 0, \\ \Psi(0) = \psi_-, \end{cases} \tag{7}$$

where $\beta \in (-\pi, \pi)$ is a constant. A unique solution $\Psi(\,\cdot\,; c) \in C^1([0,\infty))$ to (7) exists for any $c > 0$. In the second step, we will find $c > 0$ such that $\Psi(1; c) = \beta$. If there exists such $c > 0$, the pair $(\Psi(\,\cdot\,; c), c)$ is a solution to (6) with $\psi_+ = -\beta$.

Let us start the proof. Set $B = B(c) := c/(\beta - \psi_-)$ for $\beta \neq \psi_-$. Then the representation of $\Psi(s; c)$ is given as follows. For $|B| < 1$, i.e. $c < |\beta - \psi_-|$ and $n \in \mathbb{N} \cup \{0\}$,

$$\Psi(s;c) = \begin{cases} 2\arctan\left\{B + \sqrt{1-B^2}\tan\left(\dfrac{b_1}{2}s + K_1\right)\right\} \pm 2n\pi \\ \qquad\qquad (s \in (\max\{0, s_{n-1}^\pm\}, s_n^\pm)), \\ \pi + 2n\pi \\ \qquad (s = s_n^+ \text{ if } B \in (0,1) \text{ i.e. } c \in (0, \beta - \psi_-)), \\ -(\pi + 2n\pi) \\ \qquad (s = s_n^- \text{ if } B \in (-1,0) \text{ i.e. } c \in (0, -(\beta - \psi_-))), \end{cases} \tag{8}$$

where $b_1 := (\beta - \psi_-)\sqrt{1 - B^2}$,

$$K_1 := \arctan\left(\frac{\tan(\psi_-/2) - B}{\sqrt{1-B^2}}\right),$$

and $s_n^\pm = \pm 2(\pi/2 + n\pi \mp K_1)/b_1$ ($n \in \mathbb{Z}$) which are positive if $n \in \mathbb{N} \cup \{0\}$. Note that $\Psi(s;c)$ as in (8) is a C^1 function for $s > 0$. For $|B| \geq 1$, i.e. $c \geq |\beta - \psi_-|$,

$$\Psi(s;c) = \begin{cases} 2\arctan\left\{B + \sqrt{B^2-1}\left(\dfrac{2}{1-K_2e^{b_2 s}} - 1\right)\right\} \\ \qquad\qquad (|B| > 1,\ B \neq 1/\sin\psi_-), \\ \psi_- \quad (B = 1/\sin\psi_-), \\ 2\arctan\left(B - \dfrac{2B}{cs + K_3}\right) \quad (|B| = 1), \end{cases} \tag{9}$$

where $b_2 := (\beta - \psi_-)\sqrt{B^2 - 1}$,

$$K_2 := \frac{\tan(\psi_-/2) - B - \sqrt{B^2-1}}{\tan(\psi_-/2) - B + \sqrt{B^2-1}}, \quad K_3 := -\frac{2B}{\tan(\psi_-/2) - B}.$$

We remark that $\sin\psi_- = 1/B$ for $B \geq 1$ is equivalent to $\tan(\psi_-/2) = B - \sqrt{B^2-1}$ if $\psi_- \in (0, \pi/2)$ and $\tan(\psi_-/2) = B + \sqrt{B^2-1}$ if $\psi_- \in (\pi/2, \pi)$.

Using (8) and (9), we can readily derive the following lemma.

Lemma 4. *If $\psi_- \in (0,\pi)$, then $|\Psi(s;c)| < \pi$ for $s \in [0,1]$ and $c > 0$. In particular, if $c \geq |\beta - \psi_-|$, it holds for $s \in [0, \infty)$.*

Proof. For $|B| < 1$, it is verified that $b_1 s/2 + K_1 \in (-\pi/2, \pi/2)$ for $s \in [0,1]$. For $|B| \geq 1$, we can prove that $1 - K_2 e^{b_2 s} \neq 0$ and $cs + K_3 > 0$ for $s > 0$. These facts, (8), (9) and $\Psi(0;c) = \psi_- \in (0,\pi)$ imply the conclusion. Q.E.D.

In the rest of this section, our aim is to find $c > 0$ such that $\Psi(1;c) = \beta$. We first observe that for $c > 0$ the equation $\Psi' = -c\sin\Psi + \beta - \psi_-$ is equivalent to

$$\Psi' = -c\left(\sin\Psi - \frac{\beta - \psi_-}{c}\right). \tag{10}$$

By a simple calculation, we have

$$\Psi(1;c) = \beta - c\int_0^1 \sin\Psi(\sigma;c)\,d\sigma. \tag{11}$$

Here, we divide into three cases:

$$\text{(i)}\ \ 0 < \psi_- \le \beta, \quad \text{(ii)}\ \ 0 \le \beta < \psi_-, \quad \text{(iii)}\ \ \beta < 0 < \psi_-.$$

For each case, we will consider whether there exists $c > 0$ such that $\Psi(1; c) = \beta$.

The case (i): If $c \in (0, \beta - \psi_-]$ which is equivalent to $(\beta - \psi_-)/c \ge 1$, (10) implies that for $s \in (0, 1)$

$$\Psi'(s; c) \ge -c\left(1 - \frac{\beta - \psi_-}{c}\right) \ge 0,$$

so that $\Psi(s; c) \ge \Psi(0; c) = \psi_- > 0$ for $s \in [0, 1]$. It follows from this fact, Lemma 4 and (11) that for any $c \in (0, \beta - \psi_-]$

$$\Psi(1; c) = \beta - c\int_0^1 \sin\Psi(\sigma; c)\, d\sigma < \beta,$$

which means that there is no $c \in (0, \beta - \psi_-]$ such that $\Psi(1; c) = \beta$.

Let us consider the case $c \in (\beta - \psi_-, \infty)$ which is equivalent to $0 \le (\beta - \psi_-)/c < 1$. According to (9), for $\psi_- \in (0, \pi)$ satisfying $\sin\psi_- = 1/B = (\beta - \psi_-)/c$, the constant function

$$\Psi(s; c) \equiv \psi_- = \arcsin\left(\frac{\beta - \psi_-}{c}\right) \quad \text{or} \quad \pi - \arcsin\left(\frac{\beta - \psi_-}{c}\right) \tag{12}$$

is a solution of (7). By virtue of the uniqueness of the solution to (7), if $\sin\psi_- < (\beta - \psi_-)/c$, then $\sin\Psi(s; c) < (\beta - \psi_-)/c$ for $s \in [0, 1]$. Thus it follows from (10) that $\Psi'(s; c) > 0$ for $s \in (0, 1]$, so that we have $\Psi(s; c) \ge \Psi(0; c) = \psi_- > 0$ for $s \in [0, 1]$. Using this fact, Lemma 4 and (12) and applying an argument similar to that for the case $c \in (0, \beta - \psi_-]$, we can prove that if $\sin\psi_- \le (\beta - \psi_-)/c$, there is no $c \in (\beta - \psi_-, \infty)$ such that $\Psi(1; c) = \beta$. On the other hand, if $\sin\psi_- > (\beta - \psi_-)/c$, then $\sin\Psi(s; c) > (\beta - \psi_-)/c$ for $s \in [0, 1]$. From this fact and (10), we have $\Psi'(s; c) < 0$ for $s \in (0, 1]$. Thus we obtain $\Psi(1; c) < \Psi(0; c) = \psi_- \le \beta$, which implies that there is no $c \in (\beta - \psi_-, \infty)$ such that $\Psi(1; c) = \beta$.

Therefore, in the case (i), there is no $c > 0$ such that $\Psi(1; c) = \beta$.

The case (ii): Since $\psi_- \in (0, \pi)$ and $\beta < \psi_-$, we have $\sin\psi_- > (\beta - \psi_-)/c$. This fact and the uniqueness of the solution to (7) imply that $\sin\Psi(s; c) > (\beta - \psi_-)/c$ for $s > 0$. Then it follows from (10) that $\Psi'(s; c) < 0$ for $s > 0$. In proof by contradiction, suppose that there exists $c > 0$ such that $\Psi(1; c) = \beta\,(\ge 0)$. Then (11) gives

$$\int_0^1 \sin\Psi(\sigma; c)\, d\sigma = 0.$$

Thus there exists $\sigma_0(c) \in (0,1)$ such that $\sin\Psi(\sigma_0(c);c) = 0$. Since $\Psi(s;c)$ is monotone decreasing in s, we have $\Psi(1;c) < \Psi(\sigma_0(c);c) = 0$. This contradicts $\Psi(1;c) = \beta \geq 0$. Therefore, $\Psi(1;c) \neq \beta$ for any $c > 0$.

The case (iii) : Let us first prove that there exists a unique $s_+(c) > 0$ such that $\Psi(s_+(c);c) = \beta$ only if $0 < c < (\beta - \psi_-)/\sin\beta$.

If $c \in (0, -(\beta - \psi_-))$, it follows that

$$\Psi'(s;c) \leq c + \beta - \psi_- < 0 \quad \text{for} \quad s > 0. \tag{13}$$

Integrating with respect to s, we see that for $s > 0$

$$\Psi(s;c) \leq \psi_- + (c + \beta - \psi_-)s.$$

Choose $s_*(c)$ satisfying $\psi_- + (c + \beta - \psi_-)s_*(c) = \beta$. Note that $s_*(c) = (\beta - \psi_-)/(c + \beta - \psi_-) > 1$ for $c \in (0, -(\beta - \psi_-))$. Then we have $\Psi(s_*(c);c) \leq \beta$. Since $\Psi(0;c) = \psi_- > \beta$ and $\Psi(s;c)$ is monotone decreasing in s, there exists $s_+(c) \in (0, s_*(c)]$ such that $\Psi(s_+(c);c) = \beta$.

If $c \in [-(\beta - \psi_-), \infty)$, an argument similar to the case (ii) implies that

$$\Psi'(s;c) < 0 \quad \text{for} \quad s > 0. \tag{14}$$

Here, we observe that for $s > 0$

$$\sin\Psi(s;c) > \frac{\beta - \psi_-}{c}$$

is equivalent to

$$(\Psi_*(c) :=)\, 2\arctan(B + \sqrt{B^2 - 1}) < \Psi(s;c) < \pi$$

for $B = c/(\beta - \psi_-) \leq -1$ since $\Psi(s;c)$ is a C^1 solution to (7) satisfying $\Psi(0;c) = \psi_- \in (0,\pi)$. We remark that $\Psi_*(c) \in [-\pi/2, 0)$ for $c \in [-(\beta - \psi_-), \infty)$ through $-1 \leq B + \sqrt{B^2 - 1} < 0$ for $B \leq -1$. Then, if β fulfills

$$-\pi < \beta \leq \Psi_*(c),$$

there is no $s_+(c) > 0$ such that $\Psi(s_+(c);c) = \beta$. Thus we consider only the case

$$\Psi_*(c) < \beta\, (< 0). \tag{15}$$

Note that $c \in [-(\beta - \psi_-), \infty)$ and (15) imply

$$-(\beta - \psi_-) \leq c < \frac{\beta - \psi_-}{\sin\beta}. \tag{16}$$

From (9), $b_2 = (\beta - \psi_-)\sqrt{B^2 - 1} < 0$ and (15), we see that for $B \leq -1$

$$\lim_{s\to\infty} \Psi(s;c) = \Psi_*(c) < \beta.$$

Since $\Psi(0;c) = \psi_- > \beta$ and $\Psi(s;c)$ is monotone decreasing in s by virtue of (14), there exists a unique $s_+(c) > 0$ such that $\Psi(s_+(c);c) = \beta$.

Let us show that there exists $c \in (0, (\beta - \psi_-)/\sin\beta)$ such that $s_+(c) = 1$ only if $\beta \in (-\psi_-, 0)$. Integrating the equation $\Psi' = -c\sin\Psi + \beta - \psi_-$ on the interval $[0, s_+(c)]$ and setting

$$F(c) := \int_0^{s_+(c)} \sin\Psi(\sigma;c)\,d\sigma,$$

we are led to

$$(1 - s_+(c))(\beta - \psi_-) = -cF(c).$$

Taking account of $\beta - \psi_- < 0$ and (16), we have only to prove that there exists $c \in (0, (\beta - \psi_-)/\sin\beta)$ such that $F(c) = 0$. With the help of (13) and (14), we see that $\Psi'(s;c) < 0$ for $s \in (0, s_+(c))$ and $c > 0$. By means of the change of values $\phi = \Psi(s;\,\cdot\,)$ and the fact that the map $\Psi : [0, s_+(c)] \ni s \mapsto \phi \in [\beta, \psi_-]$ is the bijection, we obtain

$$F(c) = -\int_\beta^{\psi_-} \frac{\sin\phi}{-c\sin\phi + \beta - \psi_-}\,d\phi.$$

Differentiating with respect to c, we have

$$F'(c) = -\int_\beta^{\psi_-} \frac{\sin^2\phi}{(-c\sin\phi + \beta - \psi_-)^2}\,d\phi < 0. \tag{17}$$

Therefore, F is monotone decreasing in c. In addition,

$$F(0) = \frac{1}{\psi_- - \beta}\int_\beta^{\psi_-} \sin\phi\,d\phi = -\frac{\cos\psi_- - \cos\beta}{\psi_- - \beta}. \tag{18}$$

Then it follows from (18) that

$$F(0) \begin{cases} > 0 & (\beta \in (-\psi_-, 0)), \\ \leq 0 & (\beta \in (-\pi, -\psi_-]). \end{cases} \tag{19}$$

Hence, if $\beta \in (-\pi, -\psi_-]$, there is no $c \in (0, (\beta - \psi_-)/\sin\beta)$ such that $F(c) = 0$, which means that $s_+(c) \neq 1$ for any $c \in (0, (\beta - \psi_-)/\sin\beta)$.

Let us consider only the case $\beta \in (-\psi_-, 0)$. First, it follows that

$$F(-(\beta-\psi_-)) = \frac{1}{\psi_- - \beta}\int_\beta^{\psi_-}\left(1 - \frac{1}{1+\sin\phi}\right)d\phi \tag{20}$$
$$= \begin{cases} -\infty & (\beta \in (-\pi, -\pi/2]), \\ \dfrac{G(\beta;\psi_-)}{g(\beta;\psi_-)} & (\beta \in (-\pi/2, 0)), \end{cases}$$

where $g(\beta;\psi_-) := (\psi_- - \beta)(1+\tan(\psi_-/2))(1+\tan(\beta/2))$ and

$$G(\beta;\psi_-) := g(\beta;\psi_-) - 2\left(\tan\frac{\psi_-}{2} - \tan\frac{\beta}{2}\right).$$

We can prove that the equation $G(\beta;\psi_-) = 0$ with respect to β has a unique solution $\beta_* \in (-\pi/2, 0)$ and

$$G(\beta;\psi_-) \begin{cases} < 0 & (\beta \in (-\pi/2, \beta_*)), \\ = 0 & (\beta = \beta_*), \\ > 0 & (\beta \in (\beta_*, 0)). \end{cases} \tag{21}$$

Taking account of $g(\beta;\psi_-) > 0$ for $\beta \in (-\pi/2, 0)$, (20) and (21) imply

$$F(-(\beta-\psi_-)) \begin{cases} < 0 & (\beta \in (-\pi, \beta_*)), \\ = 0 & (\beta = \beta_*), \\ > 0 & (\beta \in (\beta_*, 0)). \end{cases} \tag{22}$$

Furthermore, we have

$$F\left(\frac{\beta-\psi_-}{\sin\beta}\right) = -\frac{\sin\beta}{\psi_- - \beta}\int_\beta^{\psi_-}\left(1 - \frac{\sin\beta}{\sin\beta - \sin\phi}\right)d\phi \tag{23}$$
$$= \frac{\tan\beta}{\psi_- - \beta}\left[\log\left|\frac{\tan(\phi/2) - \tan(\beta/2)}{\tan(\phi/2) - \cot(\beta/2)}\right|\right]_{\phi\to\beta}^{\phi=\psi_-}$$
$$= -\infty.$$

Note that $-\psi_- < \beta_*$ since $G(-\psi_-;\psi_-) < 0$. Then, with the help of (17), (19), (22) and (23), we find that if $\beta \in (-\psi_-, \beta_*]$, a unique c such that $F(c) = 0$ exists in $(0, -(\beta-\psi_-)]$, and if $\beta \in (\beta_*, 0)$, it exists in $(-(\beta-\psi_-), (\beta-\psi_-)/\sin\beta)$. That is, we obtain the existence of $c \in (0, (\beta-\psi_-)/\sin\beta)$ satisfying $s_+(c) = 1$.

Consequently, it follows from (i)–(iii) that there exists a unique solution (Ψ, c) of

$$\begin{cases} \Psi' = -c\sin\Psi + \beta - \psi_-, & s \in (0,1), \\ \Psi(0) = \psi_-, \quad \Psi(1) = \beta \end{cases}$$

if and only if $\beta \in (-\psi_-, 0)$. That is, there exists a unique solution (Θ, c) to (6) if and only if $\psi_+ \in (0, \psi_-)$. In addition, by virtue of (13) and (14), we have $\Theta' < 0$ for $s \in (0,1)$, which implies that the curvature of the profile curve $\mathcal{W}(0)$ is negative. Also, from an argument in (iii), the wave speed c fulfills (4) for $\psi_* = -\beta_*$. This completes the proof.

Acknowledgements. The authors are grateful to the referees for helpful comments in improving the main result. This work is partially supported by JSPS KAKENHI Grant Number (A) 18H03670.

References

[1] X. Chen and J.-S. Guo, *Motion by curvature of planar curves with end points moving freely on a line*, Math. Ann., **350** (2011), no. 2, 277–311.

[2] M. Gage, *On an area-preserving evolution equation for plane curves*, Nonlinear problems in geometry, Contemp. Math., **51** (1986), 51–62.

[3] Y. Giga and K. Ito, *Loss of convexity of simple closed curves moved by surface diffusion*, Progr. Nonlinear Differential Equations Appl., **35** (1999), 305–320.

[4] J.-S. Guo, H. Matano, M. Shimojo and C.-H. Wu, *On a free boundary problem for the curvature flow with driving force*, Arch. Ration, Mech. Anal., **219** (2016), no. 3, 1207–1272.

[5] T. Kagaya and Y. Kohsaka, Existence of non-convex traveling waves for surface diffusion of curves with constant contact angles, preprint (arXiv:1810.02049).

[6] M. Shimojo and T. Kagaya, *Exponential stability of a traveling wave for an area preserving curvature motion with two endpoints moving freely on a line*, Asymptot. Anal., **96** (2016), no. 2, 109–134.

Institute of Mathematics for Industry, Kyushu University
E-mail address: `kagaya@imi.kyushu-u.ac.jp`

Graduate School of Maritime Sciences, Kobe University
E-mail address: `kohsaka@maritime.kobe-u.ac.jp`

Advanced Studies in Pure Mathematics 85, 2020
The Role of Metrics in the Theory of Partial Differential Equations
pp. 239–258

Uniqueness problem for closed non-smooth hypersurfaces with constant anisotropic mean curvature

Miyuki Koiso

Abstract.

We study a variational problem for piecewise-smooth hypersurfaces in the $(n+1)$-dimensional Euclidean space. An anisotropic energy is the integral of an energy density that depends on the normal at each point over the considered hypersurface, which is a generalization of the area of surfaces. The minimizer of such an energy among all closed hypersurfaces enclosing the same $(n+1)$-dimensional volume is unique and it is (up to rescaling) so-called the Wulff shape. The Wulff shape and equilibrium hypersurfaces of this energy for volume-preserving variations are not smooth in general. In this article we give recent results on the uniqueness and non-uniqueness for closed equilibria. We also give nontrivial self-similar shrinking solutions of anisotropic mean curvature flow. This article is an announcement of forthcoming papers [10], [13].

§1. Introduction

An anisotropic surface energy was introduced by J. W. Gibbs (1839–1903) in order to model the shape of small crystals ([26], [27]), which is defined as follows. Let $\gamma : S^n \to \mathbb{R}_{>0}$ be a positive continuous function on the unit sphere $S^n = \{\nu \in \mathbb{R}^{n+1} \mid \|\nu\| = 1\}$ in the $(n+1)$-dimensional Euclidean space $\mathbb{R}^{n+1}$. Let X be a closed piecewise-C^2 weakly immersed hypersurface in $\mathbb{R}^{n+1}$ (the definition of piecewise-C^2 weakly immersed hypersurface will be given in §2). X will be represented as a piecewise-C^2 mapping $X : M \to \mathbb{R}^{n+1}$ from an n-dimensional oriented connected compact C^∞ manifold M into $\mathbb{R}^{n+1}$, and the unit normal vector field

Received March 3, 2019.
Revised June 18, 2019.
2010 *Mathematics Subject Classification.* 49Q10, 53C45, 53C42, 53C44.
Key words and phrases. anisotropic mean curvature, anisotropic surface energy, Wulff shape, anisotropic mean curvature flow, crystalline variational problem.

ν along X is defined on M except a set with measure zero. Then, we define the anisotropic energy of X as $\mathcal{F}_\gamma(X) = \int_M \gamma(\nu)\, dA$, where dA is the n-dimensional volume form of M induced by X. If $\gamma \equiv 1$, $\mathcal{F}_\gamma(X)$ is the usual n-dimensional volume of X. A positive continuous function on S^n will be sometimes called an integrand.

It is known that, for any positive number $V > 0$, among all closed hypersurfaces in $\mathbb{R}^{n+1}$ enclosing $(n+1)$-dimensional volume V, there exists a unique (up to translation in $\mathbb{R}^{n+1}$) minimizer $W_\gamma(V)$ of $\mathcal{F}_\gamma$ ([22]). $W_\gamma(V_0)$ for $V_0 := (n+1)^{-1}\int_{S^n}\gamma(\nu)\, dS^n$ is called the Wulff shape for γ, and we denote it by W_γ (the standard definition of the Wulff shape will be given in §2). If $\gamma \equiv 1$, W_γ is the unit sphere S^n. All $W_\gamma(V)$ are homothetic to W_γ. W_γ is convex but is not smooth in general.

Each equilibrium hypersurface of $\mathcal{F}_\gamma$ for variations that preserve the enclosed $(n+1)$-dimensional volume (we call such a variation a volume-preserving variation) has constant anisotropic mean curvature. Here the anisotropic mean curvature Λ of a piecewise-C^2 weakly immersed hypersurface X is defined at each regular point of X as (cf. [14], [21]) $\Lambda := (1/n)(-\mathrm{div}_M D\gamma + nH\gamma)$, where $D\gamma$ is the gradient of γ on S^n and H is the mean curvature of X (see §3 for details). If $\gamma \equiv 1$, then $\Lambda = H$.

We call a piecewise-C^2 equilibrium hypersurface X a CAMC (constant anisotropic mean curvature) hypersurface (see Definition 3.2 for details). A CAMC hypersurface is said to be stable if the second variation of the energy $\mathcal{F}_\gamma$ for any volume-preserving variation is nonnegative.

In the special case where $\gamma \equiv 1$, a CAMC hypersurface is a CMC (constant mean curvature) hypersurface of class C^ω ([13]). For another special γ defined by $\gamma(\nu_1,\nu_2,\nu_3) = \sqrt{|\nu_3^2-\nu_1^2-\nu_2^2|}$, ($\nu = (\nu_1,\nu_2,\nu_3) \in S^2$), $\mathcal{F}_\gamma(X)$ is the area of the surface X in $\mathbb{R}^3$ regarded as a surface in the Lorentz-Minkowski space $\mathbb{R}^3_1 := (\mathbb{R}^3, dx_1^2+dx_2^2-dx_3^2)$, and a CAMC surface is a CMC surface in $\mathbb{R}^3_1$ ([8]). These examples suggest that $\mathcal{F}_\gamma$ is a generalization of many important energy functionals for surfaces.

Because of the uniqueness of the minimizer mentioned above, it is natural to ask the following question. Is any closed CAMC hypersurface the Wulff shape (up to translation and homothety)? The answer to this question is not affirmative even in the case where $\gamma \equiv 1$ ([11], [12], [23]). However, it is expected that, if a closed CAMC hypersurface $X : M \to \mathbb{R}^{n+1}$ satisfies one of the following "good" conditions (I)–(III), the image $X(M)$ coincides with the Wulff shape (up to translation and homothety).

(I) X is an embedding, that is, X is an injective mapping.
(II) X is stable.
(III) $n = 2$ and M is homeomorphic to S^2.

In fact, surfaces which appear as natural phenomena are expected to satisfy conditions (I) and (II). If we assume that the Wulff shape W_γ is a smooth strictly convex hypersurface, any closed CAMC hypersurface X is also smooth and the statements above were already proved. In fact, if X satisfies one of (I)–(III), $X(M)$ is a homothety of W_γ (up to translations). This fact was proved by the following papers. (I): [1] for $\gamma \equiv 1$, [7] for general γ. (II): [3] for $\gamma \equiv 1$, [19] for general γ. (III): [9] for $\gamma \equiv 1$, [15] and [2] for general γ.

Note that W_γ is smooth and strictly convex if and only if γ is of class C^2 and the $n \times n$ matrix $D^2\gamma + \gamma \cdot I_n$ is positive definite at any point in S^n, where $D^2\gamma$ is the Hessian of γ on S^n and I_n is the identity matrix of size n. γ is called a convex integrand (or, simply, convex) if the homogeneous extension $\overline{\gamma} : \mathbb{R}^{n+1} \to \mathbb{R}_{\geq 0}$ of γ that is defined as $\overline{\gamma}(rX) := r\gamma(X)$ $(\forall X \in S^n,\ \forall r \geq 0)$ is a convex function (cf. [22]). γ is convex if and only if $D^2\gamma + \gamma \cdot I_n$ is positive semi-definite.

If γ has less regularity or less convexity (that is, $D^2\gamma + \gamma \cdot I_n$ is not positive definite at some points in S^n), the Wulff shape and CAMC hypersurfaces can have "edges". In §4, we give three typical examples. In Example 4.1, γ is convex, $\gamma \in C^0$ but $\gamma \notin C^1$. In Example 4.3, γ is not convex and $\gamma \in C^\infty$. In Example 4.2, γ is of class C^∞ and convex but $D^2\gamma + \gamma \cdot I_1$ degenerates at some points, and the curvature of the Wulff shape W_γ is unbounded near these singular points. However, we can show the following uniqueness result which can be applied also to γ having singular points.

Theorem 1.1 ([13]). *Assume that $\gamma : S^n \to \mathbb{R}_{>0}$ is of class C^2 and convex. Then, the image of any closed stable piecewise-C^2 CAMC hypersurface for γ whose r-th anisotropic mean curvature for γ is integrable for $r = 1, \ldots, n$ is (up to translation and homothety) W_γ.*

The assumption on the r-th anisotropic mean curvatures (Definition 3.1) in Theorem 1.1 is a natural one. In fact, each r-th anisotropic mean curvature of W_γ is $(-1)^r$ and hence it is integrable (Fact 3.1). The reason why we assume it is a technical reason. In fact, one of the key tools we use to prove Theorem 1.1 is a "Steiner-type integral formula" (Theorem 5.1) and it includes integrals of the r-th anisotropic mean curvatures. In the previous works [3], [24], [19], and [20], such an assumption was not necessary to be assumed, because the hypersurfaces studied there were compact and immersed and hence each r-th (anisotropic) mean curvature was bounded and integrable.

Theorem 1.1 is relating to the condition (II) above. As for the conditions (I) and (III), we will give some non-uniqueness results for non-convex γ. In order to derive them, we use the Cahn-Hoffman map ξ_γ for

γ. Assume that γ is of class C^1 on S^n. Then ξ_γ is the mapping $\xi_\gamma : S^n \to \mathbb{R}^{n+1}$ defined as $\xi_\gamma(\nu) = D\gamma|_\nu + \gamma(\nu)\nu$, here the tangent space $T_\nu(S^n)$ of S^n at $\nu \in S^n$ is naturally identified with the n-dimensional linear subspace of $\mathbb{R}^{n+1}$ ([13]). It is known that the Wulff shape W_γ is a subset of the image $\xi_\gamma(S^n)$. W_γ coincides with $\xi_\gamma(S^n)$ if and only if γ is convex. On the other hand, it is shown that the anisotropic mean curvature of ξ_γ for the normal ν is -1 at any regular point (Fact 3.1). Therefore, if γ is not convex, then automatically we have the non-empty set $\xi_\gamma(S^n) \setminus W_\gamma$ which consists of pieces of CAMC hypersurfaces which are not parts of W_γ (see Example 4.3, examples given in §7, and Remark 7.1). In [10], we found examples of non-convex γ for which $\xi_\gamma(S^n)$ includes some closed embedded piecewise-C^2 CAMC hypersurfaces which are not W_γ (cf. §7). These examples give the following non-uniqueness results.

Theorem 1.2 ([10]). *There exists a C^∞ function $\gamma : S^n \to \mathbb{R}_{>0}$ which is not convex such that there exist closed embedded piecewise-C^∞ CAMC hypersurfaces in $\mathbb{R}^{n+1}$ for γ each of which is not (any homothety or translation of) the Wulff shape.*

Theorem 1.3 ([10]). *There exists a C^∞ function $\gamma : S^2 \to \mathbb{R}_{>0}$ which is not convex such that there exist closed embedded piecewise-C^∞ CAMC surfaces in $\mathbb{R}^3$ with genus zero for γ each of which is not (any homothety or translation of) the Wulff shape.*

The same examples are applied to the anisotropic mean curvature flow. Let $X_t : M \to \mathbb{R}^{n+1}$ be one-parameter family of piecewise-C^2 hypersurfaces with anisotropic mean curvature Λ_t and unit normal ν_t. Assume that the Cahn-Hoffman field $\tilde{\xi}_t = \xi_\gamma \circ \nu_t$ (which is an anisotropic generalization of the unit normal vector field. see §3) along X_t is defined on M. If X_t satisfies $\partial X_t/\partial t = \Lambda_t \tilde{\xi}_t$, it is called an anisotropic mean curvature flow, which diminishes the anisotropic energy if $\Lambda_t \not\equiv 0$ ([10]). By a simple observation, one can show the following.

Theorem 1.4 ([10]). *Let c be a positive constant. Set*

$$X_t := \sqrt{2(c-t)}\,\xi_\gamma, \quad t \le c.$$

Then X_t is a self-similar shrinking solution, that is
(i) *$\partial X_t/\partial t = \Lambda_t \tilde{\xi}_t$, and*
(ii) *X_t is homothetic to ξ_γ and it shrinks as t increases.*

By using this result and by giving suitable examples, the following non-uniqueness result is proved (§7).

Theorem 1.5 ([10]). *There exists a C^∞ function $\gamma : S^n \to \mathbb{R}_{>0}$ which is not convex such that there exist closed embedded piecewise-C^2*

self-similar shrinking solutions in $\mathbb{R}^{n+1}$ *for* γ *each of which is homeomorphic to* S^n *and is not (any homothety or translation of) the Wulff shape.*

In contrast with this result, the round sphere is the only closed embedded self-similar shrinking solution of mean curvature flow in $\mathbb{R}^3$ with genus zero ([5]).

It is natural to regard the image $\xi_\gamma(S^n)$ of the Cahn-Hoffman map as the most fundamental closed piecewise-smooth CAMC hypersurface. Actually, the examples of non-trivial closed embedded CAMC hypersurfaces which prove Theorems 1.2, 1.3 are subsets of $\xi_\gamma(S^n)$ (see (§7)). From this point of view, we arrive at the following Conjecture.

Conjecture 1.1 ([10]). *Assume that* $\gamma : S^n \to \mathbb{R}_{>0}$ *is of class* C^2. *Let* $X : M \to \mathbb{R}^{n+1}$ *be a closed piecewise-*C^2 *CAMC hypersurface. We assume that the* r*-th anisotropic mean curvature of* X *for* γ *is integrable for* $r = 1, \ldots, n$. *Then, if* X *satisfies at least one of the conditions* (I), (II), (III) *above,* $X(M)$ *is a subset of a homothety of* $\xi_\gamma(S^n)$.

Similarly, the examples of non-trivial closed embedded self-similar shrinking solutions of the anisotropic mean curvature flow which prove Theorem 1.5 are subsets of $\xi_\gamma(S^n)$ (see §7). This fact leads us to the following Conjecture.

Conjecture 1.2 ([10]). *Assume that* $\gamma : S^2 \to \mathbb{R}_{>0}$ *is of class* C^2. *Then any closed embedded piecewise-*C^2 *self-similar shrinking solution of the anisotropic mean curvature flow for* γ *in* $\mathbb{R}^3$ *with genus zero is a subset of a homothety of* $\xi_\gamma(S^2)$.

We have some examples of non-convex γ for which there are convex closed piecewise-smooth CAMC hypersurfaces which are not W_γ (§7). If some of these hypersurfaces are stable, it is interesting in contrast with Theorem 1.1.

Question 1.1. *Does there exist a* C^∞ *function* $\gamma : S^n \to \mathbb{R}_{>0}$ *which is not convex such that there exist closed stable piecewise-*C^2 *CAMC hypersurfaces in* $\mathbb{R}^{n+1}$ *for* γ *each of which is not (any homothety or translation of)* W_γ*?*

We should mention preceding studies about our uniqueness problem. F. Morgan [17] proved that, if $\gamma : S^1 \to \mathbb{R}_{>0}$ is continuous and convex, any closed equilibrium rectifiable curve for $\mathcal{F}_\gamma$ in $\mathbb{R}^2$ with area constraint is (up to translation and homothety) a covering of the Wulff shape (see [18] for another proof). B. Palmer [20] proved the same conclusion as Theorem 1.1 but under the assumptions that $n = 3$, γ is of class C^3, and considered surfaces and the Wulff shape satisfy some extra assumptions (see Remark 6.1 for details). Relating to their results, we give one more observation on the relation between the Wulff shape W_γ and the Cahn-Hoffman map ξ_γ. As we stated above, if γ is of class C^1, W_γ is a subset of $\xi_\gamma(S^n)$. In [20] this property was essentially used. On the other hand, if γ is not differentiable at some points in S^n, $W_\gamma \subset \xi_\gamma(S^n)$ does not hold in general (see Example 4.1). Hence, the method used in [17] was different from that in [20].

We have an important remark on the assumption of the regularity of γ in Theorem 1.1. If $\gamma : S^n \to \mathbb{R}_{>0}$ is a convex integrand, $\gamma \in C^{1,1}$ if and only if W_γ is uniformly convex ([16]). This result implies that, if we assume that γ is of class C^2, W_γ never has straight line segment. This fact leads us to a future subject to weaken the assumption on the regularity of γ.

This article is organized as follows. In §2 we give the formulation of piecewise-C^r weakly immersed hypersurfaces and the definition of the anisotropic energy for them. We also recall the definition of the Wulff shape. In §3, we give the definitions of various anisotropic curvatures (Definition 3.1). The first variation formula of the anisotropic surface energy (Proposition 3.1) and the Euler-Lagrange equations for our variational problem (Proposition 3.2) are given. In §4, we give examples of integrands with various regularities and convexities, and their Wulff shapes and Cahn-Hoffmann maps. In §5, we give the definition of anisotropic parallel hypersurface (Definition 5.1) and give two integral formulas: a Steiner-type integral formula (Theorem 5.1) and a Minkowski-type integral formula (Theorem 5.2). They are generalizations of a well-known concept (parallel hypersurfaces) and integral formulas (Steiner formula and Minkowski formula) in the isotropic case to the anisotropic case, which are used to prove Theorem 1.1. In §6 we give an outline of the proof of Theorem 1.1. In §7 we give examples which prove Theorems 1.2, 1.3, and 1.5.

§2. Piecewise-C^r weakly immersed hypersurface and its anisotropic energy

First we recall the definition of a *piecewise-C^r weakly immersed hypersurface*, ($r \in \mathbb{N}$), defined in [13]. Let $M = \cup_{i=1}^k M_i$ be an n-dimensional oriented compact connected C^∞ manifold, where each M_i is an n-dimensional connected compact submanifold of M with piecewise-C^∞ boundary, and $M_i \cap M_j = \partial M_i \cap \partial M_j$, ($i, j \in \{1, \ldots, k\}$, $i \neq j$). We call a map $X : M \to \mathbb{R}^{n+1}$ a piecewise-C^r weak immersion (or a piecewise-C^r weakly immersed hypersurface) if X satisfies the following conditions (A1), (A2), and (A3) ($i \in \{1, \ldots, k\}$).

(A1) X is continuous, and each $X_i := X|_{M_i} : M_i \to \mathbb{R}^{n+1}$ is of class C^r.

(A2) The restriction $X|_{M_i^o}$ of X to the interior M_i^o of M_i is a C^r-immersion.

(A3) The unit normal vector field $\nu_i : M_i^o \to S^n$ along $X_i|_{M_i^o}$ can be extended to a C^{r-1}-mapping $\nu_i : M_i \to S^n$. Here, if $(u^1, \ldots, u^n)$ is a local coordinate system in M_i, then $\{\nu_i, \partial/\partial u^1, \ldots, \partial/\partial u^n\}$ gives the canonical orientation in $\mathbb{R}^{n+1}$.

The anisotropic energy of a piecewise-C^1 weak immersion $X : M \to \mathbb{R}^{n+1}$ is defined as follows. Assume that $\gamma : S^n \to \mathbb{R}_{\geq 0}$ is a nonnegative continuous function. Denote by $S(X)$ the set of all singular points of X, here a singular point of X is a point in M at which X is not an immersion. Let $\nu : M \setminus S(X) \to S^n$ be the unit normal vector field along $X|_{M \setminus S(X)}$. The anisotropic energy $\mathcal{F}_\gamma(X)$ of X is defined as

$$\mathcal{F}_\gamma(X) := \int_M \gamma(\nu)\, dA := \sum_{i=1}^{k} \int_{M_i} \gamma(\nu_i)\, dA. \tag{1}$$

Note that, since the n-dimensional Hausdorff measure of $X(S(X))$ is zero ([13]), the improper integral in the right hand side of (1) converges.

For a continuous function $\gamma : S^n \to \mathbb{R}_{>0}$, the boundary W_γ of the convex set $\tilde{W}[\gamma] := \cap_{\nu \in S^n} \{X \in \mathbb{R}^{n+1} | \langle X, \nu \rangle \leq \gamma(\nu)\}$ is called the Wulff shape for γ, where $\langle\ ,\ \rangle$ stands for the standard inner product in $\mathbb{R}^{n+1}$. We should remark that originally $\tilde{W}[\gamma]$ itself was called the Wulff shape.

§3. First variation formula, anisotropic curvatures, and anisotropic Gauss map

From now on, we assume that $\gamma : S^n \to \mathbb{R}_{>0}$ is a positive function of class C^2. Let $X : M = \cup_{i=1}^k M_i \to \mathbb{R}^{n+1}$ be a piecewise-C^2 weak immersion. The Cahn-Hoffman field $\tilde{\xi}_i$ along $X_i = X|_{M_i}$ for γ (or the anisotropic Gauss map of X for γ) is defined as $\tilde{\xi}_i := \xi_\gamma \circ \nu_i : M_i \to \mathbb{R}^{n+1}$. For any point $p \in M_i^o$, the linear map $S_p^\gamma : T_p M_i \to T_p M_i$ given by the $n \times n$ matrix $S^\gamma := -d\tilde{\xi}_i$ is called the anisotropic shape operatior of X_i. Various anisotropic curvatures of X at each regular point are defined by using the eigenvalues of S^γ as follows.

Definition 3.1 (anisotropic curvatures, cf. [7], [21]). (i) The eigenvalues of S^γ are called the anisotropic principal curvatures of X. We denote them by $k_1^\gamma, \ldots, k_n^\gamma$.

(ii) Let σ_r^γ be the elementary symmetric functions of $k_1^\gamma, \ldots, k_n^\gamma$:

$$\sigma_r^\gamma := \sum_{1 \leq l_1 < \cdots < l_r \leq n} k_{l_1}^\gamma \cdots k_{l_r}^\gamma, \quad r = 1, \ldots, n. \tag{2}$$

Set $\sigma_0^\gamma := 1$. $H_r^\gamma := \sigma_r^\gamma / {}_nC_r$ is called the r-th anisotropic mean curvature of X.

(iii) $\Lambda := H_1^\gamma$ is called the anisotropic mean curvature of X.

Remark 3.1. S^γ is not symmetric in general. However, we have the following good properties of the anisotropic curvatures.

(i) ([6]) If $A := d\xi_\gamma = D^2\gamma + \gamma \cdot I_n$ is positive definite at a point $\nu(p)$, $(p \in M_i^o)$, then all of the anisotropic principal curvatures of X at p are real.

(ii) ([13]) k_i^γ is not a real value in general. However, each H_r^γ is always a real valued function on M_i^o.

Remark 3.2. At any regular point of X, it holds that (cf. [14])

$$\Lambda = -\frac{1}{n}\mathrm{trace}_M(D^2\gamma + \gamma \cdot I_n) \circ d\nu = -\frac{1}{n}\mathrm{trace}_M\, d(\tilde{\xi}_\gamma). \tag{3}$$

Now we give the first variation formula for the anisotropic surface energy $\mathcal{F}_\gamma$.

Proposition 3.1 ([13]). *Assume that the map $X : M_0 \to \mathbb{R}^{n+1}$ satisfies* (A1), (A2), *and* (A3) *in* §2 *with $r = 2$, $X_i = X$, $M_i = M_0$, and $\nu_i = \nu$. Let $X_\epsilon : M_0 \to \mathbb{R}^{n+1}$, $(\epsilon \in J := [-\epsilon_0, \epsilon_0])$, be a variation of X, that is, $\epsilon_0 > 0$ and $X_0 = X$. Assume for simplicity that X_ϵ is of class C^∞ in ϵ. We also assume that, for each $\epsilon \in J$, the anisotropic mean curvature Λ_ϵ of X_ϵ is bounded on M_0^o. Set*

$$\delta X := \left.\frac{\partial X_\epsilon}{\partial \epsilon}\right|_{\epsilon=0}, \quad \psi := \langle \delta X, \nu \rangle.$$

Then the first variation of the anisotropic energy $\mathcal{F}_\gamma$ is given as follows.

$$\left.\frac{d\mathcal{F}_\gamma(X_\epsilon)}{d\epsilon}\right|_{\epsilon=0} = -\int_{M_0} n\Lambda\psi\, dA - \oint_{\partial M_0} \langle \delta X, R(p(\tilde{\xi}))\rangle\, d\tilde{s}, \tag{4}$$

here $d\tilde{s}$ is the $(n-1)$-dimensional volume form of ∂M_0 induced by X, N is the outward-pointing unit conormal along ∂M_0, R is the $\pi/2$-rotation on the (N, ν)-plane, and p is the projection from $\mathbb{R}^{n+1}$ to the (N, ν)-plane.

On the other hand the first variation of the $(n+1)$-dimensional volume enclosed by X_ϵ is $\delta V = \int_{M_0} \psi\, dA$ (cf. [4]). This with (4) gives the following Euler-Lagrange equations.

Proposition 3.2 (Euler-Lagrange equations, Koiso [13]. For $n = 2$, see B. Palmer [20]). *A piecewise-C^2 weak immersion $X : M = \cup_{i=1}^k M_i \to \mathbb{R}^{n+1}$ is a critical point of the anisotropic energy $\mathcal{F}_\gamma$ for volume-preserving variations if and only if the following (i) and (ii) hold.*

(i) *The anisotropic mean curvature of X is constant on $M \setminus S(X)$.*

(ii) *$\tilde{\xi}_i(\zeta) - \tilde{\xi}_j(\zeta) \in T_\zeta M_i \cap T_\zeta M_j = T_\zeta(\partial M_i \cap \partial M_j)$ holds at any point $\zeta \in \partial M_i \cap \partial M_j$, where a tangent space of a submanifold of $\mathbb{R}^{n+1}$ is naturally identified with a linear subspace of $\mathbb{R}^{n+1}$.*

In view of Proposition 3.2, we give the following definition.

Definition 3.2 ([13]). A piecewise-C^2 weak immersion $X : M = \cup_{i=1}^k M_i \to \mathbb{R}^{n+1}$ is called a hypersurface with constant anisotropic mean curvature (CAMC for short) if both of (i) and (ii) in Proposition 3.2 hold.

The Cahn-Hoffman map $\xi_\gamma : S^n \to \mathbb{R}^{n+1}$ is CAMC. More precisely, we have the following:

Fact 3.1 ([14], [13]). Since ξ_γ^{-1} gives the unit normal vector field ν_{ξ_γ} for the Cahn-Hoffman map ξ_γ, the anisotropic shape operator of ξ_γ is $S^\gamma = -d(\xi_\gamma \circ \nu_{\xi_\gamma}) = -d(\mathrm{id}_{S^n}) = -I_n$. Hence, the anisotropic principal curvatures of ξ_γ are -1, and hence each r-th anisotropic mean curvature of ξ_γ is $(-1)^r$. Particularly the anisotropic mean curvature of ξ_γ for the normal ν and that of W_γ for the outward-pointing unit normal is -1 at any regular point.

If γ is convex, we have the following striking result.

Theorem 3.1 ([13]). *Assume $\gamma : S^n \to \mathbb{R}_{>0}$ is of class C^2 and convex. Let $X : M = \cup_{i=1}^k M_i \to \mathbb{R}^{n+1}$ be a closed piecewise-C^2 CAMC hypersurface with unit normal $\nu_i : M_i \to S^n$. Then, the Cahn-Hoffman fields $\tilde{\xi}_i := \xi \circ \nu_i : M_i \to \mathbb{R}^{n+1}$, $(i = 1, \ldots, k)$, defines a C^0 map $\tilde{\xi} : M \to \mathbb{R}^{n+1}$.*

Remark 3.3. Let X be a graph $X(u_1, u_2) = (u_1, u_2, f(u_1, u_2))$ of a C^∞ function $f : D(\subset \mathbb{R}^2) \to \mathbb{R}$. Then

$$\Lambda = \frac{1}{2} \sum_{i,j=1,2} \overline{\gamma}_{x_i x_j}\Big|_{(-Df,1)} f_{u_i u_j}, \quad Df := (f_{u_1}, f_{u_2}),$$

holds, here $\overline{\gamma}$ is the homogeneous extension of γ. Hence, the equation "$\Lambda \equiv$ constant" is elliptic or hyperbolic depends on $(\overline{\gamma}_{x_i x_j}|_{(-Df,1)})_{i,j=1,2}$.

§4. Examples of integrands with various regularities and convexities, and their Wulff shapes and Cahn-Hoffmann maps

In this section, we give several examples of integrands $\gamma : S^n \to \mathbb{R}_{>0}$ with various regularities and convexities, and we study their Wulff shapes and Cahn-Hoffmann maps. Sometimes we will assume $n = 1$ or $n = 2$ for simplicity. However, one can generalize them to higher dimensional examples (for example, by rotation around a suitable straight line). Detailed explanations and computations for Examples 4.1, 4.2, and 4.3 are given in [13], Below we use the notation $\nu = (\nu_1, \ldots, \nu_{n+1})$ for $\nu \in S^n$.

Example 4.1 ($\gamma \in C^0$ and convex, but $\gamma \notin C^1$)**.** Let r, h be positive constants. Define $\gamma : S^n \to \mathbb{R}_{>0}$ as follows.

$$\gamma(\nu) = r\left(\nu_1^2 + \cdots + \nu_n^2\right)^{1/2} + h|\nu_{n+1}|, \quad \nu = (\nu_1, \ldots, \nu_{n+1}) \in S^n.$$

Then, $\gamma \in C^0(S^n)$ and convex, but γ is not differentiable at any point in the set

$$S(\gamma) := \left\{(\nu_1, \ldots, \nu_{n+1}) \in S^n \mid \nu_{n+1} = 0, \pm 1\right\}.$$

We call each point in $S(\gamma)$ a singular point of γ. The Wulff shape W_γ is the cylinder C with radius r and height $2h$ covered with two flat discs with radius r, that is

$$W_\gamma = C \cup \Gamma_+ \cup \Gamma_-,$$

where

$$C = \left\{(x_1, \ldots, x_{n+1}) \in \mathbb{R}^{n+1} \mid \left(x_1^2 + \cdots + x_n^2\right)^{1/2} = r, \ x_{n+1} \in [-h, h]\right\},$$

$$\Gamma_\pm = \left\{x = (x_1, \ldots, x_n, \pm h) \in \mathbb{R}^{n+1} \mid \left(x_1^2 + \cdots + x_n^2\right)^{1/2} \leq r\right\}.$$

The Cahn-Hoffman map $\xi_\gamma : S^n \setminus S(\gamma) \to \mathbb{R}^{n+1}$ is given as

$$\xi(p\cos\theta, \sin\theta) = (rp, \operatorname{sgn}(\sin\theta) \cdot h), \ p \in S^{n-1}, \ \theta \in (-\pi/2, 0) \cup (0, \pi/2).$$

The image $\xi_\gamma(S^n \setminus S(\gamma))$ is the union of two circles with radius r with height $\pm h$ as follows (Figure 1).

$$\xi(S^n \setminus S(\gamma)) = \left\{(x_1, \ldots, x_n, \pm h) \in \mathbb{R}^{n+1} \mid \left(x_1^2 + \cdots + x_n^2\right)^{1/2} = r\right\}.$$

Hence, this example indicates that the regularity of γ has an influence on the relation between the Wulff shape and the Cahn-Hoffmann map.

Here we give a comment about the relation between the singular points of γ and the flat faces of W_γ. The singular point $(0, \ldots, 0, 1)$ (resp.

$(0, \ldots, 0, -1))$ is the outward-pointing unit normal to the n-dimensional flat disc Γ_+ (resp. Γ_-), and each singular point $(p, 0) := (\nu_1, \ldots, \nu_n, 0)$ is the outward-pointing unit normal to the straight line segment $\sigma_p := \{(rp, x_{n+1}) \in \mathbb{R}^{n+1} \mid -h \leq x_{n+1} \leq h\}$. We call each of Γ_+, Γ_-, and σ_p a flat face of W_γ. The dimension of each flat face F coinsides with the dimension of the linear space consisting of the directions where γ is not differentiable at the normal to F.

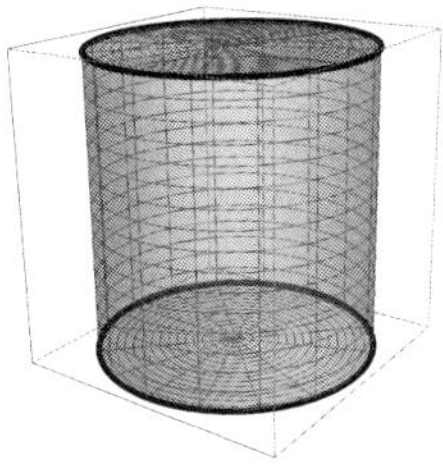

Fig. 1. The Wulff shape W_γ (a cylinder covered with two flat discs) and the image of the Cahn-Hoffman map (two thick circles at the top and at the bottom) for γ in Example 4.1.

Example 4.2 ($\gamma \in C^\infty$ and convex, but not uniformly convex)**.** Set $n = 1$. For $m \in \mathbb{N}$, define $\gamma_m(\nu) := (\nu_1^{2m} + \nu_2^{2m})^{1/(2m)}$. Then γ_m is of class C^∞ and convex. γ_m is uniformly convex if and only if $m = 1$.

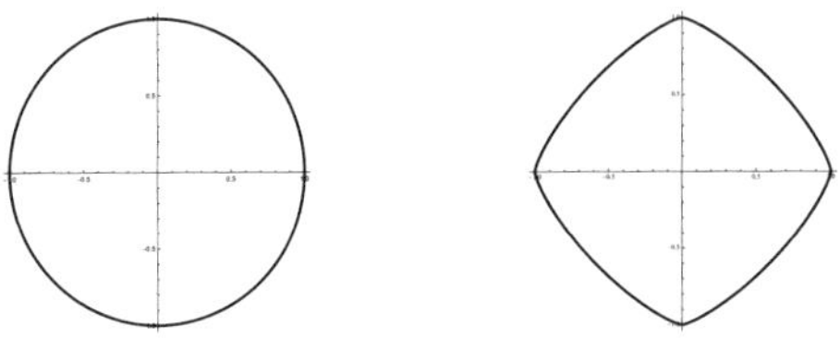

Fig. 2. Left: W_{γ_m} for $m = 1$, right: W_{γ_m} for $m = 2$, for $\gamma_m(\nu) := (\nu_1^{2m} + \nu_2^{2m})^{1/(2m)}$ in Example 4.2.

The Cahn-Hoffman map ξ_m for γ_m is represented as follows.

$$\begin{aligned}\xi_m(\cos\theta, \sin\theta) &= (\cos^{2m}\theta + \sin^{2m}\theta)^{(1/(2m))-1}(\cos^{2m-1}\theta, \sin^{2m-1}\theta) \\ &=: (f_m(\theta), g_m(\theta)).\end{aligned}$$

Moreover we have

$$\begin{aligned}A_m &:= D^2\gamma_m + \gamma_m \cdot 1 \\ &= (2m-1)\cos^{2m-2}\theta\sin^{2m-2}\theta(\cos^{2m}\theta + \sin^{2m}\theta)^{(1/(2m))-2}.\end{aligned}$$

Hence,

(i) If $m \geq 2$, $A_m = 0$ at $\theta = (1/2)\ell\pi$, $(\ell \in \mathbb{Z})$.

(ii) A_m is positive-definite on $S^1 \setminus \{(\cos\theta, \sin\theta) \mid \theta = (1/2)\ell\pi, (\ell \in \mathbb{Z})\}$.

The curvature κ_m of ξ_m with respect to the outward-pointing normal ν is represented as

$$\kappa_m(\theta) = \frac{-f_m' g_m'' + f_m'' g_m'}{((f_m')^2 + (g_m')^2)^{3/2}}$$
$$= \frac{-1}{2m-1} \cos^{-2m+2}\theta \sin^{-2m+2}\theta (\cos^{2m}\theta + \sin^{2m}\theta)^{2-\frac{1}{2m}}.$$

Hence, near each point where A_m is degenerate, κ_m is unbounded, that is,

$$\lim_{\theta(\neq \frac{\ell}{2}\pi) \to \frac{\ell}{2}\pi} \kappa_m(\theta) = -\infty, \quad \ell \in \mathbb{Z},\ m \geq 2$$

holds. However, since

$$\kappa_m\, ds = -d\theta$$

holds, we have

$$\int_0^{2\pi} \kappa_m\, ds = -\int_0^{2\pi} d\theta = -2\pi.$$

Example 4.3 ($\gamma \in C^\infty$, and γ is not convex). Set $n = 1$. Define γ as $\gamma(\nu) = 4\nu_1^3 - 3\nu_1 + 2$. Then, $\gamma \in C^\infty$, and γ is not convex. The whole of the closed curve with self-intersection in Figure 3 is the image $\hat{W}_\gamma := \xi_\gamma(S^1)$ of the Cahn-Hoffman map ξ_γ, while the closed convex solid curve that is a proper subset of $\hat{W}_\gamma$ is the Wulff shape W_γ. By computation, we obtain

$$\xi_\gamma(\cos\theta, \sin\theta) = (8\cos^2\theta \sin^2\theta + 4\cos^2\theta + 2\cos\theta - 3,$$
$$-8\cos^3\theta\sin\theta + 2\sin\theta)$$
$$=: (f(\theta), g(\theta)),$$

$$A := d\xi_\gamma = 2(-16\cos^3\theta + 12\cos\theta + 1)(-\sin\theta, \cos\theta) =: a(\theta)(-\sin\theta, \cos\theta),$$

$$\kappa_\gamma(\theta) := \frac{-f'g'' + f''g'}{((f')^2 + (g')^2)^{3/2}} = \frac{-1}{2|-16\cos^3\theta + 12\cos\theta + 1)|} = \frac{-1}{|a(\theta)|}.$$

Hence, for any $\rho \in a^{-1}(0)$,

$$\lim_{\theta(\notin a^{-1}(0)) \to \rho} \kappa_\gamma(\theta) = -\infty.$$

$a^{-1}(0) \subset S^1$ is the set of exactly six points, and they correspond to the singular points of ξ_γ (Figure 3).

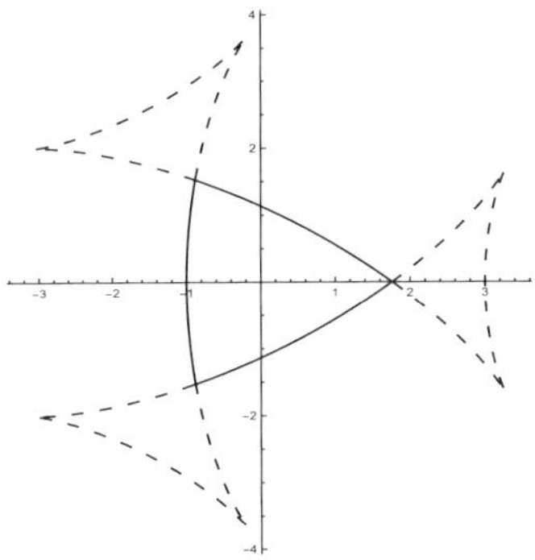

Fig. 3. The image of the Cahn-Hoffman map ξ_γ for γ in Example 4.3. The six vertices are the image of the singular points of ξ_γ. The closed convex solid curve is the Wulff shape W_γ.

§5. Steiner-type formula and Minkowski-type formula

In this section we give two useful integral formulas that are generalizations of the Steiner formula and the Minkowski formula (see [6] for smooth case). They are proved in [13] and are used to prove Theorem 1.1.

'*Anisotropic parallel hypersurface*' is a generalization of parallel hypersurface which is defined as follows.

Definition 5.1 (Anisotropic parallel hypersurface, cf. [21]). Let X be a piecewise-C^2 weak immersion. For any real number t, we call the map $X_t := X + t\tilde{\xi} : M \to \mathbb{R}^{n+1}$ the anisotropic parallel deformation of X of height t. If X_t is a piecewise-C^2 weak immersion, then we call it the anisotropic parallel hypersurface of X of height t.

The anisotropic energy $\mathcal{F}_\gamma(X_t)$ of the anisotropic parallel hypersurface $X_t := X + t\tilde{\xi}$ is a polynomial of t of degree at most n as follows.

Theorem 5.1 (Steiner-type formula [13]). *Assume that $\gamma : S^n \to \mathbb{R}_{>0}$ is of class C^2. Let $X : M = \cup_{i=1}^k M_i \to \mathbb{R}^{n+1}$ be a piecewise-C^2 weak immersion. Consider anisotropic parallel hypersurfaces $X_t = X + t\tilde{\xi} : M \setminus S(X) \to \mathbb{R}^{n+1}$, where $S(X)$ is the set of singular points of X. Then the following integral formula holds.*

$$\mathcal{F}_\gamma(X_t) = \int_M \gamma(\nu)\, dA_t = \int_M \gamma(\nu)(1 - tk_1^\gamma)\cdots(1 - tk_n^\gamma)\, dA$$
$$= \int_M \gamma(\nu) \sum_{r=0}^{n} (-1)^r t^r ({}_nC_r) H_r^\gamma\, dA,$$

where $\int_M$ *means* $\sum_{i=1}^k \int_{M_i}$.

The isotropic version of Theorem 5.1 is known as the Weyl's tube formula ([25]). The isotropic 2-dimensional version is the well-known Steiner's formula.

Next we give a generalization of the Minkowski formula.

Theorem 5.2 (Minkowski-type formula [13]). *Assume that* $\gamma : S^n \to \mathbb{R}_{>0}$ *is of class* C^2. *Assume also that* $X : M = \cup_{i=1}^k M_i \to \mathbb{R}^{n+1}$ *is a closed piecewise-*C^2 *weak immersion and that* X *satisfies the following condition.*

$$\tilde{\xi}_i(\zeta) = \tilde{\xi}_j(\zeta), \quad \forall \zeta \in \partial M_i \cap \partial M_j,$$

here $\tilde{\xi}_i := \xi_\gamma \circ \nu_i : M_i \to \mathbb{R}^{n+1}$ *is the Cahn-Hoffman field along* $X|_{M_i}$, *and the tangent space of a submanifold of* $\mathbb{R}^{n+1}$ *is naturally identified with a linear subspace of* $\mathbb{R}^{n+1}$. *Then we have the following integral formulas.*

(i)

$$\int_M (\gamma(\nu) + \Lambda \langle X, \nu \rangle)\, dA = 0.$$

(ii) *Assume that the* r*-th anisotropic mean curvature of* X *for* γ *is integrable on* M *for* $r = 1, \dots, n$. *Then,*

$$\int_M (\gamma(\nu) H_r^\gamma + \langle X, \nu \rangle H_{r+1}^\gamma)\, dA = 0, \quad r = 0, \dots, n-1,$$

holds, where H_r^γ *is the* r*-th anisotropic mean curvature of* X.

§6. Outline of the proof of Theorem 1.1

The idea is to generalize the proof of the uniqueness of stable closed CMC hypersurface in $\mathbb{R}^{n+1}$ given by [24], which was used also in [19], [20]. Since the Cahn-Hoffman field $\tilde{\xi}$ along a closed CAMC hypersurface $X : M \to \mathbb{R}^{n+1}$ is well-defined on the whole of M (Theorem 3.1), we can consider the anisotropic parallel hypersurfaces $X_t := X + t\tilde{\xi}$, ($t \in \mathbb{R}$, $|t| \ll 1$) of X. By taking suitable homotheties of X_t if necessary, we have a volume-preserving variation $Y_t = \mu(t) X_t$, ($\mu(t) > 0$, $\mu(0) = 1$) of X. By using Theorems 5.1 and 5.2, we obtain the following equality.

$$\left.\frac{d^2 \mathcal{F}_\gamma(Y_t)}{dt^2}\right|_{t=0} = \frac{-1}{n} \int_M \gamma(\nu) \sum_{1 \le i < j \le n} (k_i^\gamma - k_j^\gamma)^2\, dA.$$

Since γ is convex, all k_i^γ are real values on $M \setminus S(X)$ (Remark 3.1 (i)). Hence, if X has constant anisotropic mean curvature Λ and stable,

$k_1^\gamma = \cdots = k_n^\gamma = \Lambda/n \neq 0$ holds on $M \setminus S(X)$. Therefore, from Corollary 1 in [21], $X(M \setminus S(X)) \subset rW$ holds for some $r > 0$. Because M is closed and W has anisotropic mean curvature -1, this inclusion implies that $X(M) = (1/|\Lambda|)W$ holds. □

Remark 6.1. B. Palmer [20] proved the same conclusion as our Theorem 1.1 under some extra assumptions. First of all, [20] treats the following class of surfaces in $\mathbb{R}^3$. There, a surface Σ in $\mathbb{R}^3$ is called piecewise smooth if there exists a collection $\Gamma_1, \ldots, \Gamma_L$ of pairwise disjoint embedded C^1 curves in Σ such that $\Sigma^* := \Sigma \setminus (\Gamma_1 \cup \cdots \cup \Gamma_L)$ is a disjoint union of smooth open surfaces $\Sigma_1, \ldots, \Sigma_N$. In addition, each Σ_j together with its boundary curves is assumed to be a smooth surface with piecewise smooth boundary. In this class of surfaces, under the following assumptions (i)–(viii), [20] proved that Σ coincides with W_γ (up to homothety and translation).

(i) $n = 2$.

(ii) $\gamma : S^2 \to \mathbb{R}_{>0}$ is of class C^3 and convex.

(iii) The Wulff shape W_γ is piecewise smooth.

(iv) Σ is a closed piecewise smooth surface without self-intersection with unit normal $\nu : \Sigma^* \to S^2$.

(v) The Cahn-Hoffman field $\tilde{\xi} : \Sigma^* \to \mathbb{R}^3$ can be continuously extended to the whole of Σ.

(vi) $\nu(\Sigma^*)$ is included in the set of all unit normals to W_γ.

(vii) There exists a positive value c such that all of the principal curvatures of the Wulff shape W_γ with respect to the inward-pointing normal are bounded from below by c.

(viii) Σ is stable.

We should explain the difference between the above assumptions and the assumption of Theorem 1.1 in this paper. We study hypersurfaces in $\mathbb{R}^{n+1}$ without restriction on the dimension. Our hypersurfaces are piecewise-C^2 weak immersions from n-dimensional manifolds, which means that the principal curvatures can be unbounded unlike [20]. We assume that $\gamma : S^n \to \mathbb{R}_{>0}$ is of class C^2, not C^3. The condition (iii) implies that the principal curvatures of the Wulff shape W_γ are bounded. We do not assume such a boundedness, and in fact there are examples of W_γ which have unbounded principal curvatures (cf. Example 4.2) which we can treat. We do not need to assume that the considered surface has no self-intersection like the assumption (iv). We proved that (v) is automatically satisfied (Theorem 3.1). We do not need to assume (vi). We do not assume (vii). However, instead of it, we assume that the r-th anisotropic mean curvatures of the considered hypersurface for γ are integrable.

§7. Examples to prove Theorems 1.2, 1.3, and 1.5

The examples below prove Theorems 1.2, 1.3, and 1.5. Detailed explanation and more examples are given in [10].

As in §4, we denote points in S^1 by $\nu = (\nu_1, \nu_2) = (\cos\theta, \sin\theta)$. Define $\gamma : S^1 \to \mathbb{R}_{>0}$ as

$$\gamma(\cos\theta, \sin\theta) := \cos^6\theta + \sin^6\theta = \nu_1^6 + \nu_2^6.$$

Then the Cahn-Hoffman map $\xi_\gamma : S^1 \to \mathbb{R}^2$ for γ is represented as

$$\begin{aligned}\xi_\gamma(\cos\theta, \sin\theta) = \big(&(\cos\theta)(\cos^6\theta + 6\cos^4\theta\sin^2\theta - 5\sin^6\theta),\\ &(\sin\theta)(-5\cos^6\theta + 6\cos^2\theta\sin^4\theta + \sin^6\theta)\big).\end{aligned} \tag{5}$$

The image $\xi_\gamma(S^1)$ is shown in Figure 4 (a). Because of Fact 3.1, the closed curves in Figures 4 (c), (d), (e) are CAMC, but each of which is not (any homothety or translation of) the Wulff shape (Figure 4 (b)). Hence, these curves prove Theorems 1.2 and 1.5 for $n = 1$. We should remark that there are more closed CAMC curves for γ which are subsets of $\xi_\gamma(S^1)$. The complete collection of these curves is given in [10].

Remark 7.1. The piecewise-C^∞ closed curve in Figure 4 (f) is not CAMC, because, for the outward-pointing normal, the anisotropic curvature is -1 on the solid curve, while it is 1 on the dashed curves. Similarly, for γ given in Example 4.3, the Wulff shape W_γ is the unique closed embedded CAMC curve which is a subset of $\xi_\gamma(S^1)$.

We can construct higher dimensional examples by suitable rotations. Here we give one example. First, rename the ν_2-axis to the ν_3-axis, and consider a new ν_2-axis to obtain $\mathbb{R}^3 = \{(\nu_1, \nu_2, \nu_3)\}$. By rotating γ around the ν_3-axis, we obtain a new integrand $\gamma_1 : S^2 \to \mathbb{R}_{>0}$ represented as

$$\gamma_1(\nu_1, \nu_2, \nu_3) = (\nu_1^2 + \nu_2^2)^3 + \nu_3^6, \quad (\nu_1, \nu_2, \nu_3) \in S^2. \tag{6}$$

The Cahn-Hoffman map $\xi_{\gamma_1} : S^2 \to \mathbb{R}^3$ is given as follows (Figure 5a).

$$\begin{aligned}\xi_{\gamma_1}(\nu) = \big(&(\cos\theta)(\cos^6\theta + 6\cos^4\theta\sin^2\theta - 5\sin^6\theta)(\cos\rho),\\ &(\cos\theta)(\cos^6\theta + 6\cos^4\theta\sin^2\theta - 5\sin^6\theta)(\sin\rho),\\ &(\sin\theta)(-5\cos^6\theta + 6\cos^2\theta\sin^4\theta + \sin^6\theta)\big),\end{aligned} \tag{7}$$

$(\nu = (\cos\theta\cos\rho, \cos\theta\sin\rho, \sin\theta) \in S^2)$.

By the same way as above, we obtain closed piecewise-C^∞ CAMC surfaces and closed piecewise-C^∞ non-CAMC surfaces for γ_1 which are

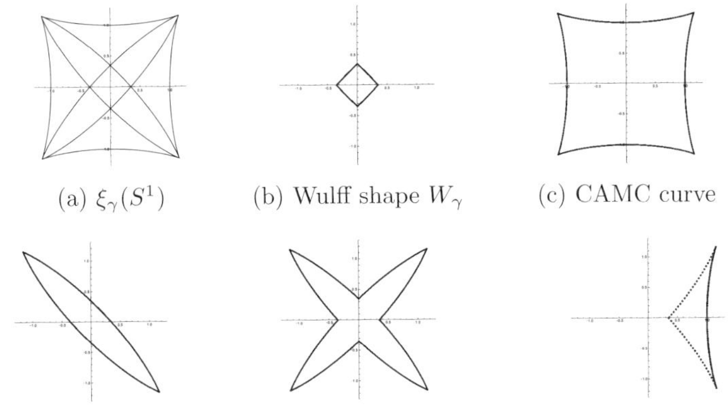

(a) $\xi_\gamma(S^1)$ (b) Wulff shape W_γ (c) CAMC curve

(d) CAMC curve (e) CAMC curve (f) not CAMC

Fig. 4. (a) is the image $\xi_\gamma(S^1)$ of the Cahn-Hoffman map given by (5). (b)–(f) are its subsets. (b)–(e): The anisotropic curvature for the outward-pointing normal is -1. (f): For the outward-pointing normal, the anisotropic curvature is -1 on the solid curve, while it is 1 on the dashed curves.

subsets of $\xi_{\gamma_1}(S^2)$ and which are not the Wulff shape W_{γ_1} (up to homothety and translation). In fact, The Wulff shape W_{γ_1} is the surface of revolution (Figure 5 (b)) given by rotating W_γ (Figure 4 (b)) around the vertical axis. The two piecewise-C^∞ closed surfaces in Figures 5 (c), 5 (d) given by rotating the closed curves in Figures 4 (c), 4 (e) around the vertical axis are CAMC.

We should remark that the piecewise-C^∞ closed surface in Figure 5 (e) given by rotating the closed curve in Figure 4 (f) is not CAMC.

Examples given in this section prove Theorems 1.2, 1.3, and 1.5.

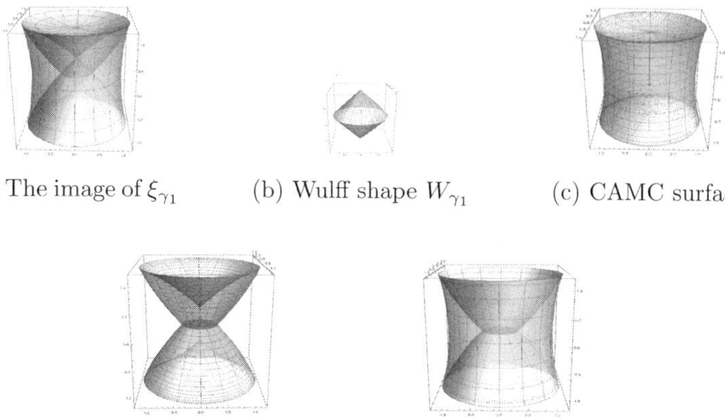

(a) The image of ξ_{γ_1} (b) Wulff shape W_{γ_1} (c) CAMC surface

(d) CAMC surface (e) not CAMC

Fig. 5. (a): The image of the Cahn-Hoffman map $\xi_{\gamma_1} : S^2 \to \mathbb{R}^3$ given by (7). (b)–(e) are its subsets. (b)–(d): The anisotropic mean curvature for the outward-pointing normal is -1. (e): The anisotropic mean curvature is -1 on the 'outer part', while it is 1 on the 'inner part'. Hence, this surface is not CAMC.

§8. Acknowledgments

The author would like to thank the referees for their valuable comments which helped to improve the manuscript.

This work was partially supported by JSPS KAKENHI Grant Number JP18H04487.

References

[1] A. D. Alexandrov, A characteristic property of spheres, *Ann. Mat. Pura Appl.*, **58** (1962), 303–315.

[2] N. Ando, Hartman-Wintner's theorem and its applications, *Calc. Var. Partial Differential Equations*, **43** (2012), 389–402.

[3] J. L. Barbosa and M. do Carmo, Stability of hypersurfaces with constant mean curvature, *Math. Z.*, **185** (1984), 339–353.

[4] J. L. Barbosa, M. do Carmo and J. Eschenburg, Stability of hypersurfaces of constant mean curvature in Riemannian manifolds, *Math. Z.*, **197** (1988), 123–138.

[5] S. Brendle, Embedded self-similar shrinkers of genus 0, *Ann. of Math.*, **183** (2016), 715–728.

[6] Y. He and H. Li, Integral formula of Minkowski type and new characterization of the Wulff shape, *Acta Mathematica Sinica, English Series*, **24** (2008), 697–704.

[7] Y. He, H. Li, H. Ma and J. Ge, Compact embedded hypersurfaces with constant higher order anisotropic mean curvatures, *Indiana Univ. Math. J.*, **58** (2009), 853–868.

[8] A. Honda, M. Koiso and Y. Tanaka, Non-convex anisotropic surface energy and zero mean curvature surfaces in the Lorentz-Minkowski space, *Journal of Math-for-Industry*, **5** (2013), 73–82.

[9] H. Hopf, Differential geometry in the large, Lecture Notes in Math., **1000**, Springer-Verlag, Berlin, 1989.

[10] Y. Jikumaru and M. Koiso, Non-uniqueness of closed embedded non-smooth hypersurfaces with constant anisotropic mean curvature, preprint, arXiv:1903.03958 [math.DG].

[11] N. Kapouleas, Complete constant mean curvature surfaces in euclidean three-space, *Ann. of Math.* (2), **131** (1990), 239–330.

[12] N. Kapouleas, Constant mean curvature surfaces constructed by fusing Wente tori, *Invent. Math.*, **119** (1995), 443–518.

[13] M. Koiso, Uniqueness of stable closed non-smooth hypersurfaces with constant anisotropic mean curvature, preprint, arXiv:1903.03951 [math.DG].

[14] M. Koiso and B. Palmer, Geometry and stability of surfaces with constant anisotropic mean curvature, *Indiana Univ. Math. J.*, **54** (2005), 1817–1852.

[15] M. Koiso and B. Palmer, Anisotropic umbilic points and Hopf's theorem for surfaces with constant anisotropic mean curvature, *Indiana Univ. Math. J.*, **59** (2010), 79–90.

[16] F. Morgan, The cone over the Clifford torus in $\mathbf{R}^4$ is Φ-minimizing, *Math. Ann.*, **289** (1991), 341–354.

[17] F. Morgan, Planar Wulff shape is unique equilibrium, *Proc. Amer. Math. Soc.*, **133** (2005), 809–813.

[18] P. B. Mucha and P. Rybka, A new look at equilibria in Stefan-type problems in the Plane, *SIAM J. Math. Anal.*, **39** (2007), 1120–1134.

[19] B. Palmer, Stability of the Wulff shape, *Proc. Amer. Math. Soc.*, **126** (1998), 3661–3667.

[20] B. Palmer, Stable closed equilibria for anisotropic surface energies: surfaces with edges, *J. Geom. Mech.*, **4** (2012), 89–97.

[21] R. C. Reilly, The relative differential geometry of nonparametric hypersurfaces, *Duke Math. J.*, **43** (1976), 705–721.

[22] J. E. Taylor, Crystalline variational problems, *Bull. Amer. Math. Soc.*, **84** (1978), 568–588.

[23] H. C. Wente, Counterexample to a conjecture of H. Hopf, *Pacific J. Math.*, **121** (1986), 193–243.

[24] H. C. Wente, A note on the stability theorem of J. L. Barbosa and M. do Carmo for closed surfaces of constant mean curvature, *Pac. J. Math.*, **147** (1991), 375–379.

[25] H. Weyl, On the volume of tubes, *Amer. J. Math.*, **61** (1939), 461–472.
[26] W. L. Winterbottom, Equilibrium shape of a small particle in contact with a foreign substrate, *Acta Metallurgica*, **15** (1967), 303–310.
[27] G. Wulff, Zur Frage der Geschwindigkeit des Wachsthums und der Auflösung der Krystallflächen, *Zeitschrift für Krystallographie und Mineralogie*, **34** (1901), 449–530.

Institute of Mathematics for Industry,
Kyushu University,
Motooka Nishi-ku,
FUKUOKA 819-0395, Japan
E-mail address: koiso@math.kyushu-u.ac.jp

Advanced Studies in Pure Mathematics 85, 2020
The Role of Metrics in the Theory of Partial Differential Equations
pp. 259–268

Stability properties of blow up solutions for critical Wave Maps

Joachim Krieger

Abstract.

We review a recent result, obtained jointly with S. Miao, on the stability properties of the blow up solutions constructed in [5], [10] for the co-rotational Wave Maps problem. This provides a partial answer to a question raised in [8], [10].

§1. The critical Wave Maps problem and its co-rotational reduction

Let $M \hookrightarrow \mathbb{R}^N$ a Riemannian manifold isometrically embedded into an ambient Euclidean space. Further, let $\mathbb{R}^{n+1}$ be Minkowski space equipped with a metric of signature $(-1, 1, \ldots)$, and let $\Box$ the corresponding d'Alembertian. We call a map $u : \mathbb{R}^{n+1} \longrightarrow M$ a *wave map*, provided it satisfies the condition $\Box u \perp TM_u$. In case $M = S^{k-1} \hookrightarrow \mathbb{R}^k$, one obtains the following system of nonlinear wave equations for the component vector u:

$$\Box u = (-|\nabla u|^2 + |u_t|^2)u$$

This equation, as well as its analogues for other targets, has received a lot of attention recently, particularly in the energy critical case $n = 2$. In this case, the natural conserved energy

$$E(u) := \int_{\mathbb{R}^2} |\nabla_{t,x} u|^2 \, dx$$

just barely fails to provide a priori control over the norms $\|\nabla_{t,x} u\|_{H^\epsilon}$, $\epsilon > 0$, which ensure strong local well-posedness according to the classical work [7]. The issue of global regularity versus singularity development

Received May 2, 2019.
2010 *Mathematics Subject Classification.* 35L05, 35B40.
Key words and phrases. critical wave equation, blow up, stability.

has attracted much attention, being related to subtle geometric properties of the target. In particular, the recent seminal work [12] provides the following.

Theorem 1.1 (Sterbenz-Tataru [12]). *Let* $u[0] = (u_0, u_1) : \mathbb{R}^2 \longrightarrow S^2 \times TS^2$ *initial data of regularity* $H^{1+} \times H^{0+}$ *with energy strictly below that of the minimum energy non-constant harmonic map* $Q : \mathbb{R}^2 \longrightarrow S^2$ *(which, up to the natural symmetries, is given by stereographic projection* $\mathbb{R}^2 \longrightarrow S^2$*). Then there exists a global in time wave map* $u : \mathbb{R}^{2+1} \longrightarrow S^2$ *of class* $H^{1+} \times H^{0+}$ *extending the data, and which scatters at time* $t = \pm\infty$.

This theorem is sharp in that the data $(Q, 0)$ corresponding to the static solution $u(t, x) = Q(x)$ clearly do not lead to a scattering solution. There remains the question as to what happens for energies $E(u) > E(Q)$. Here a key feature of Q suggests a simplification of things: in fact, $Q : \mathbb{R}^2 \to S^2$, given by stereographic projection, is a special case of a *co-rotational* wave map. This refers to maps which in terms of polar coordinates on $\mathbb{R}^2$ can be written as

$$u(t, \theta, r) = \begin{pmatrix} \cos\theta \sin U(t, r) \\ \sin\theta \sin U(t, r) \\ \cos U(t, r) \end{pmatrix}.$$

Using the fact that wave maps are critical points of the Lagrangian action functional

$$\mathcal{L}(u) = \int_{\mathbb{R}^{2+1}} \left[-|u_t|^2 + |\nabla_x u|^2\right] dtdx$$

and using polar coordinates, this reduces for co-rotational wave maps to

$$2\pi \int_{\mathbb{R}^{1+1}} \left[-U_t^2 + U_r^2 + \frac{\sin^2 U}{r^2}\right] rdrdt.$$

Looking at its critical points, it follows that the co-rotational wave maps problem reduces to the following *scalar wave equation*

$$-U_{tt} + U_{rr} + \frac{1}{r}U_r = \frac{\sin(2U)}{2r^2}. \tag{1.1}$$

In this picture, the static solution Q is then simply given by $Q(r) = 2\arctan r$. The problem (1.1) has given rise to numerous theoretical and numerical investigations in the last 30 years.

§2. Singularities for the co-rotational problem

Solutions of (1.1) which break down in finite time were observed numerically in [2], [6], and the role that Q must play in their description, provided such solutions exist, was explained in the works [3], [13]. Specifically, we quote the following result of Cote, which is in fact a special case of a more general result of his:

Theorem 2.1 (R. Cote, [3]). *Denote $Q_j(r) = Q(r) + j \cdot \pi$, $j \in \mathbb{Z}$. Then if $u(t,r)$ is a solution of (1.1) on $[0,T)$ of regularity H^1 which becomes singular at time $t = T$ in the sense that it cannot be continued continuously in the energy topology beyond T, then there exist a sequence of times $t_n \to T$, as well as a sequence of scales $\lambda_{j,n}$ and profiles Q_j, $j = 1, \ldots, J$ such that we can write*

$$U(t_n, \cdot) = \sum_{j=1}^{J} \left(Q_j \left(\frac{\cdot}{\lambda_{j,n}} \right) - Q_j(\infty) \right) + \eta(t_n, \cdot)$$

where $\eta(t,x)$ extends continuously in the energy topology beyond $t = T$. Also, the profiles asymptotically 'decouple' in the sense that for $j \neq k$

$$\lim_{n \to \infty} \left| \log \left(\frac{\lambda_{j,n}}{\lambda_{k,n}} \right) \right| = +\infty$$

and further $\lim_{n \to \infty} \frac{t_n}{\lambda_{j,n}} = +\infty$.

We note that a *continuous in time soliton resolution* analogous to the preceding theorem is still missing in general, and that the existence of multi-profile ($J \geq 2$) blow ups still appears to be open. However, the existence of finite time blow ups with *one profile* is known since the works [5], [10], [11], providing a precise dynamical description of the blow up. These works display two very different phenomena:

- In [11], the initial data $U[0]$ are of class C^∞, and the solution behaves near blow up time $t = T$, in accordance wit the previous result, as follows:

$$U(t, \cdot) = Q(\lambda(T - t) \cdot) + \eta(t, \cdot) \tag{2.1}$$

  where we have the relation $\lambda(t) \sim t^{-1} e^{\sqrt{|\log(T-t)|}}$, and $\eta(t, \cdot)$ extends continuously in the energy topology past blow up time.
- In [5], [10], the initial data $U[0]$ are of lesser regularity, $H^{1+\nu-}$ for some $\nu > 0$, which in fact can be chosen arbitrarily otherwise, and the solution decouples as

$$U(t, \cdot) = Q(\lambda(T - t) \cdot) + \eta(t, \cdot) \tag{2.2}$$

with $\lambda(t) = t^{-1-\nu}$. These solutions are in fact of regularity C^∞ in the *interior* of the light cone $r < |T-t|$, but experience a loss of regularity across the light cone $r = |T-t|$.

Observe that by choosing $\nu > 0$ large, one can make the second type of solutions have any *finite* degree of smoothness, but presumably never C^∞. In fact, we formulate here.

Conjecture. *Any finite time one bubble blow up solution which is of regularity C^∞ before the blow up admits a description* (2.1) *near blow up time T and with $\lambda(t)$ a logarithmic correction (which also includes factors of the form $e^{|\log t|^\alpha}$, $\alpha < 1$) of t^{-k}, $k \geq 1$. Only those solutions with $k = 1$ are stable under co-rotational perturbations.*

Note that the paper [11] *does establish the stability* of the first type of blow up solutions in the co-rotational context (amongst other things), while the papers [5], [10] do not make any corresponding assertion about stability. In fact at first sight one might conjecture that the very property of stability itself might force a more or less unique blow up dynamics. We next discuss a recent result, [9], obtained jointly with S. Miao, which in fact does establish stability also for the second type of blow up solution, provided the parameter $\nu > 0$ is sufficiently small. This answers a question raised in [8], [10].

§3. Stability of the blow up solutions with $\lambda(t) = t^{-1-\nu}$, $0 < \nu \ll 1$

The following result, obtained jointly with S. Miao, is proved in [9]. To formulate it, we shall identify a function $\varepsilon(r)$ on $[0,\infty)$ with the map

$$(r,\theta) \longrightarrow \varepsilon(r)e^{i\theta} \in \mathbb{C} \simeq \mathbb{R}^2.$$

Theorem 3.1 (K.-Miao [9]). *Let $\nu > 0$ be sufficiently small, and u_ν one of the blow up solutions constructed in* [5], [10] *on some interval $(0, t_0]$, with $0 < t_0 = t_0(\nu)$. Then there is $\delta_0 = \delta_0(\nu) > 0$ small enough, such that for any data perturbation $(\varepsilon_0, \varepsilon_1) \in H^4_{\mathbb{R}^2} \times H^3_{\mathbb{R}^2}$ and with*

$$\big\|(\varepsilon_0, \varepsilon_1)\big\|_{H^4_{\mathbb{R}^2} \times H^3(\mathbb{R}^2)} < \delta_0,$$

the perturbed data $u_\nu(t_0, \cdot) + \varepsilon_0, \partial_t u_\nu(t_0, \cdot) + \varepsilon_1)$ at time t_0 lead to a solution $u(t,r)$ for (1.1) *on $(0, t_0] \times \mathbb{R}_+$ with $u(t,0) = 0$ for all t and*

$$u(t,r) = Q(\lambda(t)r) + \varepsilon(t,r),\ \lambda(t) = t^{-1-\nu}.$$

where $(\varepsilon(t,\cdot),\varepsilon_t(t,\cdot))\in H^{1+\nu-}_{\mathbb{R}^2}\times H^{\nu-}_{\mathbb{R}^2}$ *and further*

$$\lim_{t\to 0}\int_0^t\left(\varepsilon_t^2+\varepsilon_r^2+\frac{\sin^2\varepsilon}{2r^2}\right)rdr=0.$$

We immediately observe some special features here:

The perturbed solution blows up at the same space-time location. This is explained by the fact that the unperturbed solution u_ν experiences a shock across the light cone, as explained earlier. Specifically, from [5], [10], one infers an expansion of u_ν of the form

$$u_\nu(t,r)=Q(\lambda(t)r)+c_1\frac{r\lambda(t)}{(t\cdot\lambda(t))^2}Q_1(a)+c_2\frac{r\lambda(t)\log(r\lambda(t))}{(t\cdot\lambda(t))^2}Q_2(a)+\cdots$$

for suitable constants $c_{1,2}$, where $a=\frac{r}{t}$, and $Q_{1,2}(a)$ admit Puiseux type expansions around $a=1$ with leading singularity (i.e. the shock) given by

$$(1-a)^{\frac{1}{2}+\nu}\log(1-a)$$

for Q_2, and without the logarithm for Q_1. It is this shock which apparently provides a certain rigidity to these blow up solutions: applying a time translation, i.e. replacing $u_\nu(t,r)$ by $u_\nu(t-\delta,r)$, will amount to perturbing the data at time $t=t_0$ by a function in $H^{1+\nu-}\times H^{\nu-}$ and no better. However, the perturbations $(\varepsilon_0,\varepsilon_1)$ considered in the theorem are of much higher regularity, avoiding a simple contradiction.

The function $\varepsilon(t,\cdot)$ *extends continuously past time* $t=0$ *according to a recent result by Duyckaerts-Jia-Kenig-Merle* [4]. Observe that the strategy for the proof of both the preceding theorem as well as the papers [5], [10], is to control the blow up solution in the forward light cone emanating from the singularity $(0,0)$, and to invoke smallness of energy outside of the light cone to prevent blow up there (in fact the co-rotational nature prevents energy concentration there), but our constructive methods have not yet been extended outside of the light cone to render the solution explicit there.

The maximum value of ν *for which stability holds is not yet known, but it is presumably fairly large, of the order of* $\frac{1}{3}$. It would be very interesting to see how the stability fails there. There is in fact numerical evidence that this sort of blow up is unstable if ν is sufficiently large, see [2].

The stability of these co-rotational blow ups under non co-rotational perturbations is a challenging open problem. We believe, however, that the techniques developed in our work may be useful in settling this ultimate stability problem.

§4. An overview of the proof of the theorem

4.1. Formulation of the equations

The proof relies on a Fourier based parametrix method, the basic idea being that passage to an adapted coordinate system with the new variables $\tau = \int_t^\infty \lambda(s)\,ds$, $R = \lambda(t)r$ leads to a nonlinear wave equation on $[\tau_0, \infty) \times \mathbb{R}_+$, $\tau_0 = \int_{t_0}^\infty \lambda(s)\,ds$, whose solution decays temporally as a consequence of temporally decaying damping factors, as well as the very properties of the parametrix used to solve it. The main difficulty one encounters is naturally tied to a resonant mode tied to the scaling symmetry of the equation, which, however, can be handled by somewhat delicate re-iteration arguments. The equation for the perturbation $\varepsilon(t, r)$, which we now think of as a function of τ, R, becomes

$$(4.1)\quad -\left(\left(\partial_\tau + \frac{\lambda'(\tau)}{\lambda(\tau)} R\partial_R\right)^2 + \frac{\lambda'(\tau)}{\lambda(\tau)}\left(\partial_\tau + \frac{\lambda'(\tau)}{\lambda(\tau)} R\partial_R\right)\right)\varepsilon - \mathcal{L}\varepsilon = \lambda^{-2} N(\varepsilon)$$

where we use the notation

$$\partial_R^2 + \frac{1}{R}\partial_R - \frac{\cos(2Q(R))}{R^2} = -\mathcal{L},$$

and the source term $N(\varepsilon)$ is given by

$$N(\varepsilon) = \frac{\cos(2u_\nu) - \cos(2Q(\lambda(t)r)}{r^2}\varepsilon + \frac{\sin(2u_\nu)}{2r^2}(\cos(2\varepsilon) - 1) + \frac{\cos(2u_\nu)}{2r^2}(\sin(2\varepsilon) - 2\varepsilon).$$

The elliptic operator $\mathcal{L}$ has a resonance at zero, given by the function $\phi_0(R) = \frac{R}{1+R^2}$. This function is also in the kernel of the operator

$$\mathcal{D} := \partial_R + \frac{1}{R} - \frac{2}{R(R^2+1)},$$

which is consistent with the fact that we have the factorisation $\mathcal{L} = \mathcal{D}^*\mathcal{D}$. It turns out that the presence of the resonance causes a certain degeneracy of the spectral measure associated with the operator $\mathcal{L}$, see [1], [10], and in order to avoid this, we shall mostly work with $\mathcal{D}\varepsilon$. Passing from the latter to ε requires using the formula

$$\varepsilon(\tau, R) = \phi(\mathcal{D}\varepsilon) + c(\tau)\phi_0(R),$$

where we have introduced the operator $\phi(g) = \phi_0(R)\int_0^R (\phi_0(s))^{-1} g(s)\,ds$. A full description of the solution $\varepsilon(\tau, R)$ then amounts to a description of the pair $\big(c(\tau), \mathcal{D}\varepsilon(\tau, R)\big)$. For $\mathcal{D}\varepsilon$, one infers the equation

(4.2)

$$\begin{aligned}
&-\left(\left(\partial_\tau + \frac{\lambda'}{\lambda}R\partial_R\right)^2 + 3\frac{\lambda'}{\lambda}\left(\partial_\tau + \frac{\lambda'}{\lambda}R\partial_R\right)\right)\mathcal{D}\varepsilon - \tilde{\mathcal{L}}\mathcal{D}\varepsilon \\
&= \lambda^{-2}\mathcal{D}\left(N(\varepsilon)\right) - \frac{4R}{(R^2+1)^2}\left(2\left(\frac{\lambda'}{\lambda}\right)^2 + \left(\frac{\lambda'}{\lambda}\right)'\right)\varepsilon \\
&\quad - \frac{\lambda'}{\lambda}\frac{4R}{(R^2+1)^2}\left(\partial_\tau + \frac{\lambda'}{\lambda}R\partial_R\right)\varepsilon - \frac{\lambda'}{\lambda}\left(\partial_\tau + \frac{\lambda'}{\lambda}R\partial_R\right)\left(\frac{4R}{(R^2+1)^2}\varepsilon\right) \\
&\quad + \left(2\left(\frac{\lambda'}{\lambda}\right)^2 + \left(\frac{\lambda'}{\lambda}\right)'\right)\mathcal{D}\varepsilon \\
&=: \lambda^{-2}\mathcal{D}\left(N(\varepsilon)\right) + \mathcal{R}(\varepsilon, \mathcal{D}\varepsilon) + \left(2\left(\frac{\lambda'}{\lambda}\right)^2 + \left(\frac{\lambda'}{\lambda}\right)'\right)\mathcal{D}\varepsilon,
\end{aligned}$$

where the new operator $\tilde{\mathcal{L}} = \mathcal{D}\mathcal{D}^*$ no longer has a resonance at zero, resulting in a much more regular spectral measure.
The evolution of the resonance in turn is described by a simple ODE, which is of course coupled to the variable $\mathcal{D}\varepsilon$:

$$\left(\partial_\tau + \frac{\lambda'}{\lambda}\right)^2 c(\tau) + \frac{\lambda'}{\lambda}\left(\partial_\tau + \frac{\lambda'}{\lambda}\right)c(\tau) + h(\tau) + \lambda^{-2}n(\tau) = 0, \tag{4.3}$$

where the functions $h(\tau), n(\tau)$ are determined by the relations

$$h(\tau) = \lim_{R\to 0} R^{-1}\mathcal{D}^*\mathcal{D}\varepsilon(\tau, R),\ n(\tau) = \lim_{R\to 0} R^{-1}N(\varepsilon)(\tau, R).$$

4.2. The Fourier Representation

Proceeding as in [1], [10], our point of departure is a distorted Fourier basis $\phi_{KST}(R,\xi)$ associated to the operator $\mathcal{L}$, and which gives the Fourier representation

$$\begin{aligned}
&f(R) = \int_0^\infty x(\xi)\phi_{KST}(R,\xi)\rho(\xi)\,d\xi,\ x(\xi) = \int_0^\infty f(R)\phi_{KST}(R,\xi)\,dR, \\
&\mathcal{L}\big(R^{-\frac{1}{2}}\phi_{KST}(R,\xi)\big) = \xi R^{-\frac{1}{2}}\phi_{KST}(R,\xi).
\end{aligned}$$

Here the spectral measure $\rho(\xi)$ is quite singular at the origin $\rho(\xi) \sim \frac{1}{\xi\log^2\xi}$ for $\xi \ll 1$, reflecting the presence of the resonance $\phi_0(R)$, but we can improve things by passing to a Fourier representation associated to the operator $\tilde{\mathcal{L}}$, for which a Fourier base is given by the functions

$$\phi(R,\xi) = \xi^{-1}\mathcal{D}\big(R^{-\frac{1}{2}}\phi_{KST}(R,\xi)\big),$$

and the corresponding spectral measure $\tilde{\rho}(\xi) = \xi\rho(\xi)$. Then setting

$$\mathcal{D}\varepsilon(\tau,R) = \int_0^\infty \overline{x}(\tau,\xi)\phi(R,\xi)\tilde{\rho}(\xi)\,d\xi,$$

one infers an integro-differential equation for the variable $\overline{x}(\tau,\xi)$, as follows, where $\mathcal{F}$ refers to the distorted Fourier transform:

(4.4)

$$\begin{aligned}
&-\left(\partial_\tau - \frac{\lambda'}{\lambda}2\xi\partial_\xi + \frac{\lambda'}{\lambda}\mathcal{K}\right)^2\overline{x} - 3\frac{\lambda'}{\lambda}\left(\partial_\tau - \frac{\lambda'}{\lambda}2\xi\partial_\xi + \frac{\lambda'}{\lambda}\mathcal{K}\right)\overline{x} - \xi\overline{x}\\
&-\left(2\left(\frac{\lambda'}{\lambda}\right)^2 + \left(\frac{\lambda'}{\lambda}\right)'\right)\overline{x}\\
&= \mathcal{F}\left(\lambda^{-2}\mathcal{D}\left(N(\varepsilon)\right)\right) + \mathcal{F}\left(\mathcal{R}(\varepsilon,\mathcal{D}\varepsilon)\right).
\end{aligned}$$

Here the important operator $\mathcal{K} = -2 - \frac{\tilde{\rho}'(\xi)\xi}{\tilde{\rho}(\xi)} + \mathcal{K}_0$, where the last operator acts by

$$\big(\mathcal{K}_0 f\big)(\xi) = \int_0^\infty \frac{\tilde{\rho}(\eta)F(\xi,\eta)}{\xi-\eta}f(\eta)\,d\eta$$

for a suitable kernel $F(\xi,\eta)$ whose fine properties we don't specify. Then the strategy is to solve (4.4) by means of a parametrix based method. Specifically, upon introducing the operator

$$\mathcal{D}_\tau := \partial_\tau - \frac{\lambda'}{\lambda}2\xi\partial_\xi - \frac{\lambda'}{\lambda}\frac{\tilde{\rho}'(\xi)\xi}{\tilde{\rho}(\xi)} - \frac{\lambda'}{\lambda} \tag{4.5}$$

we can cast the preceding equation in the following elegant form:

(4.6)

$$-\left(\mathcal{D}_\tau^2 + \frac{\lambda'}{\lambda}\mathcal{D}_\tau + \xi\right)\overline{x} = \mathfrak{R}(\tau,\overline{x}) + \mathcal{F}\left(\lambda^{-2}\mathcal{D}\left(N(\varepsilon)\right)\right) + \mathcal{F}\left(\mathcal{R}(\varepsilon,\mathcal{D}\varepsilon)\right),$$

where we omit the fine structure of the linear, time-weighted non-local terms $\mathfrak{R}(\tau,\overline{x})$. Importantly, the operator on the left admits a completely explicit homogeneous parametrix given by

(4.7)

$$\overline{x}(\tau,\xi) = \frac{\lambda(\tau)}{\lambda(\tau_0)} \frac{\tilde{\rho}^{\frac{1}{2}}\left(\frac{\lambda^2(\tau)}{\lambda^2(\tau_0)}\xi\right)}{\tilde{\rho}^{\frac{1}{2}}(\xi)} \cos\left(\lambda(\tau)\xi^{\frac{1}{2}}\int_{\tau_0}^{\tau}\lambda(u)^{-1}du\right)\overline{x}_0\left(\frac{\lambda(\tau)^2}{\lambda(\tau_0)^2}\xi\right)$$
$$+\,\xi^{-\frac{1}{2}}\frac{\tilde{\rho}^{\frac{1}{2}}\left(\frac{\lambda^2(\tau)}{\lambda^2(\tau_0)}\xi\right)}{\tilde{\rho}^{\frac{1}{2}}(\xi)} \sin\left(\lambda(\tau)\xi^{\frac{1}{2}}\int_{\tau_0}^{\tau}\lambda(u)^{-1}du\right)\overline{x}_1\left(\frac{\lambda(\tau)^2}{\lambda(\tau_0)^2}\xi\right).$$

This allows us to run a fixed point argument essentially using the norm

$$\left\|\langle\xi\rangle^{2+\kappa}\xi^{\frac{1}{2}}\langle\log\xi\rangle^{-1-\kappa}\overline{x}(\tau,\xi)\right\|_{L^2_{d\xi}(0,\infty)}$$

upon which a suitable temporal decay (by reference to τ) is imposed, and similarly for $c(\tau)$, the multiple of the resonance. For the rather intricate details of the iterative scheme which gives the final fixed point, we refer to [9].

References

[1] I. Bejenaru, J. Krieger, D. Tataru, *A codimension-two stable manifold of near soliton equivariant wave maps*, Anal. PDE, 6 (2013), no. 4, 829–857.

[2] P. Bizon, T. Chmaj, Z. Tabor, *Formation of singularities for equivariant (2+1)-dimensional wave maps into the 2-sphere*, Nonlinearity, 14 (2001), no. 5, 1041–1053.

[3] R. Cote, *On the soliton resolution for equivariant wave maps to the sphere*, Comm. Pure Appl. Math., 68 (2015), no. 11, 1946–2004.

[4] T. Duyckaerts, H. Jia, C. Kenig, F. Merle, *Universality of blow up profile for small blow up solutions to the energy critical wave map equation*, Int. Math. Res. Not. IMRN 2018, no. 22, 6961–7025.

[5] C. Gao, J. Krieger, *Optimal polynomial blow up range for critical wave maps*, Commun. Pure Appl. Anal., 14 (2015), no. 5, 1705–1741.

[6] J. Isenberg, S. Liebling, *Singularity formation in 2+1 wave maps*, J. Math. Phys., 43 (2002), no. 1, 678–683.

[7] S. Klainerman, M. Machedon, *Smoothing estimates for null forms and applications. A celebration of John F. Nash, Jr.*, Duke Math. J., 81 (1995), no. 1, 99–133 (1996).

[8] J. Krieger, *Global regularity and singularity development for wave maps*, Surveys in differential geometry. Vol. XII. Geometric flows, 167–201, Surv. Differ. Geom., 12, Int. Press, Somerville, MA, 2008.

[9] J. Krieger, S. Miao, *On stability of blow up solutions for the critical co-rotational wave maps problem*, preprint, available at https://arxiv.org/abs/1803.02706.

[10] J. Krieger, W. Schlag, D. Tataru, *Renormalization and blow up for charge one equivariant critical wave maps*, Inventiones Mathematicae, vol. 171, p. 543–615, 2008.

[11] P. Raphaël, I. Rodnianski, *Stable blow up dynamics for the critical co-rotational wave maps and equivariant Yang-Mills problems*, Publ. Math. Inst. Hautes Études Sci., 115 (2012), 1–122.

[12] J. Sterbenz, D. Tataru, *Regularity of wave-maps in dimension* $2+1$, Comm. Math. Phys., 298 (2010), no. 1, 231–264.

[13] M. Struwe, *Equivariant wave maps in two space dimensions. Dedicated to the memory of Jurgen K. Moser*, Comm. Pure Appl. Math., 56 (2003), no. 7, 815–823.

Bâtiment des Mathématiques, EPFL,
Station 8, CH-1015 Lausanne, Switzerland
E-mail address: joachim.krieger@epfl.ch

Advanced Studies in Pure Mathematics 85, 2020
The Role of Metrics in the Theory of Partial Differential Equations
pp. 269–279

Weighted $L^2 - L^2$ estimate for wave equation in R^3 and its applications

Ning-An Lai

Abstract.

In this work we establish a weighted $L^2 - L^2$ estimate for inhomogeneous wave equation in 3-D, by introducing a Morawetz multiplier with weight to the power $s(1 < s < 2)$, and then integrating on the light cones and t slice. With this weighted $L^2 - L^2$ estimate in hand, we may give a new proof of global existence for small data Cauchy problem of semilinear wave equation with supercritical power in 3-D. What is more, by combining the Huygens' principle for wave equations in 3-D, the global existence for semilinear wave equation with scale invariant damping in 3-D is established.

§1. Introduction

We are going to establish a weighted $L^2 - L^2$ estimate for the inhomogeneous wave equation with zero initial data in 3-D

$$\partial_t^2 \phi - \Delta \phi = F(t, x), \ (t, x) \in \mathbf{R}_+ \times \mathbf{R}^3. \tag{1}$$

For this purpose, we introduce a Morawetz multiplier with power $s(1 < s < 2)$ in the weight, which is showed in (11) below. After making computation in the polar coordinates, we integrate the resulted equality on the time slice and the light cones. Finally, by combining Hardy type inequalities, we establish the weighted $L^2 - L^2$ estimate to the solution of (1) with zero initial data.

As one application, we give a new proof of global existence for small data Cauchy problem of semilinear wave equation

$$\partial_t^2 \phi - \Delta \phi = |\phi|^p, \ (t, x) \in \mathbf{R}_+ \times \mathbf{R}^3, \tag{2}$$

Received March 14, 2019.
Revised June 22, 2019.
2010 *Mathematics Subject Classification.* 35L71, 35B44.
Key words and phrases. Weighted $L^2 - L^2$ estimate, wave equation, global existence, semilinear.

for $1+\sqrt{2}<p<3$. There has been a long history for the study of the large time behavior of the small data Cauchy problem of the following semilinear equation in all dimensions

$$\partial_t^2\phi-\Delta\phi=|\phi|^p,\ (t,x)\in\mathbf{R}_+\times\mathbf{R}^n. \tag{3}$$

The pioneering work is due to John [8], in which he proved global existence for the Cauchy problem (2) with smooth data for $p>1+\sqrt{2}$ and blow-up for $1<p<\sqrt{2}$ with appropriate assumption on the data. Then Strauss [20] announced a critical power $p_c(n)(n\geq 2)$ for the small data Cauchy problem of (3), which is the positive root of the following quadratic equation

$$(n-1)p^2-(n+1)p-2=0.$$

For small data Cauchy problem of (3), critical power means that if $p\in(1,p_c(n)]$ there will be no global solutions while if $p\in(p_c(n),\infty)$ the solutions exist globally in time. This conjecture has been completely solved recently (we refer the reader to [3, 4, 5, 17, 18, 19, 21, 24, 25, 26] and references therein). As the key point for global existence in [8], John established the following weighted pointwise estimate for the solution to (1) with zero data

$$\left\|t(t-r)^{p-2}\phi\right\|_{L^\infty(\mathbf{R}_+^{1+3})}\leq C_p\left\|t^p(t-r)^{p(p-2)}F\right\|_{L^\infty(\mathbf{R}_+^{1+3})}, \tag{4}$$

for $1+\sqrt{2}<p\leq 3$. Lindblad and Sogge [14] established an analogy of (4) without the weight $|t-r|$, which involves the L^p norm of the radial variables, and showed a somewhat simpler proof. We point out that both the methods in [8] and [14] heavily rely on the explicit expression of the solution of the linear wave equation in 3-D. Not much later, Georgiev, Lindblad and Sogge [3] established a weighted Strichartz estimate with simpler weights which are invariant under Lorentz rotations, and can give an elementary proof of global existence for small data Cauchy problem (3) with $p_c(n)<p\leq\frac{n+3}{n-1}$ and $n\geq 3$. In this work, based on the weighted L^2-L^2 estimate, we are going to establish another analogy of (4), which includes the weights $|t+r|$ and $|t-r|$, and L^2 norm of the angular variable. Then we may show global existence for the small data Cauchy problem (2) for $1+\sqrt{2}<p<3$, avoiding to use the explicit formula of the solution.

Another application of our weighted L^2-L^2 estimate is to prove the global existence for small data Cauchy problem of semilinear wave equation with scale invariant damping in 3-D

$$\partial_t^2\Phi-\Delta\Phi+\frac{2}{1+t}\Phi_t=|\Phi|^p,\ (t,x)\in\mathbf{R}_+\times\mathbf{R}^3. \tag{5}$$

Here, scale invariant means that, under the following scaling

$$\widetilde{\Phi}(t,x) := \Phi(\sigma(1+t)-1, \sigma x),\ \ \sigma > 0,$$

the equation

$$\partial_t^2 \Phi - \Delta\Phi + \frac{2}{1+t}\Phi_t = 0,\ \ (t,x) \in \mathbf{R}_+ \times \mathbf{R}^n \tag{6}$$

remains invariant. By using the so-called Liouville transformation

$$\phi(t,x) := (1+t)\Phi(t,x),$$

equation (5) in n dimensional space will be changed into

$$\partial_t^2 \phi - \Delta\phi = \frac{|\phi|^p}{(1+t)^{p-1}},\ \ (t,x) \in \mathbf{R}_+ \times \mathbf{R}^n. \tag{7}$$

For the small data Cauchy problem (7), we know it admits a critical power $p = p_c(n+2)$, due to the known results. We refer the reader to [1, 2, 16, 23] and references therein. However, the global existence results in high dimensions($p > p_c(n+2), n \geq 3$) have only been established under symmetrical assumption. See [2] for $n = 3$, [1] for odd and higher dimensions($n \geq 5$) and [16] for $n \geq 4$. We aim to prove global existence to the small data Cauchy problem of (5) for $p > p_c(5)$ without radial symmetrical assumption. However, compared to the Cauchy problem of (2), we have two difficulties. Firstly, we can take derivative to the equation (5) only one time at most, due to the fact $1 < p_c(5) < 2$. Secondly, an additional weight will appear in the nonlinear term after making the Liouville transformation. We overcome the first difficulty by establishing a new weighted estimate to the solution, based on Sobolev embedding and our weighted $L^2 - L^2$ estimate, while the second one is overcame by using the Strong Huygens' principle for wave equations in 3-D.

Remark 1.1. *We remark that in* [10], *Kato and Sakuraba showed another proof of global existence for the small data Cauchy problem of* (5).

Remark 1.2. *For the small data Cauchy problem of*

$$\partial_t^2 \Phi - \Delta\Phi + \frac{\mu}{1+t}\Phi_t = |\Phi|^p,\ \ (t,x) \in \mathbf{R}_+ \times \mathbf{R}^n, \tag{8}$$

it is conjectured that the critical power is $p = p_c(n+\mu)$. *There are some partial results on blow-up and lifespan estimate, see* [6, 7, 9, 12, 22] *and references therein.*

Remark 1.3. *We refer the reader to* [11] *for the detailed proof of the Theorems below.*

§2. Weighted $L^2 - L^2$ estimate for inhomogeneous wave equation in 3-D

In this section we first establish the weighted $L^2 - L^2$ estimate for the inhomogeneous wave equation with zero initial data.

Theorem 2.1. *Consider the following Cauchy problem of wave equation*

$$\begin{cases} \partial_t^2\phi - \Delta\phi = F(t,x), & (t,x) \text{ in } \mathbf{R}_+ \times \mathbf{R}^3, \\ \phi(0,x) = \phi_t(0,x) = 0, & x \in \mathbf{R}^3. \end{cases} \tag{9}$$

If we suppose that

$$\text{supp } F \subset \{|x| \le t+1\},$$

then we have

$$\begin{aligned} \sup_t \Bigg(&\left\|(t+2-r)^{s/2}\nabla_{t,x}\phi\right\|_{L^2(\mathbf{R}^3)} + \left\|(t+2-r)^{s/2}\left(\frac{\phi}{r}\right)\right\|_{L^2(\mathbf{R}^3)} \\ &+ \left\|(t+2-r)^{s/2}\left(\frac{\nabla_{S^2}\phi}{r}\right)\right\|_{L^2(\mathbf{R}^3)}\Bigg) \\ &\le C_\delta \left\|(t+2+r)^{s/2}(t+2-r)^{1/2+\delta}F\right\|_{L^2(\mathbf{R}_+\times\mathbf{R}^3)}, \end{aligned} \tag{10}$$

for any constant $\delta > 0$ and $1 < s < 2$. And C_δ denotes a positive constant depending on δ.

The proof is based on the following Morawetz type multiplier

$$\begin{aligned} X &= (t+2+r)^s(\partial_t + \partial_r) + (t+2-r)^s(\partial_t - \partial_r) \\ &\quad + \frac{1}{r}\left((t+2+r)^s - (t+2-r)^s\right) \\ &= (t+2+r)^s\left(\partial_t + \partial_r + \frac{1}{r}\right) + (t+2-r)^s\left(\partial_t - \partial_r - \frac{1}{r}\right). \end{aligned} \tag{11}$$

Remark 2.1. *We should mention that in [15], Lindblad and Sterbenz introduced the multiplier*

$$K_0^s = \frac{1}{2}(t+r)^{2s}\left(\partial_t + \partial_r\right) + \frac{1}{2}(t-r)^{2s}\left(\partial_t - \partial_r\right)$$

with s ranged as $\frac{1}{2} \le s \le 1$ and established a generalized Morawetz estimate for the electro-magnetic field with non-vanishing charge.

Remark 2.2. *In [13], the author and Zhou used the multiplier (11) with $s = 1$ to establish a similar weighted $L^2 - L^2$ estimate for the inhomogeneous wave equation, and furthermore demonstrated an elementary proof of global existence for small data Cauchy problem of (3) in high dimensions ($n \geq 4$).*

§3. Sobolev Embedding

In order to use the above weighed $L^2 - L^2$ estimate to prove global existence of semilinear wave equations, we need the following key inequality, which can be proved by integration by parts.

Theorem 3.1. *If we suppose that a smooth function satisfies*

$$supp\ \phi \subset \{|x| \leq t+1\},$$

then for $1 < s < 2$, we have

$$\left\| r^{\frac{1}{2}}(t+2+r)^{\frac{1}{2}}(t+2-r)^{\frac{s-1}{2}}\phi \right\|_{L^\infty_r L^2(S^2)} \leq C\left(\left\|(t+2-r)^{\frac{s}{2}}\phi_r\right\|_{L^2(\mathbf{R}^3)} + \left\|(t+2-r)^{\frac{s}{2}}\left(\frac{\phi}{r}\right)\right\|_{L^2(\mathbf{R}^3)} \right). \tag{12}$$

§4. Global existence for semilinear wave equation in 3-D

Consider the following Cauchy problem

$$\begin{cases} \partial_t^2\phi - \Delta\phi = |\phi|^p, & (t,x) \text{ in } \mathbf{R}_+ \times \mathbf{R}^3, \\ \phi(0,x) = \varepsilon f(x),\ \phi_t(0,x) = \varepsilon g(x), & x \in \mathbf{R}^3, \end{cases} \tag{13}$$

where $f, g \in C_0^\infty(\mathbf{R}^3)$ have compact support

$$supp\ f, g \subset \{x : |x| \leq 1\}.$$

We will provide a new method for the global existence to the above problem. By combining the inequalities (10) and (12) with $\frac{s-1}{2} = \frac{1}{p}$, and noting the Sobolev embedding theorem on S^2

$$\|\phi\|_{L^\infty(S^2)} \leq C\|\phi\|_{H^2(S^2)},$$

we come to

$$
\begin{aligned}
(14)\qquad & \left\| r^{\frac{1}{2}}(t+2+r)^{\frac{1}{2}}(t+2-r)^{\frac{1}{p}}\phi \right\|_{L_t^\infty L_r^\infty H^2(S^2)} \\
& \leq C\varepsilon + C\left\| r^{\frac{1}{2}}(t+2+r)^{\frac{1}{2}}(t+2-r)^{\frac{1}{p}}\phi \right\|^p_{L_t^\infty L_r^\infty H^2(S^2)} \\
& \quad \times \left\| (t+2+r)^{\frac{1}{2}+\frac{1}{p}-\frac{p}{2}}(t+2-r)^{-\frac{1}{2}+\delta} r^{-\frac{p}{2}} \right\|_{L_t^2 L_r^2}.
\end{aligned}
$$

Hereinafter, L_r^q denotes $L^q(r^2dr)$ for $2 \leq q < \infty$.

Then if $1+\sqrt{2} < p < 3$, we have for small enough δ

$$
\begin{aligned}
(15)\qquad M^2 & \triangleq \left\| (t+2+r)^{\frac{1}{2}+\frac{1}{p}-\frac{p}{2}}(t+2-r)^{-\frac{1}{2}+\delta} r^{-\frac{p}{2}} \right\|^2_{L_t^2 L_r^2} \\
& = C\int_0^\infty (t+2)^{-1+2\delta+\frac{2+4p-2p^2}{p}}\, dt \\
& \leq C,
\end{aligned}
$$

and hence the global existence can be established by iteration argument, based on (14).

§5. Global existence for Damped Semilinear Wave Equation in 3-D

Another application of our weighted L^2-L^2 estimate is to prove the global existence of the following Cauchy problem

$$
(16)\qquad \begin{cases} \partial_t^2\Phi - \Delta\Phi + \dfrac{2}{1+t}\Phi_t = |\Phi|^p, & (t,x) \text{ in } \mathbf{R}_+ \times \mathbf{R}^3, \\ \Phi(0,x) = \varepsilon f(x),\ \Phi_t(0,x) = \varepsilon g(x), & x \in \mathbf{R}^3, \end{cases}
$$

where $f, g \in C_0^\infty(\mathbf{R}^3)$ have compact support

$$
supp\ f, g \subset \{x : |x| \leq 1\}.
$$

By the transformation

$$
\phi(t,x) := (1+t)\Phi(t,x),
$$

problem (16) can be rewritten as

$$
(17)\qquad \begin{cases} \partial_t^2\phi - \Delta\phi = \dfrac{|\phi|^p}{(1+t)^{p-1}}, & (t,x) \text{ in } \mathbf{R}_+ \times \mathbf{R}^3, \\ \phi(0,x) = \varepsilon f(x),\ \phi_t(0,x) = \varepsilon\{f(x)+g(x)\}, & x \in \mathbf{R}^3. \end{cases}
$$

By combining the inequality (10) and the Sobolev embedding

$$H^1(S^2) \hookrightarrow L^q(S^2), \text{ for } 2 \le q < \infty,$$

we have

$$\left\| (t+2-r)^{\frac{s}{2}} \frac{\phi}{r} \right\|_{L_t^\infty L_r^2 L^q(S^2)} \le C \left\| (t+2+r)^{\frac{s}{2}} (t+2-r)^{\frac{1}{2}+\delta} F \right\|_{L^2(\mathbf{R}_+ \times \mathbf{R}^3)}. \tag{18}$$

On the other hand, by combining (10) and (12), one gets

$$\left\| r^{\frac{1}{2}} (t+2+r)^{\frac{1}{2}} (t+2-r)^{\frac{s-1}{2}} \phi \right\|_{L_t^\infty L_r^\infty L^2(S^2)} \le C_\delta \left\| (t+2+r)^{\frac{s}{2}} (t+2-r)^{\frac{1}{2}+\delta} F \right\|_{L^2(\mathbf{R}_+ \times \mathbf{R}^3)}. \tag{19}$$

Making interpolation between (18) and (19) yields

$$\left\| r^{\frac{1}{2}-\frac{3\theta}{2}} (t+2+r)^{\frac{1-\theta}{2}} (t+2-r)^{\frac{s-1+\theta}{2}} \phi \right\|_{L_t^\infty L_r^\sigma L^\beta(S^2)} \le C \left\| (t+2+r)^{\frac{s}{2}} (t+2-r)^{\frac{1}{2}+\delta} F \right\|_{L^2(\mathbf{R}_+ \times \mathbf{R}^3)}, \tag{20}$$

where

$$\frac{1}{\sigma} = \frac{\theta}{2} + \frac{1-\theta}{\infty}, \quad \frac{1}{\beta} = \frac{\theta}{q} + \frac{1-\theta}{2} \tag{21}$$

and

$$0 < \theta \ll 1.$$

In order to prove global existence for problem (17) with $p > p_c(5)$ by applying the above inequality, we also have to use the strong Huygen's principle for wave equations in 3-D. The solution for (9) can be written as

$$\phi(t,x) = C \int_0^t \int_{\mathbf{R}^3} \frac{\delta(t-\tau-|x-y|)}{t-\tau} F(\tau, y)\, dy d\tau. \tag{22}$$

On the characteristic cone$(t-\tau = |x-y|)$, it is easy to get

$$t+2-|x| = \tau + |x-y| + 2 - |x| \le 2 + \tau + |y|,$$

which yields for $\alpha > 0$ by combining (22)

$$(t+2-r)^{\alpha}\phi \le C\int_0^t\int_{\mathbf{R}^3}\frac{\delta(t-\tau-|x-y|)}{t-\tau}(\tau+2+y)^{\alpha}|F(\tau,y)|\,dyd\tau. \tag{23}$$

It holds that by combining (20)

$$\begin{aligned}&\left\|r^{\frac{1}{2}-\frac{3\theta}{2}}(t+2+r)^{\frac{1-\theta}{2}}(t+2-r)^{\frac{s-1+\theta}{2}+\alpha}\phi\right\|_{L_t^{\infty}L_r^{\sigma}L^{\beta}(S^2)}\\&\le C\left\|(t+2+r)^{\frac{s}{2}+\alpha}(t+2-r)^{\frac{1}{2}+\delta}|F|\right\|_{L^2(\mathbf{R}_+\times\mathbf{R}^3)}.\end{aligned} \tag{24}$$

Applying (24) to the Cauchy problem (17) after taking one spherical derivative, we get

$$\begin{aligned}&\left\|r^{\frac{1}{2}-\frac{3\theta}{2}}(t+2+r)^{\frac{1-\theta}{2}}(t+2-r)^{\frac{s-1+\theta}{2}+\alpha}\phi\right\|_{L_t^{\infty}L_r^{\sigma}W^{1,\beta}(S^2)}\\&\le C\varepsilon+C\Big\|(t+2+r)^{\frac{s}{2}+\alpha-p+1}(t+2-r)^{\frac{1}{2}+\delta}\\&\quad\times\|\phi\|^{p-1}_{L^{\infty}(S^2)}\|\phi\|_{H^1(S^2)}\Big\|_{L_t^2L_r^2}.\end{aligned} \tag{25}$$

Sine $q > 2$ and $0 < \theta \ll 1$, we may get from (21) that $\beta > 2$. Then Sobolev embedding

$$W^{1,\beta}(S^2) \hookrightarrow L^{\infty}(S^2)$$

and Hölder inequality yield that

$$\begin{aligned}&\left\|r^{\frac{1}{2}-\frac{3\theta}{2}}(t+2+r)^{\frac{1-\theta}{2}}(t+2-r)^{\frac{s-1+\theta}{2}+\alpha}\phi\right\|_{L_t^{\infty}L_r^{\sigma}W^{1,\beta}(S^2)}\\&\le C\varepsilon+\left\|r^{\frac{1}{2}-\frac{3\theta}{2}}(t+2+r)^{\frac{1-\theta}{2}}(t+2-r)^{\frac{s-1+\theta}{2}+\alpha}\phi\right\|^{p}_{L_t^{\infty}L_r^{\sigma}W^{1,\beta}(S^2)}\\&\quad\times\Big\|r^{-(\frac{1}{2}-\frac{3\theta}{2})p}(t+2+r)^{\frac{s}{2}+\alpha-p+1-\frac{p(1-\theta)}{2}}\\&\qquad\cdot(t+2-r)^{\frac{1}{2}+\delta-(\frac{s}{2}-\frac{1-\theta}{2}+\alpha)p}\Big\|_{L_t^2L_r^{\frac{2\sigma}{\sigma-2p}}}.\end{aligned} \tag{26}$$

Let

$$N \triangleq \left\|r^{-(\frac{1}{2}-\frac{3\theta}{2})p}(t+2+r)^{\frac{s}{2}+\alpha-p+1-\frac{p(1-\theta)}{2}}(t+2-r)^{\frac{1}{2}+\delta-(\frac{s}{2}-\frac{1-\theta}{2}+\alpha)p}\right\|_{L_t^2L_r^{\frac{2\sigma}{\sigma-2p}}},$$

and

$$s = 2-\frac{1}{p},\ \alpha = \frac{3}{2p}-\frac{1}{2},$$

then direct computation yields that for $p_c(5) < p < 3$ and small enough $\delta, \theta(\delta > p\theta)$

$$N^2 = \int_0^\infty \left(\int_0^{t+1} (t+2+r)^{\left(\frac{3}{2}+\frac{1}{p}-\frac{3p}{2}+\frac{p\theta}{2}\right)\frac{2}{1-p\theta}} \times (t+2-r)^{\left(-\frac{1}{2}+\delta-\frac{p\theta}{2}\right)\frac{2}{1-p\theta}} r^{2-\left(\frac{1}{2}-\frac{3\theta}{2}\right)\frac{2p}{1-p\theta}} dr \right)^{1-p\theta} dt \leq C \int_0^\infty (t+2)^{-1+2\delta+\frac{2+6p-4p^2}{p}} dt \leq C. \tag{27}$$

Then the global existence for Cauchy problem (17) and hence for (16) can be obtained by iteration argument.

§ Acknowledgment

The author would like to express his sincere thank to Prof. Yi Zhou for the helpful discussion and comments. He also would like to thank the anonymous reviewer for the careful reading and helpful comments.

The author is supported by Natural Science Foundation of Zhejiang Province(LY18A010008), NSFC(11771194), Postdoctoral Research Foundation of China(2017M620128, 2018T110332), the Scientific Research Foundation of the First-Class Discipline of Zhejiang Province (B)(201601).

References

[1] M. D'Abbicco and S. Lucente, NLWE with a special scale invariant damping in odd space dimension, Dynamical Systems, Differential Equations and Applications AIMS Proceedings, (2015), 312–319.

[2] M. D'Abbicco, S. Lucente and M. Reissig, A shift in the Strauss exponent for semilinear wave equations with a not effective damping, J. Differential Equations, **259**, (2015), 5040–5073.

[3] V. Georgiev, H. Lindblad and C. D. Sogge, Weighted Strichartz estimates and global existence for semilinear wave equations, Amer. J. Math., **119**, (1997), 1291–1319.

[4] R. T. Glassey, Finite-time blow-up for solutions of nonlinear wave equations, Math. Z., **177**, (1981), 323–340.

[5] R. T. Glassey, Existence in the large for $\Box u = F(u)$ in two space dimensions, Math. Z., **178**, (1981), 233–261.

[6] M. Ikeda and M. Sobajima, Life-span of solutions to semilinear wave equation with time-dependent critical damping for specially localized initial data, Math. Ann., **372**, (2018), 1017–1040.

[7] T. Imai, M. Kato, H. Takamura and K. Wakasa, The lifespan of solutions of semilinear wave equations with the scale invariant damping in two space dimensions, J. Differential Equations, **269(10)**, (2020), 8387–8424.

[8] F. John, Blow-up of solutions of nonlinear wave equations in three space dimensions, Manuscripta Math., **28**, (1979), 235–268.

[9] M. Kato, H. Takamura and K. Wakasa, The lifespan of solutions of semilinear wave equations with the scale invariant damping in one space dimension, Differential Integral Equations, **32(11-12)**, (2019), 659–678.

[10] M. Kato and M. Sakuraba, Global existence and blow-up for semilinear damped wave equations in three space dimensions, Nonlinear Analysis, **182**, (2019), 209–225.

[11] N. A. Lai, Weighted $L^2 - L^2$ estimate for wave equation and its application, arXiv:1807.05109.

[12] N. A. Lai, H. Takamura and K. Wakasa, Blow-up for semilinear wave equations with the scale invariant damping and super-Fujita exponent, J. Differential Equations, **263(9)**, (2017), 5377–5394.

[13] N. A. Lai, Y. Zhou, An elementary proof of Strauss conjecture, Journal of Functional Analysis, **267**, (2014), 1364–1381.

[14] H. Lindblad and C. D. Sogge, Long-time existence for small amplitude semilinear wave equations, Amer. J. Math., **118**, (1996), 1047–1135.

[15] H. Lindblad and J. Sterbenz, Global stability for charged-scalar fields on Minkowski space, IMRP Int. Math. Res. Pap., (2006).

[16] A. Palmieri, A global existence result for a semilinear scale-invariant wave equation in even dimension, Math. Methods Appl. Sci., **42**, (2019), 2680–2706.

[17] M. A. Rammaha, Finite-time blow up for nonlinear wave equations in high dimensions, Comm. Partial Differential Equations, **12**, (1987), 677–700.

[18] J. Schaeffer, The equation $\Box u = |u|^p$ for the critical value of p, Proc. Roy. Soc. Edinburgh Sect.A, **101**, (1985), 31–44.

[19] T. C. Sideris, Nonexistence of global solutions to semilinear wave equations in high dimensions, J. Differential Equations, **52**, (1984), 378–406.

[20] W. A. Strauss, Nonlinear scattering theory at low energy, J. Funct. Anal., **41**, (1981), 110–133.

[21] H. Takamura and K. Wakasa, The sharp upper bound of the lifespan of solutions to critical semilinear wave equations in high dimensions, J. Differential Equations, **251**, (2011), 1157–1171.

[22] Z. Tu and J. Lin, A note on the blowup of scale invariant damping wave equation with sub-Strauss exponent, arXiv:1709.00866v2.

[23] K. Wakasa, The lifespan of solutions to semilinear damped wave equations in one space dimension, Communications on Pure and Applied Analysis, **15**, (2016), 1265–1283.

[24] B. Yordanov and Q. S. Zhang, Finite time blow up for critical wave equations in high dimensions, J. Funct. Anal., **231**, (2006), 361–374.
[25] Y. Zhou, Cauchy problem for semilinear wave equations with small data in four space dimensions, J. Partial Differential Equations, **8**, (1995), 135–144.
[26] Y. Zhou, Blow up of solutions to semilinear wave equations with critical exponent in high dimensions, Chinese Ann. Math. Ser.B, **28**, (2007), 205–212.

Institute of Nonlinear Analysis and Department of Mathematics,
Lishui University,
Lishui City 323000, China
E-mail address: ninganlai@lsu.edu.cn

Advanced Studies in Pure Mathematics 85, 2020
The Role of Metrics in the Theory of Partial Differential Equations
pp. 281–289

On an obstacle problem arising in large exponent asymptotics for one dimensional fully nonlinear diffusions of power type

Qing Liu

Abstract.

In this note, we discuss the limit behavior for a fully nonlinear diffusion equation of power type in one space dimension. It turns out that, when the initial value is Lipschitz and convex, the solution converges locally uniformly to a unique limit function that is independent of the time variable as the exponent tends to infinity. We characterize the limit as the minimal solution of an obstacle problem. Such asymptotic behavior is closely related to applications in mathematical models of image denoising and collapsing sandpiles.

§1. Introduction

This short note is a simplified presentation of our work on large exponent behavior for power-type nonlinear evolution equations [17]. Although some parts of our results apply to the Euclidean space of general dimensions, in this paper we restrict ourselves in one space dimension so as to give a clearer view of the topic. Moreover, we choose to consider a heat-type problem rather than a general parabolic operator so that we can better clarify our basic idea.

The equation we are concerned with is as follows:

$$\text{(PH)} \quad \begin{cases} u_t - |u_{xx}|^{\alpha-1} u_{xx} = 0 & \text{in } \mathbb{R} \times (0, \infty), \quad (1.1) \\ u(\cdot, 0) = u_0 & \text{in } \mathbb{R}, \quad (1.2) \end{cases}$$

where $\alpha > 0$ is a given exponent and u_0 is a given Lipschitz initial value. We are interested in the limit behavior of the unique solution u_α of (PH) as $\alpha \to \infty$.

Received Nobember 13, 2018.
Revised August 20, 2019.
2010 *Mathematics Subject Classification.* Primary 35K55; Secondary 35B40, 35D40.
Key words and phrases. fully nonlinear diffusion equations, large exponent behavior, obstacle problems.

Under the assumption that u_0 is Lipschitz and convex, we show that as $\alpha \to \infty$, $u_\alpha(x,t) \to V(x)$ locally uniformly for all $(x,t) \in \mathbb{R} \times (0,\infty)$, where V is the minimal viscosity supersolution of the following obstacle problem:

$$\min\{-U_{xx} + 1,\ U - u_0\} = 0 \quad \text{in } \mathbb{R}; \tag{1.3}$$

see Theorem 3.1 for a more precise statement of our main result. Here we omit the introduction of viscosity solutions to elliptic problems like (1.3). The standard definition of viscosity solutions can be found in many references such as [10, 14].

It is worth remarking that a standard obstacle problem is usually posed in a bounded domain with the Dirichlet boundary condition [7, 13]. See also [6, 11] for well-posedness results in certain unbounded domains. In contrast to the classical setting, our problem here is quite different. Our domain is the whole space $\mathbb{R}$ and we do not assume the obstacle u_0 to be bounded. In fact, solutions of (1.3) are not unique in general, as shown in Example 3.3. Our result above amounts to saying that the large exponent limit selects the minimal solution of (1.3). It is possible to extend our results to more general equations in general dimensions; see [17].

Our analysis in this note relies on the convexity of u_0 in $\mathbb{R}$. In fact, if $u_0 \in C^2(\mathbb{R})$ is Lipschitz and convex, then it is clear that u_0 is a classical subsolution to (PH). The comparison principle then immediately implies that $u_\alpha \geq u_0$ for all $\alpha > 0$. Note that we can reach the same conclusion by approximation if u_0 is not smooth. This shows the reason why the convexity assumption of u_0 relates the limit of u_α to the obstacle problem (1.3). When u_0 is nonconvex, the situation becomes more complicated. We refer the reader to [15] for results in the nonconvex case based on careful asymptotic analysis for the initial layer.

We finally remark that our results can be used as an alternative approach to study a model describing collapsing sandpiles [4, 12]. Also, the large exponent behavior for a more general class of equations than (1.1), especially the power curvature flow, has applications in image processing [1, 3, 8, 18]. We refer the reader to [15, 17] for more details on these applications. It is also an important topic to investigate the limit behavior as $\alpha \to 0$, which was discussed in [9, 16].

Acknowledgments

This paper is based on the author's talk in the conference *The Role of Metrics in the Theory of Partial Differential Equations*, the 11th Mathematical Society of Japan Seasonal Institute, held at Hokkaido University,

July 2–13, 2018. The author would like to thank the organizers for invitation and hospitality. The work of the author was supported by JSPS Grant-in-Aid for Young Scientists, No. 16K17635 and by the grant from Central Research Institute of Fukuoka University, No. 177102.

§2. The Power-type Diffusion in One Dimension

Let us quickly review the wellposedness of (PH). Using the theory of viscosity solutions [10], one can show that for every fixed $\alpha > 0$ there exists a unique viscosity solution $u_\alpha \in C(\mathbb{R} \times [0, \infty))$ of (PH) satisfying the following growth condition: for any $T > 0$ there exists $C_T > 0$ such that

$$|u_\alpha(x, t)| \leq C_T(1 + |x|) \quad \text{for all } (x, t) \in \mathbb{R} \times [0, T].$$

In particular, the following comparison result holds.

Theorem 2.1 (Comparison principle for power-type heat equation). *Fix $\alpha \geq 1$. Suppose that $u \in USC(\mathbb{R} \times [0, \infty))$ and $v \in LSC(\mathbb{R} \times [0, \infty))$ are respectively a locally bounded subsolution and a locally bounded supersolution of* (1.1). *Assume that either $u(\cdot, 0)$ or $v(\cdot, 0)$ is Lipschitz in $\mathbb{R}$. Assume in addition that for any $T > 0$ there exists $C_T > 0$ such that*

$$|u(x, t)| + |v(x, t)| \leq C_T(1 + |x|) \quad \textit{for all } (x, t) \in \mathbb{R} \times [0, T].$$

If $u(x, 0) \leq v(x, 0)$ in $\mathbb{R}$, then $u(x, t) \leq v(x, t)$ for all $(x, t) \in \mathbb{R} \times [0, \infty)$.

We refer the reader to [15, Appendix A] for a complete proof of this result and detailed discussion on the existence of solutions of (PH) by Perron's method. Moreover, we have the following result on the Lipschitz preserving property.

Lemma 2.2 (Lipschitz regularity preserving property). *Assume that u_0 is Lipschitz in $\mathbb{R}$. Let u_α be the unique viscosity solution of (PH) for every $\alpha > 0$. Then $u_\alpha(\cdot, t)$ is Lipschitz in $\mathbb{R}$ for any $t \geq 0$. Moreover, for any $t \geq 0$,*

$$\|(u_\alpha)_x(\cdot, t)\|_{L^\infty(\mathbb{R})} \leq \|(u_0)_x\|_{L^\infty(\mathbb{R})}.$$

The proof is based on comparison between u_α and a translation of u_α in space. We omit the details here.

§3. Large exponent behavior for convex initial data

We discuss the problem under the assumption that u_0 is Lipschitz and convex. It turns out that in this case the large exponent limit of u_α is related to the obstacle problem (1.3).

Theorem 3.1 (Large exponent asymptotics). *Assume that u_0 is Lipschitz and convex in $\mathbb{R}$. Let u_α be the unique viscosity solution of (PH). Then as $\alpha \to \infty$, $u_\alpha \to V$ locally uniformly in $\mathbb{R} \times (0, \infty)$, where V is Lipschitz function in $\mathbb{R}$ and for any $x \in \mathbb{R}$,*

$$V(x) = \inf\{U(x) : \ U \textit{ is a viscosity supersolution of } (1.3)\}. \tag{3.1}$$

Remark 3.2. Under the assumptions in Theorem 3.1, the set of supersolutions of (1.3) is not empty. Indeed, since u_0 is Lipschitz, there exists $L > 0$ such that

$$u_0(x) \leq u_0(0) + L|x|$$

for all $x \in \mathbb{R}$. Let $W_a(x) = u_0(0) + L\sqrt{a^2 + x^2}$ for $x \in \mathbb{R}$ and $a > 0$. It is clear that $W_a \geq u_0$ in $\mathbb{R}$ for any $a > 0$. Moreover, we may let $a > 0$ sufficiently large such that $(W_a)_{xx} \leq 1$ in $\mathbb{R}$. In other words, W_a is a supersolution of (1.3) when $a > 0$ is taken large.

Example 3.3. Let $u_0 : \mathbb{R} \to \mathbb{R}$ be given by

$$u_0(x) = \begin{cases} 2x - 4 & \text{when } x \geq 2, \\ 0 & \text{when } -2 \leq x \leq 2, \\ -2x - 4 & \text{when } x \leq -2. \end{cases}$$

It is not difficult to see that

$$U_1(x) = \frac{1}{2}(x-1)^2, \quad U_2(x) = \frac{1}{2}(x+1)^2,$$

are both smooth solutions of (1.3). Here u_0 is not smooth, but this can be modified without influence on the non-uniqueness of solutions. Besides the above two solutions, we can actually construct infinitely many smooth solutions of (1.3) by translating the quadratic function $y = x^2$ horizontally between U_1 and U_2.

On the other hand, the least solution (supersolution) as expressed in (3.1) is

$$V(x) = \begin{cases} 2x - 4 & \text{when } x > 3, \\ (x-1)^2/2 & \text{when } 1 < x \leq 3, \\ 0 & \text{when } -1 < x \leq 1, \\ (x+1)^2/2 & \text{when } -3 < x \leq -1, \\ -2x - 4 & \text{when } x \leq -3. \end{cases}$$

To show that such V is the minimal supersolution, we argue by contradiction. Due to the symmetry, suppose that there exists another supersolution U such that $U(x_0) < V(x_0)$ for $x_0 \in (1, 3)$. Let $\tilde{V}(x) =$

$\min\{V(x), U(x)\}$ for any $x \in \mathbb{R}$. It is clear that $\tilde{V}(x_0) < V(x_0)$, $\tilde{V}(1) = 0$ and $\tilde{V}(3) = 2$.

On the other hand, a standard argument shows that $\tilde{V}$ is also a supersolution of (1.3). Since V given above satisfies $V_{xx} = 1$ in $(1, 3)$ with boundary conditions $V(1) = 0$ and $V(3) = 2$, by comparison principle for this Dirichlet problem, we deduce that $V \leq \tilde{V}$ in $(1, 3)$. This yields $V(x_0) \leq \tilde{V}(x_0)$, which is a contradiction.

In order to prove Theorem 3.1, we take the relaxed half limits of u_α: for all $(x, t) \in \mathbb{R} \times [0, \infty)$,

$$(3.2)\quad \begin{aligned}\overline{u}(x,t) &= \operatorname*{limsup^*}_{\alpha\to\infty} u_\alpha(x,t)\\ &= \lim_{\delta\to 0} \sup\left\{u_\alpha(y,s) : y \in \mathbb{R}, s \geq 0, |x-y| + |t-s| + 1/\alpha \leq \delta\right\};\end{aligned}$$

$$(3.3)\quad \begin{aligned}\underline{u}(x,t) &= \operatorname*{liminf_*}_{\alpha\to\infty} u_\alpha(x,t)\\ &= \lim_{\delta\to 0} \inf\left\{u_\alpha(y,s) : y \in \mathbb{R}, s \geq 0, |x-y| + |t-s| + 1/\alpha \leq \delta\right\}.\end{aligned}$$

By Lemma 2.2, both $\overline{u}$ and $\underline{u}$ are Lipschitz in space.

We first show that $\underline{u}$ is a supersolution of (1.3).

Proposition 3.4 (Supersolution property). *Assume that u_0 is Lipschitz and convex. Let u_α be the solution of (PH) and $\underline{u}$ be given as in* (3.3). *Then for every $t > 0$, $\underline{u}(\cdot, t)$ is a supersolution of* (1.3).

Proof. Fix an arbitrary $t_0 > 0$. Suppose by contradiction that there exist $\varphi \in C^2(\mathbb{R})$ and $x_0 \in \mathbb{R}$ such that $\underline{u}(\cdot, t_0) - \varphi$ attains at a strict minimum at $x_0 \in \mathbb{R}$ and

$$\varphi_{xx}(x_0) > 1.$$

Then for any $m > 0$, $\underline{u}(x,t) - \varphi(x) + m(t-t_0)^2$ attains a local minimum at $(x_m, t_m) \in \mathbb{R} \times (0, \infty)$ with $(x_m, t_m) \to (x_0, t_0)$ as $m \to \infty$. In particular, we can fix $m > 0$ sufficiently large to have $\varphi_{xx}(x_m) > 1$. It follows that for any $b > 0$,

$$\underline{u}(x,t) - \varphi(x) + m(t-t_0)^2 + b(x-x_m)^4 + b(t-t_m)^2$$

has a strict minimum at (x_m, t_m) in $\mathbb{R} \times (0, \infty)$. By the definition of relaxed half limit $\underline{u}$, we deduce that

$$u_\alpha(x,t) - \varphi(x) + m(t-t_0)^2 + b(x-x_m)^4 + b(t-t_m)^2$$

attains a minimum in $\mathbb{R}\times(0,\infty)$ at (x_α,t_α) with $(x_\alpha,t_\alpha)\to(x_m,t_m)$ as $\alpha\to\infty$. Since u_α is a supersolution of (PH), we have

$$-2m(t_\alpha-t_0)-2b(t_\alpha-t_m)\geq\left(\varphi_{xx}(x_\alpha)-12b(x_\alpha-x_m)^2\right)^\alpha. \tag{3.4}$$

Note that the term

$$\varphi_{xx}(x_\alpha)-12b(x_\alpha-x_m)^2$$

has a lower bound greater than 1 if $\alpha>0$ is sufficiently large. Sending $\alpha\to\infty$ in (3.4), we obtain a contradiction, since the right hand side tends to infinity while the left hand side is finite.

It remains to show that $\underline{u}(\cdot,t)\geq u_0$ in $\mathbb{R}$ for all $t\geq 0$. This actually follows from the fact that u_0 is a stationary subsolution of (PH) for all $\alpha>0$. Q.E.D.

We next prove that $\overline{u}$ is below any supersolution of (1.3).

Proposition 3.5 (Comparison with supersolutions)**.** *Assume that u_0 is Lipschitz and convex. Let u_α be the solution of (PH) and $\overline{u}$ be given as in* (3.2). *Then $\overline{u}(\cdot,t)\leq U$ in $\mathbb{R}$ holds for any $t>0$ and any viscosity supersolution U of* (1.3).

Let us first give a remark that will be used in our proof. If U is a viscosity supersolution of

$$U_{xx}\leq 1\quad\text{in }\mathbb{R}, \tag{3.5}$$

we see that $W(x):=U(x)-x^2/2$ satisfies $W_{xx}\leq 0$ in the viscosity sense. This implies the concavity of W in $\mathbb{R}$; see for instance [2, Lemma 1]. Since it is easy to see that the reverse implication also holds, we thus can equivalently define any supersolution of (3.5) to be a 1-semiconcave function U in $\mathbb{R}$, that is,

$$U(x+h)+U(x-h)\leq 2U(x)+h^2\quad\text{for all }x,h\in\mathbb{R}. \tag{3.6}$$

It follows that any supersolution of (3.5) is locally Lipschitz in $\mathbb{R}$.

Proof of Proposition 3.5. Fix $R>0$ arbitrarily. We use the standard mollification to approximate U and u_0 respectively by smooth functions U^ε and $u_{0,\varepsilon}$ with $\varepsilon>0$ small such that

$$(U_\varepsilon)_{xx}\leq 1\quad\text{in }\mathbb{R}, \tag{3.7}$$

$$\sup_{|x|\leq R}|U_\varepsilon(x)-U(x)|\leq\omega_R(\varepsilon),\quad\sup_{x\in\mathbb{R}}|u_0(x)-u_{0,\varepsilon}(x)|\leq L\varepsilon,$$

$$U_\varepsilon\geq u_{0,\varepsilon}\quad\text{in }\mathbb{R}. \tag{3.8}$$

Here ω_R denotes a modulus of continuity depending on R, and L denotes the Lipschitz constant of u_0. In addition, $u_{0,\varepsilon}$ is Lipschitz continuous with Lipschitz constant L for any $\varepsilon > 0$ small. The supersolution property (3.7) is due to the fact that the mollification preserves the inequality (3.6).

We next modify the values of U_ε outside $[-R, R]$. Let

$$q_\varepsilon(x) = \begin{cases} (U_\varepsilon)_{xx}(x), & \text{if } |x| \leq R, \\ 1, & \text{if } |x| > R. \end{cases}$$

We then integrate q_ε twice, setting, for all $x \in \mathbb{R}$,

$$\tilde{U}_\varepsilon(x) = U_\varepsilon(0) + \int_0^x \left((U_\varepsilon)_x(0) + \int_0^y q_\varepsilon(z)\, dz \right) dy.$$

Note that $\tilde{U}_\varepsilon \notin C^2(\mathbb{R})$ due to discontinuity of $(\tilde{U}_\varepsilon)_{xx}$ at $x = \pm R$. However we still have

$$(\tilde{U}_\varepsilon)_{xx} \leq 1 \quad \text{in } \mathbb{R} \tag{3.9}$$

in the viscosity sense. Moreover, by (3.8) it is obvious that

$$\tilde{U}_\varepsilon \geq U_\varepsilon \geq u_{0,\varepsilon} \quad \text{in } \mathbb{R}. \tag{3.10}$$

Since $\tilde{U}_\varepsilon$ grows quadratically at space infinity and $u_{0,\varepsilon}$ is Lipschitz, there exists $R' > R$ such that

$$\tilde{U}_\varepsilon(x) \geq u_{0,\varepsilon}(x) + L|x| \quad \text{for all } |x| \geq R'.$$

It follows that, for any $|x| \geq R'$ and $\sigma \in (1/2, 1)$,

$$\begin{aligned} \sigma\tilde{U}_\varepsilon(x) - u_{0,\varepsilon}(x) &\geq (\sigma - 1)|u_{0,\varepsilon}(0)| + ((\sigma - 1)L + \sigma L)\,|x| \\ &\geq (\sigma - 1)|u_{0,\varepsilon}(0)|. \end{aligned}$$

In view of (3.10), by choosing σ close to 1 we also have

$$\sigma\tilde{U}_\varepsilon(x) - u_{0,\varepsilon}(x) \geq (\sigma - 1) \sup_{|x| \leq R'} \left|\tilde{U}_\varepsilon(x)\right| \quad \text{for all } |x| \leq R'.$$

Since $u_{0,\varepsilon} \geq u_0 - L\varepsilon$ in $\mathbb{R}$, combining both estimates above, we deduce that, for any $\sigma \in (0, 1)$ near 1,

$$\sigma\tilde{U}_\varepsilon + C_{\varepsilon,\sigma} \geq u_0 \quad \text{in } \mathbb{R},$$

where

$$C_{\varepsilon,\sigma} = (1 - \sigma) \left(|u_{0,\varepsilon}(0)| + \sup_{|x| \leq R'} \left|\tilde{U}_\varepsilon(x)\right| \right) + L\varepsilon.$$

By the viscosity inequality (3.9), we can show that

$$(x,t) \mapsto \sigma \tilde{U}_\varepsilon(x) + \sigma^\alpha t + C_{\varepsilon,\sigma}$$

is a supersolution of (1.1). We apply Theorem 2.1 to obtain that

$$u_\alpha(x,t) \leq \sigma \tilde{U}_\varepsilon(x) + \sigma^\alpha t + C_{\varepsilon,\sigma} \quad \text{for all } (x,t) \in \mathbb{R} \times [0,\infty).$$

Letting $\alpha \to \infty$, we have

$$\overline{u}(x,t) \leq \sigma \tilde{U}_\varepsilon(x) + C_{\varepsilon,\sigma}$$

for any $(x,t) \in \mathbb{R} \times (0,\infty)$. By letting $\sigma \to 1$, we are led to

$$\overline{u}(x,t) \leq U_\varepsilon(x) + \omega_R(\varepsilon) + L\varepsilon$$

for every $(x,t) \in [-R,R] \times (0,\infty)$. We finally send $\varepsilon \to 0$ to obtain

$$\overline{u}(x,t) \leq U(x)$$

for all $(x,t) \in [-R,R] \times (0,\infty)$. We conclude the proof by noticing that $R > 0$ is arbitrary. Q.E.D.

We finally prove the main theorem.

Proof of Theorem 3.1. By Proposition 3.4 and Lemma 2.2, we deduce that for any $t \geq 0$, $\underline{u}(\cdot,t)$ is a Lipschitz supersolution of (1.3). On the other hand, Proposition 3.5 implies that

$$\overline{u}(\cdot,t) \leq \min\{\underline{u}(\cdot,t), U\}$$

for any supersolution U of (1.3), since it is easily seen that the right hand side is also a supersolution of (1.3). Then $\overline{u} = \underline{u}$ in $\mathbb{R} \times (0,\infty)$ and all of the statements in Theorem 3.1 follow easily. Q.E.D.

We refer the reader to [15] for results in the more complicated case when u_0 is not necessarily convex.

References

[1] L. Alvarez, F. Guichard, P.-L. Lions and J.-M. Morel, Axioms and fundamental equations of image processing, *Arch. Rational Mech. Anal.*, 123(3):199–257, 1993.

[2] L. Alvarez, J.-M. Lasry and P.-L. Lions, Convex viscosity solutions and state constraints, *J. Math. Pures Appl.* (9), 76(3):265–288, 1997.

[3] L. Alvarez, P.-L. Lions and J.-M. Morel, Image selective smoothing and edge detection by nonlinear diffusion. II. *SIAM J. Numer. Anal.*, 29(3):845–866, 1992.

[4] G. Aronsson, L. C. Evans and Y. Wu, Fast/slow diffusion and growing sandpiles, *J. Differential Equations*, 131(2):304–335, 1996.

[5] G. Barles and B. Perthame, Exit time problems in optimal control and vanishing viscosity method, *SIAM J. Control Optim.*, 26(5):1133–1148, 1988.

[6] A. Bensoussan and J.-L. Lions, *Applications of variational inequalities in stochastic control*, North-Holland, New York, 1982.

[7] L. Caffarelli, The obstacle problem revisited, *J. Fourier Anal. Appl.*, 4(4–5):383–402, 1998.

[8] F. Cao, *Geometric curve evolution and image processing*, volume 1805 of *Lecture Notes in Mathematics*, Springer-Verlag, Berlin, 2003.

[9] R. M. Chen and Q. Liu, A nonlinear parabolic equation with discontinuity in the highest order and applications, *J. Differential Equations*, 260(2):1200–1227, 2016.

[10] M. G. Crandall, H. Ishii and P.-L. Lions, User's guide to viscosity solutions of second order partial differential equations, *Bull. Amer. Math. Soc. (N.S.)*, 27(1):1–67, 1992.

[11] P. Daskalopoulos and P. M. N. Feehan, Existence, uniqueness, and global regularity for degenerate elliptic obstacle problems, *preprint*, 2011.

[12] L. C. Evans, M. Feldman and R. F. Gariepy, Fast/slow diffusion and collapsing sandpiles, *J. Differential Equations*, 137(1):166–209, 1997.

[13] A. Friedman, *Variational principles and free-boundary problems*, John Wiley & Sons, Inc., New York, 1982.

[14] Y. Giga, *Surface evolution equations, a level set approach*, volume 99 of *Monographs in Mathematics*, Birkhäuser Verlag, Basel, 2006.

[15] Q. Liu, Large exponent behavior for power-type nonlinear evolution equations and applications, *to appear in J. Evol. Equ.*

[16] Q. Liu, The vanishing exponent limit for motion by a power of mean curvature, *Interfaces Free Bound.*, 22(1), 51–84, 2020.

[17] Q. Liu and N. Yamada, An obstacle problem arising in large exponent limit of power mean curvature flow equation, *Trans. Amer. Math. Soc.*, 372(3), 2103–2141, 2019.

[18] R. Malladi and J. A. Sethian, Image processing via level set curvature flow, *Proc. Nat. Acad. Sci. U.S.A.*, 92(15):7046–7050, 1995.

Qing Liu,
Department of Applied Mathematics,
Fukuoka University,
Fukuoka 814-0180, Japan
E-mail address: qingliu@fukuoka-u.ac.jp

Advanced Studies in Pure Mathematics 85, 2020
The Role of Metrics in the Theory of Partial Differential Equations
pp. 291–302

Singular limit problem for the Navier–Stokes equations in a curved thin domain

Tatsu-Hiko Miura

Abstract.

We consider the incompressible Navier–Stokes equations with slip boundary conditions in a three-dimensional curved thin domain around a given closed surface. Under suitable assumptions on given data, we establish the convergence on the limit surface of the average in the thin direction of a strong solution to the bulk equations as the width of the curved thin domain tends to zero. Moreover, we characterize the limit as a unique weak solution to limit equations, which are the damped and weighted Navier–Stokes equations on the limit surface.

§1. Introduction

Fluid flows in thin domains appear in many problems of natural sciences like biology, meteorology, and ocean dynamics. In the study of the Navier–Stokes equations in a 3D thin domain we naturally expect to get the global-in-time existence of a strong solution for large data since a 3D thin domain with very small width can be considered almost 2D. It is also interesting to study a singular limit problem as the width of a thin domain tends to zero. When a thin domain degenerates into some lower dimensional set, it is important to derive limit equations on the limit set and to compare the original and limit equations in order to understand the effects of the limit set and the thin direction on these equations. Raugel and Sell [22] first proved the global existence of a strong solution to the Navier–Stokes equations under the periodic boundary conditions in a thin product domain $\Omega_\varepsilon = Q_2 \times (0,\varepsilon)$ with a rectangle Q_2 and a sufficiently small $\varepsilon > 0$. Temam and Ziane [26] generalized the result in [22] to a thin product domain $\Omega_\varepsilon = \omega \times (0,\varepsilon)$ with

---

Received November 30, 2018.
Revised April 23, 2019.
2010 *Mathematics Subject Classification.* 35B25, 35Q30, 76D05, 35R01.
Key words and phrases. Navier–Stokes equations, thin domain, singular limit problem, surface fluids.

a bounded domain ω in $\mathbb{R}^2$ and combinations of the Dirichlet, periodic, and Hodge boundary conditions. They also studied the behavior of the average in the vertical direction of a solution to the bulk Navier–Stokes equations and proved its convergence towards a unique solution to the 2D Navier–Stokes equations on ω under certain boundary conditions. Besides the works [22, 26] there is a large amount of literature on the Navier–Stokes equations in thin product domains whose boundaries are flat (see [5, 6, 7, 15, 16] and the references cited therein). However, in physical problems there are various kinds of thin domains with nonflat boundaries (see [21, 23] for examples of thin domains), so generalization of the shape of a thin domain is not only mathematically interesting but also important for applications. The first generalization was done by Temam and Ziane [27], who considered a thin spherical shell

$$\Omega_\varepsilon = \{x \in \mathbb{R}^3 \mid a < |x| < a + a\varepsilon\}, \quad a > 0$$

and established the global existence of a strong solution under the Hodge boundary conditions as well as derived limit equations on a sphere by the same argument as in [26]. Iftimie, Raugel, and Sell [8] studied the Navier–Stokes equations in a flat thin domain

$$\Omega_\varepsilon = \{(x', x_3) \in \mathbb{R}^3 \mid x' \in (0,1)^2,\, 0 < x_3 < \varepsilon g(x')\}, \quad g\colon (0,1)^2 \to \mathbb{R}$$

with a nonflat top boundary under the periodic boundary conditions on the lateral boundaries and Navier's slip boundary conditions on the top and bottom boundaries, and proved the global existence of a strong solution as well as compared the behavior of solutions to the bulk and limit equations. The existence result in [8] was generalized by Hoang [3] and Hoang and Sell [4] to a flat thin domain with nonflat top and bottom boundaries.

In the recent work [14] we considered a curved thin domain around a given 2D closed surface Γ of the form

$$\Omega_\varepsilon = \{y + rn(y) \mid y \in \Gamma,\, \varepsilon g_0(y) < r < \varepsilon g_1(y)\}, \quad g_0, g_1 : \Gamma \to \mathbb{R},$$

where n is the unit outward normal vector field of Γ, and studied the Navier–Stokes equations with Navier's slip boundary conditions

$$(1) \quad \begin{cases} \partial_t u^\varepsilon + (u^\varepsilon \cdot \nabla) u^\varepsilon - \nu \Delta u^\varepsilon + \nabla p^\varepsilon = f^\varepsilon & \text{in} \quad \Omega_\varepsilon \times (0,\infty), \\ \operatorname{div} u^\varepsilon = 0 & \text{in} \quad \Omega_\varepsilon \times (0,\infty), \\ u^\varepsilon \cdot n_\varepsilon = 0 & \text{on} \quad \Gamma_\varepsilon \times (0,\infty), \\ 2\nu P_\varepsilon D(u^\varepsilon) n_\varepsilon + \gamma_\varepsilon u^\varepsilon = 0 & \text{on} \quad \Gamma_\varepsilon \times (0,\infty), \\ u^\varepsilon|_{t=0} = u_0^\varepsilon & \text{in} \quad \Omega_\varepsilon. \end{cases}$$

Here Γ_ε is the boundary of Ω_ε with unit outward normal vector field n_ε, $P_\varepsilon a := a-(a\cdot n_\varepsilon)n_\varepsilon$ the tangential component of $a \in \mathbb{R}^3$ on Γ_ε, $\nu > 0$ the viscosity coefficient independent of ε, $\gamma_\varepsilon \geq 0$ the friction coefficient, and $D(u^\varepsilon) := \{\nabla u^\varepsilon + (\nabla u^\varepsilon)^T\}/2$ the strain rate tensor. The slip boundary conditions introduced by Navier [18] state that the fluid slips on the boundary with velocity proportional to the tangential component of the stress vector. We proved in [14] the global existence of a strong solution u^ε to (1) for large data depending on ε and derived limit equations on the limit surface Γ by showing the convergence of the average in the thin direction of u^ε to a unique weak solution of the limit equations.

The aim of [14] is not just a further generalization of the shape of a thin domain in the study of the Navier–Stokes equations, but a rigorous derivation of the surface Navier–Stokes equations by the thin film limit. Fluid flows on surfaces have attracted the interest of many researchers in mathematics, engineering, and other fields. The Navier–Stokes equations on a Riemannian manifold were introduced by Taylor [25] from a differential geometric point of view (see also [1]) and studied by Mitrea and Taylor [12] and Nagasawa [17]. When a manifold is a 2D surface in $\mathbb{R}^3$, several numerical methods for the surface Navier–Stokes and Stokes equations were developed in the recent works [19, 20, 24]. Moreover, the Navier–Stokes equations on an evolving surface have recently derived by the local conservation laws of mass and momentum [9], a global energetic variational approach [11], and the thin film limit [13], although the result of [13] was based on formal calculations. In [14] we established the first result on a rigorous derivation of the surface Navier–Stokes equations by the thin film limit, which justifies the result of [13] when a limit surface is stationary in time. The limit equations derived in [14] are in fact the damped and weighted Navier–Stokes equations on a closed surface in which the damping term and the weight function come from the friction on the boundary conditions of (1) and the nonconstant width of the curved thin domain, respectively.

In the following, we give the results of [14] on a singular limit problem for (1) that provides a rigorous derivation of the Navier–Stokes type equations on a closed surface. We also explain relations between the limit equations derived in [14] and the surface Navier–Stokes equations introduced in other works.

§2. Notations

Let us fix notations on a closed surface and a curved thin domain. Unless otherwise stated we assume that all functions given here are sufficiently smooth (see [14] for the precise regularity of each function). We denote by I_3 the identity matrix of size three and by

$$a \otimes b := \begin{pmatrix} a_1b_1 & a_1b_2 & a_1b_3 \\ a_2b_1 & a_2b_2 & a_2b_3 \\ a_3b_1 & a_3b_2 & a_3b_3 \end{pmatrix}, \quad \nabla u := \begin{pmatrix} \partial_1 u_1 & \partial_1 u_2 & \partial_1 u_3 \\ \partial_2 u_1 & \partial_2 u_2 & \partial_2 u_3 \\ \partial_3 u_1 & \partial_3 u_2 & \partial_3 u_3 \end{pmatrix}$$

the tensor product of $a = (a_1, a_2, a_3)$ and $b = (b_1, b_2, b_3)$ in $\mathbb{R}^3$ and a vector field $u = (u_1, u_2, u_3)$ on an open set in $\mathbb{R}^3$.

Let Γ be a two-dimensional closed (i.e. compact and without boundary), connected, oriented, and smooth surface in $\mathbb{R}^3$ with unit outward normal vector field n. We set $P := I_3 - n \otimes n$ on Γ and define the tangential gradient and tangential derivatives of a function η on Γ by

$$\nabla_\Gamma \eta := P\nabla\tilde{\eta}, \quad \underline{D}_i\eta := \sum_{j=1}^{3} P_{ij}\partial_j\tilde{\eta} \quad \text{on} \quad \Gamma,\ i = 1,2,3.$$

Here $\tilde{\eta}$ is an extension of η to an open neighborhood of Γ with $\tilde{\eta}|_\Gamma = \eta$. Note that the values of $\nabla_\Gamma\eta$ and $\underline{D}_i\eta$, $i = 1,2,3$ are independent of a choice of $\tilde{\eta}$ (see e.g. [2, Section 16.1]). For a (not necessarily tangential) vector field $v = (v_1, v_2, v_3)$ on Γ we define the tangential gradient matrix of v and the surface strain rate tensor by

$$\nabla_\Gamma v := \begin{pmatrix} \underline{D}_1 v_1 & \underline{D}_1 v_2 & \underline{D}_1 v_3 \\ \underline{D}_2 v_1 & \underline{D}_2 v_2 & \underline{D}_2 v_3 \\ \underline{D}_3 v_1 & \underline{D}_3 v_2 & \underline{D}_3 v_3 \end{pmatrix}, \quad D_\Gamma(v) := P\left(\frac{\nabla_\Gamma v + (\nabla_\Gamma v)^T}{2}\right)P$$

on Γ and the surface divergence of v by $\mathrm{div}_\Gamma v := \mathrm{tr}[\nabla_\Gamma v]$ on Γ. Moreover, for a matrix-valued function $A\colon \Gamma \to \mathbb{R}^{3\times 3}$ and $j = 1,2,3$ we set

$$[\mathrm{div}_\Gamma A]_j := \sum_{i=1}^{3} \underline{D}_i A_{ij} \quad \text{on} \quad \Gamma, \quad A = \begin{pmatrix} A_{11} & A_{12} & A_{13} \\ A_{21} & A_{22} & A_{23} \\ A_{31} & A_{32} & A_{33} \end{pmatrix}$$

and define $\mathrm{div}_\Gamma A := ([\mathrm{div}_\Gamma A]_1, [\mathrm{div}_\Gamma A]_2, [\mathrm{div}_\Gamma A]_3)$ on Γ.

Next we give the definitions of a curved thin domain and average operators. Let g_0 and g_1 be functions on Γ such that

$$g := g_1 - g_0 \geq c \quad \text{on} \quad \Gamma \tag{2}$$

with some constant $c > 0$. For a sufficiently small $\varepsilon > 0$ we define a curved thin domain Ω_ε and its inner and outer boundaries Γ_ε^0 and Γ_ε^1 by

$$\begin{aligned} \Omega_\varepsilon &:= \{y + rn(y) \mid y \in \Gamma,\ r \in (\varepsilon g_0(y), \varepsilon g_1(y))\} \subset \mathbb{R}^3, \\ \Gamma_\varepsilon^i &:= \{y + \varepsilon g_i(y)n(y) \mid y \in \Gamma\}, \quad i = 0,1 \end{aligned}$$

and denote by $\Gamma_\varepsilon := \Gamma_\varepsilon^0 \cup \Gamma_\varepsilon^1$ the boundary of Ω_ε. Let φ be a function on Ω_ε. We define the average of φ in the thin direction by

$$M\varphi(y) := \frac{1}{\varepsilon g(y)} \int_{\varepsilon g_0(y)}^{\varepsilon g_1(y)} \varphi(y + rn(y))\, dr, \quad y \in \Gamma.$$

We also apply M to a vector field u on Ω_ε and denote by $M_\tau u := PMu$ the averaged tangential component of u.

Finally, we define function spaces on Γ. For $\eta, \xi \in C^1(\Gamma)$ we have an integration by parts formula (see e.g. [2, Lemma 16.1])

$$\int_\Gamma (\eta \underline{D}_i \xi + \xi \underline{D}_i \eta)\, d\mathcal{H}^2 = -\int_\Gamma \eta \xi H n_i \, d\mathcal{H}^2, \quad i = 1, 2, 3,$$

where $\mathcal{H}^2$ is the two-dimensional Hausdorff measure and $H := -\mathrm{div}_\Gamma n$ is (twice) the mean curvature of Γ. Based on this formula, for $i = 1, 2, 3$ we say that $\eta \in L^2(\Gamma)$ has the weak tangential derivative η_i in $L^2(\Gamma)$ if

$$(\eta_i, \xi)_{L^2(\Gamma)} = -(\eta, \underline{D}_i \xi + \xi H n_i)_{L^2(\Gamma)} \quad \text{for all} \quad \xi \in C^1(\Gamma).$$

In this case we write $\underline{D}_i \eta = \eta_i$ and define the Sobolev space

$$H^1(\Gamma) := \{\eta \in L^2(\Gamma) \mid \underline{D}_i \eta \in L^2(\Gamma) \text{ for all } i = 1, 2, 3\}.$$

Also, for $\mathcal{X} = L^2, H^1$ and the function g on Γ given in (2) we set

$$\begin{aligned} \mathcal{X}(\Gamma, T\Gamma) &:= \{v \in \mathcal{X}(\Gamma)^3 \mid v \cdot n = 0 \text{ on } \Gamma\}, \\ \mathcal{X}_{g\sigma}(\Gamma, T\Gamma) &:= \{v \in \mathcal{X}(\Gamma, T\Gamma) \mid \mathrm{div}_\Gamma(gv) = 0 \text{ on } \Gamma\} \end{aligned}$$

and denote by $H^{-1}(\Gamma, T\Gamma)$ the dual space of $H^1(\Gamma, T\Gamma)$ (via the $L^2(\Gamma)$-inner product). We further define function spaces

$$\begin{aligned} \mathcal{R} &:= \{w(x) = a \times x + b,\ x \in \mathbb{R}^3 \mid a, b \in \mathbb{R}^3,\ w|_\Gamma \cdot n = 0 \text{ on } \Gamma\}, \\ \mathcal{K}(\Gamma) &:= \{v \in H^1(\Gamma, T\Gamma) \mid D_\Gamma(v) = 0 \text{ on } \Gamma\}. \end{aligned}$$

The space $\mathcal{R}$ describes the extrinsic infinitesimal symmetry of the embedded surface Γ, while $\mathcal{K}(\Gamma)$ represents the intrinsic one. The inclusion $\mathcal{R} \subset \mathcal{K}(\Gamma)$ holds for any closed surface. It is known that $\mathcal{R} = \mathcal{K}(\Gamma)$ if Γ is a surface of revolution or it is closed and convex, but for other closed surfaces the question whether this relation holds is still open.

§3. Main results

To state our main results we further require a few notations and assumptions. For nonnegative constants γ_ε^0 and γ_ε^1 we define the friction coefficient γ_ε on Γ_ε by $\gamma_\varepsilon := \gamma_\varepsilon^i$ on Γ_ε^i, $i=0,1$. Also, let

$$\begin{aligned}\mathcal{R}_i &:= \{w \in \mathcal{R} \mid w|_\Gamma \cdot \nabla_\Gamma g_i = 0 \text{ on } \Gamma\}, \quad i = 0,1,\\ \mathcal{R}_g &:= \{w \in \mathcal{R} \mid w|_\Gamma \cdot \nabla_\Gamma g = 0 \text{ on } \Gamma\},\\ \mathcal{K}_g(\Gamma) &:= \{v \in \mathcal{K}(\Gamma) \mid v \cdot \nabla_\Gamma g = 0 \text{ on } \Gamma\}.\end{aligned}$$

We make the following assumptions on γ_ε^0, γ_ε^1, and Γ.

Assumption 1. There exists a constant $c>0$ such that

$$\gamma_\varepsilon^0 \le c\varepsilon, \quad \gamma_\varepsilon^1 \le c\varepsilon \quad \text{for all} \quad \varepsilon \in (0,1).$$

Assumption 2. Either of the following conditions is satisfied:

(A1) There exists a constant $c>0$ such that

$$\begin{aligned}\gamma_\varepsilon^0 &\ge c\varepsilon \quad \text{for all} \quad \varepsilon \in (0,1) \quad \text{or}\\ \gamma_\varepsilon^1 &\ge c\varepsilon \quad \text{for all} \quad \varepsilon \in (0,1).\end{aligned}$$

(A2) The relation $\mathcal{K}_g(\Gamma) = \{0\}$ holds.

(A3) The relations

$$\mathcal{R}_g = \mathcal{R}_0 \cap \mathcal{R}_1, \quad \mathcal{R}_g|_\Gamma := \{w|_\Gamma \mid w \in \mathcal{R}_g\} = \mathcal{K}_g(\Gamma)$$

hold and $\gamma_\varepsilon^0 = \gamma_\varepsilon^1 = 0$ for all $\varepsilon \in (0,1)$.

These assumptions are used to show the uniform norm equivalence of the Stokes operator A_ε on Ω_ε with slip boundary conditions

$$c^{-1}\|u\|_{H^k(\Omega_\varepsilon)} \le \|A_\varepsilon^{k/2}u\|_{L^2(\Omega_\varepsilon)} \le c\|u\|_{H^k(\Omega_\varepsilon)}, \quad u \in D(A_\varepsilon^{k/2})$$

for $k=1,2$ with a constant $c>0$ independent of ε, see [14] for details. We explain in [14, Remark 1.8] when the conditions of Assumption 2 are satisfied. Here we note that the condition (A3) is imposed to consider a thin spherical shell around the unit sphere S^2 in $\mathbb{R}^3$ of the form

$$\Omega_\varepsilon = \{x \in \mathbb{R}^3 \mid 1 < |x| < 1+\varepsilon\} \quad (\Gamma = S^2,\ g_0 = 0,\ g_1 = 1),$$

which was studied in [27] under different boundary conditions. Also, the authors of [8] imposed (A3) in the study of the Navier–Stokes equations in a flat thin domain around the flat torus $\Gamma = \mathbb{T}^2$.

Let $L^2_\sigma(\Omega_\varepsilon)$ be the standard L^2-solenoidal space on Ω_ε, i.e.

$$L^2_\sigma(\Omega_\varepsilon) = \{u \in L^2(\Omega_\varepsilon)^3 \mid \operatorname{div} u = 0 \text{ in } \Omega_\varepsilon,\ u \cdot n_\varepsilon = 0 \text{ on } \Gamma_\varepsilon\}.$$

Under Assumptions 1 and 2, we define function spaces

$$\mathcal{H}_\varepsilon := \begin{cases} L^2_\sigma(\Omega_\varepsilon), & \text{if the condition (A1) or (A2) is satisfied,} \\ L^2_\sigma(\Omega_\varepsilon) \cap \mathcal{R}_g^\perp, & \text{if the condition (A3) is satisfied,} \end{cases}$$
$$\mathcal{V}_\varepsilon := H^1(\Omega_\varepsilon)^3 \cap \mathcal{H}_\varepsilon,$$

where $\mathcal{R}_g^\perp$ is the orthogonal complement of $\mathcal{R}_g$ in $L^2(\Omega_\varepsilon)^3$.

Now we present the main results of [14] on a singular limit problem for (1) in a slightly modified (but essentially equivalent) form.

Theorem 3 ([14, Theorem 1.6]). *Under Assumptions* 1 *and* 2, *let*

$$u_0^\varepsilon \in \mathcal{V}_\varepsilon, \quad f^\varepsilon \in L^\infty(0,\infty;\mathcal{H}_\varepsilon), \quad \varepsilon \in (0,1).$$

Suppose that the following conditions are satisfied:

(a) *There exist* $c > 0$, $\varepsilon_1 \in (0,1)$, *and* $\alpha \in (0,1)$ *such that*

$$\|u_0^\varepsilon\|^2_{H^1(\Omega_\varepsilon)} + \|f^\varepsilon\|^2_{L^\infty(0,\infty;L^2(\Omega_\varepsilon))} \le c\varepsilon^{-1+\alpha}$$

for all $\varepsilon \in (0,\varepsilon_1)$.

(b) *There exist*

$$v_0 \in L^2(\Gamma, T\Gamma), \quad f \in L^\infty(0,\infty;H^{-1}(\Gamma,T\Gamma))$$

such that

$$\lim_{\varepsilon\to 0} M_\tau u_0^\varepsilon = v_0 \quad \text{weakly in} \quad L^2(\Gamma,T\Gamma),$$
$$\lim_{\varepsilon\to 0} M_\tau f^\varepsilon = f \quad \text{weakly-}\star \text{ in} \quad L^\infty(0,\infty;H^{-1}(\Gamma,T\Gamma)).$$

(c) *For* $i = 0, 1$ *there exists* $\gamma^i \ge 0$ *such that* $\lim_{\varepsilon\to 0} \varepsilon^{-1}\gamma^i_\varepsilon = \gamma^i$.

Then there exists $\varepsilon_2 \in (0,1)$ *such that the Navier–Stokes equations* (1) *admit a global-in-time strong solution*

$$u^\varepsilon \in C([0,\infty);\mathcal{V}_\varepsilon) \cap L^2_{loc}([0,\infty);H^2(\Omega_\varepsilon)^3) \cap H^1_{loc}([0,\infty);\mathcal{H}_\varepsilon)$$

for each $\varepsilon \in (0,\varepsilon_2)$ *and*

$$\lim_{\varepsilon\to 0} Mu^\varepsilon \cdot n = 0 \quad \text{strongly in} \quad C([0,\infty);L^2(\Gamma)).$$

Moreover, there exists a tangential vector field

$$v \in C([0,\infty); L^2_{g\sigma}(\Gamma, T\Gamma)) \cap L^2_{loc}([0,\infty); H^1_{g\sigma}(\Gamma, T\Gamma)) \cap H^1_{loc}([0,\infty); H^{-1}(\Gamma, T\Gamma))$$

on Γ *such that*

$$\lim_{\varepsilon\to 0} M_\tau u^\varepsilon = v \quad \textit{weakly in} \quad L^2(0,T; H^1(\Gamma, T\Gamma))$$

for each $T>0$ *and* v *is a unique weak solution to*

$$(3)\quad \begin{cases} g\Big(\partial_t v + \overline{\nabla}_v v\Big) - 2\nu\left\{P\mathrm{div}_\Gamma[gD_\Gamma(v)] - \dfrac{1}{g}(\nabla_\Gamma g\otimes\nabla_\Gamma g)v\right\} \\ \qquad\qquad + (\gamma^0+\gamma^1)v + g\nabla_\Gamma q = gf \quad \textit{on} \quad \Gamma\times(0,\infty), \\ \mathrm{div}_\Gamma(gv) = 0 \quad \textit{on} \quad \Gamma\times(0,\infty), \\ v|_{t=0} = v_0 \quad \textit{on} \quad \Gamma \end{cases}$$

with an associated pressure q.

Here $\overline{\nabla}_v v := P(v\cdot\nabla_\Gamma)v$ is the covariant derivative of v along itself. For the definition of a weak solution to (3), see [14, Section 10.5].

We also give an estimate for the difference between $M_\tau u^\varepsilon$ and v as well as a result on the strong convergence of $M_\tau u^\varepsilon$.

Theorem 4 ([14, Theorems 1.7 and 10.25]). *Under the assumptions of Theorem* 3*, let* u^ε *be a global strong solution to* (1) *for* $\varepsilon\in(0,\varepsilon_2)$ *and* v *be a weak solution to* (3)*. Then*

$$(4)\quad \max_{t\in[0,T]} \|M_\tau u^\varepsilon(t) - v(t)\|^2_{L^2(\Gamma)} + \int_0^T \|\nabla_\Gamma M_\tau u^\varepsilon(t) - \nabla_\Gamma v(t)\|^2_{L^2(\Gamma)}\,dt$$
$$\le c_T\left\{\delta(\varepsilon)^2 + \|M_\tau u_0^\varepsilon - v_0\|^2_{L^2(\Gamma)} + \|M_\tau f^\varepsilon - f\|^2_{L^\infty(0,\infty;H^{-1}(\Gamma,T\Gamma))}\right\}$$

for all $T>0$*, where* $c_T>0$ *is a constant depending only on* T *and*

$$\delta(\varepsilon) := \varepsilon^{\alpha/4} + |\varepsilon^{-1}\gamma^0_\varepsilon - \gamma^0| + |\varepsilon^{-1}\gamma^1_\varepsilon - \gamma^1|.$$

In particular, for all $T>0$ *we have*

$$\lim_{\varepsilon\to 0} M_\tau u^\varepsilon = v \quad \textit{strongly in} \quad C([0,T]; L^2(\Gamma, T\Gamma)) \cap L^2(0,T; H^1(\Gamma, T\Gamma))$$

provided that $\lim_{\varepsilon\to 0}\varepsilon^{-1}\gamma^i_\varepsilon = \gamma^i$ *for* $i=0,1$ *and*

$$\lim_{\varepsilon\to 0}\|M_\tau u_0^\varepsilon - v_0\|_{L^2(\Gamma)} = \lim_{\varepsilon\to 0}\|M_\tau f^\varepsilon - f\|_{L^\infty(0,\infty;H^{-1}(\Gamma,T\Gamma))} = 0.$$

To prove Theorems 3 and 4 we consider a weak formulation for (1) satisfied by a strong solution u^ε. We use the average operator M and the change of variables formula for $\varphi\colon \Omega_\varepsilon \to \mathbb{R}$ (here $J(y,r)$ is a Jacobian)

$$\int_{\Omega_\varepsilon} \varphi(x)\,dx = \int_\Gamma \int_{\varepsilon g_0(y)}^{\varepsilon g_1(y)} \varphi(y+rn(y))J(y,r)\,dr\,d\mathcal{H}^2(y)$$

to transform the weak formulation for u^ε into that for $M_\tau u^\varepsilon$ in which we take a test function from $H^1_{g\sigma}(\Gamma, T\Gamma)$. Here we require the estimates of the strong solution u^ε (see [14, Theorem 8.4])

$$\|u^\varepsilon(t)\|^2_{H^1(\Omega_\varepsilon)} \le c\varepsilon^{-1+\alpha}, \qquad \int_0^t \|u^\varepsilon(s)\|^2_{H^2(\Omega_\varepsilon)}\,ds \le c\varepsilon^{-1+\alpha}(1+t)$$

for $t \ge 0$ to show that a residual term appearing in the weak formulation for $M_\tau u^\varepsilon$ is "sufficiently small" with respect to ε. After transformation of the weak formulation we show the energy estimate for $M_\tau u^\varepsilon$, but we cannot take $M_\tau u^\varepsilon$ as a test function of its weak formulation since it is not in $H^1_{g\sigma}(\Gamma, T\Gamma)$, i.e. $\mathrm{div}_\Gamma(gM_\tau u^\varepsilon)$ does not vanish on Γ in general. To overcome this difficultly we replace $M_\tau u^\varepsilon$ with its weighted solenoidal part $v^\varepsilon \in H^1_{g\sigma}(\Gamma, T\Gamma)$ and show the energy estimate for v^ε. Using it we derive the energy estimate for $M_\tau u^\varepsilon$ and prove its weak convergence to a weak solution of (3). We can also show the uniqueness of a weak solution v to (3) and the estimate (4) by using the weak formulations for $M_\tau u^\varepsilon$ and v as in the case of a flat bounded domain in $\mathbb{R}^2$. For details, we refer to [14, Section 10].

§4. Remarks on the limit equations

If $g \equiv 1$ and $\gamma^0 = \gamma^1 = 0$ in (3), then we have

$$(5) \qquad \partial_t v + \overline{\nabla}_v v - 2\nu P\mathrm{div}_\Gamma[D_\Gamma(v)] + \nabla_\Gamma q = f, \quad \mathrm{div}_\Gamma v = 0$$

on $\Gamma \times (0,\infty)$. It is shown in [13, Lemma 2.5] that

$$2P\mathrm{div}_\Gamma[D_\Gamma(v)] = \Delta_B v + Kv \quad \text{on} \quad \Gamma$$

for a tangential vector field v on Γ satisfying $\mathrm{div}_\Gamma v = 0$ on Γ, where Δ_B and K are the Bochner Laplacian on Γ and the Gaussian curvature of Γ. Also, since Γ is a 2D surface, the Gaussian curvature K agrees with the Ricci curvature Ric of Γ, i.e. $Kw = \mathrm{Ric}(w)$ for a tangential vector field w on Γ (see e.g. [10]). By these relations, the equations (5) read

$$(6) \qquad \partial_t v + \overline{\nabla}_v v - \nu\{\Delta_B v + \mathrm{Ric}(v)\} + \nabla_\Gamma q = f, \quad \mathrm{div}_\Gamma v = 0$$

on $\Gamma\times(0,\infty)$. The equations (6) were introduced and called the "correct" Navier–Stokes equations on a Riemannian manifold by Taylor [25] (see also [1]) and studied by Mitrea and Taylor [12] and Nagasawa [17]. Hence the original limit equations (3) can be seen as the damped and weighted Navier–Stokes equations on a closed surface.

The limit equations (5) are also related to the Navier–Stokes equations on an evolving surface derived in the recent works [9, 11, 13]. When a surface $\Gamma(t)$ evolves in time with the (scalar) normal velocity V_Γ^N, the evolution of the fluid velocity v on $\Gamma(t)$ is described by

$$(7)\qquad \begin{cases} \partial^\bullet v - 2\nu\mathrm{div}_\Gamma[D_\Gamma(v)] + \nabla_\Gamma q + qHn = f & \text{on} \quad \Gamma(t),\, t>0, \\ \mathrm{div}_\Gamma v = 0 & \text{on} \quad \Gamma(t),\, t>0, \\ v\cdot n = V_\Gamma^N & \text{on} \quad \Gamma(t),\, t>0. \end{cases}$$

Here $\partial^\bullet v = \partial_t v + (v\cdot\nabla)v$ is the material derivative of v along itself and H the mean curvature of $\Gamma(t)$. If $\Gamma(t)=\Gamma$ is stationary in time, then $V_\Gamma^N=0$ and the equations (7) reduces to the limit equations (5).

Finally, we see that different boundary conditions in a thin domain problem yield different limit equations. Temam and Ziane [27] studied the Navier–Stokes equations under the Hodge boundary conditions

$$(8)\qquad u\cdot n_\varepsilon = 0, \quad \mathrm{curl}\, u\times n_\varepsilon = 0 \quad \text{on} \quad \Gamma_\varepsilon$$

in the thin spherical shell $\Omega_\varepsilon = \{x\in\mathbb{R}^3 \mid 1<|x|<1+\varepsilon\}$. They derived limit equations on the unit sphere S^2 in $\mathbb{R}^3$ of the form

$$(9)\qquad \partial_t v + \overline{\nabla}_v v - \nu(\Delta_B v - v) + \nabla_\Gamma q = f, \quad \mathrm{div}_\Gamma v = 0$$

on $S^2\times(0,\infty)$. On the other hand, if $\Gamma=S^2$, then $K=1$ and the limit equations (5) equivalent to (6) become

$$(10)\qquad \partial_t v + \overline{\nabla}_v v - \nu(\Delta_B v + v) + \nabla_\Gamma q = f, \quad \mathrm{div}_\Gamma v = 0$$

on $S^2\times(0,\infty)$. The structures of (9) and (10) are very different by the signs of νv in the viscous terms. This is due to the difference between the Hodge boundary conditions (8) and the perfect slip boundary conditions

$$(11)\qquad u\cdot n_\varepsilon = 0, \quad 2P_\varepsilon D(u^\varepsilon)n_\varepsilon = 0 \quad \text{on} \quad \Gamma_\varepsilon.$$

Indeed, by $u\cdot n_\varepsilon = 0$ on Γ_ε we have

$$2P_\varepsilon D(u^\varepsilon)n_\varepsilon - \mathrm{curl}\, u\times n_\varepsilon = 2W_\varepsilon u \quad \text{on} \quad \Gamma_\varepsilon,$$

$$W_\varepsilon u = \begin{cases} u & \text{on} \quad \Gamma_\varepsilon^0 = \{x\in\mathbb{R}^3 \mid |x|=1\}, \\ \dfrac{-u}{1+\varepsilon} & \text{on} \quad \Gamma_\varepsilon^1 = \{x\in\mathbb{R}^3 \mid |x|=1+\varepsilon\}, \end{cases}$$

where W_ε is the Weingarten map of Γ_ε that represents the curvatures of Γ_ε. Therefore, the perfect slip boundary conditions (11) differ from the Hodge boundary conditions (8) by the curvatures of the boundary Γ_ε.

Acknowledgments. The author is grateful to Professor Yoshikazu Giga for his valuable comments on this work. The work of the author was supported by Grant-in-Aid for JSPS Fellows No. 16J02664 and the Program for Leading Graduate Schools, MEXT, Japan.

References

[1] D. G. Ebin and J. Marsden, Groups of diffeomorphisms and the motion of an incompressible fluid, Ann. of Math. (2), **92** (1970), 102–163.

[2] D. Gilbarg and N. S. Trudinger, Elliptic partial differential equations of second order, Reprint of the 1998 edition, Classics in Mathematics, Springer-Verlag, Berlin, 2001.

[3] L. T. Hoang, Incompressible fluids in thin domains with Navier friction boundary conditions (I), J. Math. Fluid. Mech., **12** (2010), 435–472.

[4] L. T. Hoang and G. R. Sell, Navier–Stokes equations with Navier boundary conditions for an oceanic model, J. Dynam. Differential Equations, **22** (2010), 563–616.

[5] C. Hu, Navier–Stokes equations in 3D thin domains with Navier friction boundary condition, J. Differential Equations, **236** (2007), 133–163.

[6] D. Iftimie, The 3D Navier–Stokes equations seen as a perturbation of the 2D Navier–Stokes equations, Bull. Soc. Math. France, **127** (1999), 473–517.

[7] D. Iftimie and G. Raugel, Some results on the Navier–Stokes equations in thin 3D domains, J. Differential Equations, **169** (2001), 281–331.

[8] D. Iftimie, G. Raugel and G. R. Sell, Navier–Stokes equations in thin 3D domains with Navier boundary conditions, Indiana Univ. Math. J., **56** (2007), 1083–1156.

[9] T. Jankuhn, M. A. Olshanskii and A. Reusken, Incompressible fluid problems on embedded surfaces: Modeling and variational formulations, Interfaces Free Bound., **20** (2018), 353–378.

[10] J. Jost, Riemannian geometry and geometric analysis, 6th ed., Universitext, Springer, Heidelberg, 2011.

[11] H. Koba, C. Liu and Y. Giga, Energetic variational approaches for incompressible fluid systems on an evolving surface, Quart. Appl. Math., **75** (2017), 359–389.

[12] M. Mitrea and M. Taylor, Navier–Stokes equations on Lipschitz domains in Riemannian manifolds, Math. Ann., **321** (2001), 955–987.

[13] T.-H. Miura, On singular limit equations for incompressible fluids in moving thin domains, Quart. Appl. Math., **76** (2018), 215–251.

[14] T.-H. Miura, Navier–Stokes equations in a curved thin domain, preprint, arXiv:1811.09816 [math.AP].
[15] I. Moise, R. Temam and M. Ziane, Asymptotic analysis of the Navier–Stokes equations in thin domains, Topol. Methods Nonlinear Anal., **10** (1997), 249–282.
[16] S. Montgomery-Smith, Global regularity of the Navier–Stokes equation on thin three-dimensional domains with periodic boundary conditions, Electron. J. Differential Equations, (1999), No. 11, 19.
[17] T. Nagasawa, Construction of weak solutions of the Navier–Stokes equations on Riemannian manifold by minimizing variational functionals, Adv. Math. Sci. Appl., **9** (1999), 51–71.
[18] C. L. M. H. Navier, Mémoire sur les lois du mouvement des fluides, Mem. Acad. R. Sci. Inst. France, **6** (1823), 389–440.
[19] I. Nitschke, A. Voigt and J. Wensch, A finite element approach to incompressible two-phase flow on manifolds, J. Fluid Mech., **708** (2012), 418–438.
[20] M. A. Olshanskii, A. Quaini, A. Reusken and V. Yushutin, A finite element method for the surface Stokes problem, SIAM J. Sci. Comput., **40** (2018), A2492–A2518.
[21] M. Prizzi, M. Rinaldi and K. P. Rybakowski, Curved thin domains and parabolic equations, Studia Math., **151** (2002), 109–140.
[22] G. Raugel and G. R. Sell, Navier–Stokes equations on thin 3D domains. I. Global attractors and global regularity of solutions, J. Amer. Math. Soc., **6** (1993), 503–568.
[23] G. Raugel, Dynamics of partial differential equations on thin domains, In: Dynamical systems, Montecatini Terme, 1994, Lecture Notes in Math., **1609**, Springer, Berlin, 1995, pp. 208–315.
[24] A. Reusken, Stream function formulation of surface Stokes equations, IMA J. Numer. Anal., published electronically, https://doi.org/10.1093/imanum/dry062.
[25] M. E. Taylor, Analysis on Morrey spaces and applications to Navier-Stokes and other evolution equations, Comm. Partial Differential Equations, **17** (1992), 1407–1456.
[26] R. Temam and M. Ziane, Navier-Stokes equations in three-dimensional thin domains with various boundary conditions, Adv. Differential Equations, **1** (1996), 499–546.
[27] R. Temam and M. Ziane, Navier-Stokes equations in thin spherical domains, In: Optimization methods in partial differential equations, South Hadley, MA, 1996, Contemp. Math., **209**, Amer. Math. Soc., Providence, RI, 1997, pp. 281–314.

Department of Mathematics
Kyoto University
Kitashirakawa Oiwake-cho, Sakyo-ku, Kyoto 606-8502, Japan
E-mail address: t.miura@math.kyoto-u.ac.jp

Advanced Studies in Pure Mathematics 85, 2020
The Role of Metrics in the Theory of Partial Differential Equations
pp. 303–313

Lower bound for the lifespan of solutions to the generalized KdV equation with low-degree of nonlinearity

Hayato Miyazaki

Abstract.

This paper is a survey article to review the Cauchy problem for the generalized KdV equation with low-degree of nonlinearity. In [5], the local well-posedness for the equation is established in an appropriate class under a non-degenerate condition for the initial data. In this paper, we give a lower bound estimate for the lifespan of the solution under the condition as a consequence of the contraction principle. The lifespan depends on the size and a quantity corresponding to the distance from zero of the initial data.

§1. Introduction

In this paper, we give a survey of recent progress on the generalized Korteweg-de Vries (KdV) equation with low-degree of nonlinearity

$$\text{(GK)} \qquad \begin{cases} \partial_t u + \partial_x^3 u = \pm |u|^\alpha \partial_x u, \ (t,x) \in \mathbb{R}^+ \times \mathbb{R}, \\ u(x,0) = u_0(x), \ x \in \mathbb{R}, \end{cases}$$

where $u = u(t,x)$ is a complex-valued unknown function and $\alpha \in (0,1)$. The equation (GK) is a lower nonlinearity version of the celebrated generalized KdV equation

$$\partial_t u + \partial_x^3 u + u^k \partial_x u = 0, \ k \in \mathbb{Z}^+. \tag{1}$$

The equation (1) on the real line or the torus has been extensively studied in mathematical and physical literatures (e.g. Chapter 7–8 in

Received January 7, 2019.
Revised June 2, 2019.
2010 *Mathematics Subject Classification.* Primary 35Q53; Secondary 35A01.
Key words and phrases. generalized KdV equation, well-posedness, lifespan of solutions.

[6]). In particular, many authors study well-posedness for (1) from the point of view of the scale invariance to the equation. A pioneer work on the local well-posedness in the Soblev spaces $H^s(\mathbb{R}) = (1-\partial_x^2)^{-\frac{s}{2}} L^2(\mathbb{R})$, $s \geqslant s_c := 1/2 - 2/k$ is given by Kenig, Ponce and Vega [4]. They prove that (1) is locally well-posed in H^s with $s > 3/4$ if $k = 1$, $s \geqslant 1/4$ if $k = 2$, $s \geqslant 1/12$ if $k = 3$ and $s \geqslant s_c$ if $k \geqslant 4$. As for subsequent works, we refer to Section 7.5 in [6].

In the case of (GK), since the nonlinearity is not Lipschitz continuous in $H^s(\mathbb{R})$ or the weighted Sobolev spaces $H^s(\mathbb{R}) \cap L^2(\langle x\rangle^r \, dx)$ for all s, $r \in \mathbb{R}$ which are basically considered as the solution space, it is not expected that (GK) has an "unique" solution in these spaces (cf. [3, Theorem 1.7]). As a first attempt to construct the unique local solution, under a non-degenerate condition introduced by Cazenave and Naumkin [1] (see (4)), Linares, the author and Ponce [5] establish the following local well-posedness result:

Theorem 1.1 (Local well-posedness, [5]). *Fix $\alpha \in (0,1)$ and $m = \left[\frac{1}{\alpha}\right] + 1$. Let $s \in \mathbb{Z}^+$ satisfy $s \geqslant 2m+4$. Assume that*

$$u_0 \in H^s(\mathbb{R}),\ \langle x\rangle^m u_0 \in L^\infty(\mathbb{R}),\ \langle x\rangle^m \partial_x^j u_0 \in L^2(\mathbb{R}),\ j = 1,2,3,4, \tag{2}$$

with

$$\|u_0\|_{H^s} + \left\|\langle x\rangle^m u_0\right\|_{L^\infty} + \sum_{j=1}^{4} \left\|\langle x\rangle^m \partial_x^j u_0\right\|_{L^2} =: \delta(u_0) =: \delta, \tag{3}$$

and

$$\inf_{x\in\mathbb{R}} \langle x\rangle^m |u_0(x)| =: \lambda(u_0) =: \lambda > 0. \tag{4}$$

Then there exists $T = T(\delta, \lambda; \alpha, s) > 0$ such that (GK) *has a unique local solution*

$$u \in C([0,T], H^s(\mathbb{R})),\ \langle x\rangle^m \partial_x^j u \in C([0,T], L^2(\mathbb{R})),\ j = 1,2,3,4 \tag{5}$$

with

$$\langle x\rangle^m u \in C([0,T], L^\infty(\mathbb{R})), \quad \partial_x^{s+1} u \in L^\infty(\mathbb{R}, L^2([0,T])), \tag{6}$$

and

$$\sup_{0\leqslant t\leqslant T} \left\|\langle x\rangle^m (u(t) - u_0)\right\|_{L^\infty} \leqslant \frac{\lambda}{2}.$$

Moreover, the map $u_0 \mapsto u(t)$ is continuous in the following sense: For any compact $I \subset [0,T]$, there exists a neighborhood V of u_0 satisfying (2) *and* (4) *such that the map is Lipschitz continuous from V into the class defined by* (5) *and* (6).

Remark 1.2. The non-degenerate condition (4) inspired by [1] comes from that $|u|^\alpha$ is not regular enough (only C^α for $\alpha \in (0,1)$). In the proof of Theorem 1.1, we also apply the Kato smoothing effect (Lemma 2.1 below) as in Linares, Ponce and Santos [7], in order to remove derivative loss of the nonlinearity.

The purpose of this paper is to give a lower bound estimate for the lifespan of the solution to (GK) through a review of the proof of Theorem 1.1. Before we state the main result, we shall give the definition of the lifespan to the solution as follows:

$$T_{\delta,\lambda} := \sup\{T \in (0,\infty];\ \text{there exists a unique solution to (GK)} \\ \text{in the class given by Theorem 1.1}\},$$

where δ and λ are defined by (3) and (4). The main result in this paper is the following:

Proposition 1.3 (Lower bound for the lifespan). *Let $u(t)$ be a corresponding maximal solution to* (GK) *on $[0,T_{\delta,\lambda})$ given by Theorem* 1.1 *under the same assumptions. Then there exists the constant $C>0$ such that*

$$(T_{\delta,\lambda}-t)^{1/2} \geqslant C\frac{\lambda(u(t))}{\delta(u(t))} \\ \times\left(1+\left(\frac{1+\lambda(u(t))}{\lambda(u(t))}\right)^{s+1-\alpha}\left(\delta(u(t))^\alpha+\delta(u(t))^{s+1}\right)\right)^{-1}$$

for any $t\in[0,T_{\delta,\lambda})$, where $\delta(u)$ and $\lambda(u)$ are defined by (3) *and* (4), *respectively.*

The rest of the paper is organized as follows: In Section 2, we summarize useful estimates. Section 3 is devoted to the partial proof of Theorem 1.1 since we employ it to show the main result. Finally, in Section 4, we prove Proposition 1.3 by the argument in [2].

At the end of this section, let us introduce some notations used throughout this paper. We set $\langle x\rangle = (1+|x|^2)^{\frac{1}{2}}$ and $[x]$ denotes the greatest integer less than or equal to x for any $x\in\mathbb{R}$. For any q, $r\geqslant 1$, $\|F\|_{L_t^qL_x^r}$ and $\|F\|_{L_t^rL_x^q}$ are defined by the usual Lebesgue spaces with respect to $(t,x)\in\mathbb{R}^+\times\mathbb{R}$, respectively. We also denote $\|F\|_{L_T^qL_x^r}$ and $\|F\|_{L_T^rL_x^q}$ by the corresponding norms restricted to the time interval $[0,T]$. Further, $\mathcal{F}$ stands for the usual Fourier transform on $\mathbb{R}$. Let $U(t)=e^{-t\partial_x^3}$ be the Airy operator defined by $U(t)=\mathcal{F}^{-1}e^{it\xi^3}\mathcal{F}$.

§2. Preliminaries

We shall start this section by some linear estimates for the Airy operator called the sharp Kato smoothing effect found in [4].

Lemma 2.1 (Kato smoothing effects). *The following estimates hold:*

$$\|\partial_x U(t) f\|_{L_x^\infty L_t^2} = \frac{1}{\sqrt{3}} \|f\|_{L^2},$$

$$\left\| \partial_x \int_0^t U(t-s) f(s)\, ds \right\|_{L_x^\infty L_t^2} \leqslant C \|f\|_{L_t^1 L_x^2}.$$

The next estimate is an important tool to treat estimates of the nonlinearity. This is established by the successive integration by parts.

Lemma 2.2 ([5]). *Let $\mu > 0$ and $r \in \mathbb{Z}^+$. Denote $\theta \in [0,1]$ with $(1-\theta)r \in \mathbb{Z}^+$. Then it holds that*

$$\left\| \langle x \rangle^{\theta\mu} \partial_x^{(1-\theta)r} f \right\|_{L^2} \leqslant C \| \langle x \rangle^{\mu} f\|_{L^2}^{\theta} \|\partial_x^r f\|_{L^2}^{1-\theta} + L.O.T.,$$

where the lower order terms L.O.T. are bounded by

$$\sum_{\substack{0 \leqslant \beta \leqslant 1, \\ (1-\beta)(r-1) \in \mathbb{Z}^+}} \left\| \langle x \rangle^{\beta(\mu-1)} \partial_x^{(1-\beta)(r-1)} f \right\|_{L^2}.$$

§3. Local well-posedness

We give a partial proof of Theorem 1.1 since we employ the estimate in the proof, in order to prove the main result.

Proof of Theorem 1.1. To simplify the exposition, we shall consider real-valued functions. Let us introduce the complete metric space

$$X_{T,M} = \Big\{ u \in C([0,T], H^s);\ \|\!|u|\!\|_{X_{T,M}} := \|u\|_{L_T^\infty H^s} + \|\langle x \rangle^m u\|_{L_T^\infty L_x^\infty} + \sum_{l=1}^{4} \left\| \langle x \rangle^m \partial_x^l u \right\|_{L_T^\infty L_x^2} + \left\| \partial_x^{s+1} u \right\|_{L_x^\infty L_T^2} \leqslant M, \sup_{0 \leqslant t \leqslant T} \left\| \langle x \rangle^m (u(t) - u_0) \right\|_{L^\infty} \leqslant \frac{\lambda}{2} \Big\}$$

equipped with the distance function $d_{X_{T,M}}(u,v) = \|\!|u-v|\!\|_{X_{T,M}}$. Here the constant M will be chosen later. Notice that we have

$$\frac{\lambda}{2} \leqslant \langle x \rangle^m |u(x,t)| \leqslant \langle x \rangle^m |u_0(x)| + \frac{\lambda}{2} \tag{7}$$

for any $(x,t)\in\mathbb{R}\times[0,T]$ as long as $u\in X_{T,M}$. Set

$$\Phi(u(t)) = U(t)u_0 \pm \int_0^t U(t-s)(|u|^\alpha \partial_x u)(s)\,ds.$$

We shall prove that Φ is a contraction map in $X_{T,M}$. Let us only show that Φ maps from $X_{T,M}$ to itself. As for the remaining proof, we refer the reader to [5]. By using Lemma 2.1 and the Leibniz rule, one has

$$\begin{aligned}(8)\qquad &\left\|\partial_x^s \Phi(u)\right\|_{L_T^\infty L_x^2} + \left\|\partial_x^{s+1}\Phi(u)\right\|_{L_x^\infty L_T^2}\\ &\leqslant C\left\|\partial_x^s u_0\right\|_{L^2} + C\sum_{j=0}^{s} \underbrace{\left\|\partial_x^j(|u|^\alpha)\partial_x^{s+1-j}u\right\|_{L_T^1 L_x^2}}_{=:A_j}.\end{aligned}$$

We shall treat A_j. A use of the Hölder inequality gives us

$$\begin{aligned}(9)\qquad A_0 &\leqslant CT^{1/2}\left\|\partial_x^{s+1}u\right\|_{L_x^\infty L_T^2}\left\|\langle x\rangle |u|^\alpha\right\|_{L_T^\infty L_x^\infty}\left\|\langle x\rangle^{-1}\right\|_{L^2}\\ &\leqslant CT^{1/2}\left\|\partial_x^{s+1}u\right\|_{L_x^\infty L_T^2}\left\|\langle x\rangle^m u\right\|_{L_T^\infty L_x^\infty}^\alpha\\ &\leqslant CT^{1/2}M^{\alpha+1}.\end{aligned}$$

The estimates for the intermediate terms A_j $(1\leqslant j\leqslant s-1)$ can be obtained by the interpolation using Lemma 2.2 between the terms in A_0 and A_s. Hence, it suffices to handle A_0 and A_s. As for A_s, it holds that

$$\begin{aligned}A_s &\leqslant CT\left(\left\||u|^{\alpha-s}|\partial_x u|^s\partial_x u\right\|_{L_T^\infty L_x^2} + \cdots + \left\||u|^{\alpha-1}|\partial_x^s u|\partial_x u\right\|_{L_T^\infty L_x^2}\right)\\ &=: CT\left(A_{s,s} + \cdots + A_{s,1}\right).\end{aligned}$$

Since the middle term $A_{s,j}$ $(2\leqslant j\leqslant s-1)$ also can be estimated by the interpolation between $A_{s,1}$ and $A_{s,s}$, we only have to estimate $A_{s,1}$ and $A_{s,s}$. Using (7) and Sobolev embedding, one reaches to

$$\begin{aligned}A_{s,1} &\leqslant C\left\|\left(\lambda^{-1}\langle x\rangle^m\right)^{1-\alpha}\partial_x u\right\|_{L_T^\infty L_x^\infty}\left\|\partial_x^s u\right\|_{L_T^\infty L_x^2}\\ &\leqslant C\lambda^{\alpha-1}\left(\left\|\langle x\rangle^m \partial_x u\right\|_{L_T^\infty L_x^2} + \left\|\langle x\rangle^m \partial_x^2 u\right\|_{L_T^\infty L_x^2}\right)\left\|\partial_x^s u\right\|_{L_T^\infty L_x^2}.\end{aligned}$$

In the similar way in the above, we have

$$A_{s,s} \leqslant C\lambda^{\alpha-s}\left(\left\|\langle x\rangle^m \partial_x u\right\|_{L_T^\infty L_x^2} + \left\|\langle x\rangle^m \partial_x^2 u\right\|_{L_T^\infty L_x^2}\right)^s\left\|\partial_x u\right\|_{L_T^\infty L_x^2}.$$

Therefore it turns out that

$$(10)\qquad A_s \leqslant CT(\lambda^{\alpha-1}+\lambda^{\alpha-s})(M^2+M^{s+1}).$$

Combining (8) with (9) and (10), we obtain

$$\|\partial_x^s\Phi(u)\|_{L_T^\infty L_x^2} + \left\|\partial_x^{s+1}\Phi(u)\right\|_{L_x^\infty L_T^2} \leqslant C\,\|\partial_x^s u_0\|_{L^2} + CT^{1/2}\left(1+\lambda^{\alpha-1}+\lambda^{\alpha-s}\right)\left(M^{\alpha+1}+M^{s+1}\right), \tag{11}$$

when $T\leqslant 1$. One also sees from Sobolev embedding that

$$\|\Phi(u)\|_{L_T^\infty L_x^2} \leqslant \|u_0\|_{L^2} + CTM^{\alpha+1}. \tag{12}$$

Let us next consider $\left\|\langle x\rangle^m\,\partial_x^l\Phi(u)\right\|_{L_T^\infty L_x^2}$ for any $l=1,\dots,4$. Note that

$$x^\beta U(t)f = U(t)(x+3t\partial_x^2)^\beta f$$

for all $\beta\in\mathbb{Z}^+$. By using the successive integration by part, we have

$$\begin{aligned}\left\|\langle x\rangle^\beta\,U(t)f\right\|_{L^2} &\leqslant C\left(\|U(t)f\|_{L^2} + \left\||x|^\beta U(t)f\right\|_{L^2}\right)\\ &\leqslant C\,\|f\|_{L^2} + C\left\|(x+3it\partial_x^2)^\beta f\right\|_{L^2}\\ &\leqslant C\,\langle t\rangle^\beta\left(\left\|\langle x\rangle^\beta f\right\|_{L^2} + \left\|\partial_x^{2\beta}f\right\|_{L^2}\right)\end{aligned} \tag{13}$$

for any $\beta\in\mathbb{Z}^+$. Thus, when $T\leqslant 1$, it follows from (13) that

$$\begin{aligned}&\left\|\langle x\rangle^m\,\partial_x^l\Phi(u)\right\|_{L_T^\infty L_x^2}\\ &\leqslant C\left(\left\|\langle x\rangle^m\,\partial_x^l u_0\right\|_{L^2} + \left\|\partial_x^{2m+l}u_0\right\|_{L^2}\right)\\ &\quad + \left(\left\|\langle x\rangle^m\,\partial_x^l(|u|^\alpha\partial_x u)\right\|_{L_T^1L_x^2} + \left\|\partial_x^{2m+l}(|u|^\alpha\partial_x u)\right\|_{L_T^1L_x^2}\right).\end{aligned} \tag{14}$$

Since $s\geqslant 2m+4$, by means of (9), (10) and $H^s\hookrightarrow \dot H^\sigma$ $(\sigma=l,\,2m+l)$, we see that

$$\begin{aligned}&\left\|\partial_x^l(|u|^\alpha\partial_x u)\right\|_{L_T^1L_x^2} + \left\|\partial_x^{2m+l}(|u|^\alpha\partial_x u)\right\|_{L_T^1L_x^2}\\ &\leqslant C\,\||u|^\alpha\partial_x u\|_{L_T^1L_x^2} + C\left\|\partial_x^s(|u|^\alpha\partial_x u)\right\|_{L_T^1L_x^2}\\ &\leqslant CTM^{\alpha+1} + CT^{1/2}M^{\alpha+1} + CT(\lambda^{\alpha-1}+\lambda^{\alpha-s})(M^2+M^{s+1}).\end{aligned} \tag{15}$$

Further, a direct computation shows

$$\begin{aligned}&\left\|\langle x\rangle^m\,\partial_x^l(|u|^\alpha\partial_x u)\right\|_{L_T^1L_x^2}\\ &\leqslant C\left\|\langle x\rangle^m\,|u|^\alpha\partial_x^{l+1}u\right\|_{L_T^1L_x^2} + \cdots + C\left\|\langle x\rangle^m\,\partial_x^l(|u|^\alpha)\partial_x u\right\|_{L_T^1L_x^2}\\ &\leqslant CT\left(\left\|\langle x\rangle^m\,|u|^\alpha\partial_x^{l+1}u\right\|_{L_T^\infty L_x^2} + \cdots + \left\|\langle x\rangle^m\,\partial_x^{l-1}(|u|^\alpha)\partial_x^2 u\right\|_{L_T^\infty L_x^2}\right.\\ &\quad\left. + \left\|\langle x\rangle^m\,|u|^{\alpha-l}|\partial_x u|^{l+1}\right\|_{L_T^\infty L_x^2} + \cdots + \left\|\langle x\rangle^m\,|u|^{\alpha-1}|\partial_x^l u|\partial_x u\right\|_{L_T^\infty L_x^2}\right)\\ &=: CT\left(\mathrm{I}_0 + \cdots + \mathrm{I}_{l-1} + \mathrm{I}_{l,1} + \cdots + \mathrm{I}_{l,l}\right).\end{aligned}$$

As in the estimation of (11), it suffices to treat I_0, $\mathrm{I}_{l,1}$ and $\mathrm{I}_{l,l}$. Firstly, we deduce that

$$\begin{aligned}\mathrm{I}_0 &\leqslant C\left\|\langle x\rangle^m u\right\|_{L_T^\infty L_x^\infty}\left\||u|^{\alpha-1}\partial_x^{l+1}u\right\|_{L_T^\infty L_x^2}\\ &\leqslant C\lambda^{\alpha-1}\left\|\langle x\rangle^m u\right\|_{L_T^\infty L_x^\infty}\left\|\langle x\rangle^{m-1}\partial_x^{l+1}u\right\|_{L_T^\infty L_x^2}\leqslant C\lambda^{\alpha-1}M^2.\end{aligned}$$

Note that when $l=4$, Lemma 2.2 is exploited. Further, we see from Sobolev embedding and $m(l-\alpha)<ml$ that

$$\begin{aligned}\mathrm{I}_{l,1} &\leqslant C\lambda^{\alpha-l}\left\|\langle x\rangle^{m(l-\alpha)}|\partial_x u|^l\right\|_{L_T^\infty L_x^\infty}\left\|\langle x\rangle^m\partial_x u\right\|_{L_T^\infty L_x^2}\\ &\leqslant C\lambda^{\alpha-l}\left(\left\|\langle x\rangle^m\partial_x u\right\|_{L_T^\infty L_x^2}+\left\|\langle x\rangle^m\partial_x^2 u\right\|_{L_T^\infty L_x^2}\right)^l\left\|\langle x\rangle^m\partial_x u\right\|_{L_T^\infty L_x^2}\\ &\leqslant C\lambda^{\alpha-l}M^{l+1}.\end{aligned}$$

On the other hand, by using Sobolev embedding again, it holds that

$$\mathrm{I}_{l,l}\leqslant C\lambda^{\alpha-1}\left\|\langle x\rangle^{m-1}\partial_x u\right\|_{L_t^\infty L_x^\infty}\left\|\langle x\rangle^m\partial_x^l u\right\|_{L_T^\infty L_x^2}\leqslant C\lambda^{\alpha-1}M^2.$$

Combining these estimates, we find

$$\left\|\langle x\rangle^m\partial_x^l(|u|^\alpha\partial_x u)\right\|_{L_T^\infty L_x^2}\leqslant C\lambda^{\alpha-1}M^2+C\lambda^{\alpha-l}M^{l+1}. \tag{16}$$

Hence, it is concluded from (14), (15) and (16) that

$$\begin{aligned}&\left\|\langle x\rangle^m\partial_x^l\Phi(u)\right\|_{L_T^\infty L_x^2}\\ &\leqslant C\left(\left\|\langle x\rangle^m\partial_x^l u_0\right\|_{L^2}+\left\|\partial_x^{2m+l}u_0\right\|_{L^2}\right)\\ &\quad+CT^{1/2}(1+\lambda^{\alpha-1}+\lambda^{\alpha-s})(M^{\alpha+1}+M^{s+1})\end{aligned} \tag{17}$$

as long as $T\leqslant 1$. Let us next handle $\|\langle x\rangle^m\Phi(u)\|_{L_T^\infty L_x^\infty}$. By using (13) and the fact $\frac{d}{dt}U(t)u_0=-\partial_x^3U(t)u_0$, together with Sobolev embedding, we obtain

$$\begin{aligned}&\left\|\langle x\rangle^m(U(t)u_0-u_0)\right\|_{L^\infty}\\ &\leqslant\left\|\int_0^t\langle x\rangle^m U(s)\partial_x^3u_0ds\right\|_{L^\infty}\\ &\leqslant CT\left(\left\|\langle x\rangle^m\partial_x^3u_0\right\|_{L^2}+\left\|\langle x\rangle^m\partial_x^4u_0\right\|_{L^2}+\|u_0\|_{H^{2m+4}}\right)\end{aligned} \tag{18}$$

whenever $T \leqslant 1$. Similarly to (18), it comes from (13) and Sobolev embedding that

$$\begin{aligned}&\left\| \langle x\rangle^m \int_0^t U(t-s)\left(|u|^\alpha \partial_x u\right)(s)ds \right\|_{L^\infty} \\ &\leqslant CT\left(\left\| \langle x\rangle^m |u|^\alpha \partial_x u \right\|_{L_T^\infty L_x^2} + \left\| \langle x\rangle^m \partial_x\left(|u|^\alpha \partial_x u\right) \right\|_{L_T^\infty L_x^2} \right) \\ &\quad + C\langle T\rangle^m \left(\left\| |u|^\alpha \partial_x u \right\|_{L_T^1 L_x^2} + \left\| \partial_x^{2m+1}\left(|u|^\alpha \partial_x u\right) \right\|_{L_T^1 L_x^2} \right).\end{aligned}$$

We here compute

$$\begin{aligned}\left\| \langle x\rangle^m |u|^\alpha \partial_x u \right\|_{L_T^1 L_x^2} &\leqslant T \left\| \langle x\rangle^m u \right\|_{L_T^\infty L_x^\infty} \left\| |u|^{\alpha-1} \partial_x u \right\|_{L_T^\infty L_x^2} \\ &\leqslant CT\lambda^{\alpha-1} \left\| \langle x\rangle^m u \right\|_{L_T^\infty L_x^\infty} \left\| \langle x\rangle^m \partial_x u \right\|_{L_T^\infty L_x^2} \\ &\leqslant CT\lambda^{\alpha-1} M^2.\end{aligned}$$

Also, Sobolev embedding provides that

$$\begin{aligned}&\left\| \langle x\rangle^m \partial_x(|u|^\alpha \partial_x u) \right\|_{L_T^\infty L_x^2} \\ &\leqslant C\left\| \langle x\rangle^m |u|^{\alpha-1}(\partial_x u)^2 \right\|_{L_T^\infty L_x^2} + \left\| \langle x\rangle^m |u|^\alpha \partial_x^2 u \right\|_{L_T^\infty L_x^2} \\ &\leqslant C\lambda^{\alpha-1} \left\| \langle x\rangle^m \partial_x u \right\|_{L_T^\infty L_x^\infty} \left\| \langle x\rangle^m \partial_x u \right\|_{L_T^\infty L_x^2} \\ &\quad + C\|u\|_{L_T^\infty L_x^\infty}^\alpha \left\| \langle x\rangle^m \partial_x^2 u \right\|_{L_T^\infty L_x^2} \\ &\leqslant C(1+\lambda^{\alpha-1})(M^2 + M^{1+\alpha}).\end{aligned}$$

Further, arguing as in (15), we deduce that

$$\left\| \partial_x^\beta(|u|^\alpha \partial_x u) \right\|_{L_T^1 L_x^2} \leqslant CT^{1/2}(1+\lambda^{\alpha-1}+\lambda^{\alpha-s})(M^{\alpha+1}+M^{s+1})$$

for some $\beta \in (0, s]$ if $T \leqslant 1$. Therefore it follows from these estimates that

$$\begin{aligned}&\left\| \langle x\rangle^m \int_0^t U(t-s)\left(|u|^\alpha \partial_x u\right)(s)\,ds \right\|_{L_T^\infty L_x^\infty} \\ &\leqslant CT^{1/2}(1+\lambda^{\alpha-1}+\lambda^{\alpha-s})(M^{\alpha+1}+M^{s+1})\end{aligned} \tag{19}$$

as long as $T \leqslant 1$. Combining (18) with (19), we see that

$$\begin{aligned}&\left\| \langle x\rangle^m \Phi(u) \right\|_{L_T^\infty L_x^\infty} \\ &\leqslant \left\| \langle x\rangle^m u_0 \right\|_{L^\infty} \\ &\quad + CT\left(\left\| \langle x\rangle^m \partial_x^3 u_0 \right\|_{L^2} + \left\| \langle x\rangle^m \partial_x^4 u_0 \right\|_{L^2} + \|u_0\|_{H^{2m+4}} \right) \\ &\quad + CT^{1/2}(1+\lambda^{\alpha-1}+\lambda^{\alpha-s})(M^{\alpha+1}+M^{s+1})\end{aligned} \tag{20}$$

whenever $T \leqslant 1$. Therefore, by using (11) and (12) together with (17) and (20), one establishes that

$$\begin{aligned}\|\!|u|\!\|_{X_T} &\leqslant C_0(1+T)\delta + CT^{1/2}(1+\lambda^{\alpha-1}+\lambda^{\alpha-s})(M^{\alpha+1}+M^{s+1}) \\ &\leqslant C_1\delta + CT^{1/2}\left(\frac{1+\lambda}{\lambda}\right)^{s-\alpha}(M^{\alpha+1}+M^{s+1})\end{aligned}$$

as long as $T \leqslant 1$, where $C_1 = 2C_0$. Therefore, taking $M = 2C_1\delta$, it holds that $\|\!|u|\!\|_{X_T} \leqslant M$ whenever $T = T(\delta,\lambda;\alpha,s)$ satisfies

$$CT^{1/2}\left(\frac{1+\lambda}{\lambda}\right)^{s-\alpha}\left(2^{\alpha+1}(C_1\delta)^{\alpha} + 2^{s+1}(C_1\delta)^{s}\right) \leqslant 1. \tag{21}$$

Finally, combining (18) with (19), when $T \leqslant 1$, we easily have

$$\begin{aligned}&\|\langle x\rangle^m(\Phi(u(t)) - u_0)\|_{L^\infty} \\ &\leqslant CT^{1/2}\delta + CT^{1/2}\left(\frac{1+\lambda}{\lambda}\right)^{s-\alpha}(M^{\alpha+1}+M^{s+1}),\end{aligned}$$

which implies

$$\sup_{0\leqslant t\leqslant T}\|\langle x\rangle^m(\Phi(u(t)) - u_0)\|_{L^\infty} \leqslant \frac{\lambda}{2}$$

as long as $T = T(\delta,\lambda;\alpha,s)$ satisfies

$$CT^{1/2}\delta + CT^{1/2}\left(\frac{1+\lambda}{\lambda}\right)^{s-\alpha}\left((2C_1\delta)^{\alpha+1} + (2C_1\delta)^{s+1}\right) \leqslant \frac{\lambda}{2} \tag{22}$$

since $M = 2C_1\delta$. Thus taking $T = T_0$ satisfying (21) and (22), $\Phi(u) \in X_{T_0,M}$ is valid. The rest of the proof is exhibited in [5]. Q.E.D.

§4. Proof of the main result

Once we obtain the local well-posedness by the contraction principle, the lower bound estimate for the lifespan of the solution is established.

Proof of Proposition 1.3. We use the technique in [2]. Since the case $T_{\delta,\lambda} = \infty$ is trivial, we may assume $T_{\delta,\lambda} < \infty$. We take an initial data $u(t)$ for any $t \in [0, T_{\delta,\lambda})$. Arguing as in the proof of Theorem 1.1, one sees that if $T - t \leqslant 1$,

$$C_1\delta(u(t)) + C(T-t)^{\frac{1}{2}}\left(\frac{1+\lambda(u(t))}{\lambda(u(t))}\right)^{s-\alpha}(M^{\alpha+1}+M^{s+1}) \leqslant M$$

and

$$C(T-t)^{\frac{1}{2}}\left(\delta(u(t))+\left(\frac{1+\lambda(u(t))}{\lambda(u(t))}\right)^{s-\alpha}(M^{\alpha+1}+M^{s+1})\right)\leqslant\frac{\lambda(u(t))}{2},$$

then $T<T_{\delta,\lambda}$. Hence we obtain $T_{\delta,\lambda}-t>1$, or

$$C_1\delta(u(t))+C(T_{\delta,\lambda}-t)^{\frac{1}{2}}\left(\frac{1+\lambda(u(t))}{\lambda(u(t))}\right)^{s-\alpha}(M^{\alpha+1}+M^{s+1})>M,$$

or

$$(T_{\delta,\lambda}-t)^{\frac{1}{2}}\left(\delta(u(t))+\left(\frac{1+\lambda(u(t))}{\lambda(u(t))}\right)^{s-\alpha}(M^{\alpha+1}+M^{s+1})\right)>\frac{\lambda(u(t))}{2C}.$$

By taking $M=2C_1\delta(u(t))$, it holds that

$$\begin{aligned}(T_{\delta,\lambda}-t)^{\frac{1}{2}}>{}&\frac{1}{C_2}\frac{\lambda(u(t))}{\delta(u(t))}\\&\times\left(1+\left(\frac{1+\lambda(u(t))}{\lambda(u(t))}\right)^{s-\alpha}\left(\delta(u(t))^{\alpha}+\delta(u(t))^{s}\right)\right)^{-1},\end{aligned}$$

because of $\delta(u)>\lambda(u)$ and $C_2>1$. This completes the proof. Q.E.D.

§ Acknowledgment

The author is very grateful to the referees for reading our manuscript carefully and giving useful suggestions.

§ Note added in proof

After the acceptance of our manuscript, we realize that we need to modify the statement of Proposition 1.3. In fact, it is required to replace s in the lower bound by $s+1$, because we need to take a sufficient condition for that Φ is a contraction map in $X_{T,M}$ into account. The condition is derived from combining estimates corresponding to (3.19), (3.20), (3.25) and (3.29) in [5].

References

[1] Thierry Cazenave and Ivan Naumkin, *Local existence, global existence, and scattering for the Schrödinger equation*, Commun. Contemp. Math., **19** (2017), no. 2, 1650038, 20, MR3611666.

[2] Thierry Cazenave and Fred B. Weissler, *The Cauchy problem for the critical nonlinear Schrödinger H^s*, Nonlinear Anal., **14** (1990), no. 10, 807–836, MR1055532.

[3] Masayuki Hayashi and Tohru Ozawa, *Well-posedness for a generalized derivative nonlinear Schrödinger equation*, J. Differential Equations, **261** (2016), no. 10, 5424–5445, MR3548257.

[4] Carlos E. Kenig, Gustavo Ponce and Luis Vega, *Well-posedness and scattering results for the generalized Korteweg-de Vries equation via the contraction principle*, Comm. Pure Appl. Math., **46** (1993), 527–620, MR1211741.

[5] Felipe Linares, Hayato Miyazaki and Gustavo Ponce, *On a class of solutions to the generalized KdV type equation*, Commun. Contemp. Math., **21** (2019), no. 7, 1850056, 21. MR4017781.

[6] Felipe Linares and Gustavo Ponce, *Introduction to nonlinear dispersive equations*, Second, Universitext, Springer, New York, 2015, MR3308874.

[7] Felipe Linares, Gustavo Ponce and Gleison N. Santos, *On a Class of Solutions to the Generalized Derivative Schrödinger Equations*, Acta Math. Sin. (Engl. Ser.), **35** (2019), no. 6, 1057–1073, MR3952703.

Advanced Science Program,
Department of Integrated Science and Technology,
National Institute of Technology,
Tsuyama College,
Tsuyama, Okayama, 708-0824, Japan
E-mail address: miyazaki@tsuyama.kosen-ac.jp

Advanced Studies in Pure Mathematics 85, 2020
The Role of Metrics in the Theory of Partial Differential Equations
pp. 315–324

On some effects of background metrics for several partial differential equations

Makoto Nakamura

Abstract.

Several semilinear partial partial differential equations are considered in homogeneous and isotropic spacetimes. The Cauchy problems of the equations are considered in Lebesgue spaces and Sobolev spaces. Dissipative and anti-dissipative properties of the spatial variance on the problems are remarked.

§1. Introduction

In this paper, we introduce some results on the Cauchy problems for semilinear partial differential equations in homogeneous and isotropic spaces. These spaces describe expanding and contracting spaces. Let us consider the following line element. For any natural number n and any fixed real numbers $(\omega^0, \ldots, \omega^n) \in (-\pi/2, \pi/2]^{1+n}$, we consider a $(1+n)$-dimensional space $\mathbb{M}^{1+n}$ defined by

$$\mathbb{M}^{1+n} := \{z \in \mathbb{C}^{1+n} \mid z^\alpha = x^\alpha e^{i\omega^\alpha},\ x^\alpha \in \mathbb{R},\ 0 \le \alpha \le n\},$$

where $\mathbb{C}$ denotes the set of complex numbers. We consider a generalization of the Einstein equations for non-Hermitian complex line elements of the form $g_{\alpha\beta}(z)dz^\alpha dz^\beta$, where $\{g_{\alpha\beta}\}_{0\le\alpha,\beta\le n}$ are complex-valued functions for $z = (z^0, \ldots, z^n) \in \mathbb{M}^{1+n}$. Under the cosmological principle, we have the solution of the generalized Einstein equations as

$$(1) \qquad g_{\alpha\beta}dz^\alpha dz^\beta = -c^2(dz^0)^2 + a(z^0)^2 q^2 \left(1 + \frac{k^2 r^2}{4}\right)^{-2} \sum_{j=1}^{n} (dz^j)^2,$$

Received December 16, 2018.
Revised April 15, 2019.
2010 *Mathematics Subject Classification.* 35Q75, 35G20, 35Q30.
Key words and phrases. semilinear Klein-Gordon equation, semilinear Shrödinger equation, semilinear diffusion equation, Navier-Stokes equations.

where $c > 0$ is the speed of light, $q(\neq 0), k \in \mathbb{C}$ are constants, $r := \{\sum_{\alpha=1}^{n} (z^\alpha)^2\}^{1/2}$, and $a(\cdot)$ is a complex-valued function which denotes the scale-function of the space. For any function f on $\mathbb{M}^{1+n}$, we consider the derivative $\partial_\alpha f(z)$ for $z \in \mathbb{M}^{1+n}$ by

$$(2)\quad \partial_\alpha f(z) := \lim_{\substack{h\to 0\\ h\in\mathbb{R}\setminus\{0\}}} \frac{f(z^0,\ldots,z^{\alpha-1},z^\alpha+he^{i\omega^\alpha},z^{\alpha+1},\ldots,z^n)-f(z)}{he^{i\omega^\alpha}}.$$

Since $z = (z^0,\ldots,z^n) \in \mathbb{M}^{1+n}$ is parametrized by $x = (x^0,\ldots,x^n) \in \mathbb{R}^{1+n}$ by the relation

$$(3)\qquad z^\alpha = x^\alpha e^{i\omega^\alpha},$$

if we put $f_*(x) := f(z)$ with (3), then we have $\partial_\alpha f(z) = e^{-i\omega^\alpha}\partial f_*(x)/\partial x^\alpha$. We consider the scale-function under the curvature $k = 0$ given by

$$(4)\qquad a(z^0) := \begin{cases} a(0)\left(1+\frac{n(1+\sigma)\partial_0 a(0) z^0}{2a(0)}\right)^{2/n(1+\sigma)} & \text{if } \sigma\neq -1,\\ a(0)\exp\left(\frac{\partial_0 a(0) z^0}{a(0)}\right) & \text{if } \sigma = -1,\end{cases}$$

where $\sigma \in \mathbb{R}$ is a constant in the equation of state in the Einstein equations. Let us consider the background spacetime $\mathbb{M}^{1+n}$ with the line element (1). We put $q = 1$ and $k = 0$ in (1). We refer to [1] on the general relativity.

As the equation of motion of the massive scalar field described by a complex-valued function $\phi = \phi(z^0,\ldots,z^n)$ with the mass m and a potential $\lambda|\phi|^{p-1}\phi^2/(p+1)$ for $\lambda \in \mathbb{C}$ and $1 \le p < \infty$, we derive the second order differential equation

$$(5)\qquad -\frac{1}{c^2}\left(\partial_0^2 + \frac{n\partial_0 a}{a}\partial_0 + \frac{m^2c^4}{\hbar^2}\right)\phi + \frac{1}{a^2}\Delta_z\phi + \lambda|\phi|^{p-1}\phi = 0,$$

where $\hbar$ is the Planck constant, $\partial_0 := \partial/\partial z^0$ and $\Delta_z := \sum_{j=1}^n \partial^2/(\partial z^j)^2$. We also show that the nonrelativistic limit of (5) yields the equation

$$(6)\qquad \pm i\frac{2m}{\hbar}\partial_0 u + \frac{1}{a^2}\Delta_z u + \lambda|uw|^{p-1}u = 0$$

with a suitable transform from ϕ to $u = u(z^0,\ldots,z^n)$ (see (9), below), where $i := (-1)^{1/2}$ and w is a weight function defined by $w(z^0) := b_0(a(0)/a(z^0))^{n/2}$ for a constant $b_0 \in \mathbb{C}$. The double-sign $\pm$ appears in same order throughout the paper.

Let us derive the equations (5) and (6). For any $\lambda \in \mathbb{C}$ and any complex-valued C^2 function ϕ on $\mathbb{M}^{1+n}$, we define the Lagrangian

$$L(\phi) := -\frac{1}{2} g^{\alpha\beta} \partial_\alpha \phi \, \partial_\beta \phi - \frac{1}{2} \left(\frac{mc}{\hbar} \right)^2 \phi^2 + \frac{\lambda}{p+1} |\phi|^{p-1} \phi^2.$$

Then with the constraint condition $\arg \delta\phi = \arg \phi$, the Euler-Lagrange equation for the action $\int_{\mathbb{M}^{1+n}} L(\phi)(-g)^{1/2} dz$ is given by

$$\text{(7)} \qquad \frac{1}{(-g)^{1/2}} \partial_\alpha \left((-g)^{1/2} g^{\alpha\beta} \partial_\beta \phi \right) - \left(\frac{mc}{\hbar} \right)^2 \phi + \lambda |\phi|^{p-1} \phi = 0,$$

which is rewritten as the equation (5). For any constant $b_0 \in \mathbb{C}$, we define a weight function $w(z^0)$ and a function $b(z^0)$ by

$$\text{(8)} \qquad w(z^0) := b_0 \left(\frac{a(0)}{a(z^0)} \right)^{n/2}, \quad b(z^0) := w(z^0) \exp\left(\mp i \frac{mc^2}{\hbar} z^0 \right),$$

where we note $b(0) = b_0$. We transform ϕ to u by the equation

$$\text{(9)} \qquad \phi(z^0, \ldots, z^n) = u(z^0, \ldots, z^n) b(z^0).$$

We assume $mz^0/\hbar \in \mathbb{R}$. Then the nonrelativistic limit ($c \to \infty$) of this equation yields (6) (see [7]).

The Cauchy problems for the semilinear Klein-Gordon equation and the semilinear Schrödinger equation in the de Sitter spacetime have been considered in [4] and [8]. We consider the semilinear complex Ginzburg-Landau type equation, the semilinear diffusion equation, and the Navier-Stokes equations in this paper.

§2. Semilinear complex Ginzburg-Landau type equation

In this section, we introduce some results on the Cauchy problem of the equation (6) without proofs. The equation (6) is rewritten as

$$\text{(10)} \qquad \pm i \frac{2m}{\hbar} \partial_t u + \frac{1}{a^2 e^{2i\omega}} \Delta u - \frac{\lambda}{e^{2i\omega}} |uw|^{p-1} u = 0$$

by a transformation (3) with $t := x^0$, $\omega^0 = 0$, $\omega := \omega^1 = \cdots = \omega^n$. We consider how the spatial variance affects the existence of the solutions. Let T_0 be the maximal existence time of the scale-function $a(\cdot)$ defined by (4). Since the equation (10) has a variable coefficient, we use a change of variable $s = s(t) := \int_0^t a(\tau)^{-2} d\tau$. We put $S_0 := s(T_0)$. We use conventions $a(s) := a(t(s))$ and $w(s) := w(t(s))$ for $s \in [0, S_0)$ as far as there is no fear of confusion. A direct computation shows

$$S_0 = \begin{cases} \dfrac{2}{a_0 a_1 (4 - n(1+\sigma))} & \text{if } \ a_1 (4 - n(1+\sigma)) > 0, \\ \infty & \text{if } \ a_1 (4 - n(1+\sigma)) \leq 0. \end{cases}$$

For $0 \leq \mu_0 < n/2$ and $0 < S \leq S_0$, we consider the Cauchy problem given by

$$(11) \quad \begin{cases} \pm i \dfrac{2m}{\hbar} \partial_s u(s,x) + \dfrac{1}{e^{2i\omega}} \Delta u(s,x) - \dfrac{\lambda a(s)^2}{e^{2i\omega}} \left(|uw|^{p-1} u \right)(s,x) = 0, \\ u(0,\cdot) = u_0(\cdot) \in H^{\mu_0}(\mathbb{R}^n) \end{cases}$$

for $(s,x) \in [0,S) \times \mathbb{R}^n$, where $H^{\mu_0}(\mathbb{R}^n)$ denotes the Sobolev space of order $\mu_0 \geq 0$. Since $u = u(t,\cdot)$ is a global solution of (10) if it exists on $[0,T_0)$, we say $u = u(s,\cdot) = u(t(s),\cdot)$ is a global solution of (11) if it exists on $[0,S_0)$.

Let us consider the well-posedness of (11). For any real numbers $2 \leq q \leq \infty$ and $2 \leq r < \infty$, we say that the pair (q,r) is admissible if it satisfies $1/r + 2/nq = 1/2$. For $\mu_0 \geq 0$ and two admissible pairs $\{(q_j, r_j)\}_{j=1,2}$, we define a function space

$$X^{\mu_0}([0,S)) := \left\{ u \in C([0,S), H^{\mu_0}(\mathbb{R}^n)); \ \max_{\mu = 0, \mu_0} \|u\|_{X^{\mu}([0,S))} < \infty \right\}$$

with a metric $d(u,v) := \|u - v\|_{X^0([0,S))}$ for u, $v \in X^{\mu_0}([0,S))$, where

$$\|u\|_{X^{\mu}([0,S))} := \begin{cases} \|u\|_{L^\infty((0,S),L^2(\mathbb{R}^n)) \cap \bigcap_{j=1,2} L^{q_j}((0,S),L^{r_j}(\mathbb{R}^n))} & \text{if } \mu = 0, \\ \|u\|_{L^\infty((0,S),\dot{H}^{\mu}(\mathbb{R}^n)) \cap \bigcap_{j=1,2} L^{q_j}((0,S),\dot{B}^{\mu}_{r_j 2}(\mathbb{R}^n))} & \text{if } \mu > 0. \end{cases}$$

Here, $\dot{H}^{\mu}(\mathbb{R}^n)$ and $\dot{B}^{\mu}_{r_j 2}(\mathbb{R}^n)$ denote the homogeneous Sobolev and Besov spaces, respectively. Since the propagator of the linear part of the first equation in (11) is written as $\exp(\pm i\hbar \exp(-2i\omega) s\Delta/2m)$, we assume $0 \leq \pm\omega \leq \pi/2$ to define it as a pseudo-differential operator. We note that the scaling critical number of p for (11) is $p(\mu_0) := 1 + 4/(n - 2\mu_0)$ when $a(\cdot) = 1$. We put

$$p_1(\mu_0) := 1 + \frac{4}{n - 2\mu_0} \cdot \left(1 + \frac{4}{n - 2\mu_0} \cdot \frac{2\mu_0}{n(1+\sigma)} \right)^{-1}$$

for $\sigma \neq -1$.

We have the following results for time-local, time-global and blowing-up solutions for the problem (11) (see [8]). The proofs are based on the

Strichartz estimates and the energy estimates for Theorem 1, the energy conservation for Corollaries 2 and 3, the energy estimates with the Gagliardo-Nirenberg inequality and the virial identities for Corollaries 4 and 5.

Theorem 1. *Let* $n \geq 1$, $\lambda \in \mathbb{C}$, $0 \leq \mu_0 < n/2$, *and* $1 \leq p \leq p(\mu_0)$. *Let* ω *satisfy* $0 \leq \pm\omega \leq \pi/2$ *and* $\omega \neq -\pi/2$. *Assume* $\mu_0 < p$ *if* p *is not an odd number. There exist two admissible pairs* $\{(q_j, r_j)\}_{j=1,2}$ *with the following properties.*

(I) (*Local solutions.*) *For any* $u_0 \in H^{\mu_0}(\mathbb{R}^n)$, *there exist* $S > 0$ *with* $S \leq S_0$ *and a unique local solution* u *of* (11) *in* $X^{\mu_0}([0, S))$. *Here,* S *depends only on the norm* $\|u_0\|_{\dot{H}^{\mu_0}(\mathbb{R}^n)}$ *when* $p < p(\mu_0)$, *while* S *depends on the profile of* u_0 *when* $p = p(\mu_0)$. *The solutions depend on the initial data continuously.*

(II) (*Small global solutions.*) *Assume that one of the following conditions from* (i) *to* (vi) *holds*: (i) $\mu_0 = 0$, $p = p(0)$, (ii) $\mu_0 > 0$, $p = p(\mu_0)$, $a_1 \geq 0$, (iii) $1 < p < p(\mu_0)$, $a_1 > 0$, $\sigma < -1$, (iv) $1 < p < p_1(\mu_0)$, $a_1 < 0$, $\sigma > -1$, (v) $p_1(\mu_0) < p < p(\mu_0)$, $a_1 > 0$, $\sigma > -1$, (vi) $\mu_0 > 0$, $1 < p < p(\mu_0)$, $a_1 > 0$, $\sigma = -1$. *If* $\|u_0\|_{\dot{H}^{\mu_0}(\mathbb{R}^n)}$ *is sufficiently small, then the solution* u *obtained in* (1) *is a global solution, namely,* $S = S_0$.

Corollary 2. *Let* $\mu_0 = 0$ *or* $\mu_0 = 1$. *Let* $\lambda > 0$. *Let* $1 \leq p < 1+4/n$ *when* $\mu_0 = 0$. *Let* $1 \leq p < 1 + 4/(n-2)$ *and* $a_1(p - 1 - 4/n) \geq 0$ *when* $\mu_0 = 1$. *For any* $u_0 \in H^{\mu_0}(\mathbb{R}^n)$, *the local solution* u *given by* (I) *in Theorem* 1 *is a global solution.*

Corollary 3. *Let* $\mu_0 = 1$, $\lambda < 0$, $a_1 \geq 0$ *and* $1 \leq p < 1 + 4/n$. *Let* $\omega = 0$ *or* $\omega = \pi/2$. *For any* $u_0 \in H^1(\mathbb{R}^n)$, *the local solution* u *given by* (I) *in Theorem* 1 *is a global solution.*

Corollary 4. *Let* $\mu_0 = 1$ *and* $\lambda < 0$. *Let* $\omega \neq 0, \pi/2$. *Put* $p_0 := 2/(\sin 2\omega)^2 - 1$. *Let* $p_0 < p \leq 1 + 4/(n-2)$. *Let* $a_1(p - 1 - 4/n) \leq 0$ *and* $S_0 = \infty$. *For any* $u_0 \in H^1(\mathbb{R}^n)$ *with negative energy*

$$\int_{\mathbb{R}^n} \frac{1}{2}|\nabla u_0(x)|^2 + \frac{\lambda a_0^2 |u_0(x)|^{p+1}}{p+1}\, dx < 0, \tag{12}$$

the solution u *given by* (I) *in Theorem* 1 *blows up in finite time.*

Corollary 5. *Let* $\mu_0 = 1$ *and* $\lambda < 0$. *Let* $\omega = 0$ *or* $\omega = \pi/2$. *Let* $1 + 4/n \leq p \leq 1 + 4/(n-2)$. *Let* $a_1 \leq 0$ *and* $S_0 = \infty$. *For any* $u_0 \in H^1(\mathbb{R}^n)$ *which satisfies* $\||x|u_0(x)\|_{L^2_x(\mathbb{R}^n)} < \infty$ *and* (12), *the solution* u *given by* (I) *in Theorem* 1 *blows up in finite time.*

Remark 6. *The results in Theorem 1 also hold for the gauge variant equation*

$$\pm i\frac{2m}{\hbar}\partial_s u(s,x)+\frac{1}{e^{2i\omega}}\Delta u(s,x)-\frac{\lambda a(s)^2}{e^{2i\omega}w(s)}|uw|^p(s,x)=0, \tag{13}$$

provided $\mu_0 < p$ when p is not an even number since the proof of Theorem 1 does not require the gauge invariance. The result (2) *in Theorem 1 especially shows that we always have small global solutions for $1<p<1+4/n$ when $a(\cdot)$ is not a constant under the conditions* (iii) *and* (vi)*. This fact follows from the strong dissipative property of the equation* (11) *caused by the spatial expansion $a_1>0$. This result is much different from the case $a(\cdot)=1$ in the following sense. In the case $a(\cdot)=1$ and $\omega=0$, some weighted spaces, for example, $(1+|x|)^{-1}L^2(\mathbb{R}^n)$, have been needed to show global solutions for $1+2/n<p<1+4/n$. For the gauge variant Schrödinger equation* (13) *with $a(\cdot)=1$ and $\omega=0$, there exist blow-up solutions for small initial data for $1<p\le 1+2/n$. In the case $a(\cdot)=1$ and $\omega=\pm\pi/4$, there exist blow-up solutions for small initial data for $1<p\le 1+2/n$.*

§3. Semilinear diffusion equation

Let us consider the special case $\pm 2m/\hbar=1$, $\omega=\pi/4$, $\lambda=-1$, $\sigma=-1$ and $a_0=1$ in (11) in Lebesgue spaces. We consider the Cauchy problem (P') given by

$$(P')\begin{cases}\dfrac{\partial u}{\partial s}(s,x)-\Delta u(s,x)=A(s)f(u)(s,x) & \text{for } (s,x)\in[0,S_0)\times\mathbb{R}^n,\\ u(0,\cdot)=u_0(\cdot),\end{cases}$$

where $f(u):=\lambda|u|^{p-1}u$ or $f(u):=\lambda|u|^p$ with $1<p<\infty$ and $\lambda\in\mathbb{C}$, and we have put $H:=a_1(=a_1/a_0)$ (the Hubble constant),

$$A(s):=(1-2Hs)^{n(p-1)/4-1} \tag{14}$$

and

$$S_0:=\begin{cases}\dfrac{1}{2H} & \text{if } H>0,\\ \infty & \text{if } H<0.\end{cases} \tag{15}$$

We define a function space X by

$$X := \begin{cases} L^\infty\left(\left(0, \frac{1}{2H}\right) \times \mathbb{R}^n\right) & \text{if } H > 0, \\ \{u \in L^\infty((0,\infty), L^1(\mathbb{R}^n)) \cap L^\infty((0,\infty)\times\mathbb{R}^n); \|u\|_X < \infty\} & \\ & \text{if } H < 0, \end{cases}$$

where if $H < 0$, we define the norm $\|u\|_X$ by

$$\|u\|_X := \max\left\{\|u\|_{L^\infty((0,\infty),L^1(\mathbb{R}^n))}, \sup_{0<s<\infty}(1+s)^{n/2}\|u(s,\cdot)\|_{L^\infty(\mathbb{R}^n)}\right\}.$$

For any real number $R > 0$, we define a metric space $X(R)$ by

$$X(R) := \{u \in X; \|u\|_X \le R\}$$

with a metric $d(u,v) := \|u - v\|_X$.

We show the existence of the solutions for small data, and the asymptotic behaviors of them. We remark that the Fujita exponent $1+2/n$ in [2] does not appear in the de Sitter spacetime (namely, $\sigma = -1$). The proof is based on the $L^p - L^q$ estimate for the operator $e^{t\Delta}$.

Theorem 7 ([6]). *Let $H \in \mathbb{R}, n \ge 1, 1 < p < \infty$.*

(I) (*Small global solutions for $H > 0$*) *Let $H > 0$. If $H^{-1/(p-1)}$ $\|u_0\|_{L^\infty}$ is sufficiently small, then there exists a unique global solution u for (P') in $X(R)$ for some $R > 0$.*

(II) (*Small global solutions for $H < 0$*) *Let $H < 0$. If $M^{1/(p-1)}$ $\|u_0\|_{L^1\cap L^\infty}$ is sufficiently small, then there exists a unique global solution u for (P') in $X(R)$ for some $R > 0$. Here, we have put*

$$M := \begin{cases} (-2H)^{n(p-1)/4-1} & \text{if } p \ge 1 + \frac{4}{n},\ H < -\frac{1}{2}, \\ (-2H)^{n(p-1)/4-1} & \text{if } p < 1 + \frac{4}{n},\ -\frac{1}{2} \le H < 0, \\ 1 & \text{if } p \ge 1 + \frac{4}{n},\ -\frac{1}{2} \le H < 0, \\ 1 & \text{if } p < 1 + \frac{4}{n},\ H < -\frac{1}{2}. \end{cases}$$

(III) (*Asymptotic behavior for $H > 0$*) *Let $H > 0$, and let u be the solution obtained in (1) for the initial datum u_0. If $u_0 \in W^{1,\infty}(\mathbb{R}^n)$, then we have*

$$\|u(s,\cdot) - v(\cdot)\|_{L^\infty} = o(1) \quad \text{as } s \nearrow \frac{1}{2H},$$

where v is a function in $W^{1,\infty}(\mathbb{R}^n)$ defined by

$$v := e^{\frac{1}{2H}\Delta}u_0 + \int_0^{\frac{1}{2H}} e^{(\frac{1}{2H}-\tau)\Delta} f(u)(\tau)A(\tau)\, d\tau. \tag{16}$$

(IV) (*Asymptotic behavior for $H < 0$*) *Let $H < 0$, and let u be the solution obtained in (2). Then we have*

$$\|u(s,\cdot) - \theta G(s,\cdot)\|_{L^r} = o(s^{-n(1-1/r)/2}) \quad \text{as } s \to \infty$$

for any r with $1 \le r \le \infty$. Here, G is the Gauss kernel defined by $G(s,x) := (4\pi s)^{-n/2}e^{-|x|^2/4s}$, and we have put

$$\theta := \int_{\mathbb{R}^n} u_0(x)\, dx + \int_0^\infty A(\tau) \int_{\mathbb{R}^n} f(u)(\tau, y)\, dy d\tau.$$

§4. Navier-Stokes equations

The Navier-Stokes equations in the spacetime with the scale-function (4) are derived by the conservation of the stress-energy tensor. We consider the Cauchy problem

$$\begin{cases} \partial_j u^j + \frac{n\partial_0 a^2}{2a^2} = 0 & \text{on } [0,T) \times \mathbb{R}^n, \\ \partial_0 u^k + f(u)^k + \frac{\partial_0 a^2}{a^2} u^k + \frac{1}{\rho}\partial^k p - \frac{\mu}{\rho}\partial_j \partial^j u^k = 0 & \text{on } [0,T) \times \mathbb{R}^n, \\ u(0,\cdot) = u_0(\cdot) & \text{on } \mathbb{R}^n \end{cases} \tag{17}$$

for $1 \le j,k \le n$, where $\partial^j := a^{-2}\partial_j$, $\partial_j\partial^j := a^{-2}\Delta$, $u = (u^1, \ldots, u^n)$, $u_0 = (u_0^1, \ldots, u_0^n)$ and $f(u)^k := u^j \partial_j u^k$. We put $t := z^0$, $x = (x^1, \ldots, x^n)$ $:= (z^1, \ldots, z^n)$. We assume μ, ρ are constants. We have the following results (see [9]).

Proposition 8. *The scale-function in (17) must satisfy $a(t) = a_0\left(1 + \frac{2a_1 t}{a_0}\right)^{1/2}$.*

Consider the problem in the Minkowski spacetime given by

$$\begin{cases} \partial_j U^j = 0 & \text{on } [0,T) \times \mathbb{R}^n, \\ \partial_t U^k + f(U)^k + \dfrac{1}{\rho}\partial^k \tilde{p} - \dfrac{\mu}{\rho}\Delta U^k = 0 & \text{on } [0,T) \times \mathbb{R}^n, \\ U(0,\cdot) = U_0(\cdot). \end{cases} \tag{18}$$

Proposition 9. *Put $\phi(x) := \frac{1}{2}a_0 a_1 \sum_{j=1}^n (x^j)^2$ for $x \in \mathbb{R}^n$. If u is the solution of (17), then the function U defined by*

$$U^j(t,x) = a(t)u^j\left(t, \frac{x}{a(t)}\right) + \frac{1}{a(t)}(\partial_j \phi)\left(\frac{x}{a(t)}\right) \tag{19}$$

for $1 \le j \le n$ is the solution of (18).

For any fixed r_0 with $1 \le r_0 < \infty$, put $\alpha(r) := \frac{n}{2}\left(\frac{1}{r_0} - \frac{1}{r}\right)$ for $1 \le r < \infty$. For $T > 0$ and r_1 with $1 \le r_1 < \infty$, we set function spaces

$$\begin{aligned}
X_0 &:= \left\{U_0 \in (L^{r_0}(\mathbb{R}^n))^n ;\ \partial_j U_0^j = 0\right\}, \\
X(T) &:= \Big\{U;\ t^{\alpha(r)} U \in BC((0,T), H(L^r(\mathbb{R}^n))^n), \\
&\qquad\qquad t^{1/2+\alpha(r_1)} \nabla U \in BC\big((0,T), \big(H(L^{r_1}(\mathbb{R}^n))^n\big)^n\big)\Big\}, \\
Y_0 &:= \left\{a_0^{-1} U_0(a_0 \cdot) - a_0^{-2} \nabla\phi(\cdot);\ U_0 \in X_0\right\}, \\
Y(T) &:= \left\{a(t)^{-1} U(t, a(t)\cdot) - a(t)^{-2} \nabla\phi(\cdot);\ U \in X(T)\right\}.
\end{aligned}$$

We put $T_0 := -\frac{2a(0)}{n(1+\sigma)\partial_0 a(0)}$ and

$$T_1 := \begin{cases} \infty & a_1 \ge 0, \\ T_0 & a_1 < 0. \end{cases}$$

We have the following result, which extends the result by Kato in [3] into expanding and contracting spaces based on the methods there.

Theorem 10. *Let $n \ge 2$. Let ρ and μ be constants with $\mu/\rho > 0$. Let $a_0 > 0$, $a_1 \in \mathbb{R}$. Let us consider the Cauchy problem* (17). *Then the following results hold.*

(I) (*Local and global solutions*) *For sufficiently small number $\varepsilon > 0$, put r_0, r_1 as $1/r_0 = 1/r_1 = 1/n - \varepsilon$, and let r be a number with $0 \le 1/r \le 1/n - \varepsilon$. Then for any $u_0 \in Y_0$, there exists a unique solution $u \in Y(T)$ of* (17), *where T with $0 < T < T_1$ can be taken of the form*

$$T = \left(\frac{C}{\|U_0\|_{(L^{r_0}(\mathbb{R}^n))^n}}\right)^{2/n\varepsilon}$$

for a constant $C > 0$ which is independent of u_0, where

$$U_0(x) := a_0 u_0\left(\frac{x}{a_0}\right) + \frac{1}{a_0}\nabla\phi\left(\frac{x}{a_0}\right)$$

and $\nabla\phi := (\partial_1, \ldots, \partial_n)\phi$. Moreover, if $a_1 < 0$ and U_0 is small such that

$$\|U_0\|_{(L^{r_0}(\mathbb{R}^n))^n} \le C T_0^{-n\varepsilon/2}$$

holds, then T can be taken as $T = T_1$. Namely, u is a unique global solution in $Y(T_1)$.

(II) (*Global solutions*) *Let* $r_0 = r_1 = n$, *and let* r *be a number with* $0 < 1/r < 1/n$. *If* $\|U_0\|_{(L^{r_0}(\mathbb{R}^n))^n}$ *is sufficiently small, then there exists a unique global solution* $u \in Y(T_1)$ *of* (17).

Acknowledgment. The author is thankful to the organizers for the invitation to the 11th MSJ-SI "The Role of Metrics in the Theory of Partial Differential Equations" at Hokkaido University. This work was supported by JSPS KAKENHI Grant Number 16H03940. The author is also thankful to the anonymous two referees for several comments to revise the paper.

References

[1] S. Carroll, *Spacetime and geometry. An introduction to general relativity*, Addison Wesley, San Francisco, CA, 2004, xiv+513 pp.

[2] H. Fujita, *On the blowing up of solutions of the Cauchy problem for* $u_t = \Delta u + u^{1+\alpha}$, J. Fac. Sci., Univ. Tokyo, Sect. I, **13** (1966), 109–124.

[3] T. Kato, *Strong* L^p*-solutions of the Navier-Stokes equation in* $\mathbb{R}^m$ *with applications to weak solutions*, Math. Z., **187** (1984), 471–480.

[4] M. Nakamura, *The Cauchy problem for semi-linear Klein-Gordon equations in de Sitter spacetime*, J. Math. Anal. Appl., **410** (2014), no. 1, 445–454.

[5] M. Nakamura, *On nonlinear Schrödinger equations derived from the nonrelativistic limit of nonlinear Klein-Gordon equations in de Sitter spacetime*, Journal of Differential Equations, **259** (2015), 3366–3388.

[6] M. Nakamura, Y. Sato, *Remarks on global solutions for the semilinear diffusion equation in the de Sitter spacetime*, to appear in Hokkaido Mathematical Journal.

[7] M. Nakamura, *Remarks on the derivation of several second order partial differential equations from a generalization of the Einstein equations*, preprint.

[8] M. Nakamura, *On the nonrelativistic limit of a semilinear field equation in a homogeneous and isotropic space*, to appear in Kyoto Journal of Mathematics.

[9] M. Nakamura, *Remarks on the Navier-Stokes equations and the elastic wave equations in homogeneous and isotropic spacetimes*, preprint.

Faculty of science, Yamagata University,
Kojirakawa-machi 1-4-12, Yamagata 990-8560, Japan
E-mail address: nakamura@sci.kj.yamagata-u.ac.jp

Advanced Studies in Pure Mathematics 85, 2020
The Role of Metrics in the Theory of Partial Differential Equations
pp. 325–336

Elastic networks, statics and dynamics

Matteo Novaga and Alessandra Pluda

Abstract.

We consider planar networks minimizing the elastic energy, we state an existence and regularity result, and we discuss some geometric properties of minimal configurations. We also consider the evolution of networks by the gradient flow of the energy, and we give a well-posedness result in the case of natural boundary conditions.

§1. Introduction

We begin by defining the mathematical objects we consider in this paper.

Definition 1.1. *A planar* network $\mathcal{N}$ *is a connected set in* $\mathbb{R}^2$, *finite union of sufficiently smooth regular curves* $\mathcal{N}^i$ *that meet at junctions. In general we allow the presence of loops and* k*-valent vertices* (*see Figure* 1). *We say that the network is regular if it has only triple junctions and the unit tangent vectors of the three concurring curves form equal angles at the junctions.*

We just defined a network as a set in the Euclidean plane, but in the following it is more convenient to *parametrize* each curve $\mathcal{N}^i$ of a network $\mathcal{N}$ with maps $\gamma^i : I \subset \mathbb{R} \to \mathbb{R}^2$. A curve is of class C^k (resp. H^k, where H^k denotes the Sobolev space of order k and exponent 2) if it admits a parametrization of class C^k (resp. H^k). We remind that a curve is regular if $|\gamma'(x)| \neq 0$ for every $x \in I$.

Received November 21, 2018.
Revised June 6, 2019.
2010 *Mathematics Subject Classification.* Primary 35A15; Secondary 53C44, 49Q10, 35K52.
Key words and phrases. networks, elastic energy, minimization problems involving curvature, geometric evolution equations, Willmore flow, parabolic systems of fourth order, singular structures.

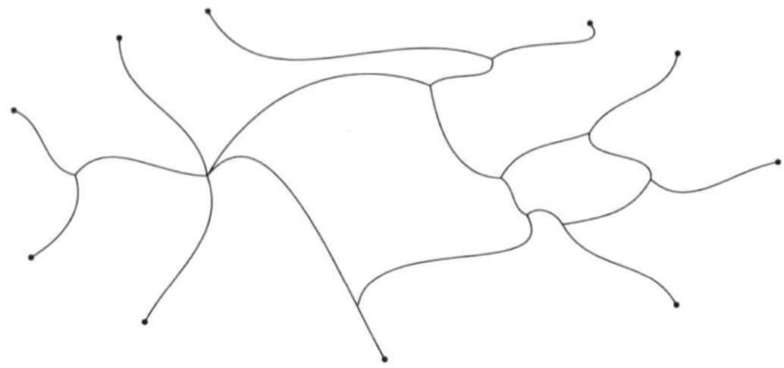

Fig. 1. A planar network.

Definition 1.2. *Let $\alpha, \beta \geq 0$ with $\alpha + \beta > 0$. The* elastic energy functional *for a network $\mathcal{N}$ is defined as*

$$E(\mathcal{N}) := \int_{\mathcal{N}} \alpha k^2 + \beta \, \mathrm{d}s = \sum_i \int_{\mathcal{N}^i} \alpha (k^i)^2 + \beta \, \mathrm{d}s^i, \tag{1}$$

where k^i is the curvature and s^i the arclength parameter of the curve $\mathcal{N}^i$.

Fixing a certain class $\mathcal{C}$ of planar networks, that is, fixing the topology of the network in the class (the number of curves and number of junctions) we are interested in the following two problems:

i) finding the minimizers of E among all networks in $\mathcal{C}$;
ii) studying the solutions of the L^2 gradient flow of the energy E with initial datum $\mathcal{N}_0$ in $\mathcal{C}$. In particular, we look for classical solution of the flow requiring that the evolving network satisfies certain boundary conditions for all times.

Looking back to the origin of elasticity theory for objects that can be modelled by one dimensional sets, we see that the first attempts regard rods and strings. That means reducing to the case of the simplest possible network: a (closed or open) curve Γ. Moreover the parameters in (1) are fixed to $\alpha = 1$ and $\beta = 0$, but the length of the curve $L(\Gamma)$ is fixed to be $\ell \in (0, \infty)$. The problem was already considered at the times of Galileo, when clearly scientists had a mechanical point of view: they looked for equations for equilibrium of moments and forces. The idea to relate the curvature of the fiber of the beam to the bending moment came only later when in 1691 Jacob Bernoulli proposed to model the bending energy of thin inextensible elastic rods with a functional involving the curvature. Finally in 1742 Daniel Bernoulli, following the principle that the potential energy of the elastic lamina must be minimal, proposed (in a letter to Euler) to attack the problem with variational techniques.

The modern formulation of Bernoulli's Problem reads as follow: find

$$\min\left\{E(\Gamma) := \int_\Gamma k^2 \,\mathrm{d}s \,\middle|\, \Gamma : [a,b] \subset \mathbb{R} \to \mathbb{R}^2,\, L(\Gamma) = \ell\right\}.$$

A natural approach to this problem is studying the critical points of the functional E, looking at the Euler–Lagrange equation, which involves four derivatives of Γ: indeed it reads as

$$2k_{ss} + k^3 - \lambda k = 0, \tag{2}$$

where k is the scalar curvature and s the arclength parameter and λ is the Lagrange multiplier associated to the length constraint. Euler was able to treat these kind of equations and divided the critical points of the functional in nine classes. He called *elastica* a solution of (2). More recently Langer and Singer contributed to the classification of the elasticae [20, 21].

We remark here that the apparently easier problem

$$\min\left\{E(\Gamma) := \int_\Gamma k^2 \,\mathrm{d}s \,\middle|\, \Gamma : [a,b] \subset \mathbb{R} \to \mathbb{R}^2,\, \Gamma \text{ closed}\right\}$$

is not well-posed. Indeed the infimum is zero and not attained. To convince yourself of this fact it is enough to consider a sequence of circles $\mathcal{C}_R$ with radius R, then $E(\mathcal{C}_R)$ goes to zero as $R \to \infty$ and the value zero is not attained. To overcome the ill-posedness, one can again fix the length of the curve, as in the original formulation of Bernoulli's problem. Other different additional constraints have been propose in the literature. Among the others, it is worth to mention:

- the length penalization (that is one asks for $\beta > 0$ in (1));
- restrict the ambient space from $\mathbb{R}^2$ to a bounded domain (confined elastica, see [14, 15]);
- defining the class $\mathcal{C}$ as the set of all embedded closed curves that enclosed a fixed area. To be more precise consider the class of all smooth, simply connected and bounded open set $\Omega \subset \mathbb{R}^2$, bounded by a Jordan curve γ, and call $A(\Omega)$ the area of Ω. One looks for

$$\min\{E(\partial\Omega) := E(\gamma) \mid A(\Omega) = A_0\}.$$

Apart from proving that the unique minimizer is the disc, it has also been established (see [7, 17]) the isoperimetric inequality

$$E^2(\partial\Omega)A(\Omega) \geq 4\pi^3.$$

Let us now consider the "opposite" case, i.e., letting $\alpha = 0$ and $\beta = 1$ in the definition of the energy functional (1). With such choice of the parameters we are looking for a network of minimal length. If we do not fix some points of the network the problem is not well-posed: indeed the length goes to zero and the network reduces to a point. So it is meaningful to consider the class of networks with n end-points of order one fixed in the plane. If we require the networks to be connected but we do not fix the topology, this gives the well-know Steiner Problem: called S a collection of n points $p_1, \dots, p_n$ in the Euclidean plane, find

$$\text{(3)} \quad \min \left\{ L(K) = \int_K 1 \, \mathrm{d}s \,\middle|\, K \text{ connected network such that } S \subset K \right\}.$$

The minimal networks are composed of straight segments meeting at triple junctions forming equal angles. The proof of existence of minimizers is considered nowadays very classical, but finding explicitly solutions is extremely challenging (even numerically) because of the lack of uniqueness due to the strong non convexity of the problem. For this reason every method to determine solutions is welcome. Motivated by the research for fast algorithms, the resolution of the Steiner problem by variational methods has recently aroused an increasing interest (see for instance [1, 2, 3, 6, 23]).

Another celebrated problem related to the minimization of the length is the isoperimetric problem. In this case we do not fix points of the network but we fix the area of the regions in which the network divides the plane, in a certain sense the topology of its complementary. More precisely an embedded compact network in $\mathbb{R}^2$ define a partition of the plane in finitely many bounded sets $\mathcal{E}_1, \dots, \mathcal{E}_n$ and an unbounded one $\mathcal{E}_0 := \mathbb{R}^2 \setminus \cup_{i=1}^n \mathcal{E}_i$. We ask the sets $\mathcal{E}_i$ to be open and we fix their area to m_i. We call $\mathcal{E}_i$ chambers and $\mathcal{E} = (\mathcal{E}_1, \dots, \mathcal{E}_n)$ cluster. Thus one want to find

$$\min \{ L(\Gamma) \,|\, \Gamma^c = \mathcal{E} = (\mathcal{E}_1, \dots, \mathcal{E}_n) \text{ with } |\mathcal{E}_i| = m_i \}.$$

For a detailed analysis of this problem see [22, 24].

§2. Minimization problems

As in the case of the length, also the minimization of the elastic energy of networks presents several possible variants. We shall present in particular three problems.

Suppose we fix a set $S \subset \mathbb{R}^2$ of n points, and $\alpha, \beta > 0$ in (1). In the same spirit of Problem (3) we look for a solution of the following problem

$$\min\{E(K) \,|\, K \text{ connected network such that } S \subset K\}.$$

The expected minimizers (for every $\alpha > 0$ and $\beta > 0$) coincide with the Steiner networks (solution of the Steiner Problem (3)), as the length is minimal and the term that involves the curvature is zero.

The second option is inspired by the isopetrimetric problem. This time we want to minimize the functional E, defined in (1), among all networks that give an n-cluster with the area of each chamber fixed. This question is widely open, the only known case is $n = 1$, proven in [7, 17].

Last but not least, we look for minimizers of the elastic energy among networks with fixed topology. For simplicity we restrict to networks of three curves that meet at two junctions, namely Theta-networks, and we fix (non zero) angles at the junctions points. More precisely we are interested in:

$$\inf\{E(\Gamma) \mid \Gamma \text{ is a Theta-network }_{(\alpha_1,\alpha_2,\alpha_3)}\}, \tag{4}$$

where a Theta-network $_{(\alpha_1,\alpha_2,\alpha_3)}$ is a network composed of three curves that meet at two triple junctions with fixed angles α_1 between γ^1 and γ^2, α_2 between γ^2 and γ^3 and, as a consequence, α_3 between γ^3 and γ^1 with $\sum_{i=1}^{3} \alpha_i = 2\pi$. Moreover each curve is regular and of class H^2.

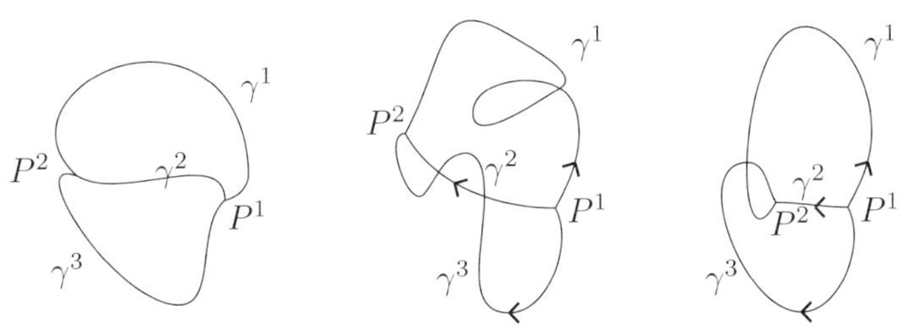

Fig. 2. Examples of Theta-networks.

Why do we fix the angles at the junctions? Suppose instead that we minimize E (with $\alpha, \beta > 0$) on the set of Theta-networks without fixing the angles. Then the infimum is zero and is clearly not attained in the class of regular networks (think for instance of a minimizing sequence given by three equal and overlapped segments whose length goes to zero).

Fixing non-zero angles at the junctions we overcome this issue, since it implies a lower bound on the energy. An embedded Theta-network can be seen as two closed curve (with a common piece). Consider two of the three curves of the Theta network that meet forming an (internal) angle of θ_1 at the junction P^1 (θ_2 at P^2 respectively), and call γ the union of these two curves. Combining the Gauss–Bonnet Theorem (in the slightly generalized version proved in [10, Appendix A]) with Hölder inequality we get

$$c := 2\pi - (\pi - \theta_1) - (\pi - \theta_2) \leq \int_\gamma |k|\, \mathrm{d}s \leq \left(\int_\gamma k^2\, \mathrm{d}s\right)^{1/2} L^{1/2}(\gamma),$$

then

$$E(\gamma) = \alpha \int_\gamma k^2\, \mathrm{d}s + \beta L(\gamma) \geq \alpha \frac{c^2}{L(\gamma)} + \beta L(\gamma) \geq 2\sqrt{\alpha\beta}\, c.$$

Hence, by fixing the angles $\theta_i \neq 0$, we get an uniform bound from below for the energy.

2.1. Existence of the minimizers

For simplicity from now on we suppose that $\alpha_1 = \alpha_2 = \alpha_3 = \frac{2\pi}{3}$ and write Theta-network for Theta-network$_{(\frac{2\pi}{3}, \frac{2\pi}{3}, \frac{2\pi}{3})}$.

First of all we notice that we can reduce to the case $\alpha = \beta = 1$. Indeed, letting

$$E_{\alpha,\beta}(\mathcal{N}) = \int_{\mathcal{N}} \alpha k^2 + \beta\, \mathrm{d}s \qquad \text{and} \qquad E_{1,1}(\mathcal{N}) = \int_{\mathcal{N}} k^2 + 1\, \mathrm{d}s,$$

and recalling the rescaling properties of the energy, we have

$$E_{\alpha,\beta}(\mathcal{N}) = \sqrt{\alpha\beta}\, E_{1,1}\left(\sqrt{\frac{\beta}{\alpha}}\mathcal{N}\right),$$

so if $\mathcal{N}_{\min}$ is a minimizer for $E_{1,1}$, then the rescaled network $\frac{\beta}{\alpha}\mathcal{N}_{\min}$ is a minimizer for $E_{\alpha,\beta}$, and vice versa. From now on for simplicity we write E for $E_{1,1}$.

We introduce the notion of "degenerate" Theta-network, that is, a network composed by two regular curves of class H^2, forming angles in pairs of $\frac{2\pi}{3}$ and $\frac{\pi}{3}$ degrees respectively, and by a curve of zero length.

Moreover, we define the functional $\overline{E}$ on networks Γ of three curves of class H^2 as follows:

$$\overline{E}(\Gamma)=\begin{cases}\displaystyle\sum_{i=1}^{3}\frac{1}{L^3(\gamma^i)}\int_0^1(\gamma^i_{xx})^2\,\mathrm{d}x+L(\gamma^i) & \text{if }\Gamma\text{ is Theta-network,}\\ \displaystyle\sum_{i=1}^{2}\frac{1}{L^3(\gamma^i)}\int_0^1(\gamma^i_{xx})^2\,\mathrm{d}x+L(\gamma^i) & \text{if }\Gamma\text{ is degenerate,}\\ +\infty & \text{otherwise.}\end{cases}$$

The two functionals E and $\overline{E}$ coincide on Theta-networks, indeed E is geometrical, invariant by reparametrization and here we simply parametrized each curve proportionally to arclength.

Theorem 1 (see [11, 10])**.** *The functional $\overline{E}$ is the relaxation of the functional E in H^2. Moreover there exists $\Gamma_{\min}$ minimizer of the functional $\overline{E}$ among 3-networks of class H^2 which is a Theta-networks, each curve $\widetilde{\gamma}$ of $\Gamma_{\min}$ is injective, of class C^∞ and solves the equation* (2)*. Hence, in particular, there exists minimizers of Problem* (4)*.*

§3. Gradient flow of the energy in the class of networks

We pass now to consider the dynamical counterpart of the problem. The questions we want to answer are the following: we fix a certain class of networks $\mathcal{C}$ and we take a network $\mathcal{N}_0$ in $\mathcal{C}$. Does there exists an evolution law that for all times $t\in[0,+\infty)$ transform the network $\mathcal{N}_0$ into a network $\mathcal{N}_t$ (still in the class C), decreasing the elastic energy along the flow? And if we want that the energy not only decrease, but that it decreases as efficiently as possible?

The evolution we have in mind can be understood as the L^2-gradient flow of the energy E where the normal velocity induces the steepest descent of the energy.

3.1. The evolution equation

How do we find an explicit expression for the velocity? Formally we derive the evolution equation by computing the first variation of E. We take a map $\gamma^i:[a,b]\to\mathbb{R}^2$ that parametrized the curve $\mathcal{N}^i$ of the network and we compute how the energy changes if we vary this parametrization by adding $\tau\psi^i$ with $\psi^i\in C^\infty([a,b],\mathbb{R}^2)$. We call the modified parametrization $\widetilde{\gamma}^i:=\gamma^i+\tau\psi^i$. If we think of ψ^i encoding the evolution in time, we can now answer the question what equations an evolution has to satisfy that decreases E most efficiently. The Cauchy

Schwarz inequality implies that the normal component of the velocity $\boldsymbol{v}^i$ of each curve $\mathcal{N}^i$ of the network is given by

$$\langle \boldsymbol{v}^i, \nu^i \rangle = \alpha(-2k^i_{ss} - (k^i)^3) + \beta k^i \qquad \text{on } [0,\infty) \times [a,b], \tag{5}$$

where ν^i is the unit normal vector of the curve γ^i. These equations are coupled with conditions at the end points of the curves $\mathcal{N}^i$. To maintain the structure (topology) of the network we have to require something on the ψ^i. For instance if two curves $\mathcal{N}^i, \mathcal{N}^j$ (parametrized by γ^i, γ^j) meet at their end points in a junction (let say $\gamma^i(a) = \gamma^j(a)$) then $\psi^i(a) = \psi^j(a)$. In the end, at each junction point we get

$$k^i = 0 \quad \text{and} \quad \sum_{i=1} \alpha(-2k^i_s \nu^i - (k^i)^2 \tau^i) + \beta \tau^i = 0 \tag{6}$$

where ν^i and τ^i are the unit normal and tangent vector, respectively. Instead at the (fixed) end point of order one the curvature has to be zero (see [4, 11, 19]).

Once we have established what the system of equations is, we want to show that there exists a solution, (at least for a shot time). The strategies to prove existence of classical solutions for all $t \in [0, +\infty)$ can be divided into two steps: local (in time) and global existence. The techniques are very different.

Before proceeding with our analysis we point out that:

- if $\alpha = 1$, $\beta = 0$ we have the Willmore flow;
- if $\alpha > 0$, $\beta > 0$ the flow is called "curve straight-shortening flow";
- if $\alpha = 0$, $\beta = 1$ we have the curve shortening flow (one dimensional Mean Curvature Flow).

Several authors considered the first two cases for the simplest network: a (closed or open) curve (see for instance [8, 12, 13, 16, 25, 27]).

3.2. Short-time existence

We shall now state the existence result for a solution of the flow.

Theorem 2 (see [19]). *Let $\mathcal{N}_0$ be a geometrically admissible initial network for the curve straight-shortening flow. Then there exists a positive time T such that within the interval $[0, T]$ the flow admits a unique solution.*

We have seen that our evolution problem is geometric, indeed both the motion equations (5) and the boundary conditions (6) are related only to geometric quantities and independent on the parametrization of the curves. But unfortunately the motion equations (5) are degenerate. Although we consider variations in any direction *only the normal movement* is specified. To turn these degenerate equations into a well-posed parabolic system of quasilinear PDEs one has to specify a suitable tangential movement. Equations (5) have the following structure:

$$\left\langle \gamma_t^i, \nu^i \right\rangle \nu^i = -A^i \nu^i = -\left\langle 2\frac{\gamma_{xxxx}^i}{|\gamma_x^i|^4} + g(\gamma_x^i, \gamma_{xx}^i, \gamma_{xxx}^i), \nu^i \right\rangle \nu^i .$$

Choosing as tangential velocity

$$T^i := \left\langle 2\frac{\gamma_{xxxx}^i}{|\gamma_x^i|^4} + g(\gamma_x^i, \gamma_{xx}^i, \gamma_{xxx}^i), \tau^i \right\rangle ,$$

we get

$$\gamma_t = -2\frac{\gamma_{xxxx}}{|\gamma_x|^4} + \tilde{f}(\gamma_{xxx}, \gamma_{xx}, \gamma_x).$$

Proceeding as for the motion equations (5) we arrive to the following system:

$$(7) \quad \begin{cases} \gamma_t^i(t,x) = -A^i(t,x)\nu^i(t,x) - T^i(t,x)\tau^i(t,x) & \\ \gamma^1(t,y) = \gamma^2(t,y) = \gamma^3(t,y) & \text{for } y \in \{0,1\} \\ \gamma_{xx}^i(t,y) = 0 & \text{for } y \in \{0,1\} \\ \sum_{i=1}^{3} \left(2k_s^i \nu^i - \mu\tau^i\right)(t,y) = 0 & \text{for } y \in \{0,1\} \\ \gamma^i(0,x) = \varphi^i(x) & \end{cases}$$

for every $t \in [0,T)\,, x \in [0,1]$ and for $i \in \{1,2,3\}$ with φ^i admissible initial parametrization. Once solved in a unique way system (7), the last step is to pass from the parametrizations back to the curve and the degenerate geometric problem. A crucial point is to check that the tangential velocity does not change the geometry of the problem, and can always be obtained by reparametrizing the curves in the right way.

3.3. Long time behavior

Recently also the long time behavior of the elastic flow of networks has been considered (see [18, 9]).

Theorem 3. *Let $\mathcal{N}_0$ be a geometrically admissible initial network. Suppose that $(\mathcal{N}(t))_{t\in[0,T_{\max})}$ is a maximal solution to the elastic flow* *with initial datum* $\mathcal{N}_0$ *in the maximal time interval* $[0, T_{\max})$ *with* $T_{\max} \in (0,\infty) \cup \{\infty\}$. *Then*

$$T_{max} = \infty$$

or at least one of the following happens:

(i) *the inferior limit of the length of at least one curve of* $\mathcal{N}(t)$ *is zero as* $t \nearrow T_{max}$.

(ii) *at one of the triple junctions*

$$\liminf_{t \nearrow T_{max}} \max \left\{ \left|\sin \alpha^1(t)\right|, \left|\sin \alpha^2(t)\right|, \left|\sin \alpha^3(t)\right| \right\} = 0,$$

where $\alpha^1(t)$, $\alpha^2(t)$ *and* $\alpha^3(t)$ *are the angles at the respective triple junction.*

We observe that existence of a global weak solution for a *second order* gradient flow for the energy $E(\mathcal{N})$, fixing the length of the curves, has been established in [26].

§4. Open problems

We list below a few open questions which we find interesting.

4.1. Statics

- Is the minimizer of Theorem 1 unique? Is it an embedded, symmetric Theta-network?
- Does Theorem 1 hold for every choice of the angles or instead there exists a minimizer of Problem (4) with a 4-point in the class of Theta-networks$_{(\alpha_1,\alpha_2,\alpha_3)}$?
- Generalize Problem (4) to networks composed of N curves; in particular find the lower semicontinuous envelope of the energy in general situations (for result in this direction we refer to [5, 28] in the case $N = 1$ in the more complicated setting of elastic clusters).

4.2. Dynamics

- Analysis of the possible singularities listed in Theorem 3.
- Definition of the flow past singularities.
- Definition of a global weak solution (for instance by variational schemes such as minimizing movements).
- Are there self-similar (for instance translating or rotating) solutions of the elastic flow?

References

[1] S. Amato, G. Bellettini and M. Paolini, Constrained BV functions on covering spaces for minimal networks and Plateau's type problems, Adv. Calc. Var., **10** (2017), 25–47.

[2] L. Ambrosio and A. Braides, Functionals defined on partitions in sets of finite perimeter. I. Integral representation and Γ-convergence, J. Math. Pures Appl., (9), **69** (1990), 285–305.

[3] L. Ambrosio and A. Braides, Functionals defined on partitions in sets of finite perimeter. II. Semicontinuity, relaxation and homogenization, J. Math. Pures Appl., (9), **69** (1990), 307–333.

[4] J. W. Barrett, H. Garcke and R. Nürnberg, Elastic flow with junctions: variational approximation and applications to nonlinear splines, Math. Models Methods Appl. Sci., **22** (2012), 1250037, 57 pp.

[5] G. Bellettini and L Mugnai, Characterization and representation of the lower semicontinuous envelope of the elastica functional, Ann. Inst. H. Poincaré Anal. Non Linéaire, **21** (2004), 839–880.

[6] M. Bonafini, G. Orlandi and E. Oudet, Variational approximation of functionals defined on 1-dimensional connected sets: the planar case, SIAM J. Math. Anal., **50** (2018), no. 6, 6307–6332.

[7] D. Bucur and A. Henrot, A new isoperimetric inequality for elasticae, J. Eur. Math. Soc., (JEMS), **19** (2017) 3355–3376.

[8] A. Dall'Acqua, C. C. Lin and P. Pozzi, Evolution of open elastic curves in $\mathbb{R}^n$ subject to fixed length and natural boundary conditions, Analysis (Berlin), **34** (2014), 209–222.

[9] A. Dall'Acqua, C. C. Lin and P. Pozzi, Flow of elastic networks: long-time existence result, Geom. Flows, **4** (2019), no. 1, 83–136.

[10] A. Dall'Acqua, M. Novaga and A. Pluda, Minimal elastic networks, arXiv:1712.09589, to appear on Indiana Univ. Math. J..

[11] A. Dall'Acqua and A. Pluda, Some minimization problems for planar networks of elastic curves, Geom. Flows, **3** (2017), 105–124.

[12] A. Dall'Acqua and P. Pozzi, A Willmore-Helfrich L^2-flow of curves with natural boundary conditions, Comm. Anal. Geom., **22**, (2014), 617–669.

[13] A. Dall'Acqua, P. Pozzi and A. Spener, The Lojasiewicz-Simon gradient inequality for open elastic curves, J. Differential Equations, **261** (2016), 2168–2209.

[14] F. Dayrens, S. Masnou and M. Novaga, Existence, regularity and structure of confined elasticae, ESAIM Control Optim. Calc. Var., **24**, (2018), 25–43.

[15] P. W. Dondl, L. Mugnai and M. Röger, Confined elastic curves, SIAM J. Appl. Math., **71** (2011), 2205–2226.

[16] G. Dziuk, E. Kuwert and R. Schätzle, Evolution of elastic curves in $\mathbb{R}^n$: existence and computation, SIAM J. Math. Anal., **33** (2002) 1228–1245.

[17] V. Ferone, B. Kawohl and C. Nitsch, The elastica problem under area constraint, Math. Ann., **365** (2016), 987–1015.

[18] H. Garcke, J. Menzel and A. Pluda, Long Time Existence of Solutions to an Elastic Flow of Networks, arXiv:1901.03246, to appear on Communications in Partial Differential Equations.

[19] H. Garcke, J. Menzel and A. Pluda, Willmore flow of planar networks, J. Differential Equations, **266** (2019), 2019–2051.

[20] J. Langer and D. A. Singer, The total squared curvature of closed curves, J. Differential Geom., **20** (1984), 1–22.

[21] J. Langer and D. A. Singer, Curve straightening and a minimax argument for closed elastic curves, Topology, **24** (1985), 75–88.

[22] F. Maggi, Sets of Finite Perimeter and Geometric Variational Problems. An introduction to geometric measure theory, Cambridge University Press, Cambridge (2012).

[23] A. Marchese and A. Massaccesi, The Steiner tree problem revisited through rectifiable G-currents, Adv. Calc. Var., **9** (2016), 19–39.

[24] F. Morgan, Geometric measure theory, A beginner's guide, Academic Press, Inc., San Diego, CA, (2000).

[25] M. Novaga and S. Okabe, Curve shortening-straightening flow for non-closed planar curves with infinite length, J. Differential Equations, **256** (2014), 1093–1132.

[26] M. Novaga and P. Pozzi, A second order gradient flow of p-elastic planar networks, SIAM J. Math. Anal., **52** (2020), no. 1, 682–708.

[27] A. Polden, Curves and Surfaces of Least Total Curvature and Fourth-Order Flows, PhD Thesis, Universität Tübingen, (1996).

[28] M. Pozzetta, A varifold perspective on the p-elastic energy of planar sets, J. Conv. Anal., **27** (2020), no. 3, 845–879.

Dipartimento di Matematica, Università di Pisa,
Largo Pontecorvo 5, 56127 Pisa, Italy
E-mail address: matteo.novaga@unipi.it

Dipartimento di Matematica, Università di Pisa,
Largo Pontecorvo 5, 56127 Pisa, Italy
E-mail address: pluda@mail.dm.unipi.it

Advanced Studies in Pure Mathematics 85, 2020
The Role of Metrics in the Theory of Partial Differential Equations
pp. 337–347

Random data Cauchy problem for the quadratic nonlinear Schrödinger equation without gauge invariance

Mamoru Okamoto

Abstract.

This article is concerned with the Cauchy problem for the quadratic nonlinear Schrödinger equation without gauge invariance. By randomizing the initial data, we prove that the Cauchy problem is almost surely locally well-posed in $H^s(\mathbb{R}^d)$ for $d \geq 5$ and $\frac{d-4}{d-3}s_c < s < s_c$, where $s_c := \frac{d}{2} - 2$ is the scaling critical regularity.

§1. Introduction

We consider the Cauchy problem for the nonlinear Schrödinger equation

$$i\partial_t u + \Delta u = |u|^2, \quad u(0,x) = \phi(x) \tag{1}$$

where $u = u(t,x) : \mathbb{R} \times \mathbb{R}^d \to \mathbb{C}$ is an unknown function, ϕ is a given initial datum. The equation (1) is invariant under the following scaling transformation:

$$u(t,x) \mapsto u^\lambda(t,x) := \lambda^2 u(\lambda^2 t, \lambda x)$$

for $\lambda > 0$. Hence, the scaling critical Sobolev regularity is $s_c := \frac{d}{2} - 2$.

Well-posedness of (1) has been studied by several researchers (see [5, 6, 9, 12, 13, 14, 15, 18, 20, 22] and references therein). In particular, (1) is locally well-posed in $H^s(\mathbb{R}^d)$ for $s \geq s_c$ and $s > -\frac{1}{4}$. Moreover, $s = -\frac{1}{4}$ is allowed in one dimension. In addition, the exponent $-\frac{1}{4}$ is sharp (up to epsilons for $d = 2, 3$). Namely, (1) is ill-posed in $H^s(\mathbb{R}^d)$ if

Received November 30, 2018.
Revised June 9, 2019.
2010 *Mathematics Subject Classification.* 35Q55.
Key words and phrases. Schrödinger equation, probabilistic well-posedness.
This work was supported by JSPS KAKENHI Grant number JP16K17624.

$d = 1, 2, 3$ and $s < -\frac{1}{4}$. The ill-posedness means that the flow map fails to be continuous.

Recently, Ikeda and Inui [9] showed nonexistence of solutions of (1) with initial data in $H^s(\mathbb{R}^d)$ satisfying a certain condition when $d \geq 4$ and $s < s_c$. Oh and Pocovnicu with the author [17] proved probabilistic well-posedness of (1) for $d = 6$, which is a corollary of the well-posedness result for the energy-critical nonlinear Schrödinger equation. To handle non-algebraic nonlinearities, they avoid thorough case-by-case analysis. In this paper, by using the case-by-case analysis, we consider the probabilistic well-posedness of (1) with low regularity data. As in [17], thanks to randomizing the initial data, we can avoid these initial data in [9] for which no solution exists.

Following the papers [1, 2, 8, 16, 23], we define the randomization. Let $\psi \in \mathcal{S}(\mathbb{R}^d)$ satisfy

$$\operatorname{supp} \psi \subset [-1, 1]^d \qquad \text{and} \qquad \sum_{n \in \mathbb{Z}^d} \psi(\xi - n) = 1 \quad \text{for any } \xi \in \mathbb{R}^d.$$

We then define the Wiener randomization of ϕ by

$$\phi^\omega := \sum_{n \in \mathbb{Z}^d} g_n(\omega) \psi(D - n) \phi, \tag{2}$$

where $\psi(D - n)\phi := \mathcal{F}^{-1}[\psi(\xi - n)\widehat{\phi}]$ and $\{g_n\}$ is a sequence of independent mean zero complex-valued random variables on a probability space $(\Omega, \mathcal{F}, P)$. In the following, we assume that the real and imaginary parts of g_n are independent and endowed with probability distributions $\mu_n^{(1)}$ and $\mu_n^{(2)}$, satisfying the following exponential moment bound:

$$\int_{\mathbb{R}} e^{\kappa x} d\mu_n^{(j)}(x) \leq e^{c\kappa^2}$$

for all $\kappa \in \mathbb{R}$, $n \in \mathbb{Z}^d$, $j = 1, 2$.

The randomization has no smoothing in terms of differentiability. However, it improves the integrability (see for example Lemma 4 in [1]).

The following is our main result.

Theorem 1.1. *Let $d \geq 5$ and let $\frac{d-4}{d-3} s_c < s < s_c$. Given $\phi \in H^s(\mathbb{R}^d)$, let ϕ^ω be its Wiener randomization defined in (2). Then, the Cauchy problem (1) with $u(0) = \phi^\omega$ is almost surely locally well-posed in $H^s(\mathbb{R}^d)$. More precisely, there exist $C, c, \gamma > 0$ such that for each $0 < T \ll 1$, there exists $\Omega_T \subset \Omega$ with $P(\Omega_T^c) \leq C \exp\Big(-\frac{c}{T^\gamma \|\phi\|_{H^s}^2}\Big)$ such that* *for each* $\omega \in \Omega_T$*, there exists a solution* $u = u^\omega \in C([-T,T]; H^s(\mathbb{R}^d))$ *to* (1) *with* $u(0) = \phi^\omega$ *in the class*

$$e^{it\Delta}\phi^\omega + C([-T,T]; H^{s_c+}(\mathbb{R}^d)) \subset C([-T,T]; H^s(\mathbb{R}^d)).$$

Set $S(t) = e^{it\Delta}$ and let $z(t) = z^\omega(t) := S(t)\phi^\omega$ denote the random linear solution with ϕ^ω as initial data. If u is a solution to (1), then the residual term $v := u - z$ satisfies the following perturbed nonlinear Schrödinger equation:

$$i\partial_t v + \Delta v = |v + z^\omega|^2, \quad v(0,x) = 0. \tag{3}$$

We use the contraction mapping theorem to find a solution to (3) (or the corresponding integral equation).

Remark 1. Once a suitable z is given, i.e., $z \in A^s$ (see (6) below), (3) is the Cauchy problem for a (deterministic) quadratic nonlinear Schrödinger equation with some additional terms. Moreover, Theorem 1.1 means that $z \in A^s \mapsto v \in C([-T,T]; H^{s_c+}(\mathbb{R}^d))$ is continuous. See Remark 1.4 in [4] for a further discussion.

In [19], the author proved the almost sure well-posedness of (1) in $H^s(\mathbb{R}^d)$ for $\frac{d-4}{d-3}s_c < s < s_c$ by performing a fixed point argument around the (unbalanced) higher order expansion as in Bényi et al. [3]. The aim in this paper is to give an alternative proof by using the unit-localized maximal function estimate in [7]. We note that the proof is applicable to other nonlinear terms u^2 and $\overline{u}^2$ (see Remark 3 below).

1.1. Notation

We summarize the notation used throughout this paper. We set $\mathbb{N}_0 := \mathbb{N} \cup \{0\}$. We denote the space of all rapidly decaying functions on $\mathbb{R}^d$ by $\mathcal{S}(\mathbb{R}^d)$.

Let $\{e_l\}$ be an orthonormal basis of $\mathbb{R}^d$ and henceforth fix our coordinate system accordingly. We define the lateral spaces $L^{p,q}_{e_l}$ for $l = 1$ with the norm

$$\|f\|_{L^{p,q}_{e_l}} := \left(\int_{\mathbb{R}_{x_1}} \left(\int_{\mathbb{R}_t} \int_{\mathbb{R}^{d-1}_{x'}} |f(t,x_1,x')|^q \, dx' dt\right)^{\frac{p}{q}} dx_1\right)^{\frac{1}{p}}$$

with analogous definitions for $l = 2, 3, \ldots, d$ and the usual modifications when $p = \infty$ or $q = \infty$.

Let θ be a smooth even function with $0 \le \theta \le 1$ and $\theta(a) = \begin{cases} 1, & \text{if } |a| \le 1, \\ 0, & \text{if } |a| \ge 2. \end{cases}$ For any $R > 0$, we set

$$\theta_{\le R}(\xi) := \theta\left(\frac{|\xi|}{R}\right), \quad \theta_R(\xi) := \theta_{\le R}(\xi) - \theta_{\le \frac{R}{2}}(\xi).$$

We define the inhomogeneous Littlewood-Paley decomposition:

$$P_1 f := \mathcal{F}^{-1}[\theta_{\le 1}\widehat{f}], \quad P_N f := \mathcal{F}^{-1}[\theta_N \widehat{f}]$$

for any $N \in 2^{\mathbb{N}}$. For any $N \in 2^{\mathbb{N}}$ and for any $l = 1, \dots, d$, we define

$$P_{N,e_l} f := \mathcal{F}^{-1}[\theta_N(\xi \cdot e_l)\widehat{f}].$$

In estimates, we use C to denote a positive constant that can change from line to line. If C is absolute or depends only on parameters that are considered fixed, then we often write $X \lesssim Y$, which means $X \le CY$. We define $X \sim Y$ to mean $C^{-1}Y \le X \le CY$. We use $a+$ and $a-$ to denote quantities $a + \varepsilon$ and $a - \varepsilon$, respectively, when $\varepsilon > 0$ is arbitrarily small and implicit constants are allowed to depend on ε.

§2. Linear estimates

In this section, we collect some linear estimates. We call (q, r) admissible if $q, r \in [2, \infty]$, $\frac{2}{q} + \frac{d}{r} = \frac{d}{2}$, and $(q, r, d) \ne (2, \infty, 2)$. Then, the following Strichartz estimate holds.

Theorem 2.1. *Let (q, r) be admissible. Then, we have*

$$\|S(t)\phi\|_{L_t^q L_x^r} \lesssim \|\phi\|_{L_x^2}.$$

For the nonlinear estimates, we use the Fourier restriction norm space.

Definition 2.2. Let $s, b \in \mathbb{R}$. The space $X^{s,b}$ is defined to be the closure of the Schwartz functions $\mathcal{S}(\mathbb{R} \times \mathbb{R}^d)$ under the norm

$$\|u\|_{X^{s,b}} := \|\langle \xi \rangle^s \langle \tau + |\xi|^2 \rangle^b \mathcal{F}_{t,x}[u](\tau, \xi)\|_{L^2_{\tau,\xi}}.$$

Here, we collect some properties of $X^{s,b}$ (see [21] for the proof). When $b > \frac{1}{2}$, $X^{s,b} \hookrightarrow C(\mathbb{R}; H^s(\mathbb{R}))$ holds. Moreover, for $b \in (\frac{1}{2}, 1)$, $\rho \in (0, \frac{3}{2} - b)$, and $T \in (0, 1)$, we have

$$\left\| \theta_{\le T}(t) \int_0^t S(t - t') f(t') dt' \right\|_{X^{s,b}} \lesssim T^\rho \|f\|_{X^{s,b-1+\rho}}. \tag{4}$$

In addition, the transfer principle yields the following : For any admissible pair (q, r) and $b > \frac{1}{2}$, we have $\|u\|_{L^q_t L^r_x} \lesssim \|u\|_{X^{0,b}}$.

Remark 2. Since $\|u\|_{L^2_{t,x}} = \|u\|_{X^{0,0}}$, for any admissible pair (q, r) with $q > 2$, an interpolation shows that $\|u\|_{L^{q-}_t L^{r-}_x} \lesssim \|u\|_{X^{0,\frac{1}{2}-}}$.

The randomization has the same Sobolev regularity as the initial data (see, for example, [1]).

Lemma 2.3. *Given $\phi \in H^s(\mathbb{R}^d)$, let ϕ^ω be its Wiener randomization defined in* (2). *Then, there exist $C, c > 0$ such that for all $\lambda > 0$,*

$$P\Big(\|\phi^\omega\|_{H^s} > \lambda\Big) \leq C \exp\bigg(-c\frac{\lambda^2}{\|\phi\|_{H^s}^2}\bigg).$$

The following lemma shows an improvement of the Strichartz estimates upon the randomization of initial data (See [1]).

Lemma 2.4. *Given $\phi \in L^2(\mathbb{R}^d)$, let ϕ^ω be its Wiener randomization defined in* (2). *Then, given finite $q, r \geq 2$, there exist $C, c > 0$ such that*

$$P\Big(\|S(t)\phi^\omega\|_{L^q_t L^r_x([-T,T]\times\mathbb{R}^d)} > \lambda\Big) \leq C \exp\bigg(-c\frac{\lambda^2}{T^{\frac{2}{q}}\|\phi\|_{L^2}^2}\bigg)$$

for all $T > 0$ and $\lambda > 0$.

Ionescu and Kenig [10, 11] proved the local smoothing estimate in $L^{\infty,2}_{e_l}$ and the maximal function estimate in $L^{2,\infty}_{e_l}$. Because Sobolev's embedding $W^{\frac{d-2}{4},\frac{2d}{d-1}}(\mathbb{R}^d) \hookrightarrow L^4(\mathbb{R}^d)$ yields that

$$\|P_N S(t) f\|_{L^4_{t,x}} \lesssim N^{\frac{d-2}{4}}\|f\|_{L^2},$$

an interpolation argument shows the following bound.

Lemma 2.5. *Let $d \geq 3$, $2 \leq p \leq q \leq \infty$ with $\frac{1}{p} + \frac{1}{q} = \frac{1}{2}$, and $l \in \{1, 2, \ldots, d\}$. Then, we have*

$$\|S(t)P_N f\|_{L^{p,q}_{e_l}} \lesssim N^{\frac{d}{p}-\frac{1}{2}}\|f\|_{L^2}, \qquad \|S(t)P_{N,e_l} f\|_{L^{q,p}_{e_l}} \lesssim N^{\frac{d}{q}-\frac{1}{2}}\|f\|_{L^2}.$$

Accordingly, the transfer principle yields that

$$\|P_N u\|_{L^{p,q}_{e_l}} \lesssim N^{\frac{d}{p}-\frac{1}{2}}\|u\|_{X^{0,b}}, \qquad \|P_{N,e_l} u\|_{L^{q,p}_{e_l}} \lesssim N^{\frac{d}{q}-\frac{1}{2}}\|u\|_{X^{0,b}}$$

for $2 \leq p \leq q \leq \infty$ with $\frac{1}{p} + \frac{1}{q} = \frac{1}{2}$, $l \in \{1, 2, \ldots, d\}$, and $b > \frac{1}{2}$. Because

$$\|P_{N,e_l} u\|_{L^{\infty,2}_{e_l}} \lesssim N^{\frac{1}{2}}\|u\|_{L^2_{t,x}} = N^{\frac{1}{2}}\|u\|_{X^{0,0}},$$

an interpolation argument as in Remark 2 leads to

$$\|P_{N,e_l}u\|_{L_{e_l}^{\frac{4}{\varepsilon},\frac{4}{2-\varepsilon}}} \lesssim N^{-\frac{1}{2}+}\|u\|_{X^{0,\frac{1}{2}-}}. \tag{5}$$

From the proof of (4.9) in [11] (see also Lemma 5.4 in [7]), we have the following unit-scale maximal function estimate:

Lemma 2.6. *Let $d \geq 3$. For all $n \in \mathbb{Z}^d$ with $|n| \geq 10$ and for each $l = 1, 2, \dots, d$, we have*

$$\|S(t)\psi(D-n)\phi\|_{L_{e_l}^{2,\infty}} \lesssim |n|^{\frac{1}{2}}\|\psi(D-n)\phi\|_{L^2}.$$

We define the norm as follows:

$$\begin{aligned}\|z\|_{A^s} &:= \|z\|_{X^{s,\frac{1}{2}+}} + \|\langle\nabla\rangle^s z\|_{L_{t,x}^{\frac{d+2}{2}+}} \\ &\quad + \left(\sum_{N\subset 2^{\mathbb{N}_0}} N^{2s}\left(\sum_{l=1}^{d} N^{-\frac{1}{2}+\frac{\varepsilon}{2}}\|P_N z\|_{L_{e_l}^{\frac{4}{2-\varepsilon},\frac{4}{\varepsilon}}}\right)^2\right)^{\frac{1}{2}}.\end{aligned} \tag{6}$$

For sufficiently small $\gamma > 0$, we set

$$\Omega_T := \left\{\omega \in \Omega \colon \|\theta_{\leq T}S(t)\phi^\omega\|_{A^s} \leq T^{-\frac{\gamma}{2}}\right\}.$$

Then, Lemmas 2.3, 2.4, and 2.6 (see also the poof of Proposition 5.3 in [7]) show that there exist $C, c > 0$ such that $P(\Omega_T^c) \leq C\exp\left(-\frac{c}{T^\gamma\|\phi\|_{H^s}^2}\right)$ for $T \in (0,1)$.

§3. Proof of Theorem 1.1

We employ the following bilinear estimate proved by Tao [20]:

Theorem 3.1. *Let N_1, N_2, and N_3 be dyadic numbers. Then, we have*

$$\left\|P_{N_3}(P_{N_1}f\overline{P_{N_2}g})\right\|_{X^{0,-\frac{1}{2}+}} \lesssim N_{\max}^{-\frac{1}{2}+}N_{\min}^{\frac{d-3}{2}}\|P_{N_1}f\|_{X^{0,\frac{1}{2}+}}\|P_{N_2}g\|_{X^{0,\frac{1}{2}+}},$$

where $N_{\max} := \max(N_1, N_2, N_3)$ and $N_{\min} := \min(N_1, N_2, N_3)$.

We are now in position to prove Theorem 1.1.

Proof of Theorem 1.1. By (4) and taking $\gamma > 0$ sufficiently small, it suffices to show that

$$\left\|P_{N_3}(P_{N_1}F\overline{P_{N_2}G})\right\|_{X^{s_c+,-\frac{1}{2}+}} \lesssim N_{\max}^{0-}\left(\|v\|_{X^{s_c+,\frac{1}{2}+}} + \|z\|_{A^s}\right)^2 \tag{7}$$

for $N_1, N_2, N_3 \in 2^{\mathbb{N}_0}$, $\max(N_1, N_2) \gtrsim N_3$, and $F, G \in \{z, v\}$.

The estimate for $v\overline{v}$ is essentially the same as that for the deterministic setting. For reader's convenience, we give the proof. When $N_1 \sim N_2 \gtrsim N_3$, Theorem 3.1 yields that

$$\begin{aligned}
&\left\| P_{N_3}(P_{N_1} v \overline{P_{N_2} v}) \right\|_{X^{s_c+,-\frac{1}{2}+}} \\
&\lesssim N_1^{-\frac{1}{2}+} N_3^{2s_c+\frac{1}{2}+} \|P_{N_1} v\|_{X^{0,\frac{1}{2}+}} \|P_{N_2} v\|_{X^{0,\frac{1}{2}+}} \\
&\lesssim N_1^{0-} \|P_{N_1} v\|_{X^{s_c+,\frac{1}{2}+}} \|P_{N_2} v\|_{X^{s_c+,\frac{1}{2}+}}.
\end{aligned}$$

When $N_3 = N_{\max}$, Theorem 3.1 yields that

$$\begin{aligned}
&\left\| P_{N_3}(P_{N_1} v \overline{P_{N_2} v}) \right\|_{X^{s_c+,-\frac{1}{2}+}} \\
&\lesssim \min(N_1, N_2)^{s_c+\frac{1}{2}} N_3^{s_c-\frac{1}{2}+} \|P_{N_1} v\|_{X^{0,\frac{1}{2}+}} \|P_{N_2} v\|_{X^{0,\frac{1}{2}+}} \\
&\lesssim N_3^{0-} \|v\|_{X^{s_c+,\frac{1}{2}+}} \|v\|_{X^{s_c+,\frac{1}{2}+}}.
\end{aligned}$$

Hence, we have (7) for $v\overline{v}$.

We consider the estimate for $z\overline{z}$. Here, we observe the decomposition

$$P_N F = P_{N,e_1} P_N F + \sum_{k=2}^{d} P_{N,e_k} \left(\prod_{j=1}^{k-1} (1 - P_{N,e_j}) \right) P_N F$$

and that the operators $\prod_{j=1}^{k-1}(1 - P_{N,e_j})$ are disposable. When $N_1 \sim N_3 \gtrsim N_2$, by Hölder's inequality, we have

$$\begin{aligned}
\left\| P_{N_3}(P_{N_1} z \overline{P_{N_2} z}) \right\|_{X^{s_c+,-\frac{1}{2}+}} &\le N_3^{s_c+} \left\| P_{N_3}(P_{N_1} z \overline{P_{N_2} z}) \right\|_{L^2} \\
&\lesssim \sum_{l=1}^{d} N_3^{s_c+} \|P_{N_1,e_l} z\|_{L_{e_l}^{\frac{4}{\varepsilon},\frac{4}{2-\varepsilon}}} \|P_{N_2} z\|_{L_{e_l}^{\frac{4}{2-\varepsilon},\frac{4}{\varepsilon}}} \\
&\lesssim N_3^{s_c+} N_1^{-s-\frac{1}{2}+\frac{d}{4}\varepsilon} N_2^{-s+\frac{1}{2}-\frac{\varepsilon}{2}} \|z\|_{A^s}^2 \\
&\lesssim N_1^{-s+s_c-\frac{1}{2}+} N_2^{-s+\frac{1}{2}} \|z\|_{A^s}^2 \\
&\lesssim N_1^{0-} \|z\|_{A^s}^2,
\end{aligned}$$

provided that $s > \max(s_c - \frac{1}{2}, \frac{s_c}{2})$ and $\varepsilon > 0$ is sufficiently small.

When $N_1 \sim N_2 \gtrsim N_3$, we use the duality argument and (5) to have

$$\begin{aligned}
&\left\|P_{N_3}(P_{N_1}z\overline{P_{N_2}z})\right\|_{X^{s_c+,-\frac{1}{2}+}} \\
&\leq N_3^{s_c+} \sup_{\|w\|_{X^{0,\frac{1}{2}-}}=1} \left| \int_{\mathbb{R}\times\mathbb{R}^d} P_{N_1}z\overline{P_{N_2}zP_{N_3}w}\,dtdx \right| \\
&\lesssim N_3^{s_c+} \sup_{\|w\|_{X^{0,\frac{1}{2}-}}=1} \sum_{l=1}^{d} \|P_{N_1}z\|_{L^2_{t,x}} \|P_{N_2}z\|_{L_{e_l}^{\frac{4}{2-\varepsilon},\frac{4}{\varepsilon}}} \|P_{N_3,e_l}w\|_{L_{e_l}^{\frac{4}{\varepsilon},\frac{4}{2-\varepsilon}}} \\
&\lesssim N_1^{-2s+\frac{1}{2}+} N_3^{s_c-\frac{1}{2}+} \|z\|_{A^s}^2 \\
&\lesssim N_1^{0-} \|z\|_{A^s}^2,
\end{aligned}$$

provided that $s > \max(\frac{1}{4}, \frac{s_c}{2})$ and $\varepsilon > 0$ is sufficiently small.

Next, we focus on the estimate for $z\overline{v}$. We note that $v\overline{z}$ is similarly handled. First, we consider the case $N_1 \sim N_2 \gtrsim N_3$. The duality, Hölder's inequality, and Remark 2 imply that

$$\begin{aligned}
&\left\|P_{N_3}(P_{N_1}z\overline{P_{N_2}v})\right\|_{X^{s_c+,-\frac{1}{2}+}} \\
&\lesssim N_3^{s_c+} \sup_{\|w\|_{X^{0,\frac{1}{2}-}}=1} \left| \int_{\mathbb{R}\times\mathbb{R}^d} P_{N_1}z\overline{P_{N_2}vP_{N_3}w}\,dtdx \right| \\
&\lesssim N_3^{s_c+} \sup_{\|w\|_{X^{0,\frac{1}{2}-}}=1} \|P_{N_1}z\|_{L_{t,x}^{\frac{d+2}{2}+}} \|P_{N_2}v\|_{L_{t,x}^{\frac{2(d+2)}{d}}} \|P_{N_3}w\|_{L_{t,x}^{\frac{2(d+2)}{d}-}} \\
&\lesssim N_1^{-s-s_c+} N_3^{s_c+} \|z\|_{A^s} \|v\|_{X^{s_c+,\frac{1}{2}+}}.
\end{aligned}$$

Hence, (7) follows from $s > 0$.

Second, we consider the case $N_1 \sim N_3 \geq N_2$. Then, we divide the proof into two cases: $N_2 \leq N_1^{\frac{1}{d-3}}$ or $N_2 \geq N_1^{\frac{1}{d-3}}$.

When $N_2 \leq N_1^{\frac{1}{d-3}}$, Theorem 3.1 yields that

$$\begin{aligned}
\left\|P_{N_3}(P_{N_1}z\overline{P_{N_2}v})\right\|_{X^{s_c+,-\frac{1}{2}+}} &\lesssim N_1^{s_c-\frac{1}{2}+} N_2^{s_c+\frac{1}{2}} \|P_{N_1}z\|_{X^{0,\frac{1}{2}+}} \|P_{N_2}v\|_{X^{0,\frac{1}{2}+}} \\
&\lesssim N_1^{-s+s_c-\frac{1}{2}+} N_2^{\frac{1}{2}+} \|z\|_{X^{s,\frac{1}{2}+}} \|v\|_{X^{s_c+,\frac{1}{2}+}} \\
&\lesssim N_1^{-s+\frac{d-4}{d-3}s_c+} \|z\|_{A^s} \|v\|_{X^{s_c+,\frac{1}{2}+}} \\
&\lesssim N_1^{0-} \|z\|_{A^s} \|v\|_{X^{s_c+,\frac{1}{2}+}}
\end{aligned}$$

provided that $s > \frac{d-4}{d-3}s_c$.

When $N_2 \geq N_1^{\frac{1}{d-3}}$, the duality, Hölder's inequality, and Remark 2 imply that

$$\begin{aligned}
&\left\| P_{N_3}(P_{N_1} z \overline{P_{N_2} v}) \right\|_{X^{s_c+,-\frac{1}{2}+}} \\
&\lesssim N_3^{s_c+} \sup_{\|w\|_{X^{0,\frac{1}{2}-}}=1} \left| \int_{\mathbb{R}\times\mathbb{R}^d} P_{N_1} z \overline{P_{N_2} v P_{N_3} w}\, dtdx \right| \\
&\lesssim N_3^{s_c+} \sup_{\|w\|_{X^{0,\frac{1}{2}-}}=1} \|P_{N_1} z\|_{L_{t,x}^{\frac{d+2}{2}+}} \|P_{N_2} v\|_{L_{t,x}^{\frac{2(d+2)}{d}}} \|P_{N_3} w\|_{L_{t,x}^{\frac{2(d+2)}{d}-}} \\
&\lesssim N_1^{-s+s_c+} N_2^{-s_c-} \|z\|_{A^s} \|v\|_{X^{s_c+,\frac{1}{2}+}} \\
&\lesssim N_1^{-s+\frac{d-4}{d-3}s_c+} \|z\|_{A^s} \|v\|_{X^{s_c+,\frac{1}{2}+}}.
\end{aligned}$$

Hence, (7) follows from $s > \frac{d-4}{d-3}s_c$.

Third, we consider the case $N_1 \leq N_2 \sim N_3$. The same calculation as in $N_1 \sim N_3 \geq N_2$ yields that (7) is valid for $s > 0$. Q.E.D.

Remark 3. The same result as in Theorem 1.1 holds for u^2. Moreover, we can improve Theorem 1.1 for $\overline{u}^2$, because $N_{\max}^{-\frac{1}{2}+} N_{\min}^{\frac{d-3}{2}}$ in Theorem 3.1 is replaced by $N_{\max}^{-\frac{3}{2}+} N_{\min}^{\frac{d-1}{2}}$. More precisely, by dividing the proof into two cases $N_2 \leq N_1^{\frac{3}{d-1}}$ or $N_2 \geq N_1^{\frac{3}{d-1}}$, the bilinear estimate for $\overline{zv}$ with $N_1 \sim N_3 \geq N_2$ holds provided that $s > \frac{d-4}{d-1}s_c$. We therefore obtain the almost sure local well-posedness for $\overline{u}^2$ when $\max(s_c - \frac{1}{2}, \frac{s_c}{2}, \frac{d-4}{d-1}s_c) < s < s_c$.

Acknowledgements. The author would like to thank the anonymous referee for careful reading and useful comments.

References

[1] Á. Bényi, T. Oh and O. Pocovnicu, Wiener randomization on unbounded domains and an application to almost sure well-posedness of NLS, *Excursions in harmonic analysis.* **4**, 3–25, Appl. Numer. Harmon. Anal., Birkhäuser/Springer, Cham, 2015.

[2] Á. Bényi, T. Oh and O. Pocovnicu, On the probabilistic Cauchy theory of the cubic nonlinear Schrödinger equation on $\mathbb{R}^d$, $d \geq 3$, *Trans. Amer. Math. Soc.* Ser. B, **2** (2015), 1–50.

[3] Á. Bényi, T. Oh and O. Pocovnicu, Higher order expansions for the probabilistic local Cauchy theory of the cubic nonlinear Schrödinger equation on $\mathbb{R}^3$, *Trans. Amer. Math. Soc.* Ser. B, **6** (2019), 114–160.

[4] Á. Bényi, T. Oh and O. Pocovnicu, On the probabilistic Cauchy theory for nonlinear dispersive PDEs, *Landscapes of time-frequency analysis*, 1–32, Appl. Numer. Harmon. Anal., Birkhäuser/Springer, Cham, 2019.

[5] T. Cazenave and F. B. Weissler, The Cauchy problem for the critical nonlinear Schrödinger equation in H^s, *Nonlinear Anal.*, **14** (1990), no. 10, 807–836.

[6] J. E. Colliander, J.-M. Delort, C. E. Kenig and G. Staffilani, Bilinear estimates and applications to 2D NLS, *Trans. Amer. Math. Soc.*, **353** (2001), no. 8, 3307–3325.

[7] B. Dodson, J. Luhrmann and D. Mendelson, Almost sure local well-posedness and scattering for the 4D cubic nonlinear Schrödinger equation, *Adv. Math.*, **347** (2019), 619–676.

[8] H. Hirayama and M. Okamoto, Random data Cauchy problem for the nonlinear Schrödinger equation with derivative nonlinearity, *Discrete Contin. Dyn. Syst.*, **36** (2016), no. 12, 6943–6974.

[9] M. Ikeda and T. Inui, Some non-existence results for the semilinear Schrödinger equation without gauge invariance, *J. Math. Anal. Appl.*, **425** (2015), no. 2, 758–773.

[10] A. Ionescu and C. E. Kenig, *Low-regularity Schrödinger maps*, Differential Integral Equations, **19** (2006), no. 11, 1271–1300.

[11] A. Ionescu and C. E. Kenig, *Low-regularity Schrödinger maps. II. Global well-posedness in dimensions* $d \geq 3$, Comm. Math. Phys., **271** (2007), no. 2, 523–559.

[12] T. Iwabuchi and K. Uriya, Ill-posedness for the quadratic nonlinear Schrödinger equation with nonlinearity $|u|^2$, *Commun. Pure Appl. Anal.*, **14** (2015), no. 4, 1395–1405.

[13] C. E. Kenig, G. Ponce and L. Vega, Quadratic forms for the 1-D semilinear Schrödinger equation, *Trans. Amer. Math. Soc.*, **348** (1996), no. 8, 3323–3353.

[14] N. Kishimoto, A remark on norm inflation for nonlinear Schrödinger equations, *Commun. Pure Appl. Anal.*, **18** (2019), no. 3, 1375–1402.

[15] N. Kishimoto and K. Tsugawa, Local well-posedness for quadratic nonlinear Schrödinger equations and the "good" Boussinesq equation, *Differential Integral Equations*, **23** (2010), no. 5–6, 463–493.

[16] J. Lührmann and D. Mendelson, Random data Cauchy theory for nonlinear wave equations of power-type on $\mathbb{R}^3$, *Comm. Partial Differential Equations*, **39** (2014), no. 12, 2262–2283.

[17] T. Oh, M. Okamoto and O. Pocovnicu, On the probabilistic well-posedness of the nonlinear Schrödinger equation equations with non-algebraic nonlinearities, *Discrete Contin. Dyn. Syst.* A, **39** (2019), no. 6, 3479–3520.

[18] M. Okamoto, Norm inflation for the generalized Boussinesq and Kawahara equations, *Nonlinear Anal.*, **157** (2017), 44–61.

[19] M. Okamoto, Remarks on the probabilistic well-posedness for quadratic nonlinear Schrödinger equations, RIMS Kôkyûroku Bessatsu, **B74** (2019), 47–64.

[20] T. Tao, Multilinear weighted convolution of L^2-functions, and applications to nonlinear dispersive equations, *Amer. J. Math.*, **123** (2001), no. 5, 839–908.
[21] T. Tao, Nonlinear dispersive equations. Local and global analysis., *CBMS Regional Conference Series in Mathematics*, **106** (2006).
[22] Y. Tsutsumi, L^2-solutions for nonlinear Schrödinger equations and nonlinear groups, *Funkcial. Ekvac.*, **30** (1987), no. 1, 115–125.
[23] T. Zhang and D. Fang, Random data Cauchy theory for the generalized incompressible Navier-Stokes equations, *J. Math. Fluid Mech.*, **14** (2012), no. 2, 311–324.

Division of Mathematics and Physics,
Faculty of Engineering,
Shinshu University,
4-17-1 Wakasato, Nagano City 380-8553, Japan
E-mail address: m_okamoto@shinshu-u.ac.jp

Advanced Studies in Pure Mathematics 85, 2020
The Role of Metrics in the Theory of Partial Differential Equations
pp. 349–358

Numerical analysis of constrained total variation flows

Koya Sakakibara

Abstract.

In this paper, we consider total variation flow with values constrained to a Riemannian manifold (constrained total variation flow), which appears in several fields of mathematical sciences. In particular, we consider the constrained total variation flow on the space of piecewise constant functions and construct a numerical scheme based on the minimizing movement scheme. We present a mathematical result on the convergence of our numerical scheme and a result of the numerical experiment. This paper is based on the joint work with Yoshikazu Giga (The University of Tokyo), Kazutoshi Taguchi (ARISE analytics Inc.), and Masaaki Uesaka (Arithmer Inc./The University of Tokyo), and a full version of this paper has been published in [7].

§1. Introduction

We consider the constrained total variation flow for a function $u\colon \Omega\times [0,T)\to M$:

$$(\mathrm{TVF};u_0)\qquad \begin{cases} \dfrac{\partial u}{\partial t} = -\pi_u\left(-\nabla\cdot\left(\dfrac{\nabla u}{|\nabla u|}\right)\right) & \text{in } \Omega\times(0,T),\\ \left(\dfrac{\nabla u}{|\nabla u|}\right)\cdot\nu^{\Omega}=0 & \text{on } \partial\Omega\times(0,T),\\ u|_{t=0}=u_0 & \text{in } \Omega, \end{cases}$$

where $\Omega\subset\mathbb{R}^k$ is a bounded Lipschitz domain, $\partial\Omega$ denotes the boundary of Ω, ν^{Ω} denotes the unit outward normal vector of $\partial\Omega$, M is a manifold

Received December 31, 2018.
Revised June 17, 2019.
2010 *Mathematics Subject Classification.* Primary 65M12; Secondary 35K45, 35K55.
Key words and phrases. total variation flow, manifold constraint, minimizing movement scheme.

embedded into $\mathbb{R}^l$, $\pi_p\colon T_p\mathbb{R}^l(=\mathbb{R}^l) \to T_pM$ is the orthogonal projection at $p \in M$, and $T > 0$.

Constrained total variation flows have appeared in several fields of applications. When $M = S^2$, this flow is used for color image denoising with preserving brightness [15]. When $M = SO(3)$, this flow appears as an essential prototype of a continuum model for the time evolution of grain boundaries in a crystal [8, 9]. When $M - SPD(3)$, the space of all symmetric positive definite three-dimensional matrices, this flow is proposed for denoising DT-MRI [3, 4, 13, 17]. In these applications, the spatially discretized total variation flows, the total variation flows on the space of piecewise constant functions, play essential roles. Therefore, in this paper, we consider the spatially discretized total variation flow and construct a numerical scheme to solve it.

The spatially discretized constrained total variation flow can be regarded as a gradient flow; therefore, one can use the minimizing movement scheme to compute it, which can be summarized as follows (see [2] for detail).

Let $H = (H, \langle\cdot,\cdot\rangle_H)$ be a real Hilbert space, E be a submanifold of H, and $\mathcal{F}$ be an extended real-valued functional on H. Moreover, let $I := [0, T)$ be a time interval and $\{t^{(n)}\}_{n=0}^N$ be a uniform time partition of I, i.e., $t^{(n)}$ is defined as $t^{(n)} = n\tau$, where $\tau = \lfloor T/N \rfloor$. Here, $\lfloor x \rfloor$ $(x \in \mathbb{R})$ denotes the floor function which returns the highest integer less than or equal to x as its value. Then, a sequence $\{u_\tau^{(n)}\}_{n=0}^N$ in E is defined by the following procedure:

(1) $u_\tau^{(0)} := u_0$.

(2) For $n \geq 1$, $u_\tau^{(n)}$ is defined as the minimizer of the following optimization problem:

$$\text{Minimize } \Phi(u, u_\tau^{(n-1)}) \text{ subject to } u \in E,$$

where

$$\Phi(u, v) := \tau\mathcal{F}(u) + \frac{1}{2}\|u - v\|_H^2, \quad u, v \in E.$$

From the viewpoint of numerical computation, it is not easy to solve the optimization problem in the above scheme because it is categorized into Riemannian optimization problems. Theory of "smooth" Riemannian optimization problem is well-studied, and we refer to [1] for its systematic explanation. Conversely, the theory of "non-smooth" Riemannian optimization problem is still scarcely explored as a subfield of the theory of Riemannian optimization problems. Here, we point out that the optimization problem derived by the minimizing movement scheme of constrained total variation flow is a non-smooth Riemannian optimization problem. We refer to [10, 11, 16, 18] as its reference. Moreover, the theory of non-smooth Riemannian optimization problem is generalized to theory of non-smooth and non-convex optimization with a separable structure in [12]. Although there are several studies in this direction, it is under development compared with the linearly constrained problem. Therefore, in this paper, we propose another numerical scheme, which includes a *linearly* constrained optimization problem in each time step instead of the Riemannian constraint problem.

§2. Preliminaries

2.1. Function spaces

Definition 2.1. A family $\Omega_\Delta = \{\Omega_\alpha\}_{\alpha\in\Delta}$ of subsets of Ω is called a *rectangular partition* of Ω if the following three conditions are satisfied:

(1) $\mathcal{H}^k\left(\Omega\setminus\bigcup_{\alpha\in\Delta}\Omega_\alpha\right)=0$;

(2) $\mathcal{H}^k(\Omega_\alpha\cap\Omega_\beta)=0$ for all $(\alpha,\beta)\in\Delta\times\Delta$ with $\alpha\neq\beta$;

(3) For each $\alpha\in\Delta$, there exists a rectangle R_α in $\mathbb{R}^k$ such that $\Omega_\alpha=\Omega\cap R_\alpha$ holds.

Here, $\mathcal{H}^k$ denotes the k-dimensional Hausdorff measure.

In the following, we fix a rectangular partition $\Omega_\Delta = \{\Omega_\alpha\}_{\alpha\in\Delta}$ of Ω, and construct a numerical scheme for spatially discretized problem based on Ω_Δ.

Define the set of edges $e(\Delta)$ as follows:

$$e(\Delta)=\left\{\gamma=\{\alpha,\beta\}\subset\Delta \mid \mathcal{H}^{k-1}(\partial\Omega_\alpha\cap\partial\Omega_\beta)\neq 0,\ \alpha\neq\beta\right\}.$$

For $\gamma=\{\alpha,\beta\}\in e(\Delta)$, we take a bijection $\mathrm{Sign}_\gamma\colon\gamma\to\{\pm1\}$, and define $E_\gamma:=\partial\Omega_\alpha\cap\partial\Omega_\beta$. Moreover, we set

$$E\Omega_\Delta=\{E_\gamma\}_{\gamma\in e(\Delta)},\quad \mathrm{Sign}_{e(\Delta)}=\{\mathrm{Sign}_\gamma\}_{\gamma\in e(\Delta)}.$$

Define a set of piecewise $\mathbb{R}^l$-valued constant functions as follows:

$$H_\Delta=\left\{U\in L^2(\Omega;\mathbb{R}^l)\mid U|_{\Omega_\alpha} \text{ is constant for each } \Omega_\alpha\in\Omega_\Delta\right\}.$$

H_Δ is a closed subspace of the Hilbert space $L^2(\Omega;\mathbb{R}^l)$, endowed with an inner product

$$\langle X,Y\rangle_{H_\Delta}=\langle X,Y\rangle_{L^2(\Omega;\mathbb{R}^l)},\quad X,Y\in H_\Delta.$$

Moreover, we define a space $H_{E\Omega_\Delta}$ of piecewise $\mathbb{R}^l$-valued constant maps on $\bigcup E\Omega_\Delta$ associated with Ω_Δ by

$$H_{E\Omega_\Delta} = \left\{ U \in L^2\left(\bigcup E\Omega_\Delta; \mathbb{R}^l\right) \,\middle|\, U|_{E_\gamma} \text{ is constant for each } E_\gamma \in E\Omega_\Delta \right\}.$$

Define a set of piecewise M-valued constant functions as follows:

$$M_\Delta = \{U \in L^2(\Omega; M) \mid U|_{\Omega_\alpha} \text{ is constant for each } \Omega_\alpha \in \Omega_\Delta\}.$$

M_Δ is a non-convex set, and regarded as a submanifold of H_Δ. For $u \in M_\Delta$, denote the tangent space of M_Δ at u by $T_u M_\Delta$:

$$T_u M_\Delta = \{X \in L^2(\Omega; \mathbb{R}^l) \mid X|_{\Omega_\alpha} \text{ is constant in } T_{u|_{\Omega_\alpha}} M \text{ for each } \Omega_\alpha \in \Omega_\Delta\}.$$

We regard M_Δ as a submanifold of H_Δ. Then, the orthogonal projection in M_Δ is naturally induced from the one in M as follows. For $u \in M_\Delta$, we define the orthogonal projection $P_u \colon H_\Delta \to T_u M_\Delta$ by

$$(P_u X)(x) := \pi_{u(x)}(X(x)) \quad \text{for a.e. } x \in \Omega.$$

2.2. Constrained discrete total variation flows

The discrete total variation functional $\boldsymbol{TV}_\Delta \colon H_\Delta \to \mathbb{R}$ is defined as

$$\boldsymbol{TV}_\Delta(u) = \sum_{\gamma \in e(\Delta)} \|(\boldsymbol{D}_\Delta u)^\gamma\|_{\mathbb{R}^l} \mathcal{H}^{k-1}(E_\gamma),$$

where $\boldsymbol{D}_\Delta^{\mathrm{Sign}_{e(\Delta)}} = \boldsymbol{D}_\Delta \colon H_\Delta \to H_{E\Omega_\Delta}$ is a discrete gradient defined as

$$\begin{aligned} \boldsymbol{D}_\Delta u &= \sum_{\{\alpha,\beta\} \in e(\Delta)} (\boldsymbol{D}_\Delta u)^{\{\alpha,\beta\}} \mathbf{1}_{E_{\{\alpha,\beta\}}} \\ &= \sum_{\{\alpha,\beta\} \in e(\Delta)} \left(\mathrm{Sign}_{\{\alpha,\beta\}}(\alpha) u^\alpha + \mathrm{Sign}_{\{\alpha,\beta\}}(\beta) u^\beta \right) \mathbf{1}_{E_{\{\alpha,\beta\}}} \end{aligned}$$

for $u = \sum_{j \in \Delta} u^j \mathbf{1}_{\Omega_j} \in H_\Delta$.

Definition 2.2. Let $u_0 \in M_\Delta$, and $I := [0, T)$. A map $u \in W^{1,2}(I; M_\Delta)$ is said to be a solution to the *discrete model* of (TVF;u_0) if u satisfies

$$(\mathrm{DTVF};u_0) \qquad \begin{cases} \dfrac{\mathrm{d}u}{\mathrm{d}t} \in -P_{u(t)} \partial \boldsymbol{TV}_\Delta(u(t)) & \text{for a.e. } t \in (0, T), \\ u|_{t=0} = u_0, \end{cases}$$

where $\partial \boldsymbol{TV}_\Delta$ denotes the subgradient of $\boldsymbol{TV}_\Delta$ in H_Δ.

The unique existence of global-in-time solution to (DTVF;u_0) have been proved in [5, 6, 14].

§3. Numerical scheme

Let $\tau > 0$ be a time step size, and $N(I,\tau) := \lfloor N/\tau \rfloor$. Define a set of time $I(\tau) = \{t^{(n)} \mid n = 0, 1, \dots, N(I,\tau)\}$ as follows:

$$t^{(n)} = \begin{cases} n\tau & \text{if } n = 0, 1, \dots, N(I,\tau) - 1, \\ T & \text{if } n = N(I,\tau). \end{cases}$$

We propose the following numerical scheme for (DTVF;u_0) based on the minimizing movement scheme [2]:

Algorithm. Let $u_0 \in M_\Delta$. Define $\{u_\tau^{(n)}\}_{n=0}^{N(I,\tau)} \subset M_\Delta$ by the following procedure.

(1) $u_\tau^{(0)} = u_0$.

(2) For $n \geq 1$, define $u_\tau^{(n)}$ as follows:

$$X_\tau^{(n-1)} := \operatorname*{argmin}_{X \in T_{u_\tau^{(n-1)}} M_\Delta} \Phi_{\mathrm{loc}}^\tau(X; u_\tau^{(n-1)}),$$

$$u_\tau^{(n)} := \mathrm{Exp}_{u_\tau^{(n-1)}} X_\tau^{(n-1)},$$

where

$$\Phi_{\mathrm{loc}}^\tau(X; u_\tau^{(n-1)}) := \tau \boldsymbol{TV}_\Delta(u_\tau^{(n-1)} + X) + \frac{1}{2}\|X\|_{H_\Delta}^2, \quad X \in T_{u_\tau^{(n-1)}} M_\Delta.$$

$\mathrm{Exp}_u \colon T_u M_\Delta \to M_\Delta$ is the exponential map at $u \in M_\Delta$ defined as

$$X(x) \longmapsto \exp_{u(x)}(X(x)) \quad \text{for a.e. } x \in X.$$

Concerning the above-proposed algorithm, we have the following propositions and the theorem.

Proposition 3.1. *$\{u_\tau^{(n)}\}_{n=0}^{N(I,\tau)}$ and $\{X_\tau^{(n)}\}_{n=0}^{N(I,\tau)-1}$ in the above algorithm are well-defined. Moreover, they satisfy*

$$X_\tau^{(n)} \in -\tau P_{u_\tau^{(n)}} \partial \boldsymbol{TV}_\Delta(u_\tau^{(n)} + X_\tau^{(n)}), \quad \|X_\tau^{(n)}\|_{H_\Delta} \leq \tau\, \mathrm{Lip}(\boldsymbol{TV}_\Delta)$$

for $n = 0, 1, \dots, N(I,\tau) - 1$.

Proposition 3.2 (Energy decay). *Let M be a compact C^2 manifold in $\mathbb{R}^l$, $u_0 \in M_\Delta$, and $\{u_\tau^{(n)}\}_{n=0}^{N(I,\tau)}$ be a sequence generated by our algorithm. If the time step size $\tau > 0$ satisfies $\tau\, \mathrm{Curv}(M)\, \mathrm{Lip}(\boldsymbol{TV}_\Delta) \leq 1$, then $\boldsymbol{TV}_\Delta(u_\tau^{(n+1)}) \leq \boldsymbol{TV}_\Delta(u_\tau^{(n)})$ holds for all $n = 0, 1, \dots, N(I,\tau) - 1$. Namely, our algorithm satisfies the energy-decaying property.*

Theorem 3.3 (Error estimate). *Let M be a path-connected and compact C^2 submanifold in $\mathbb{R}^l$. Fix two initial data $u_0^1, u_0^2 \in M_\Delta$, and let $u \in W^{1,2}(I; M_\Delta)$ be a solution of* (DTVF; u_0^1), *and $\{u_\tau^{(n)}\}_{n=0}^{N(I,\tau)}$ be a sequence generated by our algorithm with initial value u_0^2. Then, there exist constants C_0, C_1, C_2 independent of $t^{(n)}$ and τ such that the following error estimate holds:*

$$\|u(t^{(n)}) - u_\tau^{(n)}\|_{H_\Delta}^2 \leq \mathrm{e}^{C_0 t^{(n)}} \left(\|u_0^1 - u_0^2\|_{H_\Delta}^2 + t^{(n)}(C_1\tau + C_2\tau^2)\right)$$

for all $t^{(n)} \in I(\tau)$.

Concerning the proofs of the above propositions and theorem, see [7].

§4. Numerical experiment

In this section, we show a numerical result for the case where $M = SO(3)$. We adopt the Euler angle (or Euler axis) representations to describe each element in $SO(3)$. Given $R = (R_{i,j})_{i,j-1}^3 \in SO(3)$, its Euler angle representation is given by Rodrigues' rotation formula:

$$R = \cos\theta I + (1 - \cos\theta) e \otimes e + \sin\theta [e]_x,$$

where $\theta \in [0, 2\pi)$ and $e := (e_1, e_2, e_3) \in S^2$ denote the Euler axis of R and $[e]_x$ is the cross product matrix of e. They are explicitly given as follows:

$$\begin{aligned}
\theta &= \arccos\left(\frac{R_{1,1} + R_{2,2} + R_{3,3} - 1}{2}\right), \\
(e_1, e_2, e_3) &= \left(\frac{R_{3,2} - R_{2,3}}{2\sin\theta}, \frac{R_{1,3} - R_{3,1}}{2\sin\theta}, \frac{R_{2,1} - R_{1,2}}{2\sin\theta}\right), \\
[e]_x &= \begin{pmatrix} 0 & -e_3 & e_2 \\ e_3 & 0 & -e_1 \\ -e_2 & e_1 & 0 \end{pmatrix}.
\end{aligned}$$

Since the Euler axis e is in S^2, e is represented as the spherical Euler angles $(\phi, \psi) \in [0, \pi) \times [0, 2\pi)$. Therefore, R is represented using three parameters $(\theta, \phi, \psi) \in [0, 2\pi) \times [0, \pi) \times [0, 2\pi)$.

We use the following initial data u with the Euler angles $\theta, \phi, \psi \colon [0, 1] \times [0, 1] \to \mathbb{R}$:

$$\theta = \sum_{i,j=0}^{2} \boldsymbol{\theta}_{i,j} \mathbf{1}_{I_i \times J_j}, \quad \phi = \sum_{i,j=0}^{2} \boldsymbol{\phi}_{i,j} \mathbf{1}_{I_i \times J_j}, \quad \psi = \sum_{i,j=0}^{2} \boldsymbol{\psi}_{i,j} \mathbf{1}_{I_i \times J_j},$$

where

$$\boldsymbol{\theta} = \begin{pmatrix} 0.35\pi & 0.2\pi & 0.55\pi \\ 0.81\pi & 0.64\pi & 0.4\pi \\ 0.1\pi & 0.7\pi & 0.3\pi \end{pmatrix}, \quad \boldsymbol{\phi} = \begin{pmatrix} 0.4\pi & 0.5\pi & 0.7\pi \\ 0.5\pi & 0.3\pi & 0.4\pi \\ 0.6\pi & 0.3\pi & 0.4\pi \end{pmatrix},$$

$$\boldsymbol{\psi} = \begin{pmatrix} 0.2\pi & 0.25\pi & 0.3\pi \\ 0.25\pi & 0.225\pi & 0.2\pi \\ 0.3\pi & 0.2\pi & 0.35\pi \end{pmatrix},$$

$$I_0 = \left(0, \frac{2}{5}\right), \quad I_1 = \left(\frac{2}{5}, \frac{3}{5}\right), \quad I_2 = \left(\frac{3}{5}, 1\right),$$

$$J_0 = \left(0, \frac{1}{5}\right), \quad J_1 = \left(\frac{1}{5}, \frac{4}{5}\right), \quad J_2 = \left(\frac{4}{5}, 1\right).$$

Figure 1 depicts the result of the numerical experiment under the above settings. Although there does not exist any benchmark test for $SO(3)$-valued total variation flow, we can see that our proposed numerical scheme works well because the facet-preserving phenomenon is observed and the numerical solution finally reaches the constant state.

Acknowledgements

We would like to thank the anonymous referee for his/her critical comments and suggestions which helped us to improve the quality of this paper.

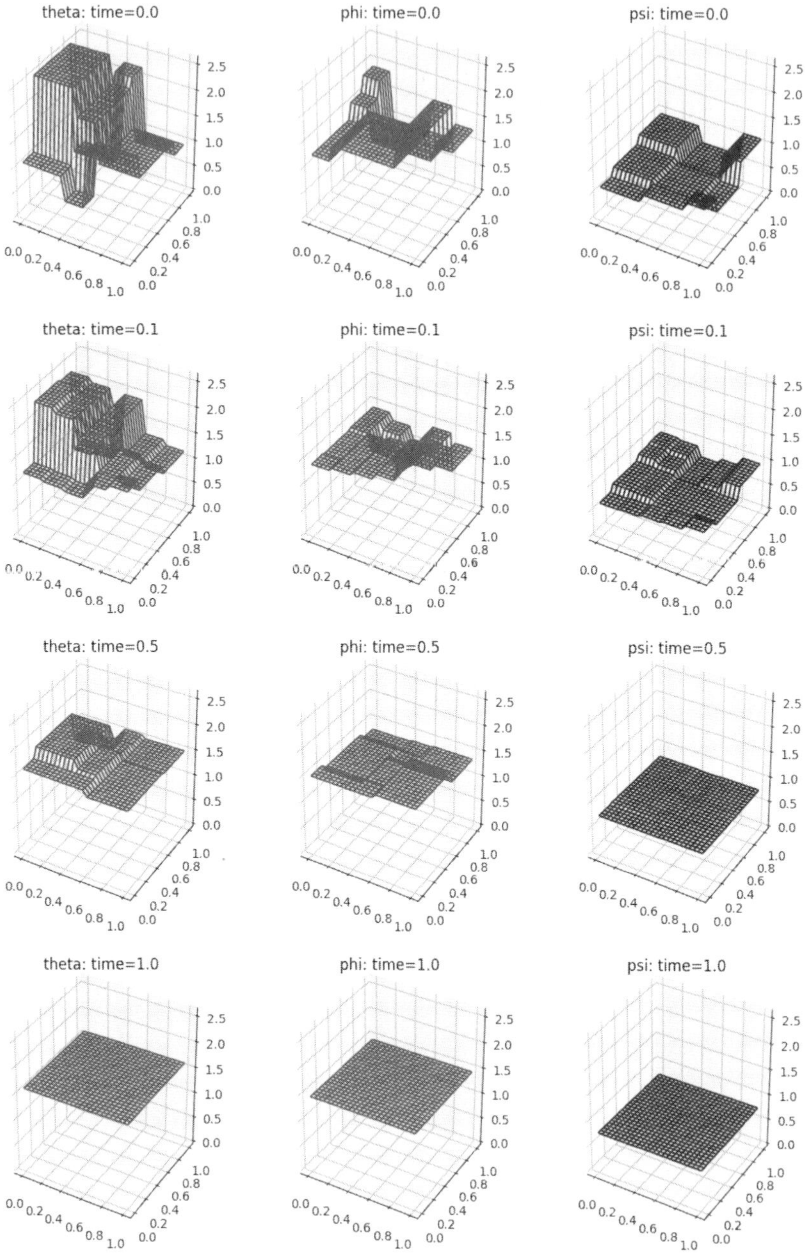

Fig. 1. Numerical results of $SO(3)$-valued total variation flow at $t = 0.0, 0.1, 0.5, 1.0$.

References

[1] P-A Absil, Robert Mahony and Rodolphe Sepulchre, *Optimization algorithms on matrix manifolds*, Princeton University Press, 2009.

[2] Luigi Ambrosio, Nicola Gigli and Giuseppe Savaré, *Gradient flows: in metric spaces and in the space of probability measures*, Springer Science & Business Media, 2008.

[3] Peter J Basser, James Mattiello and Denis LeBihan, MR diffusion tensor spectroscopy and imaging, *Biophysical journal*, 66(1):259–267, 1994.

[4] Oddvar Christiansen, Tin-Man Lee, Johan Lie, Usha Sinha and Tony F Chan, Total variation regularization of matrix-valued images, *International Journal of Biomedical Imaging*, 2007, 2007.

[5] Yoshikazu Giga and Ryo Kobayashi, On constrained equations with singular diffusivity, *Methods and Applications of Analysis*, 10(2):253–278, 2003.

[6] Yoshikazu Giga, Hirotoshi Kuroda and Noriaki Yamazaki, An existence result for a discretized constrained gradient system of total variation flow in color image processing, *Interdisciplinary Information Sciences*, 11(2):199–204, 2005.

[7] Yoshikazu Giga, Koya Sakakibara, Kazutoshi Taguchi and Masaaki Uesaka, A new numerical scheme for discrete constrained total variation flows and its convergence, /*Numerische Mathematik*/ (2020, published online), 37 pp.

[8] Ryo Kobayashi and James A Warren, Modeling the formation and dynamics of polycrystals in 3D, *Physica A: Statistical Mechanics and its Applications*, 356(1):127–132, 2005.

[9] Ryo Kobayashi, James A Warren and W Craig Carter, A continuum model of grain boundaries, *Physica D: Nonlinear Phenomena*, 140(1–2):141–150, 2000.

[10] Artiom Kovnatsky, Klaus Glashoff and Michael M Bronstein, MADMM: a generic algorithm for non-smooth optimization on manifolds, In *European Conference on Computer Vision*, pages 680–696, Springer, 2016.

[11] Rongjie Lai and Stanley Osher, A splitting method for orthogonality constrained problems, *Journal of Scientific Computing*, 58(2):431–449, 2014.

[12] Guoyin Li and Ting Kei Pong, Global convergence of splitting methods for nonconvex composite optimization, *SIAM Journal on Optimization*, 25(4):2434–2460, 2015.

[13] Xavier Pennec, Pierre Fillard and Nicholas Ayache, A Riemannian framework for tensor computing, *International Journal of Computer Vision*, 66(1):41–66, 2006.

[14] Kazutoshi Taguchi, On discrete one-harmonic map flows with values into an embedded manifold on a multi-dimensional domain, *Advances in Mathematical Sciences and Applications*, 27(1):81–113, 2018.

[15] Bei Tang, Guillermo Sapiro and Vicent Caselles, Color image enhancement via chromaticity diffusion, *IEEE Transactions on Image Processing*, 10(5):701–707, 2001.

[16] Yu Wang, Wotao Yin and Jinshan Zeng, Global convergence of ADMM in nonconvex nonsmooth optimization, *Journal of Scientific Computing*, 78(1):29–63, 2019.
[17] Andreas Weinmann, Laurent Demaret and Martin Storath, Total variation regularization for manifold-valued data, *SIAM Journal on Imaging Sciences*, 7(4):2226–2257, 2014.
[18] Junyu Zhang, Shiqian Ma and Shuzhong Zhang, Primal-dual optimization algorithms over Riemannian manifolds: an iteration complexity analysis, */Mathematical Programming/* (2019), 46 pp.

Okayama University of Science; RIKEN iTHEMS
E-mail address: ksakaki@xmath.ous.ac.jp

Advanced Studies in Pure Mathematics 85, 2020
The Role of Metrics in the Theory of Partial Differential Equations
pp. 359–367

The blow-up curve of solutions for semilinear wave equations with Dirichlet boundary conditions in one space dimension

Tetsuya Ishiwata and Takiko Sasaki

Abstract.

We consider a blow-up curve for the one dimensional wave equation. Merle–Zaag [5] showed that there is a possibility that the blow-up curve for $\partial_t^2 u - \partial_x^2 u = |u|^{p-1}u$ is not differentiable if the sign of the solution changes. To show the result, they use the variational structure of the equation. In this paper, we study the blow-up curve $\partial_t^2 u - \partial_x^2 u = |\partial_t u|^{p-1}\partial_t u$ which does not have the variational structure. We state that the blow-up curve is not differentiable if the initial data is an odd function which satisfy some conditions. Next, we show the key of the proof. This paper is an announcement of our forthcoming paper.

§1. Introduction

In this paper, we consider

$$(1)\qquad \begin{cases} \partial_t^2 u - \partial_x^2 u = 2^p|\partial_t u|^{p-1}\partial_t u, & x \in \mathbb{R}, \quad t \geq 0, \\ u(x,0) = u_0(x), \quad \partial_t u(x,0) = u_1(x), & x \in \mathbb{R}, \end{cases}$$

where $u_0, u_1 \in C^4(\mathbb{R})$. In what follows, we assume that $p > 1$ is a positive constant such that the function $|v|^{p-1}v$ $(v \geq 0)$ is of class C^4. That is, $p = 2, 3$ or $p \geq 4$.

Received March 8, 2019.
Revised June 28, 2019.
2010 *Mathematics Subject Classification.* 35B44, 35L71, 35B09.
Key words and phrases. blow-up, wave equation, positive solutions.
Tetsuya Ishiwata is supported by JSPS KAKENHI Grant Numbers 15H03632 and 15K13461.
Takiko Sasaki is supported by JSPS KAKENHI Grant Numbers 16H07288 and 18K13447.

It is well known that the solution of (1) blows-up in finite time if the initial data is large enough. In this paper, we consider the following function

$$T(x) = \sup\{t \geq 0 \mid |\partial_t u(x,t)| < \infty\} \quad \text{for} \quad x \in \mathbb{R}.$$

We call the function T and its graph blow-up curve. Since the region

$$C = \{(x,t) \in \mathbb{R} \times \mathbb{R}^+ \mid u(x,t) \text{ is a classical solution}\}$$

depends only on $\partial_t u$ for this problem, thus we defined the blow-up curve as the above. If C depends on other quantities, we should change the definition of the blow-up curve. Take the case of $\partial_t^2 u - \partial_x^2 u = |u|^{p-1}u$. In this case, only u controls the region C. Then, we define the blow-up curve as

$$\tilde{T}(x) = \sup\{t \geq 0 \mid |u(x,t)| < \infty\} \quad \text{for} \quad x \in \mathbb{R}.$$

Before stating our results, we recall some previous studies on the blow-up curve. The blow-up curve has been studied from the view points of whether or not it can be differentiable. We consider the following nonlinear wave equation:

$$\partial_t^2 u - \partial_x^2 u = F(u, \partial_t u, \partial_x u).$$

The first work was done by Caffarelli and Friedman [1]. They consideblack the case $F = |u|^{p-1}u$. They defined $\tilde{T}$ as the blow-up curve since u controls the region C. They showed $\tilde{T} \in C^1(\mathbb{R})$ if the initial conditions are large and smooth enough. Godin [2, 3] showed $\tilde{T} \in C^1(\mathbb{R})$ in the case $F = e^u$.

Moreover, Merle and Zaag [5] extended the results of [1] to more general initial data and furthermore, they also showed that there is a possibility that the blow-up curve is not differentiable when we consider a sign-changing solution. Hamza and Zaag [4] extended the results of [5] to the solutions of the damped wave equations. The variational structure of the wave equation was one of the keys of their proof in [5]. Hence, it is not clear from [5] whether the non-differentiability of the blow-up curve is due to the variational structure.

On the other hand, in the case that the nonlinearity involves the derivative of the solution, there are few results on the blow-up curve. Ohta and Takamura [6] consideblack the case $F = (\partial_t u)^2 - (\partial_x u)^2$. In this case, we can obtain the exact solution of the wave equation. By using the exact solution, they showed that there are many types of the blow-up curve. Sasaki [7] consideblack the case $F = |\partial_t u|^{p-1}\partial_t u$ and showed the blow-up curve becomes smooth if the initial conditions are large and smooth enough.

In this paper, we consider when the blow-up curves are not differentiable in the case (1) which does not have the variational structure.

§2. Main results

Let us first give as an restriction on (1). On the analogy of [5], we expect that the blow-up curve is not differentiable if $\partial_t u$ is an odd function. Hence, we only focus on the case that the solution of (1) is an odd function. Then, (1) can be rewritten as

$$(2) \qquad \begin{cases} \partial_t^2 u - \partial_x^2 u = 2^p |\partial_t u|^{p-1} \partial_t u, & x > 0, \quad t > 0, \\ \partial_t u(0,t) = 0, & t > 0, \\ u(x,0) = u_0(x), \quad \partial_t u(x,0) = u_1(x), & x > 0. \end{cases}$$

Moreover, we rewrite (2) to a system of wave equations whose nonlinear term does not involve the derivative of the solution. We introduce ϕ and ψ which are defined as

$$\phi = \partial_t u + \partial_x u, \quad \psi = \partial_t u - \partial_x u.$$

Then, (2) is rewritten as

$$(3) \qquad \begin{cases} D_- \phi = |\phi + \psi|^{p-1}(\phi + \psi), & x > 0, \quad t > 0, \\ D_+ \psi = |\phi + \psi|^{p-1}(\phi + \psi), & x > 0, \quad t > 0, \\ (\phi + \psi)(0,t) = 0, & t > 0, \\ \phi(x,0) = f(x), \quad \psi(x,0) = g(x), & x > 0, \end{cases}$$

where $D_\mp = \partial_t \mp \partial_x$, $f(x) = u_1(x) + \partial_x u_0(x)$ and $g(x) = u_1(x) - \partial_x u_0(x)$. By using $\phi + \psi$, the blow-up curve T is written as

$$T(x) = \sup\{t \geq 0 \mid |(\phi + \psi)(x,t)| < \infty\} \quad \text{for} \quad x > 0.$$

Due to the Dirichlet boundary condition, $\phi + \psi$ does not blow-up at $x = 0$. In what follows, we only consider the blow-up curve for $x > 0$.

Remark 1. Under some assumptions, the blow-up curve T becomes continuously differentiable for $x > 0$. Moreover, we can show $|T'(x)| < 1$ for $x > 0$. In this paper, we study the limit of $T'(x)$ as x goes to 0.

To state the main result, we introduce some symbols. Let Q^*, R^*, S^* and T^* be positive constants and set

$$
\begin{aligned}
&B_{R_1,R_2} = \{x \in \mathbb{R} \mid R_1 < x < R_2\},\\
&K_-(x_0,t_0) = \bigl\{(x,t) \in (\mathbb{R}^+)^2 \bigm| |x_0 - x| < t_0 - t\bigr\} \quad (x_0, t_0 \geq 0),\\
&K_1 = \bigcup_{x \in B_{Q^*+T^*,Q^*+T^*+R^*}} K_-(x,T^*), \quad K_2 = K_-(0,S^*),\\
&K = K_1 \cup K_2,
\end{aligned}
$$

where R_1 and R_2 are positive constants. In what follows, we assume $Q^* + T^* < S^* < Q^* + 2T^*$ holds. See Figure 1. The rectangle is K_1 and the isosceles right triangle is K_2.

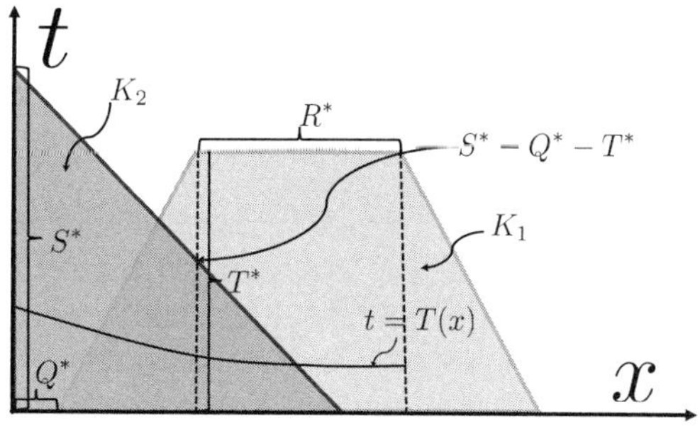

Fig. 1. The dependence domain.

To make certain of being the blow-up curve in $K_1 \cup K_2$, we consider the following system of ODEs.

$$
\begin{cases}
\dfrac{d}{dt}\tilde{\phi} = |\tilde{\phi} + \tilde{\psi}|^p, & t > 0,\\
\dfrac{d}{dt}\tilde{\psi} = |\tilde{\phi} + \tilde{\psi}|^p, & t > 0,\\
\tilde{\phi}(0) = \gamma, \quad \tilde{\psi}(0) = 0,
\end{cases}
\tag{4}
$$

where γ is a positive constant. Then, there exists a positive constant $\tilde{T}$ depending only on γ such that

$$(\tilde{\phi} + \tilde{\psi})(t) \to \infty \quad \text{as} \quad t \to \tilde{T}.$$

We suppose the following assumptions.

(A1) $\tilde{T} < S^* - Q^* - T^*$.

(A2) $\begin{cases} f \geq \gamma, \quad g \geq -\gamma \quad \text{in} \quad B_{0,Q^*+R^*+2T^*}, \\ g \geq 0 \quad \text{in} \quad B_{Q^*,Q^*+R^*+2T^*}. \end{cases}$

(A3) $f, g \in C^4(B_{0,Q^*+R^*+2T^*})$.

(A4) There exists a positive constant ε_0 such that

$$\gamma^p \geq 2(1+\varepsilon_0)(|\partial_x f| + |\partial_x g|) \quad \text{in} \quad B_{Q^*,Q^*+R^*+2T^*}.$$

(A5.1) $\partial_x f > 0$, $\partial_x g > 0$ in B_{0,S^*}.

(A5.2) $\partial_x^2 f$, $\partial_x^2(f+g) \geq 0$ in B_{0,S^*}.

(A5.3) $\partial_x^3 f$, $\partial_x^3 g \geq 0$ in B_{0,S^*}.

(A6) $\partial_x(f-g) \geq 0$ in B_{0,S^*}.

We now state the main result of this paper.

Theorem 1. *Let Q^*, R^*, S^* and T^* be positive constants satisfying $Q^* + T^* < S^* < Q^* + 2T^*$. Assume $p = 2$, 3 or $p \geq 4$ and* (A1)–(A6) *hold true. Then, there exists a unique $C^1(B_{0,Q^*+R^*+T^*})$ function T such that $0 < T(x) < S^*$ $(x \in B_{0,Q^*+R^*+T^*})$ and a unique $(C^{3,1}(\Omega))^2$ solution (ϕ, ψ) of* (3) *satisfying*

$$\phi(x,t),\ \psi(x,t) \to \infty \quad as \quad t \uparrow T(x) \tag{5}$$

for any $x \in B_{0,Q^*+R^*+T^*}$. Here,

$$\Omega = \{(x,t) \in \mathbb{R}^2 \mid x \in B_{0,Q^*+R^*+T^*},\ 0 < t < T(x)\}.$$

Moreover,

$$T'(x) \to -1 \quad as \quad x \downarrow 0. \tag{6}$$

Remark 2. Uesaka [8] consideblack the blow-up curve for the following system of the nonlinear wave equations:

$$\begin{cases} \partial_t^2 u_1 - c_1^2 \partial_x^2 u_1 = F_1(u_1, u_2), \\ \partial_t^2 u_2 - c_2^2 \partial_x^2 u_2 = F_2(u_1, u_2). \end{cases} \tag{7}$$

In [8], he only showed the Lipschitz continuity of the blow-up curve under suitable initial conditions. It is an open problem when the blow-up curve for (7) becomes smooth and when it is not differentiable.

Remark 3. Since $\partial_t u$ is an odd function, the blow-up curve T becomes an even function. Therefore, we have

$$T'(x) \to 1 \quad (x \uparrow 0) \quad \text{and} \quad T'(x) \to -1 \quad (x \downarrow 0).$$

In the final section, we state some keys of the proof of Theorem 1. We can prove Theorem 1 by showing the continuous differentiability of T for $x > 0$ and applying the Implicit Function Theorem to an approximation of T.

§3. Idea of the proof of Theorem 1

3.1. Continuous differentiability of the blow-up curve

In this section, we state the key proposition of the proof as follows:

Proposition 1. *Under the assumptions of Theorem* 1, *there exists a unique* $C^1(B_{0,Q^*+R^*+T^*})$ *function* T *such that* $0 < T(x) < S^*$ $(x \in B_{0,Q^*+R^*+T^*})$ *and a unique* $(C^{3,1}(\Omega))^2$ *solution* (ϕ, ψ) *of* (3) *satisfying*

$$\phi(x,t),\ \psi(x,t) \to \infty \quad as \quad t \uparrow T(x) \quad and \quad |T'(x)| < 1 \tag{8}$$

for any $x \in B_{0,Q^*+R^*+T^*}$. *Here,*

$$\Omega = \{(x,t) \in \mathbb{R}^2 \mid x \in B_{0,Q^*+R^*+T^*},\ 0 < t < T(x)\}.$$

The idea of the proof of Proposition 1 is showing that ϕ and ψ have the property like blowing-up solutions of ordinary differential equations. The following lemmas play an important roles in showing the property.

Lemma 1. *Assume* (A2) *and* (A5.1). *Then, we have*

$$\phi \geq 0, \quad \phi + \psi \geq 0 \quad in \quad \Omega. \tag{9}$$

Lemma 2. *Assume* (A1)–(A6) *and* $\tilde{x} \in B_{0,Q^*+R^*+T^*}$. *Then, there exist constants* ε_0, $r > 0$ *which depend on* $\tilde{x}$ *such that*

$$\partial_t\phi \geq (1+\varepsilon_0)|\partial_x\phi|, \quad \partial_t\psi \geq (1+\varepsilon_0)|\partial_x\psi| \quad in \quad B_{\tilde{x}}, \tag{10}$$

where $B_{\tilde{x}} = \{(x,t) \in \Omega \mid |x-\tilde{x}|^2 + |t - T(\tilde{x})|^2 < r\}$.

Remark 4. We notice that the constant ε_0 which appears in (10) depends on $\tilde{x}$. In [7], we can take ε_0 so as not to depend on $\tilde{x}$ by assuming that there exists ε_0 which does not depend on $\tilde{x}$ such that the (10) holds at $t = 0$. That is, we can assume (A4) holds for all x in [7]. However, due to the condition $(\phi + \psi)(0,t) = 0$, we cannot take the constant ε_0 so as not to depend on $\tilde{x}$ if $\tilde{x}$ is near 0 This makes difficult to prove Proposition 1.

By Lemma 2, we obtain the following Lemma.

Lemma 3. *Assume* (A1)–(A6) *and* $\tilde{x} \in B_{0,Q^*+R^*+T^*}$. *Then, there exist* C_1 *and* C_2 *which depend on* $\tilde{x}$ *such that*

$$C_1(\phi+\psi)^p \le \partial_t\phi, \partial_t\psi \le C_2(\phi+\psi)^p, \tag{11}$$

$$C_1(T(x)-t)^{-1/(p-1)} \le (\phi+\psi)(x,t) \le C_2(T(x)-t)^{-1/(p-1)} \tag{12}$$

in $B_{\tilde{x}}$.

Lemma 3 means (ϕ, ψ) has the property like blowing-up solutions of ordinary differential equations in some sense. By using Lemmas 2 and 3, we can show that there exists a positive constant α which depends on $\tilde{x}$ such that there exist constants $\tilde{C}_1$ and $\tilde{C}_2$ which depend on α such that

$$|\alpha| < 1, \quad \phi(x,t) \approx V_{1,\alpha}(x,t) \quad \text{and} \quad \psi(x,t) \approx V_{2,\alpha}(x,t),$$

if $(x,t) \in \Omega$ is sufficiently close to $(\tilde{x}, T(\tilde{x}))$. Here,

$$V_{j,\alpha}(x,t) = \tilde{C}_j(\alpha(x-\tilde{x}) - (t - T(\tilde{x})))^{-1/(p-1)} \quad \text{for} \quad j = 1, 2.$$

We also have

$$\partial_\tau V_{1,\alpha} = \frac{1}{1-\alpha}(V_{1,\alpha}+V_{2,\alpha})^p \quad \text{and} \quad \partial_\tau V_{2,\alpha} = \frac{1}{1+\alpha}(V_{1,\alpha}+V_{2,\alpha})^p,$$

where $\partial_\tau = \partial_t + \alpha\partial_x$. In this sense, the solution has the property like blowing-up solutions of the ordinary differential equations. By using this fact, we can show Proposition 1.

Remark 5. The constant α is actually $T'(\tilde{x})$.

3.2. Proof of the non-differentiability of the blow-up curve

In this section, we state the idea of the proof of (6). The proof is based on the Implicit Function Theorem. Let $n \in \mathbb{N}$ and M_n be given by

$$M_n = \sup_{x \in \overline{B_{0,S^*}}} \phi(x, T(x) - 1/2n) \quad \text{for} \quad n \in \mathbb{N}.$$

Since $(\phi, \psi) \in \left(C^{3,1}(\Omega)\right)^2$ and

$$\partial_t\phi(x,t) > 0 \quad \text{and} \quad \phi(x,t) \to \infty \quad \text{as} \quad t \to T(x) \quad \text{for} \quad (x,t) \in \Omega,$$

there exists an implicit function $E_n \in C^3(\overline{B_{0,S^*}})$ such that,

$$\begin{cases} \phi(x, E_n(x)) = M_n & \text{for} \quad x \in \overline{B_{0,S^*}}, \\ |T(x) - E_n(x)| < \dfrac{1}{n} & \text{for} \quad x \in B_{0,S^*}. \end{cases} \tag{13}$$

By applying the Implicit Function Theorem, we get

$$E_n'(x) = -\frac{\partial_x \phi}{\partial_t \phi}(x, E_n(x)) \quad \text{for} \quad x \in \overline{B_{0,S^*}}. \tag{14}$$

Moreover, by using Proposition 1, we can prove the uniform boundedness and the uniform equicontinuity of E_n'. Therefore, there exists $\{n_k\} \subset \{n\}$ such that there exists $\tilde{E} \in C(\overline{B_{0,S^*}})$ such that

$$E_{n_k}' \to \tilde{E} \ \text{ as } \ k \to \infty \ \text{ uniformly in } \ \overline{B_{0,S^*}} \ \text{ and } \ \tilde{E} = T' \ \text{ in } \ B_{0,S^*}$$

by using the Ascoli and Arzelà Theorem. On the other hand, by using (14) and the fact $(\phi + \psi)(0, t) = 0$, we have

$$E_{n_k}'(0) = -\frac{\partial_x \phi}{\partial_t \phi}(0, E_{n_k}(0)) = -\frac{\partial_t \phi - |\phi + \psi|^{p-1}(\phi + \psi)}{\partial_t \phi}(0, E_{n_k}(0)) = -1$$

for all k. Therefore, we have

$$T'(x) - \tilde{E}'(x) \to \tilde{E}'(0) = -1 \quad \text{as} \quad x \downarrow 0.$$

Remark 6. We notice that (6) comes from the propagation speed of ϕ. By Proposition 1, we have

$$\psi(x, t), \ (\phi + \psi)(x, t) \to \infty \quad \text{as} \quad t \uparrow T(x)$$

for $x \in B_{0,Q^*+R^*+T^*}$. Therefore, we also have the implicit function like E_n for ψ and $(\phi + \psi)$.

However, the implicit function for ϕ only has uniform boundedness and uniform equicontinuity. This means the blowing-up of ϕ is dominant in this situation. Therefore, we have (6).

References

[1] L. A. Caffarelli and A. Friedman, The blow-up boundary for nonlinear wave equations, Trans. Amer. Math. Soc., **297** (1986), 223–241.

[2] P. Godin, The blow-up curve of solutions of mixed problems for semilinear wave equations with exponential nonlinearities in one space dimension. II, Ann. Inst. H. Poincaré Anal. Non Linéaire, **17** (2000), 779–815.

[3] P. Godin, The blow-up curve of solutions of mixed problems for semilinear wave equations with exponential nonlinearities in one space dimension. I, Calc. Var. Partial Differential Equations, **13** (2001), 69–95.

[4] M. A. Hamza and H. Zaag, Blow-up behavior for the Klein-Gordon and other perturbed semilinear wave equations, Bull. Sci. Math., **137** (2013), 1087–1109.

[5] F. Merle and H. Zaag, Existence and classification of characteristic points at blow-up for a semilinear wave equation in one space dimension, Amer. J. Math., **134** (2012), 581–648.
[6] M. Ohta and H. Takamura, Remarks on the blow-up boundaries and rates for nonlinear wave equations, Nonlinear Anal., **33** (1998), 693–698.
[7] T. Sasaki, Regularity and singularity of the blow-up curve for a wave equation with a derivative nonlinearity, Adv. Differential Equations, **23** (2018), 373–408.
[8] H. Uesaka, The blow-up boundary for a system of semilinear wave equations, Further progress in analysis, World Sci. Publ., Hackensack, NJ, (2009), 845–853.

Tetsuya Ishiwata
Department of Mathematical Sciences, Shibaura Institute of Technology
307 Fukasaku, Minuma-ku, Saitama 337-8570, Japan

Takiko Sasaki
Current address
(1) *National Institute of Technology, Ibaraki College,*
866 Nakane, Hitachinaka-shi, Ibaraki-ken 312-8508, Japan
E-mail address: tsasaki@ibaraki-ct.ac.jp
(2) [*Cross-Appointment*]
Mathematical Institute, Tohoku University,
Aoba, Sendai 980-8578, Japan

Advanced Studies in Pure Mathematics 85, 2020
The Role of Metrics in the Theory of Partial Differential Equations
pp. 369–378

Local energy decay and the vector field method

Jacob Sterbenz

Abstract.

We discuss recent work concerning stability of solutions to linear and nonlinear wave equations on asymptotically flat space times. This is joint work with Jason Metcalfe, Jesus Oliver, and Daniel Tataru.

§1. Introduction

Recall that for the scalar wave equation $\Box = -\partial_t^2 + \sum_{i=1}^n \partial_i^2$ on Minkowski space $\mathbb{R}^{n+1}$, solutions to the initial value problem:

$$\Box\phi = 0, \qquad \phi(0,x) = f(x), \quad \partial_t\phi(0,x) = g(x), \tag{1}$$

exhibit the conservation of energy:

$$E(t) = E(0), \qquad E(t) = \frac{1}{2}\int_{\mathbb{R}^n} (\phi_t^2 + |\nabla_x\phi|^2)(t,x)\, dx, \tag{2}$$

assuming $(f,g) \in \dot{H}_x^1 \times L_x^2$.

Solutions to (1) also exhibit transport type behavior. This can be seen by taking a geometrical optics ansatz $\phi_\lambda = a_\lambda e^{i\lambda u}$, where u solves eikonal equation $|u_t| = |\nabla_x u|$, and to first order a_λ solves the amplified transport equation $La_\lambda = (\frac{1}{2}\Box u)a_\lambda$. Here L is the null vector field $L = u_t\partial_t - \nabla_x u \cdot \nabla_x$ whose integral curves are rays of the form $\gamma(s) = (s, x_0 + \omega s)$ for various directions $\omega \in \mathbb{S}^{n-1} \subseteq \mathbb{R}^n$. This ansatz implies that any initially localized solution to (1) will eventually move energy away from spatially compact sets. In particular, one can show that localizing the energy (2) leads to a spacetime bound:

$$\int_0^\infty \int_{|x|\leq R} (\phi_t^2 + |\nabla_x\phi|^2)(t,x)\, dxdt \leq C(R)E(0). \tag{3}$$

Received December 1, 2018.
Revised April 23, 2019.
2010 *Mathematics Subject Classification.* 35L10.
Key words and phrases. Morawetz estimate, Klainerman vector field method.

Estimates of this type have a long history, and were first introduced in the work of Morawetz [19, 20, 21]. Notice that (3) is essentially optimal if one considers only energy class initial data. On $\mathbb{R}^{3+1}$ this can be seen directly by taking an infinite sum of incoming spherical wave packets of the form $\phi = \sum_i a_i\phi_0(t-i,x)$, where $\phi_0 = |x|^{-1}(\chi(t-|x|)-\chi(t+|x|))$, where $\chi(s)$ is a smooth cutoff on $\frac{1}{2}\leq s\leq 2$, and where $\sum_i a_i^2 = 1$.

On the other hand if one restricts the previous sum to the single wave packet ϕ_0, then one has much better bounds than (3). In fact, for reasonably smooth initial data one has the explicit solution formula for $t>0$ on $\mathbb{R}^{3+1}$:

(4)
$$\phi(t,x) = \frac{1}{4\pi t}\int_{|y-x|=t} g(y)\,dS + \frac{1}{4\pi t^2}\int_{|y-x|=t}\big(f+(y-x)\cdot\nabla f\big)(y)\,dS.$$

In particular when the data $f,g\in\mathcal{S}(\mathbb{R}^3)$ are Schwartz class functions we see the solution exhibits the behavior:

(5)
$$|\phi(t,x)| \leq C(N,f,g)(1+|t|+|x|)^{-1}\big(1+\big||t|-|x|\big|\big)^{\;N},$$

for every $N>0$.

A robust way to interpolate between estimates (3) and (5) is given by Klainerman's vector field method, see [8] and [9]. One starts with the observation that the vector fields $\mathbb{L} = \{\partial_t, \partial_i, x^j\partial_i - x^i\partial_j, t\partial_i + x^i\partial_t, t\partial_t + x^i\partial_i\}$ satisfy $[\Box, X] = c(X)\Box$ for a constant $c(X)$ depending on X. Thus, if $\Box\phi = 0$, then one has the conserved energy (2) for $X^I\phi$, where X^I is some finite product of elements in $\mathbb{L}$. Via rescaled Sobolev embeddings this gives:

(6)
$$|\nabla\phi(t,x)| \leq C(f,g)(1+|t|+|x|)^{-1}\big(1+\big|t-|x|\big|\big)^{-\frac{1}{2}},$$

where $C(f,g)$ is the initial L^2 of $\nabla X^I\phi$ for all multiindices $|I|\leq 2$.

Now consider the wave equation $\Box_g = \nabla^\alpha\nabla_\alpha$ on a curved space time $(\mathcal{M},g)$, where g is some Lorentzian metric which satisfies an asymptotic flatness condition of the form $|\partial^I(g-m)| \lesssim (1+|x|)^{-|I|-\gamma}$ for some $\gamma>0$, and where m is the usual Minkowski metric. In this case there is no useful analog of (4) for studying global problems, and instead one must work with the operator $\Box_g$ more directly.

In order to get uniform decay estimates one must assume more than generic asymptotic conditions for g. Detailed local properties of $(\mathcal{M},g)$ must also be known. This can be seen by again looking at the geometrical optics ansatz $\phi_\lambda = a_\lambda e^{i\lambda u}$ where $La_\lambda = (\frac{1}{2}\Box u)a_\lambda$ (to first order) and $L = -\nabla u$ is the null generator of $u = const.$ Such approximate solutions show that the long time behavior of null geodesics in $(\mathcal{M}, g)$ will determine the local decay of solutions to $\Box_g \phi = 0$, at least in the high frequency regime. Now there are mainly two cases to consider:

First, if every null geodesic of $(\mathcal{M}, g)$ leaves all spatially compact sets in a finite amount of time, then under general asymptotic conditions estimates such as (3) continue to hold. For example see the works [2], [7], [18], and the results described below.

Second, if $(\mathcal{M}, g)$ happens to have spatially trapped null geodesics, it can be shown that the precise estimate (3) fails. However, for many of the metrics which arise in applications one can often recover a weakened version[1] such as:

$$\int_0^\infty \int_{|x|\leq R} \phi^2(t,x)\, dxdt \leq C(R)E(0). \tag{7}$$

Bounds of this type have been established for stationary black holes in a number of works such as [5], [24], [25], and recently [6] which covers all subextremal Kerr metrics.

With some additional structure on $(\mathcal{M}, g)$, estimates such as (7) are enough to bootstrap the local errors generated by commutators $[\Box_g, X]$ when X are suitably chosen vector fields which are asymptotic to the elements of $\mathbb{L}$ listed above. This ultimately leads to decay estimates like (6) for both linear and nonlinear waves on a variety of interesting backgrounds. See for example the works [13], [14], [15], [16], [22], and the results described below.

§2. Results on local energy decay

Here we discuss the results of [17] which is joint work with Jason Metcalfe and Daniel Tataru. The goal of this project is to provide general conditions under which estimate (3) holds for solutions to the perturbed wave equation:

$$P(t,x,D) = \Box_{A,g} + V(t,x), \quad \Box_{A,g} = (D_\alpha + A_\alpha)g^{\alpha\beta}(D_\beta + A_\beta) \tag{8}$$

where A and V are complex valued, and $D_\alpha = (1/i)\partial_\alpha$. The metric g and the coefficients A, V are assumed to satisfy optimal asymptotic flatness condition of the form

$$|\!|\!| (g-m, A, V) |\!|\!|_{AF} < \infty \tag{9}$$

[1] Roughly speaking one needs to take into account the amplification factor $\frac{1}{2}\Box u$ in the transport equation for a_λ. This measures the spreading of nearby null geodesics, and reasonable local energy decay can still be established if a_λ is negative along trapped null geodesics.

where $m = (-1, 1, 1, 1)$ and:

$$\| (h, A, V) \|_{AF} = \| h \|_2 + \| \langle x\rangle A \|_1 + \| \langle x\rangle^2 V \|_0,$$
$$\| q \|_k = \sum_{|I|\le k} \|\langle x\rangle^{|I|}\partial^I q\|_{l^1 L^\infty}.$$

The key additional assumptions we make here are that g is nontrapping for null geodesics, and that (g, A, V) is either *stationary* (i.e. time independent), or in certain cases that $\partial_t(g, A, V)$ is sufficiently small in the AF norm. For the strongest theorems to hold we also need to enforce a reality condition on (A, V) which implies that P is (almost) symmetric.

To measure local energy decay in an optimal way we use the dyadic norms of Metcalfe-Tataru:

$$\| u \|_{LE} = \sup_{k>0} 2^{-\frac{1}{2}k} \| u \|_{L^2(|x|\approx 2^k)}, \quad \| u \|_{LE^1} = \Big\| (\partial u, \langle x\rangle^{-1} u) \Big\|_{LE},$$
$$\| F \|_{LE^*} = \sum_{k>0} 2^{\frac{1}{2}k} \| F \|_{L^2(|x|\approx 2^k)}.$$

Then we say that *local energy decay holds for P* if:

$$\| u \|_{LE^1[0,T]} + \| \partial u \|_{L^\infty L^2[0,T]} \lesssim \| \partial u(0) \|_{L^2} + \| Pu \|_{LE^* + L^1 L^2[0,T]}. \tag{10}$$

With this notation one of the main results of [17] reads as follows:

Theorem 1. a) *Let P be stationary, nontrapping and asymptotically flat. Then local energy decay holds for P if and only if P has no negative eigenvalues, real resonances, or zero eigenvalues/resonances.*

b) *Let P be stationary, symmetric, nontrapping and asymptotically flat. Then local energy decay holds for P if and only if P has no negative eigenvalues or zero eigenvalues/resonances.*

The various obstructions described in the above theorem are:

i) An "eigenvalue" corresponds to an L^2 solution ϕ_ω to the reduced wave equation $P_\omega = P|_{D_t=\omega}$ for some $\omega \in \mathbb{C}$. We say ω is "negative" if $\Im(\omega) < 0$. Note that even when P_ω is symmetric, the eigenvalue problem here is quadratic, so there is no reason for ω to be real. Negative eigenvalues give rise to highly localized finite energy solutions $e^{i\omega t}\phi_\omega$ with exponentially growing local energy forward in time.

ii) A "real resonance" is a solution to $P_\omega \phi_\omega = 0$ with $\omega \in \mathbb{R} \setminus \{0\}$, which satisfies the outgoing Sommerfeld condition:

$$2^{-j/2} \|(\partial_r + i\omega)\phi_\omega\|_{L^2(|x|\approx 2^j)} \to 0.$$

After multiplication of $e^{i\omega t}\phi_\omega$ by a cutoff $\chi(t-r)$ supported where $t-r>0$, these correspond to approximate waves which spread in a timelike subset of $|x|\le t$. They are the most subtle obstruction to the local energy decay estimate.

iii) Finally, by a "zero eigenvalues/resonance" we simply mean a stationary solution to $P_0\phi_0=0$ with finite LE^1 norm when $0\le t\le 1$. These are associated with a root space at zero energy which leads to finite energy solutions with no time oscillation but which grow at some algebraic rate in time.

The importance of the symmetry assumption in part b) of Theorem 1 is that it automatically eliminates real (nonzero) resonances through an energy bootstrap argument. Each of i)–iii) represents a distinct obstruction to the dispersive estimate (10), and by Theorem these are the only possible obstructions.

In [17] we also prove a version of Theorem 1 part b) which includes the case when $\| \partial_t(g,A,V)\|_{AF}\le\epsilon$. Moreover this can be done in the context where one drops the assumption of no negative eigenvalues:

Theorem 2. *Let P be symmetric, slowly varying, nontrapping, asymptotically flat, and satisfying the following uniform estimate at zero energy*:

$$\|u\|_{\dot H^1}\le K_0\|P_0u\|_{\dot H^{-1}}.$$

Then there is a direct sum decomposition of the energy space into invariant subspaces

$$\mathcal{E}=S^0+S^++S^-,$$

with S^+ and S^- of equal finite dimension, so that the wave flow on the subspaces S^+, S^0 and S^- is as follows:

a) *Solutions in S^+ are spatially localized and grow exponentially forward in time.*

b) *Solutions in S^0 are uniformly bounded and satisfy local energy decay.*

c) *Solutions in S^- are spatially localized and decay exponentially forward in time.*

One can think of this result as a linear stable manifold theorem for symmetric adiabatic operators $P(t,x,D)$ which holds whenever there is a spectral gap around zero energy. This provides a quantitative version of the spectral theory for Klein-Gordon equations established in [10], and also an infinite dimensional analog of the classical "roughness theory" of nonautonomous ODE [4].

The key technical step in all of the above theorems is to prove a local energy decay estimate for solutions to $P\phi=F$, where ϕ is assumed to be compactly supported in time. This restriction allows one to truncate the time frequency variable into low, medium, and high frequencies which can then be analyzed using separate arguments. One can finally remove the time truncation with various limiting arguments depending on the context.

§3. Results on the vector field method

Here we discuss [23] which is joint work with Jesus Oliver. In this project we start by *assuming* a weaker version of (10) which is:

$$\sup_{0\leq t\leq T} \|\partial\phi(t)\|_{H^s_x} + \|\phi\|_{WLE^s[0,T]} \leq C_s\big(\|\partial\phi(0)\|_{H^s_x} + \|\Box_g\phi\|_{(WLE^{*,s}+L^1_tH^s_x)[0,T]}\big), \tag{11a}$$

$$\sup_{0\leq t\leq T} \|\partial\phi(t)\|_{H^s_x}+\|\phi\|_{LE^s[0,T]} \leq C_s\big(\|\partial\phi(0)\|_{H^s_x} + \|(\phi,\partial_t\phi)\|_{L^2\times H^s(r<R_0)[0,T]} + \|\Box_g\phi\|_{LE^{*,s}[0,T]}\big), \tag{11b}$$

where the norms are defined as follows:

$$\|\,\phi\,\|_{LE^s[0,T]} = \sum_{|J|\leq s}\big(\|\,\langle x\rangle^{-1}\partial^J\phi\,\|_{LE[0,T]} + \|\,\partial\partial^J\phi\,\|_{LE[0,T]}\big),$$

$$\|\,F\,\|_{LE^{*,s}[0,T]} = \sum_{|J|\leq s}\|\,\partial^J F\,\|_{LE^*[0,T]},$$

$$\|\phi\|_{WLE^s[0,T]} = \sum_{|J|\leq s}\big(\|\langle x\rangle^{-1}\partial^J\phi\|_{LE[0,T]} + \|\partial\partial^J\phi\|_{LE(r>R_0)[0,T]}\big),$$

$$\|\,F\,\|_{WLE^{*,s}[0,T]} = \sum_{|J|\leq s}\big(\|\,\partial^J F\,\|_{LE^*[0,T]} + \|\,\partial\partial^J F\,\|_{L^2(r<R_0)[0,T]}\big).$$

Here R_0 is some sufficiently large radius which in applications is fixed so the exterior $|x|>R_0$ does not contain trapped null rays. Note that the s index in LE^s is shifted here by one unit compared to (10) to make the notation more uniform. We also note that the space time $(\mathcal{M},g)$ may only be defined in the spatial variable for $|x|\geq r_0>0$, which is the case for black holes.

The other main consideration of [23] is that we replace the condition (9) with the weaker assumption that there exists an outgoing optical function $u=u(t,x)$ satisfying the two conditions:

I) (Weak Asymptotic Flatness) We have that (u,x) forms a uniform set of coordinates in the sense that $C^{-1}<u_t<C$, and u has the symbol bounds:

$$\|\,(\tau_-\partial_t)^i(\tau_x\tau_0\partial_x)^J(\partial_t u-1,\partial_i u+\omega^i)\,\|_{\ell^1_r L^\infty[0,\infty)} < \infty, \tag{13}$$

for all $(i, J) \in \mathbb{N} \times \mathbb{N}^3$, where $\tau_x = \langle x \rangle$, $\tau_- = \langle u \rangle$, $\tau_+ = \langle (t, x) \rangle$, $\tau_0 = \tau_- \tau_+^{-1}$, and where $\omega^i = x^i \tau_x^{-1}$.

II) (Outgoing Radiation Condition) First define symbol classes $\mathcal{Z}^k$:

$$\| q \|_{k,N} := \sum_{i+|J|\le N} \| \tau_0^{-k} (\tau_- \partial_u^b)^i (\tau_x \partial_x^b)^J q \|_{\ell_r^1 L^\infty} + \sum_{i+|J|\le N} \| \tau_0^{-k} (\tau_- \partial_u^b)^i (\tau_x \partial_x^b)^J q \|_{\ell_u^1 \ell_r^1 L^\infty(\frac{1}{2}t<r<2t)}.$$

Here $\partial^b = (\partial_u^b, \partial_x^b)$ denotes the (u, x) coordinate derivatives. Then for the inverse metric $g^{\alpha\beta}$ in (u, x) coordinates one has the symbol bounds:

$$g^{\alpha\beta} - h^{\alpha\beta} \in \mathcal{Z}^0, \qquad \sqrt{|g|}\, g^{iu} + \omega^i \in \mathcal{Z}^{\frac{1}{2}}, \tag{14a}$$

$$g^{ui} - \omega^i \omega_j g^{uj} \in \mathcal{Z}^1, \qquad g^{uu} \in \mathcal{Z}^2, \tag{14b}$$

where $h^{uu} = 0$, $h^{ui} = -\omega^i$, and $h^{ij} = \delta^{ij}$ is the Minkowski metric in Bondi coordinates (u, x).

In the regions $|x| < (1-\epsilon)t$ and $|x| > (1+\epsilon)t$, for $\epsilon > 0$, the asymptotic bounds (14) are consistent with (9). On the other hand in the region $\big| t - |x| \big| \le \epsilon t$ these conditions are consistent with the asymptotics of the metrics in the stability of Minkowski space, see [3], [1], and [11].

The main result of [23] is to show that by assuming (13) and (14), the estimates (11a) and (11b) imply uniform decay rates for both linear and nonlinear problems. On the linear side we have:

Theorem 3 (Peeling Estimates for the Homogeneous Wave Equation). *Given any $k \ge 1$ there exists an integer $N = N(k)$ depending (linearly) on k such that if $\Box_g \phi = 0$ one has:*

$$\sum_{i+|J|\le k} \Big\| \tau_+^{\frac{3}{2}} \tau_0^{\frac{1}{2}} (\tau_- \partial_u^b)^i (\tau_x \partial_x^b)^J \phi \Big\|_{L^\infty[0,T]} \le C_k \sum_{|J|\le N} \| \tau_x (\tau_x \partial_x)^J \partial\phi(0) \|_{L_x^2}. \tag{15}$$

Estimates similar to (15) are shown to hold for solutions to the inhomogeneous wave equation $\Box_g \phi = F$, with optimal norms on F. This leads to a nonlinear result:

Theorem 4 (Peeling Estimates for Solutions to Null Form Systems). *Suppose quadratic forms $\mathcal{N}$ are given which satisfy:*

$$\big| (\tau_- \partial_u^b)^i (\tau_x \partial_x^b)^J \partial_\phi^k \mathcal{N}^{\alpha\beta} \big| \le c_{i,J,k}(|\phi|),$$
$$\big| (\tau_- \partial_u^b)^i (\tau_x \partial_x^b)^J \partial_\phi^k \mathcal{N}^{uu} \big| \le c_{i,J,k}(|\phi|) \tau_0,$$

where $c_{i,J,k}(|\phi|)$ are a series of constants that are locally bounded in the quantity $|\phi|$. Let ϕ be a solution to the system of semilinear equations:

$$\Box_g \phi = \mathcal{N}^{\alpha\beta}(t,x,\phi)\partial_\alpha\phi\partial_\beta\phi,$$

Then for sufficiently small and well localized initial data the solution is global and enjoys the peeling estimates:

$$\sum_{i+|J|\leq k} \left\| \tau_+^{\frac{3}{2}}\tau_0^{\frac{1}{2}}(\tau_-\partial_u^b)^i(\tau_x\partial_x^b)^J\phi \right\|_{L^\infty[0,T]} \leq C_k\epsilon.$$

The proof of these two theorems uses a graduated series of norms $\|\cdot\|_{S^a}$ which are, roughly speaking, the local energy decay norm LE^s with some additional weighs of the form $\tau_+^a\tau_0$ for $\partial\phi$ and τ_+^a for $\partial_x^b\phi$. Estimates in the space S^a come at the expense of local errors which are roughly $\| \tau_+^{a-1}X^I\phi\|_{L^2(r\leq R)[0,T]}$ for $X\in\widetilde{\mathbb{L}}=\mathbb{L}\setminus\{t\partial_i+x^i\partial_t\}$. Such errors can be fed back into the local energy decay estimate (11a) to close the argument.

To put these results in context, we remark that Theorems 3 and 4 can be seen as sharp generalizations of previous results of Luk ([13] and [14]) for the specific case of slowly rotating Kerr backgrounds. Theorem 3 is also closely related to recent work of Moschidis [22], which applies to more general backgrounds which are similar to the ones defined by conditions (13) and (14). One of the main differences between the method used to establish Theorem 3 and the technique of [22] is our use of *time dependent* weights as well as a sequence of *graded norms*. The norms are specifically designed to handle nonlinear problems, and allows one to prove Theorem 4 in essentially the same way as one proves Theorem 3. To our knowledge Theorem 4 is the first quadratic nonlinear result on black holes with large ergoregions, which combined with the results in [6] includes the case of all subextreamal Kerr metrics.

Finally, we remark that the norms used to produce Theorems 3 and 4 are also compatible with systems of nonlinear wave equations exhibiting the weak null condition of Lindblad-Rodnianski [12]. This aspect of our method, and its relation to spacetimes satisfying certain bounds on their Ricci curvature, will be the subject of future work.

References

[1] L. Bieri, *An extension of the stability theorem of the Minkowski space in general relativity*, J. Differential Geom., **86** (2010), no. 1, 17–70.

[2] J.-F. Bony and D. Häfner, *The semilinear wave equation on asymptotically Euclidean manifolds*, Comm. Partial Differential Equations, **35** (2010), no. 1, 23–67.

[3] D. Christodoulou and S. Klainerman, *The global nonlinear stability of the Minkowski space*, Princeton Mathematical Series, 41. Princeton University Press, Princeton, NJ, 1993. x+514 pp.

[4] W. A. Coppel, *Dichotomies in stability theory*, Lecture Notes in Mathematics, **629**. Springer-Verlag, Berlin-New York, 1978.

[5] M. Dafermos and I. Rodnianski, *The red-shift effect and radiation decay on black hole spacetimes*, Commun. Pure Appl. Math., **62** (7), 859–919.

[6] M. Dafermos, I. Rodnianski and Y. Shlapentokh-Rothman, *Decay for solutions of the wave equation on Kerr exterior spacetimes* III: *The full subextremal case* $|a| < M$, Ann. of Math. (2), **183** (2016), no. 3, 787–913.

[7] M. Keel, H. Smith and C. Sogge, *Almost global existence for some semilinear wave equations*, Dedicated to the memory of Thomas H. Wolff. J. Anal. Math., **87** (2002), 265–279.

[8] S. Klainerman, *Uniform decay estimates and the Lorentz invariance of the classical wave equation*, Comm. Pure Appl. Math., **38** (1985), no. 3, 321–332.

[9] S. Klainerman, *The null condition and global existence to nonlinear wave equations*, Nonlinear systems of partial differential equations in applied mathematics, Part 1 (Santa Fe, N.M., 1984), 293–326, Lectures in Appl. Math., 23, Amer. Math. Soc., Providence, RI, 1986.

[10] H. Langer, B. Najman and C. Tretter, *Spectral theory of the Klein-Gordon equation in Pontryagin spaces*, Comm. Math. Phys., **267** (2006), no. 1, 159–180.

[11] H. Lindblad, *On the asymptotic behavior of solutions to the Einstein vacuum equations in wave coordinates*, Comm. Math. Phys., **353** (2017), no. 1, 135–184.

[12] H. Lindblad and I. Rodnianski, *The global stability of Minkowski spacetime in harmonic gauge*, Ann. of Math. (2), **171** (2010), no. 3, 1401–1477.

[13] J. Luk, *A vector field method approach to improved decay for solutions to the wave equation on a slowly rotating Kerr black hole*, Anal. PDE, **5** (2012), no. 3, 553–625.

[14] J. Luk, *The null condition and global existence for nonlinear wave equations on slowly rotating Kerr spacetimes*, J. Eur. Math. Soc. (JEMS), **15** (2013), no. 5, 1629–1700.

[15] J. Metcalfe and C. Sogge, *Hyperbolic trapped rays and global existence of quasilinear wave equations*, Invent. Math., **159** (2005), no. 1, 75–117.

[16] J. Metcalfe and C. Sogge, *Global existence of null-form wave equations in exterior domains*, Math. Z., **256** (2007), no. 3, 521–549.

[17] J. Metcalfe, J. Sterbenz and D. Tataru, *Local energy decay for scalar fields on time dependent non-trapping backgrounds*, to appear in AJM.

[18] J. Metcalfe and D. Tataru, *Global parametrices and dispersive estimates for variable coefficient wave equations*, Math. Ann., **353** (2012), no. 4, 1183–1237.

[19] C. S. Morawetz, *Exponential decay of solutions of the wave equation*, Comm. Pure Appl. Math., **19** (1966), 439–444.

[20] C. S. Morawetz, *Time decay for the nonlinear Klein-Gordon equations*, Proc. Roy. Soc. Ser. A, **306** (1968), 291–296.

[21] C. S. Morawetz, *Decay for solutions of the exterior problem for the wave equation*, Comm. Pure Appl. Math., **28** (1975), 229–264.

[22] G. Moschidis, *The r^p-weighted energy method of Dafermos and Rodnianski in general asymptotically flat spacetimes and applications*, Ann. PDE, **2** (2016), no. 6.

[23] J. Oliver and J. Sterbenz, *A Vector Field Method for Radiating Black Hole Spacetimes*, `arXiv:1705.10714`

[24] J. Marzuola, J. Metcalfe, D. Tataru and M. Tohaneanu, *Strichartz estimates on Schwarzschild black hole backgrounds*, Comm. Math. Phys., **293** (2010), no. 1, 37–83.

[25] D. Tataru and M. Tohaneanu, *A local energy estimate on Kerr black hole backgrounds*, Int. Math. Res. Not., **2011** (2008), no. 2, 248–292.

Department of Mathematics,
University of California, San Diego (UCSD)
9500 Gilman Drive, Dept 0112
La Jolla, CA 92093-0112 USA
E-mail address: jsterben@math.ucsd.edu

Advanced Studies in Pure Mathematics 85, 2020
The Role of Metrics in the Theory of Partial Differential Equations
pp. 379–389

Remark on global existence of solutions to the 1D compressible Euler equation with time-dependent damping

Yuusuke Sugiyama

Abstract.

In this paper, we consider the 1D compressible Euler equation with the damping coefficient $\lambda/(1+t)^{\mu}$. Under the assumption that $0 \leq \mu < 1$ and $\lambda > 0$ or $\mu = 1$ and $\lambda > 2$, we prove that solutions exist globally in time, if initial data are small C^1 perturbation near constant states. In particular, we remove the conditions on the limit $\lim_{|x|\to\infty}(u(0,x), v(0,x))$, assumed in previous results.

§1. Introduction

In this paper, we consider the following Cauchy problem of the compressible Euler equation with time-dependent damping

$$
\begin{cases} u_t - v_x = 0, \\ v_t + p(u)_x = -\dfrac{\lambda}{(1+t)^{\mu}} v, \\ (u(0,x), v(0,x)) = (1 + \varepsilon\phi(x), \varepsilon\psi(x)). \end{cases} \tag{1}
$$

Here $x \in \mathbb{R}$ is the Lagrangian spatial variable and $t \in \mathbb{R}_+$ is time. $u = u(t,x)$ and $v = v(t,x)$ are the real valued unknown functions, which stand for the specific volume and the fluid velocity. ε is a small positive constant. Throughout this paper, we assume that $\lambda \geq 0$ and $\mu \geq 0$. In the case with $\lambda \equiv 0$, the equations in (1) are the compressible Euler, which is a fundamental model for the compressible inviscid fluids. In the case with $\mu = 0$ and $\lambda > 0$ (constant damping), this system describes

Received November 27, 2018.
Revised May 8, 2019.
2010 *Mathematics Subject Classification.* Primary 35Q31; Secondary 35L40, 35A01.
Key words and phrases. damped p-system, method of characteristics, time-dependet damping.

the flow of fluids in porous media. We assume that the flow is barotropic ideal gases. Namely the pressure p satisfies that

$$p(u) = \frac{u^{-\gamma}}{\gamma} \quad \text{for} \quad \gamma > 1. \tag{2}$$

For initial data, in order to avoid the singularity of p', we assume that there exists constant $\delta_0 > 0$ such that for all $x \in \mathbb{R}$

$$u(0, x) \geq \delta_0. \tag{3}$$

The local existence and the uniqueness theorem for (1) with C_b^1 initial data is proved by Douglis [3] and Hartman and Winter [4] (see also Majda [11] and Courant and Lax [2]) as follows.

Proposition 1. *If $u(0, \cdot), v(0, \cdot) \in C_b^1(\mathbb{R})$ and (3) is assumed for some $\delta_0 > 0$, then (1) has a local and unique solution (u, v) satisfying*

$$(u, v) \in C_b^1([0, T] \times \mathbb{R}) \times C_b^1([0, T] \times \mathbb{R})$$

and

$$u(t, x) \geq \delta_0/2 \quad on \quad [0, T] \times \mathbb{R}$$

for some $T = T(\|u(0, \cdot)\|_{C_b^1(\mathbb{R})}, \|v(0, \cdot)\|_{C_b^1(\mathbb{R})}, \delta_0)$.

In [3, 4], the local existence and uniqueness theorem is proved in a trapezoid domain. Joining the solutions, one can construct the unique solution on $[0, T] \times \mathbb{R}$ as pointed out in Majda [11]. We note that the assumption (3) is satisfied if ε is sufficiently small for fixed $\phi \in C_b^1(\mathbb{R})$. Before recalling known results, we give notations. We set $c = \sqrt{-p'(u)}$ and $\eta = \int_u^\infty c(\xi)\, d\xi = \frac{2}{\gamma - 1} u^{-(\gamma-1)/2}$ and define Riemann invariants as follows:

$$\begin{aligned} r &= v - \eta + \frac{2}{\gamma - 1}, \\ s &= v + \eta - \frac{2}{\gamma - 1}. \end{aligned} \tag{4}$$

For the solution constructed by Proposition 1, we define its lifespan by

$$\begin{aligned} T^* = \sup\Big\{T > 0 \mid \sup_{t \in [0, T)} &\|(u, v)(t)\|_{L^\infty} + \|(u_t, v_t)(t)\|_{L^\infty} \\ &+ \|(u_x, v_x)(t)\|_{L^\infty} + \|p'(u(t))\|_{L^\infty} < \infty\Big\}. \end{aligned} \tag{5}$$

We note that the boundedness of $p'(u)$ means the positivity of u. From the definition of T^*, one may consider that there are several types of the breakdown of solutions such as the divergence of u, v, p' and time or space derivative of the solution. However the divergence of u, v and p' does not occur in the small data regime, since we can show that solutions are small if initial data are small (see Lemma 1). Hence, for the classical solution of (1) with small data, "blow-up" means always the divergence of the time or space derivative of the solution.

In the case with $\lambda \equiv 0$ (no damping case), for more general 2×2 strictly hyperbolic system including the 1D Euler equation, sufficient conditions for the blow-up (the formation singularity) has been studied by many mathematicians (e.g. Lax [10], Zabusky [20] and Klainerman and Majda [9]). Applying these blow-up results, we can show that if r_x or s_x is negative at some point, then the derivative blow-up occurs under suitable assumptions on $r(0,x)$ and $s(0,x)$. Namely there are blow-up solutions even if initial data are small perturbations near constant states. While if $r_x(0,x), s_x(0,x) \geq 0$ and ε is suitably small, the global solution exists (see Remark 2).

In the case with $\mu = 0$ and $\lambda > 0$ (constant damping case), Hsiao and Liu [6] has proved that classical solutions exist globally in time, if initial data are small perturbations near constant states and that small solutions asymptotically behave like that to the following porous media system as $t \to \infty$:

$$\begin{cases} \bar{u}_t = -p(\bar{u})_{xx}, \\ \bar{v} = -p(\bar{u})_x. \end{cases}$$

After Hsiao and Liu's work, many improvements and generalizations of this work have been investigated (see Hsiao and Liu [7], Nishihara [15], Hsiao and Serre [8], Marcati and Nishihara [14] and Mei [12]). We note that, in the above papers for the 1D Euler equation with constant damping, they assume the existence of the limit $\lim_{x\to\pm\infty}(u_0(x), v_0(x))$ and that the convergence rate of the limit is sufficiently fast in order to show the global existence of solutions via L^2 energy estimates.

In [16, 17, 18], Pan has found thresholds of μ and λ separating the existence and the nonexistence of global solution of (1) in small data regime. Namely, in the case with $0 \leq \mu < 1$ and $\lambda > 0$ or $\mu = 1$ and $\lambda > 2$, Pan [17] has proved that solutions of (1) exist globally in time, if initial data are small and compact perturbations of constant states. While, in the case with $\mu > 1$ and $\lambda > 0$ or $\mu = 1$ and $0 \leq \lambda \leq 2$, Pan has proved that solutions of (1) can blow up under some conditions on initial data in [16, 18]. In [1], Cui, Yin, Zhang and Zhu have proved that the global solution of (1) with $0 \leq \mu < 1$ and $\lambda > 0$ asymptotically behaves like that to the corresponding porous media system having a time-dependent coefficient, if $\varepsilon > 0$ is small. In [1, 16, 17, 18], they also assume that the convergence rate of the limit $\lim_{x\to\pm\infty}(u_0(x), v_0(x))$ is sufficiently fast, since their method is based on the L^2 energy method. In [19], the author has shown the global existence with small $\varepsilon > 0$, assuming the existence of the one side limit $\lim_{x\to-\infty}(u_0(x), v_0(x))$. In this paper, we show the global existence without any assumption on the behavior of initial data at spacial infinity, only assuming that initial data are small perturbation near constant states. The difference between the proofs of the main theorem of this paper and [19] is the definition of the Riemann invariant. The Riemann invariant used in [19] can not be small, when (u, v) is close to $(0, 1)$. The proof of the main theorem of this paper is somewhat simpler than that in [19].

Now we state the main theorem of this paper.

Theorem 1. *Let $\gamma > 1$, $(\phi, \psi) \in C_b^1(\mathbb{R})$. Suppose that $0 \leq \mu < 1$ and $\lambda > 0$ or $\mu = 1$ and $\lambda > 2$. There exists a number $\varepsilon_0 > 0$ such that if $0 < \varepsilon \leq \varepsilon_0$, then the Cauchy problem (1) has a global unique solution satisfying the following decay estimate:*

$$\|(r_t(t), s_t(t))\|_{L^\infty} + \|(r_x(t), s_x(t))\|_{L^\infty} \leq C\varepsilon(1+t)^{-\mu}. \tag{6}$$

As we discuss in Remark 2, the time decay rate of L^∞-norm of the time and space derivatives of the global solution of (1) with $\lambda = 0$ is -1. Therefore, one can say that the time decay rate (6) is close to that with no damping case, when the decay in the damping coefficient becomes fast. However, in the case that the convergence rate of the limit $\lim_{x\to\pm\infty}(u_0(x), v_0(x))$ is sufficiently fast, the diffusion phenomena implies that the slower the coefficient in the damping term decays, the faster the solution does. One may expect that if initial data do not converge sufficiently fast to a constant state at spatial infinity, the solution behaves like that of the equation of hyperbolic-type, not parabolic.

Remark 1. *Let $\gamma > 1$ and $(\phi, \psi) \in C_b^1(\mathbb{R})$. Suppose that $\mu > 1$ and $\lambda > 0$ or $\mu = 1$ and $0 \leq \lambda \leq 2$. In the same way as in the proof of Theorem 1, we can shown that there exists a number $\varepsilon_0 > 0$ such that if $0 < \varepsilon \leq \varepsilon_0$, then*

$$T^* \geq C\varepsilon^{-1} \quad \text{for} \quad \mu > 1 \quad \text{and} \quad \lambda \geq 0, \tag{7}$$

$$T^* C\varepsilon^{-\frac{2}{2-\lambda}} \quad \text{for} \quad \mu = 1 \quad \text{and} \quad 0 \leq \lambda < 2, \tag{8}$$

$$T^* \geq e^{\frac{C}{\varepsilon}} \quad \text{for} \quad \mu = 1 \quad \text{and} \quad \lambda = 2. \tag{9}$$

The sharp upper and lower estimates of the lifespan are given in the author's paper [19].

Notations

For $\Omega \subset \mathbb{R}^n$, $C_b^1(\Omega)$ is the set of bounded and continuous functions whose first partial derivatives are also bounded on Ω. The norm of $C_b^1(\Omega)$ is $\|f\|_{C_b^1(\Omega)} = \|f\|_{L^\infty(\Omega)} + \|(\partial_{x_1} f, \ldots, \partial_{x_n} f)\|_{L^\infty(\Omega)}$. When $\Omega = \mathbb{R}$, for abbreviation, we denote $\|\cdot\|_{L^\infty(\Omega)}$ by $\|\cdot\|_{L^\infty}$.

§2. Preliminary

First, we introduce some useful identities for the Riemann invariant, which are based on Lax's formulas in [10]. For $c = \sqrt{-p'(u)}$, the plus and minus characteristic curves are solutions to the following deferential equations:

$$\frac{dx_\pm}{dt}(t) = \pm c(u(t, x_\pm(t))).$$

For the solution (u, v) constructed in Proposition 1, $c(u(t,x))$ is continuous on $[0, T^*) \times \mathbb{R}$. Furthermore, for arbitrarily fixed $T' \in (0, T^*)$, $c(u(t,x))$ is Lipschitz continuous with x for all $t \in [0, T']$, since u_x and $c'(u)$ are uniformly bounded on $\mathbb{R} \times [0, T']$ from the definition of the lifespan T^*. Therefore, if we take initial data $x_\pm(t^*) = x^* \in \mathbb{R}$ with $t^* \in [0, T^*)$, the solution $x_\pm(t)$ uniquely exists on $[0, T^*)$ from a standard theorem for the existence and uniqueness for ordinary differential equations.

We can easily check that Riemann invariants r and s (see (4) for their definitions) are solutions to the following 1st order hyperbolic system

$$\begin{cases} \partial_- r = -\dfrac{\lambda}{2(1+t)^\mu}(r+s), \\ \partial_+ s = -\dfrac{\lambda}{2(1+t)^\mu}(r+s), \end{cases} \tag{10}$$

where $\partial_\pm = \partial_t \pm c\partial_x$. We put $A(t) = \exp(\int_0^t \frac{\lambda}{2(1+\tau)^\mu} d\tau)$. These equations can be written as

$$\begin{cases} \partial_-(A(t)r) = -\dfrac{\lambda A(t)}{2(1+t)^\mu} s, \\ \partial_+(A(t)s) = -\dfrac{\lambda A(t)}{2(1+t)^\mu} r. \end{cases} \tag{11}$$

While, differentiating the equations in (10) with x, from the identity $s_x - r_x = 2\eta_x = -2cu_x$, we have

$$
(12) \qquad \begin{cases} \partial_- r_x = \dfrac{c'}{2c} r_x(s_x - r_x) - \dfrac{\lambda}{2(1+t)^\mu}(r_x + s_x) \\ \partial_+ s_x = \dfrac{c'}{2c} r_x(r_x - s_x) - \dfrac{\lambda}{2(1+t)^\mu}(r_x + s_x). \end{cases}
$$

Multiplying the both side of the equations in (12) by $A(t)\sqrt{c}$, since

$$
\partial_- \sqrt{c} = \frac{c'}{2\sqrt{c}} s_x \quad \text{and} \quad \partial_+ \sqrt{c} = \frac{c'}{2\sqrt{c}} r_x,
$$

we have

$$
(13) \qquad \begin{cases} \partial_- y = -A(t)^{-1} \dfrac{\gamma+1}{4} u^{\frac{\gamma-3}{4}} y^2 - \dfrac{\lambda q}{2(1+t)^\mu}, \\ \partial_+ q = -A(t)^{-1} \dfrac{\gamma+1}{4} u^{\frac{\gamma-3}{4}} q^2 - \dfrac{\lambda y}{2(1+t)^\mu}, \end{cases}
$$

where $y = A(t)\sqrt{c} r_x$ and $q = A(t)\sqrt{c} s_x$. Now we rewrite (13) as integral equalities. We define $\theta_\gamma(u)$ as follows:

$$
\theta_\gamma(u) = \begin{cases} \dfrac{4}{3-\gamma} u^{\frac{3-\gamma}{4}} - \dfrac{4}{3-\gamma} & \text{for} \quad \gamma \neq 3, \\ \log u \quad \text{for} \quad \gamma = 3. \end{cases}
$$

Since it holds that

$$
(14) \qquad \sqrt{c} s_x(t, x_-(t)) = \frac{d}{dt} \theta_\gamma(u(t, x_-(t))),
$$

from integration by parts, we have

$$
\begin{aligned}
(15) \qquad & \int_0^t \frac{q(t, x_-(\tau))}{(1+\tau)^\mu} \, d\tau \\
&= \int_0^t \frac{A(\tau)}{(1+\tau)^\mu} \left(\frac{\mu}{(1+\tau)} - \frac{\lambda}{2(1+\tau)^\mu} \right) \theta_\gamma(\tau, x_-(\tau)) \, d\tau \\
&\quad + \left(\frac{A(t)\theta_\gamma(t, x_-(t))}{(1+t)^\mu} - \theta_\gamma(0, x_-(0)) \right).
\end{aligned}
$$

From the first equation in (13) and (15), y can be written on the minus characteristic curve through (t, x) as follows:

$$
\begin{aligned}
(16)\quad y(t,x) &= y(0,x_-(0)) \\
&\quad - \int_0^t \frac{\lambda A(\tau)}{2(1+\tau)^\mu}\left(\frac{\mu}{(1+\tau)} - \frac{\lambda}{2(1+\tau)^\mu}\right)\theta_\gamma(\tau,x_-(\tau))\,d\tau \\
&\quad - \frac{\lambda}{2}\left(\frac{A(t)\theta_\gamma(t,x_-(t))}{(1+\tau)^\mu} - \theta_\gamma(0,x_-(0))\right) \\
&\quad - \int_0^t A(\tau)^{-1}\frac{\gamma+1}{4}u^{\frac{\gamma-3}{4}}y^2(\tau,x_-(\tau))\,d\tau.
\end{aligned}
$$

In the same way as above, we can obtain the similar identity for q. Next, we introduce key inequalities which control u and v.

Lemma 1. *Let $\gamma > 1$, $\mu \geq 0$ and $\lambda \geq 0$. The following estimate holds for C^1 solution of (1) constructed by Proposition 1*

$$\|r(t)\|_{L^\infty} + \|s(t)\|_{L^\infty} \leq \|r(0)\|_{L^\infty} + \|s(0)\|_{L^\infty}$$

for $t \in [0, T^*)$.

Proof. The proof of this lemma is almost same as in Lemma 7 in [19]. We consider the characteristic curves $x_-(\cdot)$ and $x_+(\cdot)$ through (t, x). Namely, we solve the characteristic equations with initial data $x_\pm(t) = x$. As mentioned above, the characteristic curves exist on $[0, t]$, since it holds that $u, v \in C_b^1([0, T^*) \times \mathbb{R})$ and that $p'(u)$ is bounded on $[0, t] \times \mathbb{R}$ from Proposition 1. From (11), r and s can be written by

$$
\begin{aligned}
A(t)r(t,x) &= r(0,x_-(0)) - \int_0^t \frac{\lambda A(\tau)s(\tau,x_-(\tau))}{2(1+\tau)^\mu}\,d\tau \\
A(t)s(t,x) &= s(0,x_+(0)) - \int_0^t \frac{\lambda A(\tau)r(\tau,x_+(\tau))}{2(1+\tau)^\mu}\,d\tau.
\end{aligned}
$$

We put $\Phi(t) = \|r(t)\|_{L^\infty} + \|s(t)\|_{L^\infty}$. Summing up the above equations and taking L^∞-norm, we have that

$$A(t)\Phi(t) \leq \Phi(0) + \int_0^t \frac{\lambda A(\tau)\Phi(\tau)}{2(1+\tau)^\mu}\,d\tau.$$

By the Gronwall inequality, we have

$$\Phi(t) \leq \Phi(0).$$

Then we have the desired estimate. Q.E.D.

The following fundamental inequality plays an important role in the proof of the decay estimate (6).

Lemma 2. *Suppose that* $0 \le \mu < 1$ *and* $\lambda > 0$ *or* $\mu = 1$ *and* $\lambda > 2$. *Then it holds that*

$$A^{-1}(t)\int_0^t \frac{A(s)}{(1+s)^{2\mu}}\,ds \le C(1+t)^{-\mu}.$$

Proof. In the case that $\mu = 1$, $A(t) = (1+t)^{\lambda/2}$, from which, straightforward computation yields the desired estimate. In the case that $\mu < 1$, since

$$\frac{2(1+t)^\mu}{\lambda}\frac{d}{dt}A(t) = A(t),$$

from the integration by parts, we have that

$$\begin{aligned} A^{-1}(t)\int_0^t \frac{A(s)}{(1+s)^{2\mu}}\,ds &= \frac{2}{\lambda}\left(\frac{1}{(1+t)^\mu} - A^{-1}(t)\right) \\ &\quad + \frac{2\mu}{\lambda}A^{-1}(t)\int_0^t \frac{A(s)}{(1+s)^{\mu+1}}\,ds. \end{aligned}$$

For the third term in the right hand side of the above equality, we see that

$$\begin{aligned} A^{-1}(t)\int_0^t \frac{A(s)}{(1+s)^{\mu+1}}\,ds &= A^{-1}(t)\left(\int_0^{\frac{t}{2}} + \int_{\frac{t}{2}}^t\right)\frac{A(s)}{(1+s)^{\mu+1}}\,ds \\ &\le C\left(A^{-1}(t)A\left(\frac{t}{2}\right) + \frac{1}{(1+t)^\mu}\right) \\ &\le C_1\left(\exp(-C_2(1+t)^{1-\mu}) + \frac{1}{(1+t)^\mu}\right). \end{aligned}$$

Hence we have the desired estimate. Q.E.D.

§3. Proof of Theorem 1

From Lemma 1, we have that

$$\|v(t)\|_{L^\infty} + \|u^{-(\gamma-1)/2}(t) - 1\|_{L^\infty} \le C\varepsilon.$$

Hence we see that

$$(1+C\varepsilon)^{-\frac{2}{\gamma-1}} \le u(t,x) \le (1-C\varepsilon)^{-\frac{2}{\gamma-1}},$$

which implies $\|u(t)-1\|_{L^\infty} \le C\varepsilon$, if $\varepsilon > 0$ is sufficiently small. Applying the above estimates to (16), we have that

$$|y(t,x)| \le C\left(\varepsilon + \int_0^t \frac{\varepsilon A(s)}{(1+s)^{2\mu}}\,ds + \frac{\varepsilon A(t)}{(1+t)^{\mu}} + \int_0^t A^{-1}(s)y^2(t,x_-(s))\,ds\right).$$

Multiplying the both side of the above inequality by $(1+t)^\mu/A(t)$ and putting $Y(T) = \sup_{[0,T]}(1+t)^\mu\|r_x(t)\|_{L^\infty}$, from the definition of y and Lemma 2, we see that

$$Y(T) \le C_1\varepsilon + C_2Y^2(T).$$

Since $Y(0) \le C_3\varepsilon$, we have the uniform estimate $Y(T) \le 2(C_1+C_3)\varepsilon$ for $T \in [0,T^*)$, if ε is sufficiently small. Similarly we can obtain the same uniform estimate of s_x as for r_x, which implies the global existence $(T^* = \infty)$ and the decay estimate

$$\|(r_x,s_x)(t)\|_{L^\infty} \le C\varepsilon(1+t)^{-\mu}.$$

The decay estimate for (r_t,s_t) can be shown by (12) and Lemma 1.

Remark 2. *Here we discuss the global existence and a decay estimate of solutions of* (1) *with* $\lambda = 0$. *In order to consider them, we review some formula of* r_x *and* s_x *in the case that* $\lambda = 0$ (*cf.* [10]). *Now we note that* $A(t) \equiv 1$ *if* $\lambda = 0$. *Solving the differential equation* (13), *we have that*

$$y(t,x_-(t)) = \frac{1}{\dfrac{1}{y(0,x_-(0))} + \displaystyle\int_0^t \frac{\gamma+1}{4}u^{\frac{\gamma-3}{4}}(\tau,x_-(\tau))\,d\tau}.$$

Therefore we have that if $r_x(0,x)$, $s_x(0,x) \ge 0$ *and* ε *is suitably small, then the* (1) *has a global* C_b^1 *solution such that*

$$\|(r_t,s_t)(t)\|_{L^\infty} + \|(s_x,r_x)(t)\|_{L^\infty} \le C(1+t)^{-1}.$$

§ Acknowledgments

The author would like to thank the referee for carefully reading of the manuscript. The research is supported by Grant-in-Aid for Young Scientists Research (B), No. 16K17631.

References

[1] H. Cui, H. Yin, J. Zhang and L. Zhu, Convergence to nonlinear diffusion waves for solutions of Euler equations with time-depending damping, J. Differential Equations, **264** (2018), 4564–4602.

[2] R. Courant and P. D. Lax, On nonlinear partial differential equations with two independent variables, Comm. Pure Appl. Math., **2** (1949), 255–273.

[3] A. Douglis, Some existence theorems for hyperbolic systems of partial differential equations in two independent variables, Comm. Pure Appl. Math., **5** (1952), 119–154.

[4] P. Hartman and A. Winter, Hyperbolic partial differential equations, American Journal of Math., **74** (1952), 834–864.

[5] F. Hou and H. Yin, On the global existence and blowup of smooth solutions to the multi-dimensional compressible Euler equations with time-depending damping, Nonlinearity, **30** (2017), 2485–2517.

[6] L. Hsiao and T. P. Liu, Convergence to nonlinear diffusion waves for solutions of a system of hyperbolic conservation laws with damping, Comm. Math. Phys., **143** (1992), 599–605.

[7] L. Hsiao and T. P. Liu, Nonlinear diffusive phenomena of nonlinear hyperbolic systems, Chinese Ann. Math. Ser. B, **14** (1993), 465–480.

[8] L. Hsiao and D. Serre, Global existence of solutions for the system of compressible adiabatic flow through porous media, SIAM J. Math. Anal., **27** (1996), 70–77.

[9] S. Klainerman and A. Majda, Formation of singularities for wave equations including the nonlinear vibrating string, Comm. Pure Appl. Math., **33** (1980), 241–263.

[10] P. D. Lax, Development of Singularities of solutions of nonlinear hyperbolic partial differential equations, J. Math. Phys., **5** (1964), 611–613.

[11] A. Majda, Compressible fluid flow and systems of conservation laws in several space variables, Springer, Appl. Math. Sci., 1984.

[12] M. Mei, Best asymptotic profile for hyperbolic p-system with damping, SIAM J. Math. Anal., **42** (2010), 1–23.

[13] R. C. MacCamy, V. J. Mizel, Existence and nonexistence in the large of solutions of quasilinear wave equations, Arch. Ration. Mech. Anal., **25** (1967), 299–320.

[14] P. Marcati and K. Nishihara, The $L^p - L^q$ estimates of solutions to one-dimensional damped wave equations and their application to the compressible flow through porous media, J. Differential Equations, **191** (2003), 445–469.

[15] K. Nishihara, Asymptotic behavior of solutions of quasilinear hyperbolic equations with linear damping, J. Differential Equations, **137** (1997), 384–395.

[16] X. Pan, Remarks on 1-D Euler equations with time-decayed damping, arXiv:1510.08115.

[17] X. Pan, Global existence of solutions to 1-d Euler equations with time-dependent damping, Nonlinear Anal., **132** (2016), 327–336.

[18] X. Pan, Blow up of solutions to 1-d Euler equations with time-dependent damping, J. Math. Anal. Appl., **442** (2016), no. 2, 435–445.
[19] Y. Sugiyama, Singularity formation for the 1D compressible Euler equation with valuable damping coefficient, Nonlinear Anal., **170** (2018), 615–628.
[20] N. J. Zabusky, Exact solution for the vibrations of a nonlinear continuous model string, J. Math. Phys., **3** (1962), 1028–1039.
[21] H. Zheng, Singularity formation for the compressible Euler equations with general pressure law, J. Math. Anal. Appl., **438** (2016), 59–72.

School of Engineering, The University of Shiga Prefecture
2500, Hassaka-cho, Hikone-City, Shiga 522-8533 Japan
E-mail address: sugiyama.y@e.usp.ac.jp

Advanced Studies in Pure Mathematics 85, 2020
The Role of Metrics in the Theory of Partial Differential Equations
pp. 391–405

Short time blow-up by negative mass term for semilinear wave equations with small data and scattering damping

Ning-An Lai, Nico Michele Schiavone and Hiroyuki Takamura

Abstract.

In this paper we study blow-up and lifespan estimate for solutions to the Cauchy problem with small data for semilinear wave equations with scattering damping and negative mass term. We show that the negative mass term will play a dominant role when the decay of its coefficients is not so fast, thus the solutions will blow up in a finite time. What is more, we establish a lifespan estimate from above which is much shorter than the usual one.

§1. Introduction

We consider the Cauchy problem with small data for the semilinear wave equations with scattering damping and negative mass term

$$(1) \quad \begin{cases} u_{tt} - \Delta u + \dfrac{\mu_1}{(1+t)^\beta} u_t - \dfrac{\mu_2}{(1+t)^{\alpha+1}} u = |u|^p, & \text{in } \mathbf{R}^n \times (0,T), \\ u(x,0) = \varepsilon f(x), \quad u_t(x,0) = \varepsilon g(x), & x \in \mathbf{R}^n, \end{cases}$$

where $\mu_1 \geq 0$, $\mu_2 > 0$, $\alpha < 1$, $\beta > 1$, $p > 1$, $n \in \mathbf{N}$, $T > 0$ and $\varepsilon > 0$ is a "small" parameter. This problem comes from the recent interest in the "wave-like" or "heat-like" behaviour of semilinear wave equations with variable coefficients damping. The Cauchy problems with small data for

$$u_{tt} - \Delta u = |u|^p \quad \text{and} \quad u_t - \Delta u = |u|^p$$

Received November 28, 2018.
Revised May 21, 2019.
2010 *Mathematics Subject Classification.* primary 35L71; secondary 35B44.
Key words and phrases. wave equation, semilinear, damping, mass, blow-up, lifespan.

admit critical powers, respectively, the so-called Strauss exponent $p_S(n)$ and the Fujita exponent $p_F(n)$ (see [13] and [3]), where for "critical power" of a problem we mean the exponent p_c such that its small data solutions blow up for $1 < p \le p_c$ and exist globally in time for $p > p_c$. It is of recent interest the Cauchy problem with small data for

$$u_{tt} - \Delta u + \frac{\mu}{(1+t)^\beta} u_t = |u|^p. \tag{2}$$

If the Cauchy problem (2) admits a critical power related to $p_S(n)$, then we say it has a "wave-like" behaviour, while if it is related to $p_F(n)$, then we say it admits a "heat-like" behaviour. Generally speaking, if the decay rate β of the damping coefficients is large enough, then the damping term seems to have no influence and then we get a "wave-like" behaviour; otherwise, we get a "heat-like" behaviour. According to the different value of β, we recover four cases (overdamping, effective, scaling invariant, scattering), based on the works by Wirth [20–22] (see also [2, 4–6, 8, 9, 16–19] and references therein).

On the other hand, people are paying more attention to the Cauchy problem for

$$u_{tt} - \Delta u + \frac{\mu_1}{1+t} u_t + \frac{\mu_2^2}{(1+t)^2} u = |u|^p,$$

which includes scale-invariant damping and mass in the mean time. In some sense, this model describes the interplay between the damping and mass. For this problem, the quantity

$$\delta := (\mu_1 - 1)^2 - 4\mu_2^2$$

plays an important role to the behaviour of the solutions. We refer the reader to [10–12] and references therein.

Naturally, we want to consider the corresponding problem with scattering damping and mass term. Very recently, the authors [7] studied the Cauchy problem (1) with fast decay rate in the coefficients of the mass term, thus, $\alpha > 1$, in which we proved that the blow-up results and the upper bound lifespan estimates are the same as that of the semilinear wave equations with scattering damping but without mass term, see [9]. This implies that the negative mass term seems to have no influence on the behaviour of the solutions. In this work, we are devoted to studying the case $\alpha < 1$. Our motivation to study a negative mass term is simply a mathematical interest by [7], but one may refer to the introduction of Yagdjian and Galstian [23] which mentions its physical background. From our main result listed in Theorem 1 below, it seems that the negative mass term will affect the qualitative properties of the small data solutions of the Cauchy problem (1), due to two reasons: firstly, the non-existence of global energy solutions can be established for all $p > 1$ and $n \geq 1$; moreover, the upper bound of the lifespan is smaller than the usual one and it looks like a log-type with respect to ε.

§2. Definitions and Main Result

First of all, let us introduce energy solutions of Cauchy problem (1).

Definition 2.1. *We say that u is an energy solution for problem* (1) *over $\mathbf{R}^n \times [0,T)$ if*

$$u \in C\big([0,T), H^1(\mathbf{R}^n)\big) \cap C^1\big([0,T), L^2(\mathbf{R}^n)\big) \cap L^p_{loc}\big(\mathbf{R}^n \times (0,T)\big)$$

satisfies $u(x,0) = \varepsilon f(x)$ in $H^1(\mathbf{R}^n)$, $u_t(x,0) = \varepsilon g(x)$ in $L^2(\mathbf{R}^n)$ and

$$\begin{aligned}(3)\qquad &\int_{\mathbf{R}^n} u_t(x,t)\phi(x,t)\,dx - \int_{\mathbf{R}^n} \varepsilon g(x)\phi(x,0)\,dx \\ &+ \int_0^t ds \int_{\mathbf{R}^n} \{-u_t(x,s)\phi_t(x,s) + \nabla u(x,s)\cdot\nabla\phi(x,s)\}\,dx \\ &+ \int_0^t ds \int_{\mathbf{R}^n} \left\{\frac{\mu_1}{(1+s)^\beta}u_t(x,s) - \frac{\mu_2}{(1+s)^{\alpha+1}}u(x,s)\right\}\phi(x,s)\,dx \\ =& \int_0^t ds \int_{\mathbf{R}^n} |u(x,s)|^p\phi(x,s)\,dx\end{aligned}$$

for all test functions $\phi \in C_0^\infty(\mathbf{R}^n \times [0,T))$ and for all $t \in [0,T)$.

Theorem 1. *Let $\alpha < 1$, $n \geq 1$ and $p > 1$. Assume that both $f \in H^1(\mathbf{R}^n)$ and $g \in L^2(\mathbf{R}^n)$ are non-negative, at least one of them does not vanish identically. Suppose that u is an energy solution of* (1) *on $\mathbf{R}^n \times [0,T)$ that satisfies, for some $R \geq 1$,*

$$(4)\qquad \operatorname{supp} u \subset \big\{(x,t) \in \mathbf{R}^n \times [0,T) \colon |x| \leq t + R\big\}.$$

Then, there exists a constant $\varepsilon_0 = \varepsilon_0(f,g,R,n,p,\mu_1,\mu_2,\alpha,\beta) > 0$ such that T has to satisfy

$$T \leq 3\zeta(C\varepsilon)$$

for $0 < \varepsilon \leq \varepsilon_0$, where $\zeta = \zeta(\bar{\varepsilon})$ is the larger solution to the equation

$$(5)\qquad \bar{\varepsilon}\zeta^{\frac{2}{p-1}-n+\frac{1+\alpha}{4}} \exp\left(\frac{2}{1-\alpha}\sqrt{\mu_2 \exp\left(\frac{\mu_1}{1-\beta}\right)}\zeta^{\frac{1-\alpha}{2}}\right) = 1$$

and C is a positive constant independent of ε.

Remark 2.1. *Let us make some observations:*

- *The assumption* (4) *can be replaced by* $\mathrm{supp}\{f,g\} \subset \{x \in \mathbf{R}^n\colon |x| \le R\}$ *when* $n = 1,2$, *or* $p \le n/(n-2)$ *for* $n \ge 3$. *This fact is established by local existence of such an energy solution. See Appendix in the last section.*
- *Since letting* $\varepsilon \to 0$ *we have* $\zeta \to +\infty$ *in* (5), *it is not difficult to see that* $T \le c[\log(1/\varepsilon)]^{2/(1-\alpha)}$ *for some constant* $c > 0$ *follows from* (5). *In fact, this inequality is trivial when the exponent* $\delta := 2/(p-1) - n + (1+\alpha)/4$ *of the first* ζ *is non-negative, while* ζ^δ *can be absorbed by square root of the exponential term when* $\delta < 0$.
- *It is an open question the optimality of the upper bound of the lifespan in Theorem* 1.

§3. Kato's type lemma

In order to prove our theorem, we need a slightly different version of the improved Kato's lemma introduced in [15].

Lemma 3.1. *Let* $p > 1$ *and* $0 \le \widetilde{T}_0 < T$ *be positive constants. Suppose that* $A \in C^1([\widetilde{T}_0, T))$, $B \in C^1([0,T))$, $m \in C^1([0,T))$ *are strictly positive functions,* $B = B(t)$ *is decreasing and that* $m(t)$ *is bounded by two constants* $\overline{m}, \underline{m} > 0$ ($\underline{m} \le m(t) \le \overline{m}$) *for* $t \ge 0$. *Define the function*

$$h(t) := B(t)^{1/2} A(t)^{(p-1)/2-\delta}, \tag{6}$$

where δ *is a constant such that* $0 < \delta < (p-1)/2$ *and* $h'(t) \ge 0$ *for* $t \ge \widetilde{T}_0$.

Assume that $F \in C^2([0,T))$ *satisfies*

$$F(t) \ge A(t) \quad \text{for } t \ge \widetilde{T}_0, \tag{7}$$

$$\{m(t)F'(t)\}' \ge B(t)|F(t)|^p \quad \text{for } t \ge 0, \tag{8}$$

$$F(0), F'(0) \ge 0, \quad F(0) + F'(0) > 0. \tag{9}$$

If $F'(0) = 0$, *suppose that there exists a time* $\widetilde{t} > 0$ *such that*

$$F(\widetilde{t}) \ge 2F(0). \tag{10}$$

Define the time

$$\widetilde{T}_1 := \begin{cases} \overline{m}\,\underline{m}^{-1} F(0)/F'(0) & \text{if } F'(0) \ne 0, \\ \widetilde{t} & \text{if } F'(0) = 0. \end{cases}$$

Then, for $\widetilde{T} \geq \max\{\widetilde{T}_0, \widetilde{T}_1\}$ we have $T \leq 3\widetilde{T}$ assuming that

$$\widetilde{T}\, h(\widetilde{T}) A(\widetilde{T})^{\delta} \geq \delta^{-1}\, \overline{m} \sqrt{(p+1)/\underline{m}}\,. \tag{11}$$

Proof. First of all, let us prove that $F(t), F'(t) > 0$ for $t > 0$. We need to consider two cases according to the initial conditions (9) on F.

Case 1: $F'(0) > 0$. From (8) it follows $F'(t) \geq m(0)F'(0)m(t)^{-1} > 0$, and then $F(t) \geq F(0) + m(0)F'(0) \int_0^t m(s)^{-1} ds > 0$ for $t > 0$.

Case 2: $F'(0) = 0$. Then $F(0) > 0$. It follows from (8) evaluated in $t = 0$ that $\{mF'\}'(0) > 0$, which implies $m(t)F'(t) > m(0)F'(0) = 0$ for small $t > 0$. Hence, the fact that $\{mF'\}'(t) \geq 0$ for $t \geq 0$ from (8) yields that $m(t)F'(t) > 0$, that is $F'(t) > 0$, and so $F(t) > F(0) > 0$ for $t > 0$.

Moreover, observe that

$$F(t) \geq 2F(0) \quad \text{for } t \geq \widetilde{T}_1. \tag{12}$$

Indeed, if $F'(0) = 0$, it follows by the hypothesis (10) and by the fact that F is increasing. If $F'(0) > 0$, by (8) and because m is bounded, we have $F(t) \geq F(0) + \underline{m}\overline{m}^{-1} F'(0)t$, from which (12) follows.

Multiplying (8) by $m(t)F'(t) > 0$, we get

$$\left(\frac{1}{2}\{m(t)F'(t)\}^2\right)' \geq m(t)B(t)|F(t)|^p F'(t) \quad \text{for } t > 0.$$

From this inequality, the positivity of F and the facts that B is decreasing and m is bounded, it follows that

$$\begin{aligned} \frac{1}{2}F'(t)^2 &\geq \frac{1}{2}\overline{m}^{-2}\, \underline{m}^2 F'(0)^2 + \overline{m}^{-2}\, \underline{m} B(t) \int_0^t F(s)^p F'(s)\, ds \\ &\geq \frac{\overline{m}^{-2}\underline{m}}{p+1} B(t) F(t)^p \{F(t) - F(0)\} \quad \text{for } t \geq 0, \end{aligned}$$

and so, using equation (12), we get

$$F'(t) \geq \overline{m}^{-1} \sqrt{\underline{m}/(p+1)}\, B(t)^{1/2} F(t)^{(p+1)/2} \quad \text{for } t \geq \widetilde{T}_1. \tag{13}$$

Now, fix $\widetilde{T} := \max\{\widetilde{T}_0, \widetilde{T}_1\}$ and define the function

$$H(t) := \int_{\widetilde{T}}^t h(s)\, ds = \int_{\widetilde{T}}^t B(s)^{1/2} A(s)^{(p-1)/2-\delta}\, ds \quad \text{for } t \geq \widetilde{T}.$$

Because $0 < \delta < (p-1)/2$, from inequality (13) and from (7) we obtain

$$F'(t)/F(t)^{1+\delta} \geq \overline{m}^{-1} \sqrt{\underline{m}/(p+1)}\, B(t)^{1/2} A(t)^{(p-1)/2-\delta} \quad \text{for } t \geq \widetilde{T}.$$

Integrating this inequality on $[2\widetilde{T}, t]$ we get

$$\frac{1}{\delta}\left(\frac{1}{F(2\widetilde{T})^\delta} - \frac{1}{F(t)^\delta}\right) \geq \frac{1}{\overline{m}}\sqrt{\frac{\underline{m}}{p+1}}\int_{2\widetilde{T}}^t B(s)^{1/2}A(s)^{(p-1)/2-\delta}\,ds.$$

Neglecting the second term on the left-hand side, from (7) evaluated in $t = 2\widetilde{T}$ and recalling the definition of H, we have

$$A(2\widetilde{T})^{-\delta} \geq F(2\widetilde{T})^{-\delta} \geq \delta\,\overline{m}^{-1}\sqrt{\underline{m}/(p+1)}\,[H(t) - H(2\widetilde{T})] \tag{14}$$

for $t \geq 2\widetilde{T}$. Since h is increasing, we get

$$H(2\widetilde{T}) \geq h(\widetilde{T})\int_{\widetilde{T}}^{2\widetilde{T}} ds = \widetilde{T}h(\widetilde{T}). \tag{15}$$

Observe moreover that A is increasing, in fact

$$h'(t) = h(t)\big\{B'(t)/(2B(t)) + [(p-1)/2 - \delta]A'(t)/A(t)\big\},$$

and so, because $h, A, B > 0$, $h' \geq 0$ and $B' \leq 0$, we get $A' \geq 0$. Then, by equation (15), hypothesis (11) and the monotonicity of A, we have

$$A(2\widetilde{T})^\delta H(2\widetilde{T}) \geq A(\widetilde{T})^\delta\,\widetilde{T}\,h(\widetilde{T}) \geq \delta^{-1}\,\overline{m}\sqrt{(p+1)/\underline{m}}.$$

Inserting this inequality in (14) we obtain $2H(2\widetilde{T}) \geq H(t)$. Since $H''(t) = h'(t) \geq 0$ we have also, integrating two times on $[2\widetilde{T}, t]$, that $H(t) \geq H(2\widetilde{T}) + H'(2\widetilde{T})\{t - 2\widetilde{T}\}$. These two relations give us the estimate

$$t \leq 2\widetilde{T} + H(2\widetilde{T})/H'(2\widetilde{T}) \quad \text{for } t > 2\widetilde{T}.$$

Finally, observe that

$$v\,(H(t)/H'(t))' = 1 - H(t)H''(t)/(H'(t))^2 \leq 1$$

from which, integrating on $[\widetilde{T}, 2\widetilde{T}]$, we get

$$H(2\widetilde{T})/H'(2\widetilde{T}) \leq H(\widetilde{T})/H'(\widetilde{T}) + 2\widetilde{T} - \widetilde{T} = \widetilde{T},$$

and so we have $t \leq 3\widetilde{T}$. Therefore the proof of the lemma is completed.

§4. Proof of Theorem 1

Following the idea in [7] and [9], we introduce the multiplier

$$m(t) := \exp\left(\mu_1 \frac{(1+t)^{1-\beta}}{1-\beta}\right).$$

Clearly, $1 \geq m(t) \geq m(0) > 0$ for $t \geq 0$. Let us define the functional

$$F_0(t) := \int_{\mathbf{R}^n} u(x,t)\,dx,$$

and then $F_0(0) = \varepsilon \int_{\mathbf{R}^n} f(x)\,dx$, $F_0'(0) = \varepsilon \int_{\mathbf{R}^n} g(x)\,dx$ are non-negative and do not both equal to zero, due to the hypothesis for the initial data.

Taking into account of (4), we choose the test function $\phi = \phi(x,s)$ in the definition of energy solution (3) such that it satisfies $\phi \equiv 1$ in $\{(x,s) \in \mathbf{R}^n \times [0,t] : |x| \leq s+R\}$, to get

$$\begin{aligned}
&\int_{\mathbf{R}^n} u_t(x,t)dx - \int_{\mathbf{R}^n} u_t(x,0)\,dx + \int_0^t ds \int_{\mathbf{R}^n} \frac{\mu_1}{(1+s)^\beta} u_t(x,s)\,dx \\
&= \int_0^t \int_{\mathbf{R}^n} \frac{\mu_2}{(1+s)^{\alpha+1}} u(x,s)\,dx + \int_0^t ds \int_{\mathbf{R}^n} |u(x,s)|^p\,dx,
\end{aligned}$$

which yields, by taking derivative with respect to t,

$$F_0''(t) + \frac{\mu_1}{(1+t)^\beta} F_0'(t) = \frac{\mu_2}{(1+t)^{\alpha+1}} F_0(t) + \int_{\mathbf{R}^n} |u(x,t)|^p\,dx. \tag{16}$$

Multiplying both sides of (16) with $m(t)$ yields

$$\{m(t)F_0'(t)\}' = m(t)\frac{\mu_2}{(1+t)^{\alpha+1}} F_0(t) + m(t)\int_{\mathbf{R}^n} |u(x,t)|^p\,dx. \tag{17}$$

Integrating the previous equation twice on $[0,t]$, we obtain

$$\begin{aligned}
F_0(t) \geq\ & F_0(0) + F_0'(0)\int_0^t \frac{ds}{m(s)} + \mu_2 \int_0^t \frac{ds}{m(s)} \int_0^s \frac{m(\sigma)F_0(\sigma)}{(1+\sigma)^{\alpha+1}}\,d\sigma \\
&+ \int_0^t \frac{ds}{m(s)} \int_0^s m(\sigma)\,d\sigma \int_{\mathbf{R}^n} |u(x,\sigma)|^p\,dx \quad \text{for } t \geq 0.
\end{aligned} \tag{18}$$

By a comparison argument, we observe $F_0(t) > 0$ for $t > 0$, and consequently also $F_0'(t) > 0$ for $t > 0$ by an integration of (17). In fact, if $F_0(0) > 0$, then F_0 is strictly positive for at least small times. Supposing that t_0 is the smallest zero time of F_0, calculating (18) in t_0 we get a contradiction. If $F_0(0) = 0$, then $F_0'(0) > 0$ and so $F_0'(t) > 0$

for at least small time; due to the fact that F_0 is strictly increasing we then conclude that it is positive for at least small time $t > 0$. Supposing that $t_0 > 0$ is the smallest zero point of F_0, calculating (18) in t_0 we get again a contradiction.

Moreover observe that if $F_0'(0) = 0$, neglecting the last term on the right-hand side of (18), and noting that F is increasing and m is bounded, we have

$$F_0(t) \geq F_0(0) + 2^{-1}m(0)\mu_2 F_0(0)(1+t)^{-\max\{0,\alpha+1\}}t^2$$

and so $F_0(\widetilde{t}) \geq 2F_0(0)$, if we choose $\widetilde{t} = \widetilde{t}(\mu_1, \mu_2, \alpha, \beta) > 0$ such that

$$2^{-1}m(0)\mu_2(1+\widetilde{t})^{-\max\{0,\alpha+1\}}\,\widetilde{t}^2 = 1.$$

Now we need estimates for $\{mF_0'\}'$ and F_0. Neglecting the first term on the right-hand side of (17) and applying Hölder's inequality for the last term, there exists $C_1 = C_1(n,p,R) > 0$ such that, for $t \geq 0$,

$$\{m(t)F_0'(t)\}' \geq m(0)C_1(1+t)^{-n(p-1)}|F_0(t)|^p =: B(t)|F_0(t)|^p. \tag{19}$$

Fix $t_0 > 0$ to be chosen later and consider the auxiliary function

$$J(t) = \varepsilon J_0 + \varepsilon J_1(t-t_0) + m(0)\mu_2 \int_{t_0}^t ds \int_{t_0}^s \frac{J(\sigma)}{(1+\sigma)^{\alpha+1}}\,d\sigma \quad \text{for } t \geq t_0,$$

where $J_0 := \frac{1}{2}\|f\|_{L^1(\mathbf{R}^n)}$, $J_1 := \frac{1}{2}m(0)\,\|g\|_{L^1(\mathbf{R}^n)}$. By the similar way as above, we get by comparison argument that $F_0(t) \geq J(t)$ for $t \geq t_0$. Setting for the simplicity $c := m(0)\mu_2$, $q := 1-\alpha$, the function J satisfies

$$J''(t) = c(1+t)^{q-2}J(t) \quad \text{for } t \geq t_0,$$

with $J(t_0) = \varepsilon J_0$, $J'(t_0) = \varepsilon J_1$. One can check that the solution of this ordinary differential equation is

$$J(t) = \varepsilon c_+ J_+(t) + c_- J_-(t)$$

where, setting $B^+_{1/q} := I_{1/q}$ and $B^-_{1/q} := K_{1/q}$ the modified Bessel functions respectively of the first and second kind with order $1/q$, we have

$$J_\pm(t) := (1+t)^{1/2} B^\pm_{1/q}\left(2\frac{\sqrt{c}}{|q|}(1+t)^{q/2}\right),$$

$$\begin{aligned} c_\pm := \pm\frac{2}{q}(1+t_0)^{-1/2}\left[(1+t_0)J_1 - J_0\right] B^\mp_{1/q}\left(\frac{2\sqrt{c}}{|q|}(1+t_0)^{q/2}\right) \\ + J_0\frac{2\sqrt{c}}{|q|}(1+t_0)^{(q-1)/2} B^\mp_{1+1/q}\left(\frac{2\sqrt{c}}{|q|}(1+t_0)^{q/2}\right). \end{aligned}$$

Observe that $c_+ > 0$ at least for $t_0 > 0$ (independent of ε) large enough. From the asymptotic expansions of the modified Bessel functions (see Section 9.7 in [1]), when $t > 0$ is large we have that

$$J_\pm(t) = \frac{\sqrt{\pi^{\mp 1} q}}{2c^{1/4}}(1+t)^{\frac{1}{2}-\frac{q}{4}} \exp\left(\pm 2\frac{\sqrt{c}}{q}(1+t)^{\frac{q}{2}}\right)\left[1 + O\left((1+t)^{-\frac{q}{2}}\right)\right].$$

Consequently, we can find constants $C_2, T_1 > 0$ independent of ε, such that, for every $t \geq T_1$, the following estimate holds:

$$F_0(t) \geq \varepsilon C_2 (1+t)^{\frac{1+\alpha}{4}} \exp\left(2\frac{\sqrt{m(0)\mu_2}}{1-\alpha}(1+t)^{\frac{1-\alpha}{2}}\right) =: A(t). \tag{20}$$

Thanks to estimates (19) and (20), we can apply Lemma 3.1. Fix $\delta := (p-1)/4$ and, using the definition (6) of h, observe that

$$\begin{aligned} h'(t) &= B^{1/2}(t) A^{(p-1)/2-\delta}(t)(1+t)^{-1} \\ &\quad \times \left\{-\frac{n(p-1)}{2} + \left(\frac{p-1}{2} - \delta\right)\left[\frac{1+\alpha}{4} + \sqrt{m(0)\mu_2}(1+t)^{\frac{1-\alpha}{2}}\right]\right\}, \end{aligned}$$

so we can find a time $T_2 = T_2(n, p, \mu_1, \mu_2, \alpha, \beta) \geq 0$ such that $h'(t) > 0$ for $t \geq T_2$. Then we can choose $\widetilde{T}_0 = \max\{T_1, T_2\}$.

Let us set $\widetilde{T} \equiv \zeta - 1$ and $C = C_3[\delta m(0)\sqrt{C_1}/(2\sqrt{p+1})]^{2/(p-1)}$, where $\zeta \equiv \zeta(\overline{\varepsilon})$ is the larger solution to (5) with $\overline{\varepsilon} = C\varepsilon$. There exists $\varepsilon_0 > 0$ such that, for $0 < \varepsilon \leq \varepsilon_0$, we have $\widetilde{T} \equiv \zeta - 1 \geq \max\{\widetilde{T}_0, \widetilde{T}_1\}$, where $\widetilde{T}_1$, independent of ε, is defined as in the statement of the Lemma. We can also suppose $\widetilde{T} \geq 1$, so that $\zeta - 1 \geq \zeta/2$. Therefore, one can check that (11) holds and so the maximal existence time T of F_0 satisfies $T \leq 3\widetilde{T} \leq 3\zeta$. The proof of the Theorem 1 is completed.

§5. Appendix

In this section we are going to show the local existence and finite speed of propagation property for energy solution to our problem, as stated in the second point of Remark 2.1. In the following, the positive constant C may vary from line to line. We assume that $p \leq n/(n-2)$ when $n \geq 3$.

Let us denote the function space

$$\begin{aligned} B_{T,K} := \{&\phi \in C\left([0,T), H^1(\mathbf{R}^n)\right) \cap C^1\left([0,T), L^2(\mathbf{R}^n)\right) : \\ &\operatorname{supp}\phi \subset \{(x,t) \in \mathbf{R}^n \times [0,T) \colon |x| \leq t + R\},\ \|\phi\|_{B_{T,K}} \leq K\}, \end{aligned}$$

where T, R, K are fixed positive constants and

$$\|\phi\|_{B_{T,K}} := \sup_{t\in[0,T)} E_\phi^{1/2}(t), \quad E_\phi(t) := \frac{1}{2}\int_{\mathbf{R}^n} (\phi_t^2 + |\nabla\phi|^2)\, dx.$$

It can be proved that $B_{T,K}$ is a Banach space.

Consider the following Cauchy problem for $v \in B_{T,K}$

$$(21)\quad \begin{cases} u_{tt} - \Delta u = |v|^p + m^2(t)v - b(t)v_t =: F_v(x,t), & \text{in } \mathbf{R}^n \times (0,T), \\ u(x,0) = \varepsilon f(x), \quad u_t(x,0) = \varepsilon g(x), & x \in \mathbf{R}^n, \end{cases}$$

where we set for the simplicity

$$m^2(t) = \frac{\mu_2}{(1+t)^{\alpha+1}}, \quad b(t) = \frac{\mu_1}{(1+t)^\beta}.$$

However, all the calculations below are trivially generalized for any $m^2, b \in C([0,T))$.

We want to show that the map

$$M\colon v \mapsto u = Mv, \ v \in B_{T,K},$$

is a contraction. Note that, for $v \in B_{T,K}$, by Gagliardo-Nirenberg inequality and Poincaré inequality, we have

$$\|v\|_{L^{2p}(\mathbf{R}^n)} \le C\|v\|_{L^2(\mathbf{R}^n)}^{1-\theta(2p)} \|\nabla v\|_{L^2(\mathbf{R}^n)}^{\theta(2p)}, \quad \theta(2p) := n\left(\frac{1}{2} - \frac{1}{2p}\right),$$

for $p \le n/(n-2)$ when $n \ge 3$, and

$$\|v\|_{L^2(\mathbf{R}^n)} \le C(t+R)\|\nabla v\|_{L^2(\mathbf{R}^n)},$$

which imply

$$(22)\quad \|v\|_{L^{2p}(\mathbf{R}^n)} \le C(t+R)^{1-\theta(2p)}\|\nabla v\|_{L^2(\mathbf{R}^n)} \le C(t+R)^{1-\theta(2p)} E_v^{1/2}.$$

In particular, for fixed $T > 0$, we can check that

$$F_v(x,t) \in L^2\left(\mathbf{R}^n \times [0,T)\right).$$

Let us start proving that the map M is onto. Firstly, we show the finite speed propagation of the energy solution, i.e.

$$\operatorname{supp} u \subset \left\{(x,t) \in \mathbf{R}^n \times [0,T) \colon |x| \le t + R\right\},$$

by using the density argument similarly to [14]. By density of $C_0^\infty(\mathbf{R}^n)$ in $L^2(\mathbf{R}^n)$, we can approximate the energy data f, g by sequences of smooth and compactly supported functions $\{f_m\}_{m\in\mathbf{N}}, \{g_m\}_{m\in\mathbf{N}}$ in the energy norm $H^1(\mathbf{R}^n)$ and $L^2(\mathbf{R}^n)$ respectively. Noting that $F_v(x,t)$ has compact support, we can find also a sequence of smooth and compactly supported functions $\{F_{v,m}\}_{m\in\mathbf{N}}$ converging to F_v in the norm $L^2(\mathbf{R}^n\times[0,T))$. Let u_m be the smooth solution of the problem

$$\begin{cases} (u_m)_{tt}-\Delta u_m = F_{v,m}(x,t) & \text{in } \mathbf{R}^n\times(0,T) \\ u(x,0)=\varepsilon f_m(x), \quad u_t(x,0)=\varepsilon g_m(x) & \text{in } \mathbf{R}^n. \end{cases} \tag{23}$$

Fix $(x_0,t_0)\in\mathbf{R}^n\times(0,T)$ with $|x_0|\geq t_0+R$ and set

$$C_{(x_0,t_0)} := \{(x,t)\in\mathbf{R}^n\times[0,T) \colon |x-x_0|\leq t_0-t\},$$

the backward cone with vertex in (x_0,t_0). Then, denote the energy on a time-section of the cone as

$$E_{t_0-t}(t,u(t)) := \frac{1}{2}\int_{B_{t_0-t}(x_0)} (u_t^2+|\nabla u|^2)\,dx, \tag{24}$$

where $B_r(x_0):=\{x\in\mathbf{R}^n : |x-x_0|\leq r\}$. The standard space-time divergence form yields a local energy inequality

$$E^{1/2}_{t_0-t}(t,u_m(t)) \leq E^{1/2}_{t_0}(0,u_m(0)) + \int_0^t \|F_{v,m}(\cdot,s)\|_{L^2(B_{t_0-s}(x_0))}\,ds.$$

Applying the previous inequality to the difference u_m-u_n of two solutions of (23), we have that $\{u_m(\cdot,t)\}_{m\in\mathbf{N}}$ is a Cauchy sequence in the norm (24) uniformly in $t\in[0,t_0]$. Hence the limit u is an energy solution satisfying

$$E^{1/2}_{t_0-t}(t,u(t)) \leq E^{1/2}_{t_0}(0,u(0)) + \int_0^t \|F_v(\cdot,s)\|_{L^2(B_{t_0-t}(x_0))}\,ds,$$

which gives us the fact that

$$f(x)\equiv g(x)\equiv 0 \quad \text{in } C_{x_0,t_0}\cap\{t=0\} \quad \text{and} \quad F_v\equiv 0 \text{ in } C_{(x_0,t_0)}$$

and Poincaré inequality imply

$$u\equiv 0 \text{ in } C_{(x_0,t_0)}.$$

Next, we show that $\|Mv\|_{B_{T,K}}\leq K$. It is easy to obtain

$$\frac{\partial}{\partial t}\left(\frac{1}{2}(u_t^2+|\nabla u|^2)\right) = \operatorname{div}(u_t\nabla u)+|v|^p u_t + m^2(t)vu_t - b(t)v_tu_t. \tag{25}$$

Exploiting (22) we get the estimate

$$\begin{aligned}\int_{\mathbf{R}^n} |v|^p |u_t|\, dx &\le \left(\int_{\mathbf{R}^n} |v|^{2p}\, dx\right)^{1/2} \sqrt{2} E_u^{1/2}(t) \\ &\le C(t+R)^{p\{1-\theta(2p)\}} E_v^{p/2}(t) E_u^{1/2}(t).\end{aligned}$$

Moreover, it is trivial that

$$\int_{\mathbf{R}^n} |v||u_t|\, dx \le C(t+R) E_v^{1/2}(t) E_u^{1/2}(t)$$

and

$$\int_{\mathbf{R}^n} |v_t||u_t|\, dx \le 2E_v^{1/2}(t) E_u^{1/2}(t).$$

Integrating (25) over $\mathbf{R}^n \times [0,t]$ and using the divergence theorem and the estimates above, we obtain

$$E_u(t) \le E_u(0) + C \int_0^t a_K(s) E_u(s)^{1/2}\, ds,$$

where

$$a_K(t) := K^p (t+R)^{p\{1-\theta(2p)\}} + K(t+R) m^2(t) + K b(t),$$

which yields, by Bihari's inequality, that for some positive constant γ

$$\begin{aligned}E_u(t)^{1/2} &\le E_u^{1/2}(0) + C \int_0^t a_K(s)\, ds \\ &\le E_u^{1/2}(0) + C \max\{K, K^p\} T(1+T)^\gamma.\end{aligned}$$

Hence we can choose K large enough and T small enough such that

$$E_u^{1/2}(0) \le \frac{K}{2} \quad \text{and} \quad C\max\{K, K^p\} T(1+T)^\gamma \le \frac{K}{2},$$

and then $E_u^{1/2}(t) \le K$.

Finally, we can prove the contraction of the map M in a similar way. Fixed $v_1, v_2 \in B_{T,K}$, let

$$u_1 = Mv_1, \quad u_2 = Mv_2 \quad \text{and} \quad \overline{u} = u_1 - u_2, \quad \overline{v} = v_1 - v_2.$$

We have that $\overline{u}$ satisfies the problem

$$\begin{cases} \overline{u}_{tt} - \Delta \overline{u} = |v_1|^p - |v_2|^p + m^2(t)\overline{v} - b(t)\overline{v}_t, & \text{in } \mathbf{R}^n \times (0,T), \\ u(x,0) = u_t(x,0) \equiv 0, \quad x \in \mathbf{R}^n, \end{cases}$$

and the equation

$$(26)\quad \frac{\partial}{\partial t}\left(\frac{1}{2}(\overline{u}_t^2+|\nabla\overline{u}|^2)\right) = \operatorname{div}(\overline{u}_t\nabla\overline{u})+(|v_1|^p-|v_2|^p)\,\overline{u}_t+m^2(t)\overline{v}\overline{u}_t-b(t)\overline{v}_t\overline{u}_t.$$

Observe that, by (22), it holds

$$\begin{aligned}\int_{\mathbf{R}^n}\big||v_1|^p-|v_2|^p\big|\,|\overline{u}_t|\,dx &\le C\int_{\mathbf{R}^n}|v_1-v_2|(|v_1|+|v_2|)^{p-1}|\overline{u}_t|\,dx\\ &\le C\big\||\overline{v}|(|v_1|+|v_2|)^{p-1}\big\|_{L^2}\|\overline{u}_t\|_{L^2}\\ &\le C\|\overline{v}\|_{L^{2p}}\left(\|v_1\|_{L^{2p}}+\|v_2\|_{L^{2p}}\right)^{p-1}\|\overline{u}_t\|_{L^2},\\ &\le CK^{p-1}(t+R)^{p\{1-\theta(2p)\}}E_{\overline{v}}^{1/2}E_{\overline{u}}^{1/2}.\end{aligned}$$

Then, integrating (26) on $\mathbf{R}^n\times[0,t]$, exploiting again the Bihari's inequality and proceeding similarly as above, we reach the estimate

$$\|\overline{u}\|_{B_{T,K}}\le C\max\{1,K^{p-1}\}T(1+T)^\gamma\|\overline{v}\|_{B_{T,K}},$$

from which, choosing T small enough, we infer that M is a contraction. The proof of the desired local existence is now completed.

§ Acknowledgement

The first author is partially supported by Natural Science Foundation of Zhejiang Province(LY18A010008), NSFC(11771194), high level talent project of Lishui City (2016RC25), the Scientific Research Foundation of the First-Class Discipline of Zhejiang Province(B)(201601). The third author is partially supported by the Grant-in-Aid for Scientific Research (B)(No.18H01132), Japan Society for the Promotion of Science.

References

[1] M. Abramowitz and I. A. Stegun, Handbook of mathematical functions with formulas, graphs, and mathematical table, Vol. 2172, Courier Corporation, 1965.

[2] M. D'Abbicco, S. Lucente and M. Reissig, A shift in the Strauss exponent for semilinear wave equations with a not effective damping, J. Differ. Equ., **259**, 5040–5073 (2015).

[3] H. Fujita, On the blowing up of solutions of the Cauchy problem for $u_t=\Delta u+u^{1+\alpha}$, J. Fac. Sci. Univ. Tokyo Sect. I, **13**, 109–124 (1966).

[4] K. Fujiwara, M. Ikeda and Y. Wakasugi, Estimates of lifespan and blow-up rate for the wave equation with a time-dependent damping and a power-type nonlinearity, Funkcial. Ekvac., **62**(2), 157–189 (2019).

[5] M. Ikeda and T. Inui, The sharp estimate of the lifespan for the semilinear wave equation with time-dependent damping, Differ. Integral Equ., **32**, 1/2, 1–36 (2019).

[6] M. Ikeda and Y. Wakasugi, Global well-posedness for the semilinear damped wave equation with time dependent damping in the overdamping case, Proc. Amer. Math. Soc., **148**(1), 157–172 (2020).

[7] N.-A. Lai, N. M. Schiavone and H. Takamura, Wave-like blow-up for semilinear wave equations with scattering damping and negative mass term, M. D'Abbicco, M. Ebert, V. Georgiev and T. Ozawa, eds., "New Tools for Nonlinear PDEs and Application", Trends in Mathematics, 217–240, Birkhäuser (2019).

[8] N.-A. Lai, H. Takamura and K. Wakasa, Blow-up for semilinear wave equations with the scale invariant damping and super-Fujita exponent, J. Differ. Equ., **263**(9), 5377–5394 (2017).

[9] N.-A. Lai and H. Takamura, Blow-up for semilinear damped wave equations with subcritical exponent in the scattering case, Nonlinear Anal., **168**, 222–237 (2018).

[10] N. Nascimento, A. Palmieri and M. Reissig, Semi-linear wave models with power non-linearity and scale-invariant time-dependent mass and dissipation, Math. Nachr., **290**(11–12), 1779–1805 (2017).

[11] A. Palmieri, Global existence of solutions for semi-linear wave equation with scale-invariant damping and mass in exponentially weighted spaces, J. Math. Anal. Appl., **461**(2), 1215–1240 (2018).

[12] A. Palmieri and M. Reissig, A competition between Fujita and Strauss type exponents for blow-up of semi-linear wave equations with scale-invariant damping and mass, J. Differ. Equ., **266**(2–3), 1176–1220 (2019).

[13] W. A. Strauss, Nonlinear scattering theory at low energy, J. Funct. Anal., **41**(1), 110–133 (1981).

[14] M. Struwe, Semilinear wave equations, Bulletin of the American Mathematical Society, **26**(1), 53–85 (1992).

[15] H. Takamura, Improved Kato's lemma on ordinary differential inequality and its application to semilinear wave equations, Nonlinear Anal. Theory Methods Appl., **125**, 227–240 (2015).

[16] Z. Tu and J. Lin, A note on the blowup of scale invariant damping wave equation with sub-Strauss exponent, arXiv:1709.00866 (2017).

[17] K. Wakasa and B. Yordanov, On the nonexistence of global solutions for critical semilinear wave equations with damping in the scattering case, Nonlinear Anal. TMA, **180**, 67–74 (2019).

[18] Y. Wakasugi, Critical exponent for the semilinear wave equation with scale invariant damping, Fourier analysis, Birkhäuser, Cham, 375–390 (2014).

[19] C. Wang and M. Liu, Global existence for semilinear damped wave equations in relation with the Strauss conjecture, Discrete Contin. Dyn. Syst., **40**(2), 709–724 (2020).

[20] J. Wirth, Solution representations for a wave equation with weak dissipation, Math. Methods Appl. Sci., **27**, 101–124 (2004).

[21] J. Wirth, Wave equations with time-dependent dissipation. I. Non-effective dissipation, J. Differ. Equ., **222**, 487–514 (2006).

[22] J. Wirth, Wave equations with time-dependent dissipation. II. Effective dissipation, J. Differ. Equ., **232**, 74–103 (2007).

[23] K. Yagdjian and A. Galstian, The Klein-Gordon equation in anti-de Sitter spacetime, Rend. Semin. Mat. Univ. Politec. Torino, **67**(2), 271–292 (2009).

Institute of Nonlinear Analysis and Department of Mathematics,
Lishui University,
Lishui City 323000, China
E-mail address: ninganlai@lsu.edu.cn

Department of Mathematics, University of Rome "La Sapienza",
Piazzale Aldo Moro 5, 00185 Roma, Italy
E-mail address: schiavone@mat.uniroma1.it

Mathematical Institute, Tohoku University,
Aoba, Sendai 980-8578, Japan
E-mail address: hiroyuki.takamura.a1@tohoku.ac.jp

Advanced Studies in Pure Mathematics 85, 2020
The Role of Metrics in the Theory of Partial Differential Equations
pp. 407–416

Separability of Maxwell equation in rotating black hole spacetime and its geometric aspects

Tsuyoshi Houri, Norihiro Tanahashi and Yukinori Yasui

Abstract.

Recently a new formalism for perturbations of Maxwell's equations on the background of the Kerr-NUT-(A)dS spacetime was proposed, with which the equations are reduced to a equation of motion of a scalar field that can be solved by separation of variables. In this formalism the differential operators that commute with the operators of the equations of motion, called symmetry operators, played a key role to establish the separable structure. In this work we propose a method to reproduce these commuting symmetry operators in terms of the geometric quantities associated to the hidden symmetry of the background spacetime.

§1. Introduction

Black holes are basic objects in the gravitational theory, and their various aspects have been studied intensively. Particularly perturbative dynamics of fields in black hole spacetime has attracted much interest since it governs astrophysical phenomena around black holes and also shows intriguing mathematical properties. In this context, the separability of the perturbation equations plays a key role since it reduces the perturbation equations given as partial differential equations to a set of ordinary differential equations that may be solved analytically.

In this work we focus on spacetimes with rotating black holes [1, 2], in which the perturbation equations become complicated due to the rotation. For the Kerr black holes in four dimensions, Teukolsky made a significant progress in his seminal work, which showed that the equations of motion for the gravitational and Maxwell field can be separated

Received November 30, 2018.
Revised May 13, 2019.
2010 *Mathematics Subject Classification.* 35Q61, 83C22, 83C57.
Key words and phrases. gravitation, black holes, linear perturbations, separability, integrability.

if they are expressed in the so-called Newman-Penrose formalism [3, 4]. However, this formalism works only in the four dimensions and generalization to higher dimensions remained unaccomplished for a long time.

Recently a breakthrough about this problem was made by Lunin [5] by introducing new variables to express the vector potential, which is the dynamical variables of the Maxwell field. In terms of these new variables, the perturbation equations can be separated for all the propagating modes, and also this formalism can be generalized to the higher-dimensional case straightforwardly. In this work symmetries of the rotating black hole spacetime played an important role. It is known that the Kerr spacetime and its higher-dimensional generalization have hidden symmetries signified by the existence of nontrivial symmetric tensor satisfying $\nabla_{(a}K_{bc)} = 0$. This tensor is called the Killing tensor, and it yields a constant of motion $K^{ab}p_a p_b$, where p^a is the momentum vector, that is preserved on a geodesic. In Lunin's work, this tensor was used to introduce special coordinates that helped separating the variables of the perturbation equations.

Krtouš, Frolov and Kubizňák made a step forward about clarifying the significance of the hidden symmetries [6] in this problem. They focused on the off-shell Kerr-NUT-AdS metric in $2n$ dimensions

$$(1)\qquad g_{ab}\,dx^a\,dx^b = \sum_{\mu=1}^{n} \frac{U_\mu}{X_\mu}\,dx_\mu^2 + \sum_{\mu=1}^{n} \frac{X_\mu}{U_\mu}\left(\sum_{k=0}^{n-1} A_\mu^{(k)}\,d\psi_k\right)^2,$$

where U_μ, X_μ and $A_\mu^{(k)}$ are certain functions of x^μ. This metric is the most general one admitting the principal tensor h_{ab} which is an anti-symmetric generalization of the Killing tensor defined by

$$(2)\qquad \nabla_c h_{ab} = g_{ca}\xi_b - g_{cb}\xi_a, \qquad \xi_a = \frac{1}{2n-1}\nabla^b h_{ba},$$

where ∇_a is the covariant derivative (Levi-Civita connection) of g_{ab}. Then they showed that, by expressing the vector potential expressed with a scalar Z and a constant β as

$$(3)\qquad \mathsf{A}^a = B^{ab}\nabla_b Z, \qquad B^{ac}(g_{cb} - \beta h_{cb}) = \delta^a_b,$$

Maxwell's equation in the Lorenz gauge $\nabla_a \mathsf{A}^a = \nabla_a(B^{ab}\nabla_b Z) = 0$ can be reduced to a single scalar equation given by

$$(4)\qquad \Box Z + 2\beta\xi_c B^{cd}\nabla_d Z = 0,$$

where $\Box \equiv g^{ab}\nabla_a\nabla_b$ is the d'Alembertian operator. In our work, we call it the Lunin-Krtouš-Frolov-Kubizňák (LKFK) equation. They also showed that these equations can be expressed using differential operators whose commutators vanish, which makes it possible to solve the equations by separation of variables. This result suggests that the separability of the perturbation equation and the hidden symmetry of the spacetime are intimately related with each other, although their precise correspondence still remains unclear.

In our work, we shed new light on this problem based on studies about the hidden symmetry of the spacetime and the separability. One technique useful here is the Eisenhart-Duval lift [7, 8], in which the original spacetime is supplemented with additional dimensions as

$$(5)\ ds^2 = \tilde{g}_{AB}\, dx^A dx^B = g_{ab}\, dx^a\, dx^b + 2qA_a\, dx^a\, du + 2dudv - 2V du^2,$$

where g_{ab} is the original spacetime metric and u, v are the coordinates for the additional spacetime dimensions. By suitably choosing the metric functions A^a, V, the derivative operator $\Box + 2\beta\xi_c B^{cd}\nabla_d$ in (4) can be expressed simply as the d'Alembertian $\tilde{\Box} = \tilde{g}^{AB}\tilde{\nabla}_A\tilde{\nabla}_B$ in terms of the covariant derivative $\tilde{\nabla}_A$ associated with $\tilde{g}_{AB}$. Hence, with this technique we can reduce the perturbation equation (4) to a massless Klein-Gordon equation by absorbing the additional term $2\beta\xi_c B^{cd}\nabla_d$ into the structure of the higher-dimensional spacetime.

It turns out that Eq. (5) falls into the standard form of the metric whose geodesics is completely integrable [9], hence it is associated with the Killing tensors $\tilde{K}^{(i)}_{AB}$ corresponding to the constants of motion on those geodesics. Using these Killing tensors we can construct differential operators $\tilde{\mathcal{K}}^{(i)} \equiv \tilde{\nabla}_A \tilde{K}^{AB}_{(i)} \tilde{\nabla}_B$, and it can be shown that these operators commute with the d'Alembertian $\left[\tilde{\Box}, \tilde{\mathcal{K}}^{(i)}\right] = 0$ and also that they commutes with each other $\left[\tilde{\mathcal{K}}^{(i)}, \tilde{\mathcal{K}}^{(j)}\right] = 0$. It implies that $\tilde{\mathcal{K}}^{(i)}$ act as commuting symmetry operators that maps a solution of (4) to another one, which plays a key role in separation of variables. It can be further confirmed that the differential operators $\tilde{\mathcal{K}}^{(i)}$ coincide with the commuting derivative operators found in [6], where only the explicit form of the operators in terms of the coordinates were given. Hence we found a way to give a geometric interpretation to the commuting symmetry operators found by [6].

§2. Construction of commuting symmetry operators

2.1. Eisenhart-Duval lift of a classical mechanical system

In the Hamiltonian formulation, the motion of a relativistic particle with mass m and charge q in the presence of vector and scalar potentials

A_a and V on a D-dimensional spacetime (M, g_{ab}), on which the line element is given by $ds^2 = g_{ab}dx^a dx^b$, is governed by the Hamiltonian

$$H = \frac{1}{2} g^{ab}(p_a - qA_a)(p_b - qA_b) + V \tag{6}$$

with the constraint $H = -m^2$. For such a system, we may take the Eisenhart-Duval lift by considering the $(D+2)$-dimensional manifold $\tilde{M} = M \times R^2$ with the metric

$$d\tilde{s}^2 = \tilde{g}_{AB}\, dx^A\, dx^B = g_{ab}\, dx^a\, dx^b + 2qA_a\, dx^a\, du + 2dudv - 2V du^2, \tag{7}$$

where $\tilde{g}_{ab} = g_{ab}(x^a)$, $\tilde{g}_{au} = qA_a(x^a)$, $\tilde{g}_{uv} = 1$, $\tilde{g}_{uu} = -2V(x^a)$. The geodesic Hamiltonian on the uplifted spacetime $(\tilde{M}, \tilde{g}_{AB})$ is written as

$$\tilde{H} = \frac{1}{2}\tilde{g}^{AB} p_A p_B = \frac{1}{2} g^{ab}(p_a - qA_a p_v)(p_b - qA_b p_v) + V p_v^2 + p_u p_v, \tag{8}$$

where $\tilde{g}^{ab} = g^{ab}$, $\tilde{g}^{av} = -qg^{ab}A_b$, $\tilde{g}^{uv} = 1$, $\tilde{g}^{vv} = q^2 A_a A^a + 2V$. Since g_{ab}, A_a and V are functions of x^a only, $\partial/\partial u$ and $\partial/\partial v$ are Killing vector fields; hence, the corresponding momenta p_u and p_v are constants. Putting $\tilde{H} = 0$, $p_u = m^2$ and $p_v = 1$, we obtain the equations of motion for the Hamiltonian (6) satisfying $H = -m^2$ from (8). Thus null geodesics on the uplifted spacetime $(\tilde{M}, \tilde{g}_{AB})$ project onto the solutions of the original system in M.

2.2. Application to quantum mechanical systems

In quantum mechanics, via the Dirac quantization $p_a \to -i\nabla_a$, the equation of motion corresponding to the system of the Hamiltonian (6) is obtained as

$$\mathcal{H}\psi \equiv \Big[g^{ab}(\nabla_a - iqA_a)(\nabla_b - iqA_b) - 2V\Big]\psi = -m^2\psi, \tag{9}$$

where ∇_a is the Levi-Civita connection on (M, g_{ab}). A remarkable fact is that solutions to the equation of motion (9) are reproduced from a particular class of the solutions to the massless Klein-Gordon equation on the uplifted spacetime $(\tilde{M}, \tilde{g}_{AB})$,

$$\tilde{\Box}\Phi = 0, \tag{10}$$

where $\tilde{\Box} \equiv \tilde{g}^{AB}\tilde{\nabla}_A\tilde{\nabla}_B$ and $\tilde{\nabla}_A$ are the d'Alembertian and Levi-Civita connection on $(\tilde{M}, \tilde{g}_{AB})$. Actually $\tilde{\Box}$ is calculated from (8) as

(11)
$$\Box_{\tilde{g}} = \Box_g - 2qA^a \frac{\partial}{\partial x^a}\frac{\partial}{\partial v} - q(\nabla_a A^a)\frac{\partial}{\partial v} + (q^2 A_a A^a + 2V)\left(\frac{\partial}{\partial v}\right)^2 + 2\frac{\partial}{\partial u}\frac{\partial}{\partial v},$$

where $\Box \equiv g^{ab}\nabla_a\nabla_b$ and ∇_a are the d'Alembertian and Levi-Civita connection on (M, g_{ab}). Then restricting the solutions to the massless Klein-Gordon equation (10) to the specific ones of the form $\Phi = e^{-i\frac{m^2}{2}u+iv}\psi(x^\rho)$, we have

$$\tilde{\Box}\Phi = e^{-i\frac{m^2}{2}u+iv}\big(\mathcal{H}+m^2\big)\psi, \tag{12}$$

where $\mathcal{H}$ is the one appeared in (9). For simplicity we set $m = 0$ henceforth. Thus, instead of solving the equation of motion (9), we may solve the massless Klein-Gordon equation (10) on the uplifted spacetime $(\tilde{M}, G_{AB})$. This fact is the key in constructing symmetry operators of the equation of motion (9), as shown in the subsequent sections.

2.3. Separability and commuting symmetry operators

Generally, the separability of the differential equation $\mathcal{O}\psi = 0$ is related to the existence of first- and/or second-order differential operators $\mathcal{K}^{(i)}$ satisfying the commutation relations $[\mathcal{O}, \mathcal{K}^{(i)}] = 0$, $[\mathcal{K}^{(i)}, \mathcal{K}^{(j)}] = 0$, where the square bracket $[\ ,\ ]$ denotes the commutator $[A, B] \equiv AB - BA$. The operators satisfying the first condition are called symmetry operators since they map a solution of $\mathcal{O}\psi = 0$ to another solution, and the second condition implies that they commute with each other as differential operators. If we are able to solve the differential equation $\mathcal{O}\psi = 0$ by separation of variables, we have separation constants κ and they are related to second-order symmetry operators by $\mathcal{K}^{(i)}\psi = \kappa^{(i)}\psi$.

2.4. Construction of symmetry operators

In the Eisenhart-Duval lift, symmetry operators of the equation of motion (9) can be constructed from symmetry operators of the massless Klein-Gordon equation (10). To show it, we begin with the second-order differential operator $\tilde{\mathcal{K}}$ on $\tilde{M}$,

$$\tilde{\mathcal{K}} = \tilde{\nabla}_A \tilde{K}^{AB}\tilde{\nabla}_B, \tag{13}$$

where $\tilde{K}_{AB}$ is a rank-2 symmetric tensor on $\tilde{M}$. For $\tilde{\mathcal{K}}$ to be a symmetry operator of the massless Klein-Gordon equation (10), i.e.,

$$[\tilde{\Box}, \tilde{\mathcal{K}}] = 0, \tag{14}$$

$\tilde{K}_{AB}$ must be a Killing tensor on $(\tilde{M}, \tilde{g}_{AB})$, which obeys the Killing equation $\tilde{\nabla}_{(A}\tilde{K}_{BC)} = 0$, and must satisfy, in addition, the anomaly-free condition

$$\tilde{\nabla}^A\left(\tilde{K}_{[A}{}^C\tilde{R}_{B]C}\right) = 0, \tag{15}$$

where $\tilde{R}_{AB}$ is the Ricci tensor of $(\tilde{M}, \tilde{g}_{AB})$ [10].

Now, we suppose that the Killing tensor $\tilde{K}^{AB}$ are independent of the coordinates u and v, $\partial_u \tilde{K}^{AB} = 0$, $\partial_v \tilde{K}^{AB} = 0$. Then, applying the symmetry operator $\tilde{\mathcal{K}}$ on $\tilde{M}$ to the wave function of the form $\Phi = e^{-i\frac{m^2}{2}u+iv}\psi(x^\rho)$, we obtain the differential operator $\mathcal{K}$ on M by

$$\tilde{\mathcal{K}}\Phi = e^{iv}\mathcal{K}\psi. \tag{16}$$

Without loss of generality, we may express $\tilde{K}^{AB}$ as

$$
\begin{aligned}
(\tilde{K}^{AB}) &= \begin{pmatrix} \tilde{K}^{ab} & \tilde{K}^{au} & \tilde{K}^{av} \\ \tilde{K}^{ub} & \tilde{K}^{uu} & \tilde{K}^{uv} \\ \tilde{K}^{vb} & \tilde{K}^{vu} & \tilde{K}^{vv} \end{pmatrix} \\
&= \begin{pmatrix} K^{ab} & L^a & -qK^{ac}A_c + M^a \\ L^b & W & -qL^cA_c + T + 2WV \\ -qK^{bc}A_c + M^b & -qL^cA_c + T + 2WV & q^2K^{cd}A_cA_d - 2qM^cA_c + F \end{pmatrix}.
\end{aligned}
$$

Here, using the fact that $\tilde{K}^{AB}$ is a Killing tensor on $\tilde{M}$, we can show that K^{ab} and L^a are Killing tensor and Killing vector on M, and also that W is a constant and $\nabla_a M^a = 0$ [11]. Using $\nabla_a M^a = 0$ among them, we find

$$\mathcal{K}\psi = \nabla_a^{(A)} K^{ab} \nabla_b^{(A)} \psi + 2iM^a \nabla_a^{(A)} \psi - F\psi, \tag{17}$$

where $\nabla_a^{(A)} \equiv \nabla_a - iqA_a$. Since $[\tilde{\Box}, \tilde{K}]\Phi = e^{iv}[\mathcal{H}, \mathcal{K}]\psi$, we obtain $[\mathcal{H}, \mathcal{K}] = 0$, which means that $\mathcal{K}$ is a symmetry operator $\mathcal{K}$ of the equation of motion (9).

For commutativity of two symmetry operators $\left[\tilde{\mathcal{K}}^{(i)}, \tilde{\mathcal{K}}^{(j)}\right] = 0$ additional conditions similar to (15) must be satisfied, which must be examined separately [12].

To summarize, the procedure for obtaining commuting symmetry operators $\mathcal{K}^{(i)}$ for the equation of motion (9) is given as follows:

(1) Read out g_{ab}, A_a and V from the equation of motion (9).

(2) Using these quantities, construct the uplifted metric $\tilde{g}_{AB}$ (7).

(3) Find Killing tensors $\tilde{K}^{(i)}_{AB}$ on $(\tilde{M}, \tilde{g}_{AB})$ such that the components are independent of u and v.

(4) Check if $\tilde{K}^{(i)}_{AB}$ satisfy the anomaly-free condition (15), and also the commutativity conditions.

(5) If they hold, construct the commuting symmetry operators $\tilde{\mathcal{K}}^{(i)}$ by (13). Then it follows that $[\tilde{\Box}, \tilde{\mathcal{K}}^{(i)}] = 0 = [\tilde{\mathcal{K}}^{(i)}, \tilde{\mathcal{K}}^{(j)}]$.

(6) Finally, obtain commuting symmetry operators $\mathcal{K}^{(i)}$ from $\tilde{\mathcal{K}}^{(i)}$ via (16) and (17). Then $[\mathcal{H}, \mathcal{K}^{(i)}] = 0 = [\mathcal{K}^{(i)}, \mathcal{K}^{(j)}]$.

§3. Separability and symmetry operators of Maxwell's equations on the Kerr-NUT-(A)dS spacetime in four dimensions

To illustrate the procedure to construct the symmetry operators summarized in Section 2.4, we apply it to the LKFK equation in four-dimensional case. The Teukolsky equation in four dimensions and the LKFK equations can be treated in a similar manner.

3.1. Carter form of the Kerr-NUT-(A)dS metric

The Kerr metric in the Boyer-Lindquist coordinates (t, r, θ, ϕ) is

(18)
$$ds^2 = -\frac{\Delta}{\Sigma}\left(dt - a\sin^2\theta d\phi\right)^2 + \frac{\Sigma dr^2}{\Delta} + \Sigma d\theta^2 + \frac{\sin^2\theta}{\Sigma}\left(adt - (r^2+a^2)d\phi\right)^2,$$

where $\Delta = r^2 - 2mr + a^2$, $\Sigma = r^2 + a^2\cos^2\theta$. Performing the coordinate transformation $p^2 = a^2\cos^2\theta$, $\tau = t - a\phi$, $\sigma = -\frac{\phi}{a}$, we obtain the Kerr metric in the Carter form

$$\begin{aligned} ds^2 = &-\frac{\mathcal{Q}(r)}{r^2+p^2}(d\tau - p^2 d\sigma)^2 + \frac{r^2+p^2}{\mathcal{Q}(r)}dr^2 \\ &+ \frac{r^2+p^2}{\mathcal{P}(p)}dp^2 + \frac{\mathcal{P}(p)}{r^2+p^2}(d\tau + r^2 d\sigma)^2, \end{aligned} \tag{19}$$

where $\mathcal{Q}(r) = r^2 - 2mr + a^2$, $\mathcal{P}(p) = a^2 - p^2$. The metric (19) solves the vacuum Einstein equation with cosmological constant $R_{ab} = \lambda g_{ab}$ if and only if $\mathcal{Q}(r)$ and $\mathcal{P}(p)$ are given by $\mathcal{Q}(r) = -\frac{\lambda}{3}r^4 + \epsilon r^2 - 2mr + a^2$, $\mathcal{P}(p) = -\frac{\lambda}{3}p^4 - \epsilon p^2 + 2lp + a^2$ with constants ϵ, a, m and l. This solution is called the Kerr-NUT-(A)dS metric. The metric (19) with $\mathcal{Q}(r)$ and $\mathcal{P}(p)$ replaced by arbitrary functions depending only on r and p is called the off-shell metric. In the following calculation, we will work with the off-shell metric without using imposing the Einstein equation.

It is important to mention that the off-shell metric in the Carter form fits into Benenti's canonical form for metrics admitting the separability of the Hamilton-Jacobi equation for geodesics [9]. Actually, the components of the inverse metric in the coordinate basis are written in the canonical form

$$\begin{aligned} &g^{rr} = \frac{\mathcal{Q}}{r^2+p^2}, \quad g^{pp} = \frac{\mathcal{P}}{r^2+p^2}, \quad g^{\tau\tau} = -g^{rr}\frac{r^4}{\mathcal{Q}^2} + g^{pp}\frac{p^4}{\mathcal{P}^2}, \\ &g^{\tau\sigma} = g^{rr}\frac{r^2}{\mathcal{Q}^2} + g^{pp}\frac{p^2}{\mathcal{P}^2}, \quad g^{\sigma\sigma} = -g^{rr}\frac{1}{\mathcal{Q}^2} + g^{pp}\frac{1}{\mathcal{P}^2}, \end{aligned} \tag{20}$$

and we can easily construct a Killing tensor on the Kerr spacetime. Following Benenti's construction, the Killing tensor turns out to be given by (20) with g^{ab}, g^{rr}, g^{pp} replaced to K^{ab}, K^{rr}, K^{pp}, where

$$K^{rr} = \frac{p^2\mathcal{Q}}{r^2+p^2}, \qquad K^{pp} = -\frac{r^2\mathcal{P}}{r^2+p^2}. \tag{21}$$

3.2. Symmetry operators of the LKFK equations

In the LKFK formulation in four dimensions, Maxwell's equation (4) can be written in the form

$$\mathcal{H}^{(L)}\psi \equiv \Big[g^{ab}(\nabla_a - iqA_a^{(L)})(\nabla_b - iqA_b^{(L)}) - 2V^{(L)}\Big]\psi = 0, \tag{22}$$

where g_{ab} is the off-shell metric (19), and $A^{(L)}$ and $V^{(L)}$ are given by

$$\begin{aligned} qA_a^{(L)}dx^a &= \frac{i\beta^2 r}{\Xi_r}dr - \frac{i\beta^2 p}{\Xi_p}dp - \frac{i\beta}{r^2+p^2}\left(\frac{\mathcal{Q}}{\Xi_r} - \frac{\mathcal{P}}{\Xi_p}\right)d\tau + \frac{i\beta}{r^2+p^2}\left(\frac{p^2\mathcal{Q}}{\Xi_r} + \frac{r^2\mathcal{P}}{\Xi_p}\right)d\sigma, \\ V^{(L)} &= \frac{1}{r^2+p^2}\left(\frac{\beta^2 r\mathcal{Q}'}{2\Xi_r} + \frac{\beta^4 r^2\mathcal{Q}}{\Xi_r^2} - \frac{\beta^2 p\mathcal{P}'}{2\Xi_p} + \frac{\beta^4 p^2\mathcal{P}}{\Xi_p^2}\right), \end{aligned} \tag{23}$$

where $\Xi_r = 1 - \beta^2 r^2$, $\Xi_p = 1 + \beta^2 p^2$. This can be solved by separation of variables by using the ansatz $\psi = e^{-i\omega\tau}e^{im\sigma}R(r)\Theta(p)$ and we obtain

$$\begin{aligned} &\Xi_r\mathcal{Q}\frac{d}{dr}\left(\frac{\mathcal{Q}}{\Xi_r}\frac{dR}{dr}\right) + \left[(\omega r^2 + m)^2 - \frac{2i\mathcal{Q}}{\beta\Xi_r}(\omega + \beta^2 m) - \kappa^{(L)}\mathcal{Q}\right]R = 0, \\ &\Xi_p\mathcal{P}\frac{d}{dp}\left(\frac{\mathcal{P}}{\Xi_p}\frac{d\Theta}{dp}\right) + \left[-(\omega p^2 - m)^2 + \frac{2i\mathcal{P}}{\beta\Xi_p}(\omega + \beta^2 m) + \kappa^{(L)}\mathcal{P}\right]\Theta = 0, \end{aligned} \tag{24}$$

where $\kappa^{(L)}$ is a separation constant.

The uplifted metric $\tilde{g}_{AB}$ on $\tilde{M}$ is obtained from (7) with (19) and (23). As explained before, the uplifted metric $\tilde{g}_{AB}$ fits into Benenti's canonical form

$$\begin{aligned} &\tilde{g}^{ab} = g^{ab}, \qquad \tilde{g}^{rv} = -g^{rr}\frac{i\beta^2 r}{\Xi_r}, \qquad \tilde{g}^{pv} = g^{pp}\frac{i\beta^2 p}{\Xi_p}, \\ &\tilde{g}^{\tau v} = -g^{rr}\frac{i}{2\beta}\frac{1+\beta^2 r^2}{\Xi_r\mathcal{Q}} + g^{pp}\frac{i}{2\beta}\frac{1-\beta^2 p^2}{\Xi_p\mathcal{P}}, \quad \tilde{g}^{\sigma v} = g^{rr}\frac{i\beta}{2}\frac{1+\beta^2 r^2}{\Xi_r\mathcal{Q}} - g^{pp}\frac{i\beta}{2}\frac{1-\beta^2 p^2}{\Xi_p\mathcal{P}}, \\ &\tilde{g}^{uv} = 1 = g^{rr}\frac{r^2}{\mathcal{Q}} + g^{pp}\frac{p^2}{\mathcal{P}}, \qquad \tilde{g}^{vv} = \frac{g^{rr}}{\mathcal{Q}}\frac{d}{dr}\left(\frac{\beta^2 r}{\Xi_r}\mathcal{Q}\right) - \frac{g^{pp}}{\mathcal{P}}\frac{d}{dp}\left(\frac{\beta^2 p}{\Xi_p}\mathcal{P}\right), \end{aligned} \tag{25}$$

and the Killing tensor is given by the expression same as the metric (25) with $\tilde{g}^{AB}$, g^{ab}, g^{rr}, g^{pp} replaced to $\tilde{K}^{AB}$, K^{ab}, K^{rr}, K^{pp}, respectively. Then we obtain the symmetry operator of the LKFK equations according to (17) as

(26)
$$\mathcal{K}^{(L)}\psi = \frac{p^2}{r^2+p^2}\left[\mathcal{Q}\partial_r^2 + \left(\mathcal{Q}' + \frac{2\beta^2 r\mathcal{Q}}{\Xi_r}\right)\partial_r - \frac{1}{\mathcal{Q}}(r^2\partial_\tau - \partial_\sigma)^2 + \frac{2}{\beta}\frac{1}{\Xi_r}(\partial_\tau - \beta^2\partial_\sigma)\right]\psi$$
$$- \frac{r^2}{r^2+p^2}\left[\mathcal{P}\partial_p^2 + \left(\mathcal{P}' - \frac{2\beta^2 p\mathcal{P}}{\Xi_p}\right)\partial_p + \frac{1}{\mathcal{P}}(p^2\partial_\tau + \partial_\sigma)^2 - \frac{2}{\beta}\frac{1}{\Xi_p}(\partial_\tau - \beta^2\partial_\sigma)\right]\psi.$$

Using the multiplicative separation ansatz $\psi = e^{-i\omega\tau}e^{im\sigma}R(r)\Theta(p)$ and the separated equations (24), we find

$$\mathcal{K}^{(L)}\psi = \kappa^{(L)}\psi. \tag{27}$$

It can be confirmed by direct calculations that the anomaly-free condition (15) is satisfied by the Killing tensor constructed above [11], hence we have the commutation relations

$$\left[\mathcal{H}^{(L)}, \mathcal{K}^{(L)}\right] = 0. \tag{28}$$

§4. Outlook

An obvious next step for this work is to inspect the uplifted metrics corresponding to different perturbations. In four dimensions Maxwell's equations can be reduced to the Teukolsky equation and also to the LKFK equations, and these two equations lead to different uplifted metrics. Since these two equations should describe the same time evolution of the Maxwell fields, there would be intertwining operators connecting the solutions of these solutions. If they exist, it is expected that the uplifted metrics corresponding to the different equations of motion are related to each other by some transformations. It would be fruitful to investigate such possible connection between the uplifted spacetimes by clarifying their geometric properties such as the curvature invariants and Weyl tensors.

Also, for both equations the uplifted metric is intimately related to the symmetry of the base spacetime. For example, the vector field for the LKFK equation is given by $A_a \sim \xi^b B_{ba}$, where ξ^b is a Killing vector and the polarization tensor B_{ab} is constructed from the metric and the principal tensor corresponding to the hidden symmetry of the base spacetime. Geometric properties of this kind could provide useful information of properties of the uplifted metrics and also the vector potential ansatz that reduced the Maxwell equation to a scalar field equation. Pursuing this idea would be another future tasks for this study.

References

[1] R. P. Kerr, Gravitational field of a spinning mass as an example of algebraically special metrics, Phys. Rev. Lett., **11** (1963) 237.

[2] R. C. Myers and M. J. Perry, Black Holes in Higher Dimensional Space-Times, Annals Phys., **172** (1986) 304.

[3] S. A. Teukolsky, Rotating black holes: Separable wave equations for gravitational and electromagnetic perturbations, Phys. Rev. Lett., **29** (1972) 1114.

[4] S. A. Teukolsky, Perturbations of a rotating black hole. 1. Fundamental equations for gravitational electromagnetic and neutrino field perturbations, Astrophys. J., **185** (1973) 635.

[5] O. Lunin, Maxwell's equations in the Myers-Perry geometry, JHEP, **12** (2017) 138.

[6] P. Krtouš, V. P. Frolov and D. Kubizňák, Separation of Maxwell equations in Kerr-NUT-(A)dS spacetimes, Nucl. Phys., **B934** (2018) 7.

[7] L. P. Eisenhart, Dynamical Trajectories and Geodesics, Annals of Mathematics, **30** (1928) 591.

[8] C. Duval, G. Burdet, H. P. Kunzle and M. Perrin, Bargmann Structures and Newton cartan Theory, Phys. Rev. D, **31** (1985) 1841.

[9] S. Benenti, Separation of variables in the geodesic Hamilton-Jacobi equation, in Symplectic geometry and mathematical physics (Aix-en-Provence, 1990), vol. 99 of Progr. Math., pp. 1–36. Birkhäuser Boston, Boston, MA, 1991.

[10] B. Carter, Killing Tensor Quantum Numbers and Conserved Currents in Curved Space, Phys. Rev., **D16** (1977) 3395.

[11] T. Houri, N. Tanahashi and Y. Yasui, On symmetry operators for the Maxwell equation on the Kerr-NUT-(A)dS spacetime, Class. Quant. Grav., **37** no. 1 (2020) 015011.

[12] I. Kolar and P. Krtous, Weak electromagnetic field admitting integrability in Kerr-NUT-(A)dS spacetimes, Phys. Rev., **D91** (2015) 124045.

National Institute of Technology, Maizuru College,
Kyoto 625-8511, Japan
E-mail address: `t.houri@maizuru-ct.ac.jp`

Institute of Mathematics for Industry, Kyushu University,
Fukuoka 819-0395, Japan
E-mail address: `tanahashi@imi.kyushu-u.ac.jp`

Institute for Fundamental Sciences, Setsunan University,
Osaka 572-8508, Japan
E-mail address: `yukinori.yasui@mpg.setsunan.ac.jp`

Advanced Studies in Pure Mathematics 85, 2020
The Role of Metrics in the Theory of Partial Differential Equations
pp. 417–428

Traveling fronts in balanced bistable reaction-diffusion equations

Masaharu Taniguchi

Abstract.

This paper is an announcement of a forthcoming paper [21]. We study axially asymmetric traveling fronts to reaction-diffusion equations with balanced bistable nonlinear terms. We prove such a solution as the balanced limit of a pyramidal traveling front of an imbalanced bistable reaction-diffusion equation by preserving the ratio of a major axis and a minor axis of the cross section to be a constant. When the constant ratio is not one, the limit functions become axially asymmetric traveling fronts to balanced bistable reaction-diffusion equations.

§1. Introduction

This paper is an announcement of a forthcoming paper [21]. In this paper we study a reaction-diffusion equation

$$\text{(1)} \qquad \begin{aligned} &\frac{\partial u}{\partial t} = \Delta u - G'(u), && \boldsymbol{x} \in \mathbb{R}^n,\ t > 0, \\ &u(\boldsymbol{x}, 0) = u_0(\boldsymbol{x}), && \boldsymbol{x} \in \mathbb{R}^n, \end{aligned}$$

where $n \geq 3$ is a given integer, and given $u_0 \in X$. Here X is the set of bounded and uniformly continuous functions from $\mathbb{R}^n$ to $\mathbb{R}$ with the norm

$$\|u_0\| = \sup_{\boldsymbol{x} \in \mathbb{R}^n} |u_0(\boldsymbol{x})|.$$

The standing assumption on G is as follows. Let $G \in C^2[-1, 1]$ satisfy

$$G(1) = 0,\ G(-1) = 0,\ G'(1) = 0,\ G'(-1) = 0,\ G''(1) > 0,\ G''(-1) > 0,$$
$$G(s) > 0 \qquad \text{if} \quad -1 < s < 1.$$

Received November 28, 2018.
Revised May 15, 2019.
2010 *Mathematics Subject Classification.* 35C07, 35B20, 35K57.
Key words and phrases. traveling front, reaction-diffusion equation, asymmetric, balanced.

See Figure 1. For $G(s) = (1 - s^2)^2/4$ and $-G'(s) = s - s^3$, (1) is called the Allen–Cahn equation, the scalar Ginzburg–Landau equation or the Nagumo equation.

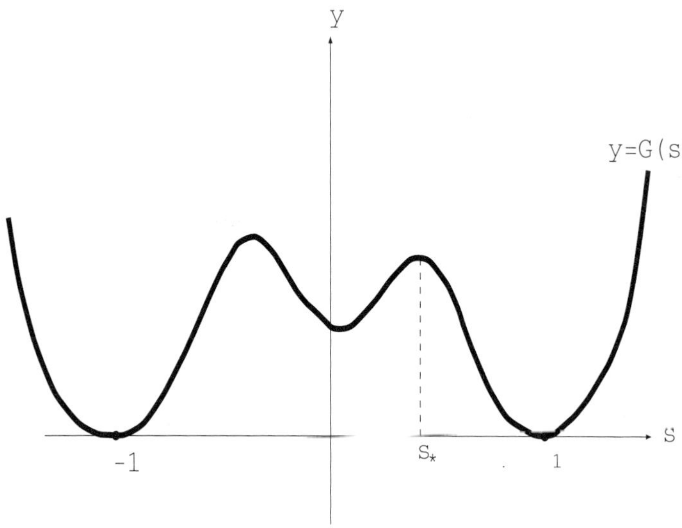

Fig. 1. The graph of G.

A function $v(\boldsymbol{x}', x_n)$ is called a traveling front if it satisfies

$$\Delta v + c\frac{\partial v}{\partial x_n} - G'(v) = 0, \qquad (\boldsymbol{x}', x_n) \in \mathbb{R}^n$$

with some constant c. Here we choose the x_n-axis as its traveling direction. In general, one might choose plural traveling directions for one traveling front. A traveling front is called axisymmetric if it is axisymmetric for a suitable choice of a traveling direction. Then planar traveling fronts are axisymmetric. It is called axially asymmetric if it is axially asymmetric for any choice of traveling direction in this paper.

The reaction term is called balanced when $G(1) = G(-1)$ and is called imbalanced when $G(1) \neq G(-1)$. When the reaction term is imbalanced, multi-dimensional traveling fronts including axisymmetric ones and axially asymmetric ones have been studied by [7, 8, 9, 10, 11, 12, 13, 14, 15, 16, 17, 18, 19, 20, 23] and so on. In this case, the propagation is mainly driven by the imbalance of the reaction kinetics and the curvature effect of an *interface*. Here a level set of a solution is often called an interface.

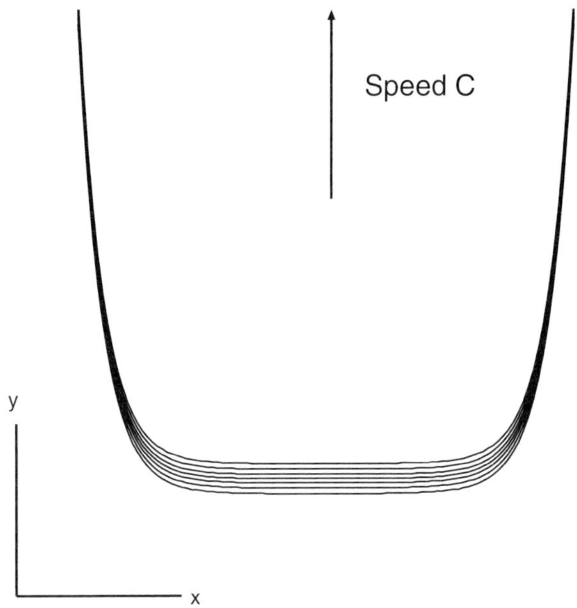

Fig. 2. A level set of an axisymmetric traveling front in $\mathbb{R}^2$.

When the reaction term is balanced, one has no driven force caused by the reaction kinetics and the propagation is mainly driven by the curvature effect of an interfaces and is also driven by interaction between portions of an interface. For Equation (1), axisymmetric traveling fronts have been studied by Chen, Guo, Hamel, Ninomiya and Roquejoffre [3] in $\mathbb{R}^n$. See Figure 2 for a level set of an axisymmetric traveling front in $\mathbb{R}^2$. See del Pino, Kowalczyk and Wei [5] for a stationary solution, that is a traveling front with speed zero, related with De Giorgi's conjecture. See [6] for a traveling wave solution with two non-planar fronts and for a traveling wave solution with non-convex fronts.

§2. Axially asymmetric traveling fronts in [21]

In this chapter we study an axially asymmetric traveling front solution to a balanced reaction-diffusion equation (1). This axially asymmetric traveling front solution is monotone decreasing in the traveling axis x_n and travels with any given positive speed.

Let $c \in (0, \infty)$ be arbitrarily fixed. Let s_* be the largest zero point of G' in $(-1, 1)$, that is, $s_* \in (-1, 1)$ is defined by

$$s_* = \min\{s_0 \in (-1,1) \bigm| -G'(s) > 0 \quad \text{if } s_0 < s < 1\}.$$

We fix θ_0 with $s_* < \theta_0 < 1$ and have $-G'(\theta_0) > 0$.

Let

$$1 \le \alpha_2 \le \cdots \le \alpha_{n-1} \tag{2}$$

be arbitrarily given and let $\zeta > 0$ be arbitrarily given. We put

$$\boldsymbol{\alpha}' = (1, \alpha_2, \ldots, \alpha_{n-1}). \tag{3}$$

For any k with

$$0 < k < \sqrt{G''(-1)}\,,$$

let

$$f_k(s) = -G'(s) + k\sqrt{2G(s)}\,, \quad s \in \mathbb{R},$$
$$F_k(s) = \int_{-1}^{s} f_k(\sigma)\,\mathrm{d}\sigma.$$

Let $\boldsymbol{\alpha}'$ be given by (3) with (2). In [21] we define

$$U(\boldsymbol{x}', x_n; \boldsymbol{\alpha}') = \lim_{k \to 0} V_k(\boldsymbol{x}', x_n + z_k; \boldsymbol{A}'(k)) \tag{4}$$

for all $(\boldsymbol{x}', x_n)$ in any compact set in $\mathbb{R}^n$, where $\boldsymbol{A}'(k)$ is a vector in $[0, \infty)^{n-1}$ and z_k is a real number. Here V_k is a pyramidal traveling front that satisfies a profile equation with c for an imbalanced reaction-diffusion equation:

$$\Delta V_k + c\frac{\partial V_k}{\partial x_n} + f_k(V_k) = 0, \qquad (\boldsymbol{x}', x_n) \in \mathbb{R}^n.$$

Here V_k satisfies

$$\frac{\partial V_k}{\partial x_n} < 0 \qquad \text{for all } (\boldsymbol{x}', x_n) \in \mathbb{R}^n,$$
$$\frac{\partial V_k}{\partial x_j} > 0 \qquad \text{if } x_j > 0,\ 1 \le j \le n-1,$$

and it also satisfies

$$V_k(\mathbf{0}', z_k) = \theta_0,$$

$$V_k(0, \ldots, 0, \overset{i}{\smile}{r_i(\boldsymbol{A}'(k))}, 0, \ldots, 0, \zeta + z_k) = \theta_0, \qquad 2 \leq i \leq n-1,$$

$$\frac{r_i(\boldsymbol{A}'(k))}{r_1(\boldsymbol{A}'(k))} = \alpha_i, \qquad 2 \leq i \leq n-1.$$

See [21] for the details. Here a symbol $\overset{\overset{i}{\smile}}{r_i(\boldsymbol{A}'(k))}$ implies that $r_i(\boldsymbol{A}'(k))$ is the i-th component. See Figure 3 for a level set of a pyramidal traveling front with speed c for an imbalanced equation.

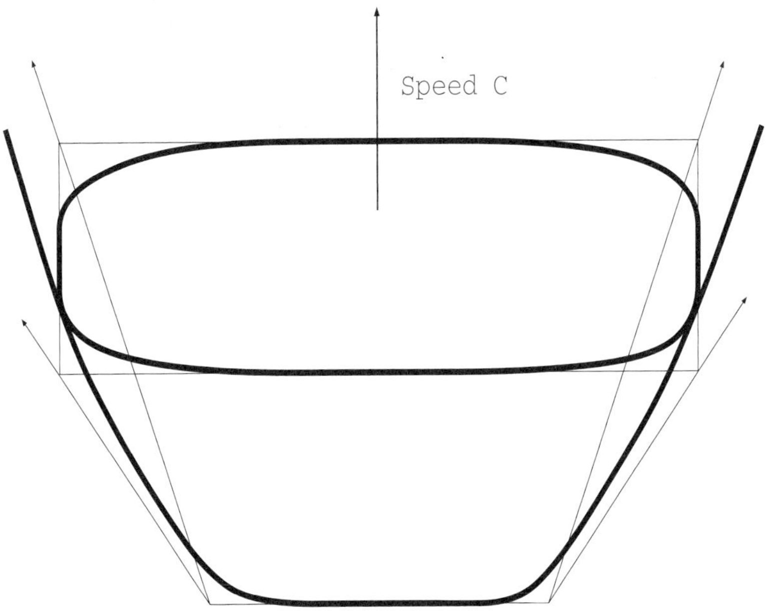

Fig. 3. A level set of a pyramidal traveling front for an imbalanced equation.

Then $U(\boldsymbol{x}; \boldsymbol{\alpha}')$ satisfies $U(\mathbf{0}; \boldsymbol{\alpha}') = \theta_0$. Let R_j be given by

$$U(0, \ldots, 0, \overset{\overset{j}{\smile}}{R_j}, 0, \ldots, 0, \zeta; \boldsymbol{\alpha}') = \theta_0 \tag{5}$$

for $1 \le j \le n-1$. See Figure 5. We often write $U(\boldsymbol{x}; \boldsymbol{\alpha}')$ simply as $U(\boldsymbol{x})$. Then $U(\boldsymbol{x}', x_n - ct) = U(\boldsymbol{x}', x_n - ct; \boldsymbol{\alpha}')$ becomes a traveling front to Equation (1) and satisfies the following properties.

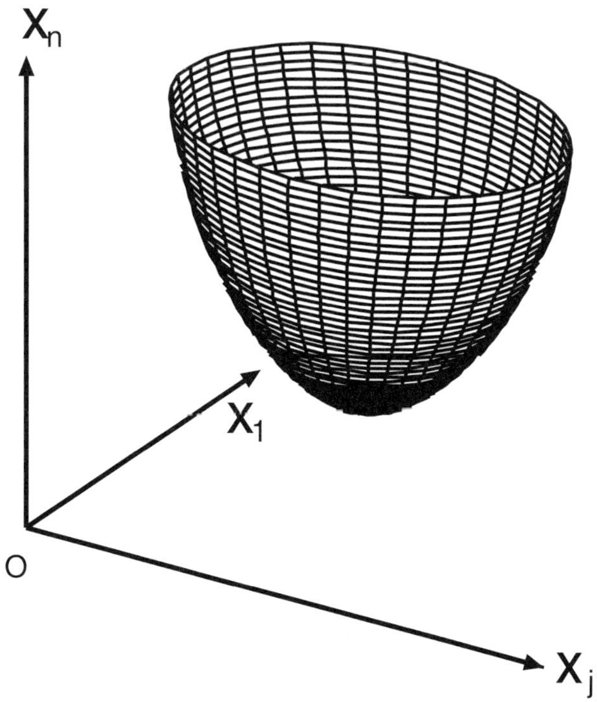

Fig. 4. A level set $\{U(\boldsymbol{x}', x_n) = \theta_0\}$ of U.

Theorem 1 ([21]). *Let* $c > 0$ *be arbitrarily given. Let* $U(\boldsymbol{x}) = U(\boldsymbol{x}; \boldsymbol{\alpha}')$ *and* R_j *be given by* (4) *and* (5), *respectively. Then one has*

$U(\mathbf{0}) = \theta_0 \in (-1,1)$ *and*

$$\Delta U + c\frac{\partial U}{\partial x_n} - G'(U) = 0, \qquad (\boldsymbol{x}', x_n) \in \mathbb{R}^n, \tag{6}$$

$$U(x_1, \ldots, x_{j-1}, -x_j, x_{j+1}, \ldots, x_n) = U(x_1, \ldots, x_{j-1}, x_j, x_{j+1}, \ldots, x_n) \quad \text{for } 1 \le j \le n-1,$$

$$\frac{\partial U}{\partial x_n}(\boldsymbol{x}) < 0 \qquad \text{if } \boldsymbol{x} \in \mathbb{R}^n,$$

$$\frac{\partial U}{\partial x_j}(\boldsymbol{x}) > 0 \qquad \text{if } \boldsymbol{x} \in \mathbb{R}^n,\ x_j > 0,\ 1 \le j \le n-1,$$

$$\frac{R_j}{R_1} = \alpha_j, \qquad 1 \le j \le n-1.$$

For every $\theta \in (-1,1)$, *one has*

$$\inf_{\boldsymbol{x} \in \mathbb{R}^n} \left\{ |\nabla U(\boldsymbol{x})| \,\middle|\, U(\boldsymbol{x}) = \theta \right\} > 0, \tag{7}$$

and can define $q_\theta(\boldsymbol{x}') \in \mathbb{R}$ *by* $U(\boldsymbol{x}', q_\theta(\boldsymbol{x}')) = \theta$ *for all* $\boldsymbol{x}' \in \mathbb{R}^{n-1}$. *Here* q_θ *belongs to* $C^1(\mathbb{R}^{n-1})$.

As in Theorem 1, a level set $\{\boldsymbol{x} \in \mathbb{R}^n \mid U(\boldsymbol{x}) = \theta\}$ is given by a graph of a function q_θ that is defined on the entire space $\mathbb{R}^{n-1}$. When $(\alpha_1, \ldots, \alpha_{n-1}) \ne (1, \ldots, 1)$, $\{\boldsymbol{x}' \in \mathbb{R}^{n-1} \mid U(\boldsymbol{x}', \zeta; \boldsymbol{\alpha}') = \theta_0\}$ is a compact set with a major axis and a minor axis and is not a disk. Thus $U(\boldsymbol{x})$ is an axially asymmetric traveling front solution with the θ_0 level set whose cross section at $x_n = \zeta > 0$ is a compact set that is different from a disk. Now the author introduces the following conjecture.

Conjecture 1. *When* $\boldsymbol{\alpha} = (\alpha_2, \ldots, \alpha_{n-1}) = (1, \ldots, 1)$ *in* (2), U *in Theorem* 1 *is axisymmetric with respect to the* x_n*-axis, and equals the traveling front studied by* [3].

It is an interesting problem to study whether the conjecture stated above holds true or not. This problem should be studied in a near future.

§3. Discussions

Equation (1) and a mean curvature flow are closely related as is seen in [2] for instance. The cross section of $\{U(\boldsymbol{x}', x_n) = \theta_0\}$ at $x_n = \zeta$ in Figure 5 will be related to *Angenent's Ovals* in a curvature flow. Let κ be the curvature of a curve in $\mathbb{R}^2$ and let θ be the angle between the tangent vector of the curve and the x-axis. Let the curve develop

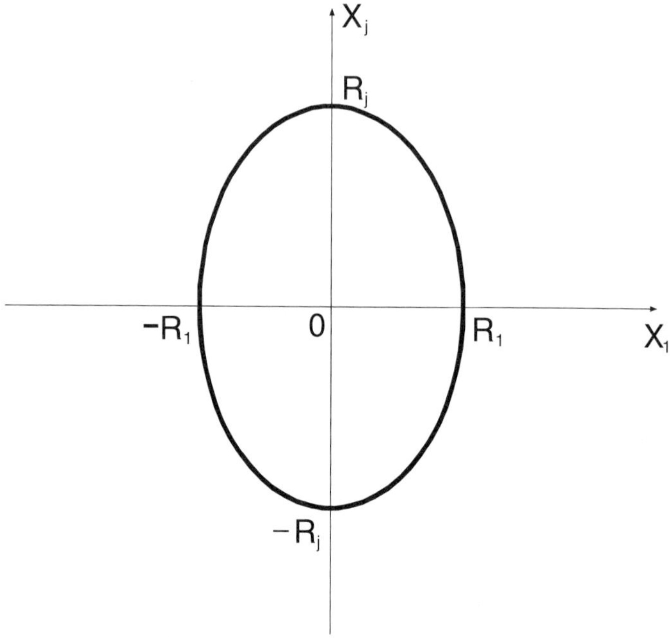

Fig. 5. The cross section of $\{U(\boldsymbol{x}', x_n) = \theta_0\}$ at $x_n = \zeta$.

with time following to the curvature flow. Then we have the following equation

$$\frac{\partial \kappa}{\partial t} = \kappa^2 \frac{\partial^2 \kappa}{\partial \theta^2} + \kappa^3.$$

This equation has the following solutions

$$\kappa(\theta, t) = \sqrt{\lambda\left(\cos 2\theta - \coth(2\lambda t)\right)}, \qquad \theta \in \mathbb{R},\ -\infty < t < 0.$$

Here $\lambda > 0$ is an arbitrarily given number. These solutions are called *Angenent's Ovals* or *Paper Clip Solutions*. See [1, 4] for these solutions.

For a mean curvature flow, Wang [22] studied an axially asymmetric traveling front that lies between two parallel planes in $\mathbb{R}^n$ for $n \geq 3$. The mean curvature flow of a graph in $\mathbb{R}^n$ is given by

$$\frac{w_t}{\sqrt{1 + |\mathrm{D}w|^2}} = \frac{1}{n-1} \sum_{i=1}^{n-1} \mathrm{D}_i \left(\frac{\mathrm{D}_i w}{\sqrt{1 + |\mathrm{D}w|^2}} \right). \tag{8}$$

Here $\mathrm{D}_i = \partial/\partial x_i$ and $\mathrm{D} = (\mathrm{D}_1, \ldots, \mathrm{D}_{n-1})$. A traveling front with speed c is given by

$$w(\boldsymbol{x}, t) = \psi(\boldsymbol{x}') + ct.$$

Here $\boldsymbol{x} = (\boldsymbol{x}', x_n) \in \mathbb{R}^n$. The profile equation is given by

$$\frac{c}{\sqrt{1 + |\mathrm{D}\psi|^2}} = \frac{1}{n-1} \sum_{i=1}^{n-1} \mathrm{D}_i \left(\frac{\mathrm{D}_i \psi}{\sqrt{1 + |\mathrm{D}\psi|^2}} \right) \tag{9}$$

Wang [22] studied a traveling front solution ψ to (9).

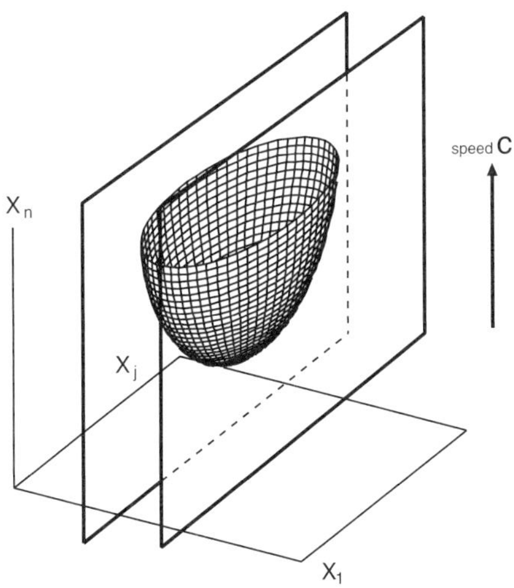

Fig. 6. A level set of an axially asymmetric traveling front in a mean curvature flow.

The relation between an axially asymmetric traveling front in Theorem 1 and that in [22] will be an interesting problem. When

$$-G'(u) = \frac{1}{\varepsilon^2}(u - u^3) \tag{10}$$

for $\varepsilon > 0$, a level set $\{\boldsymbol{x} \mid u(\boldsymbol{x}, t) = 0\}$ is approximated by the graph of $x_n = w(\boldsymbol{x}', t)$ in the limit of $\varepsilon \to 0$ as is seen in [2]. Here $u(\boldsymbol{x}, t)$ is the solution of (1) and $w(\boldsymbol{x}', t)$ is the solution of (8). See also [6] for related arguments.

Now the author conjectures that U_ε in Theorem 1 converges to ψ to (9) in Wang [22] as follows.

Conjecture 2. *Let U_ε be as in Theorem 1 for G given by (10). Then, as $\varepsilon \to 0$, a level set $\{\boldsymbol{x} \in \mathbb{R}^n \mid U_\varepsilon(\boldsymbol{x}) = 0\}$ converges to the graph of $x_n = \psi(\boldsymbol{x}')$ uniformly on any compact set in $\mathbb{R}^n$. Here ψ is a solution to (9).*

An axially asymmetric traveling front in a mean curvature flow in [22] lies between two parallel planes, and has a shape as is seen in Figure 6. A level set of an axially asymmetric traveling front in Theorem 1 is defined on the whole $\mathbb{R}^{n-1}$. The reason of the difference is as follows. In a mean curvature flow, a solution propagates only by the curvature effect. While, a solution propagates by the curvature effect and by the interaction between portions of an interface in a balanced reaction-diffusion equation. For any fixed $x_n > 0$, U can be close to $G(1)$ in $\{(\boldsymbol{x}', x_n) \mid U(\boldsymbol{x}', x_n) > \theta_0\}$, while U cannot be close to $G(-1)$ in $\{(\boldsymbol{x}', x_n) \mid U(\boldsymbol{x}', x_n) > \theta_0\}$. Then, portions of an interface attract each other, and the portions have to be apart from each other, and an interface of a traveling front cannot lie between two planes. This shows a sharp contrast between an axially asymmetric traveling front in a balanced reaction-diffusion equation and that in a mean curvature flow.

§ Acknowledgements

The author would like to express his sincere gratitude to Professor Sigurd Angenent of University of Wisconsin, Madison, Professor Hirokazu Ninomiya of Meiji University and Professor Xu-Jia Wang of Australian National University for stimulating discussions. This work is supported by JSPS Grant-in-Aid for Scientific Research (C) Grant Number 26400169.

References

[1] S. Angenent, *Shrinking doughnuts*, Nonlinear diffusion equations and their equilibrium states, 3 (Gregynog, 1989), Progr. Nonlinear Differential Equations Appl., Birkhäuser Boston, Boston, MA, **7** (1992), 21–38.

[2] X. Chen, *Generation and propagation of interfaces in reaction-diffusion equations*, J. Differential Equations, **96** (1992), 116–141.

[3] X. Chen, J.-S. Guo, F. Hamel, H. Ninomiya and J.-M. Roquejoffre, *Traveling waves with paraboloid like interfaces for balanced bistable dynamics*, Ann. Inst. H. Poincaré Anal. Non Linéaire, **24** (2007), no. 3, 369–393.

[4] P. Daskalopoulos, R. Hamilton and N. Sesum, *Classification of compact ancient solutions to the curve shortening flow*, J. Differential Geom., **84** (2010), no. 3, 455–464.

[5] M. del Pino, M. Kowalczyk and J. Wei, *On De Giorgi's conjecture in dimension $N \geq 9$*, Annals of Math., **174** (2011), 1485 1569.
[6] M. del Pino, M. Kowalczyk and J. Wei, *Traveling waves with multiple and nonconvex fronts for a bistable semilinear parabolic equation*, Comm. Pure Appl. Math., **66** (2013), 481–547.
[7] F. Hamel, R. Monneau and J.-M. Roquejoffre, *Existence and qualitative properties of multidimensional conical bistable fronts*, Discrete Contin. Dyn. Syst., **13** (2005), no. 4, 1069–1096.
[8] F. Hamel, R. Monneau and J.-M. Roquejoffre, *Asymptotic properties and classification of bistable fronts with Lipschitz level sets*, Discrete Contin. Dyn. Syst., **14** (2006), no. 1, 75–92.
[9] M. Haragus and A. Scheel, *Corner defects in almost planar interface propagation*, Ann. Inst. H. Poincaré Anal. Non Linéaire, **23** (2006), no. 3, 283–329.
[10] Y. Kurokawa and M. Taniguchi, *Multi-dimensional pyramidal traveling fronts in the Allen–Cahn equations*, Proc. Roy. Soc. Edinburgh Sect. A, **141** (2011), 1031–1054.
[11] W.-M. Ni and M. Taniguchi, *Traveling fronts of pyramidal shapes in competition-diffusion systems*, Netw. Heterog. Media, **8** (2013), no. 1, 379–395.
[12] H. Ninomiya and M. Taniguchi, *Existence and global stability of traveling curved fronts in the Allen–Cahn equations*, J. Differential Equations, **213** (2005), no. 1, 204–233.
[13] H. Ninomiya and M. Taniguchi, *Global stability of traveling curved fronts in the Allen–Cahn equations*, Discrete Contin. Dyn. Syst., **15** (2006), no. 3, 819–832.
[14] M. Taniguchi, *Traveling fronts of pyramidal shapes in the Allen–Cahn equations*, SIAM J. Math. Anal., **39** (2007), no. 1, 319–344.
[15] M. Taniguchi, *The uniqueness and asymptotic stability of pyramidal traveling fronts in the Allen–Cahn equations*, J. Differential Equations, **246** (2009), 2103–2130.
[16] M. Taniguchi, *Pyramidal traveling fronts in the Allen–Cahn equations*, RIMS Kôkyûroku, **1651** (2009), 92–109.
[17] M. Taniguchi, *Multi-dimensional traveling fronts in bistable reaction-diffusion equations*, Discrete Contin. Dyn. Syst., **32** (2012), 1011–1046.
[18] M. Taniguchi, *An $(N-1)$-dimensional convex compact set gives an N-dimensional traveling front in the Allen–Cahn equation*, SIAM J. Math. Anal., **47** (2015), no. 1, 455–476.
[19] M. Taniguchi, *Convex compact sets in $\mathbb{R}^{N-1}$ give traveling fronts of cooperation-diffusion systems in $\mathbb{R}^N$*, J. Differential Equations, **260** (2016), 4301-4338.
[20] M. Taniguchi, "Traveling front solutions in reaction-diffusion equations," submitted.

[21] M. Taniguchi, *Axially asymmetric traveling fronts in balanced bistable reaction-diffusion equations*, Ann. Inst. H. Poincaré Anal. Non Linéaire, **36** (2019), 1791–1816.

[22] X.-J. Wang, *Convex solutions to the mean curvature flow*, Ann. of Math., **173** (2011), 1185–1239.

[23] Z.-C. Wang, *Traveling curved fronts in monotone bistable systems*, Discrete Contin. Dyn. Syst., **32** (2012), 2339–2374.

Research Institute for Interdisciplinary Science, Okayama University
3-1-1, Tsushimanaka, Kita-ku, Okayama City, 700-8530, Japan
E-mail address: taniguchi-m@okayama-u.ac.jp

Advanced Studies in Pure Mathematics 85, 2020
The Role of Metrics in the Theory of Partial Differential Equations
pp. 429–439

An existence theorem for Brakke flow

Yoshihiro Tonegawa

Abstract.

Given any closed rectifiable set of codimension one with locally finite measure, [6] proved an existence result for Brakke's mean curvature flow starting from such set. After reviewing the definition and the result in [6], an outline of the proof is given.

§1. The initial data for the mean curvature flow

Given a compact smooth hypersurface $\Gamma_0 \subset \mathbb{R}^{n+1}$, one can show the existence of a family of smooth hypersurfaces $\{\Gamma_t\}_{t>0}$ for a while, where the velocity of motion of Γ_t at each point $x \in \Gamma_t$ is given by the mean curvature of Γ_t at that point. This is called the mean curvature flow (abbreviated by MCF). For example, given a sphere as the initial hypersurface, the MCF shrinks the sphere. As the radius decreases, the mean curvature increases, and in finite time, the radii of shrinking spheres become zero. This is a typical picture of the MCF and singular behavior occurs in finite time in general for any compact initial hypersurface. If the MCF is written locally as a graph, the PDE which describes this motion is similar to the well-known heat equation, since the velocity corresponds to the time-derivative of the graph and the mean curvature to the Laplacian of the graph (the mean curvature is the sum of principal curvatures) and they are supposed to be equal. It is also well-known that the heat equation has a very strong smoothing effect. Even if the initial data is very rough, the solution gets smoothed out immediately and becomes C^∞. Putting these facts together, it is natural to ask: What is the roughest set with which one can find the MCF starting from it?

Received October 30, 2018.
Revised July 1, 2019.
2010 *Mathematics Subject Classification.* Primary 53E10; Secondary 49Q05.
Key words and phrases. mean curvature flow, varifold, geometric measure theory.

There are a few well-known facts about the MCF immediately relevant to such question. First of all, the MCF has the area-decreasing property. To describe this more precisely, let us introduce some notation. Let $\mathcal{H}^n$ be the n-dimensional Hausdorff measure, which measures the hypersurface area of a given set if the set is a C^1 hypersurface in $\mathbb{R}^{n+1}$. Suppose that we have a family of hypersurfaces $\{\Gamma_t\}$ which is the MCF. Then we can check that

$$\frac{d}{dt}\mathcal{H}^n(\Gamma_t) = -\int_{\Gamma_t} |h_{\Gamma_t}|^2 \, d\mathcal{H}^n \leq 0 \tag{1}$$

and this is what is meant by the area-decreasing property. Here, h_{Γ_t} is the mean curvature vector of Γ_t. For a later use, the following characterization of the mean curvature vector is very important. For a nice C^2 hypersurface Γ, we have the following formula:

$$\int_{\Gamma} \mathrm{div}_{\Gamma} g \, d\mathcal{H}^n = -\int_{\Gamma} h_{\Gamma} \cdot g \, d\mathcal{H}^n \tag{2}$$

for any $g \in C_c^1(\mathbb{R}^{n+1};\mathbb{R}^{n+1})$. The symbol $\mathrm{div}_{\Gamma} g$ is the divergence of g restricted to the tangent space of Γ.

The property (1) shows that the hypersurface area is like the energy of the problem. For any problem of this kind, it is usually the case that one looks at the "energy class" for the initial data. Thus it is reasonable to restrict the class to the sets Γ_0 with $\mathcal{H}^n(\Gamma_0) < \infty$, or at least locally so, that is, for any $R > 0$, $\mathcal{H}^n(\Gamma_0 \cap \{x : |x| \leq R\}) < \infty$. The analogy with a solution u of the heat equation $u_t = \Delta u$ is

$$\frac{d}{dt}\int |\nabla u|^2 = -\int 2(\Delta u)^2.$$

This is very close in spirit to (1), since, if the hypersurface is a graph of u, then $\mathcal{H}^n(\Gamma) = \int \sqrt{1 + |\nabla u|^2} \approx \int (1 + \frac{1}{2}|\nabla u|^2)$ if $|\nabla u|$ is small. Since we do not need to assume $\nabla u \in L^2$ for the solution of the heat equation, the energy class may still be too restrictive, perhaps. Still, asking $\mathcal{H}^n(\Gamma_0) < \infty$ (or locally finite) is certainly a reasonable restriction. The next point to think about is whether it is reasonable to ask that Γ_0 be closed. In general, it is possible that the closure of Γ_0 is much bigger than Γ_0. All one needs to do is to add countable dense points to Γ_0 to have the closure of Γ_0 being $\mathbb{R}^{n+1}$. There seem to be some points to reflect on, but again it seems reasonable to ask Γ_0 being closed, at least to make the situation simpler. Still, we have the following well-known fact about such set Γ_0 of locally finite measure from geometric measure theory. In general, Γ_0 can be decomposed into two mutually disjoint sets, one is the so-called countably n-rectifiable set, and the other is a purely unrectifiable set (see [8, Section 13]). A set is called countably n-rectifiable if it is contained in a countable union of C^1 n-dimensional manifolds except for a null set with respect to $\mathcal{H}^n$. A purely unrectifiable set is the one which does not contain any countably n-rectifiable set of positive $\mathcal{H}^n$ measure, and it is a fractal-like irregular set. It may be possible to define some analogue of mean curvature by certain smoothing procedure for such irregular set, but the L^2 norm of any approximate mean curvature should be very large, and such irregular set will be wiped out immediately. Thus it is reasonable to consider only closed countably n-rectifiable sets as the initial data for MCF.

We are also interested in proving a time-global existence result for such initial data. If the hypersurface Γ_t is given as a boundary of time-evolving domain Ω_t, i.e., $\partial\Omega_t = \Gamma_t$, which may be regarded as a two-phase case, there are numerous approaches to obtain generalized time-global solutions such as the level-set method [2, 3] and the phase-field method [4] just to name a few. The new feature of the main existence theorem of [6] to be discussed in the following is that more general multi-phase case with a general initial data as described above can be handled compared to the existing results. For a more complete description of the relevant results, see for example [7].

§2. Definition of Brakke flow

To describe the MCF which starts from closed countably n-rectifiable set, we need the notion of Brakke flow [1]. Here, for brevity of presentation, I describe the Brakke flow without introducing the notion of varifold[1]. It is a weak solution of MCF and the idea of Brakke flow is that one makes a transition from the family of sets $\{\Gamma_t\}_{t\geq 0}$ to that of Radon measures $\{\mathcal{H}^n\lfloor_{\Gamma_t}\}_{t\geq 0}$.

Definition 2.1. *For $0 < T < \infty$ and open set $U \subset \mathbb{R}^{n+1}$, a family of Radon measures $\{\mu_t\}_{t\in[0,T]}$ is a Brakke flow in U if the following assumptions are all satisfied.*

(a) *For a.e. $t \in [0,T]$, there exist an $\mathcal{H}^n$-measurable countably n-rectifiable set $\Gamma_t \subset U$ and an $\mathcal{H}^n$-measurable function $\theta_t : \Gamma_t \to \mathbb{N}$ (where $\mathbb{N}$ is the set of natural numbers) such that $\mu_t = \theta_t\,\mathcal{H}^n\lfloor_{\Gamma_t}$, that is, μ_t is an n-dimensional Hausdorff measure restricted to Γ_t and weighted by a $\mathbb{N}$-valued function θ_t.*

[1]The Brakke flow is usually stated in the language of varifolds. On the other hand, the definition stated in this paper is equivalent.

(b) $\sup_{t\in[0,T]} \mu_t(K) < \infty$ *for all compact set* $K \subset U$.

(c) *For a.e.* $t \in [0,T]$, μ_t *has a generalized mean curvature vector* h_{μ_t} *such that*

$$\int_0^T \int_K |h_{\mu_t}|^2 \, d\mu_t dt < \infty$$

for any compact set $K \subset U$.

(d) *For any* $0 \le t_1 < t_2 \le T$ *and* $\phi \in C_c^1(U \times [0,T]; \mathbb{R}^+)$, *we have*

$$\int_U \phi \, d\mu_t \Big|_{t=t_1}^{t_2} \le \int_{t_1}^{t_2} \int_U (\nabla\phi - \phi \, h_{\mu_t}) \cdot h_{\mu_t} + \frac{\partial \phi}{\partial t} \, d\mu_t dt.$$

If Γ_t is a smooth MCF, by setting $\mu_t = \mathcal{H}^n \lfloor_{\Gamma_t}$ (so that $\theta_t = 1$) and regarding h_{μ_t} as the usual mean curvature vector of Γ_t, (a)–(d) are all satisfied. In fact, (d) is satisfied with equality in place of inequality. The generalized mean curvature vector h_{μ_t} is a $\mathbb{R}^{n+1}$-valued $\mathcal{H}^n$ measurable function such that

$$\int_U \mathrm{div}_{\Gamma_t} g \, d\mu_t = - \int_U h_{\mu_t} \cdot g \, d\mu_t$$

for all $g \in C_c^1(U; \mathbb{R}^{n+1})$. The symbol $\mathrm{div}_{\Gamma_t} g$ is the divergence of g restricted the approximate tangent space of Γ_t which exists $\mathcal{H}^n$ a.e. on Γ_t. This h_{μ_t} coincides with the classical one when Γ_t is smooth. If we assume that θ_t is equal to 1 for a.e. $t \in [0,T]$, so that $\mu_t = \mathcal{H}^n \lfloor_{\Gamma_t}$ for a.e. $t \in [0,T]$, a partial regularity result [1, 5, 9] shows that Γ_t is C^∞ MCF for a.e. $t \in [0,T]$ and $\mathcal{H}^n$ a.e. in U. For the precise statement, see [9]. The function θ_t is called multiplicity and it helps describe the somewhat undesirable "folding" of surface measures due to possible collisions.

§3. Existence theorem of [6]

Assume that $\Gamma_0 \subset \mathbb{R}^{n+1}$ is a closed countably n-rectifiable set with locally finite $\mathcal{H}^n$ measure. Additionally assume that there exists $c_1 \ge 0$ such that $\int_{\Gamma_0} \exp(-c_1|x|) \, d\mathcal{H}^n(x) < \infty$ and that $\mathbb{R}^{n+1} \setminus \Gamma_0$ is not connected. As I mentioned before, Γ_0 can be an irregular set in $\mathbb{R}^{n+1}$, even with some lower dimensional pieces, which can grow exponentially near infinity. Next, choose $E_{0,1}, \ldots, E_{0,N}$ which are non-empty, mutually disjoint open sets in $\mathbb{R}^{n+1}$ such that $\mathbb{R}^{n+1} \setminus \Gamma_0 = \bigcup_{i=1}^N E_{0,i}$, and $N \ge 2$. With the assumption of $\mathbb{R}^{n+1} \setminus \Gamma_0$ being non-connected, we can choose such a set of open sets. Here, $N \ge 2$ is arbitrary, but finite. We may regard this process of assigning a number to each connected component as "labeling". If $\mathbb{R}^{n+1} \setminus \Gamma_0$ has N connected components, it is natural to label them from 1 to N. But it is not necessary to differentiate them if one wishes. If there are countably many connected components, we just label these components from 1 to N, for some N. So the labeling of $\mathbb{R}^{n+1} \setminus \Gamma_0$ is somewhat arbitrary. Since the $\mathcal{H}^n$ measure of $\mathbb{R}^{n+1} \setminus \Gamma_0$ is locally finite, note that we always have $\bigcup_{i=1}^N \partial E_{0,i} = \Gamma_0$, that is, Γ_0 is the topological boundary of these open sets. Depending on how to put the labeling, some portion of Γ_0 may be surrounded by open sets with the same labeling. It tuned out that such portion is likely to vanish instantly at $t = 0$ by the way the MCF is constructed. In fact, the reason for assuming the non-connectedness of $\mathbb{R}^{n+1} \setminus \Gamma_0$ is that, if connected, our method of existence proof should produce trivial solution of instant vanishing at $t = 0$.

After choosing such open sets $\{E_{0,i}\}_{i=1}^N$, the conclusion of [6] is the following.

Theorem 3.1. *There exist $\{E_i(t)\}_{t\geq 0}$ for each $i = 1, \ldots, N$ and a Brakke flow $\{\mu_t\}_{t\geq 0}$ on $\mathbb{R}^{n+1}$ with the following properties.*

(a) *$E_1(t), \ldots, E_N(t)$ are open and mutually disjoint sets in $\mathbb{R}^{n+1}$ for all $t \geq 0$.*

(b) *$E_i(0) = E_{0,i}$ for $i = 1, \ldots, N$.*

(c) *$\mu_0 = \mathcal{H}^n \lfloor_{\Gamma_0}$.*

(d) *Define $d\mu = d\mu_t\, dt$ which is a Radon measure on $\mathbb{R}^{n+1} \times [0, \infty)$, and define $\Gamma_t = \mathbb{R}^{n+1} \setminus \bigcup_{i=1}^N E_i(t)$. Then for all $t > 0$, we have $\Gamma_t = \{x \in \mathbb{R}^{n+1} : (x,t) \in \operatorname{spt}\mu\}$.*

(e) *For each $t \geq 0$, $\int_{\mathbb{R}^{n+1}} \exp(-c_1|x|)\, d\mu_t(x) < \infty$.*

(f) *Each $E_i(t)$ moves continuously (resp. 1/2-Hölder continuously) with respect to the Lebesgue measure locally in space and time for $t \geq 0$ (resp. $t > 0$).*

Here, $\operatorname{spt}\mu$ is the support of μ as a measure on $\mathbb{R}^{n+1} \times [0, \infty)$. These $E_1(t), \ldots, E_N(t)$ may be considered as a set of moving domains starting from $E_{0,1}, \ldots, E_{0,N}$. μ_t is a Brakke flow starting from the initial hypersurface measure $\mathcal{H}^n \lfloor_{\Gamma_0}$, as explained in the previous section (with $U = \mathbb{R}^{n+1}$). Note that Γ_t may be considered as a time-slice of moving boundaries and (d) claims that they coincide with the space-time support of the Brakke flow μ. Due to (f), the Lebesgue measure of $E_i(t)$ does not jump suddenly in time. Due to (d), Γ_t stays non-empty and μ_t stays non-zero unless all but one $E_i(t)$ becomes empty. It is possible that Γ_t may not approach to Γ_0 as $t \to 0+$, or μ_t may not be continuous at $t = 0$. This is because some portion of Γ_0 may vanish instantly at $t = 0$. The possibility is implicitly reflected in (d) where

$\Gamma_t = \{x \in \mathbb{R}^{n+1} : (x,t) \in \operatorname{spt}\mu\}$ is not claimed for $t=0$. Thus the attainment of initial condition for general initial data may be appropriately understood through the continuity of domains $E_i(t)$ more than through the measure μ_t.

§4. Some idea of proof

The proof of the existence theorem involves a construction of time-discrete approximate MCF and certain compactness theorems of varifolds, with many estimates. Here, for the brevity, I sketch some idea of how to construct the time-discrete approximate MCF. Since Γ_0 is assumed to be only countably n-rectifiable, we need to cook up some reasonable approximate quantity which should reduce to the usual mean curvature if Γ_0 is smooth. Let $\Gamma \subset \mathbb{R}^{n+1}$ be a closed countably n-rectifiable set with $\mathcal{H}^n(\Gamma) < \infty$ for simplicity. Let me explain how an approximate mean curvature vector for Γ is defined in [6]. Let $\varepsilon > 0$ be small and let $\Phi_\varepsilon(x) = (2\pi\varepsilon^2)^{-\frac{n+1}{2}}\exp(-\frac{|x|^2}{2\varepsilon^2})$. We have $\int_{\mathbb{R}^{n+1}} \Phi_\varepsilon(x)\,dx = 1$ and $\lim_{\varepsilon\to 0+}\Phi_\varepsilon = \delta_0$ (delta function). In [6], we also truncate Φ_ε so that it has a compact support, but I skip it for simplicity. We use (2) to motivate the definition of approximate mean curvature vector. Recall that $\operatorname{div}_\Gamma g = \operatorname{tr}(T_x\Gamma \circ \nabla g) = \sum_{i,j=1}^{n+1}(T_x\Gamma)_{ij}\frac{\partial g_i}{\partial x_j}$, where $g = (g_1,\ldots,g_{n+1})$ and $T_x\Gamma : \mathbb{R}^{n+1} \to T_x\Gamma$ is the matrix representing the orthogonal projection from $\mathbb{R}^{n+1}$ to the tangent space $T_x\Gamma$ of Γ at x. One important fact about $\mathcal{H}^n$ measurable countably n-rectifiable set with locally finite $\mathcal{H}^n$ measure is the existence of approximate tangent space, which has been already mentioned when the generalized mean curvature was introduced. More precisely, for $\mathcal{H}^n$ a.e. $x \in \Gamma$, there exists a unique n-dimensional subspace $T_x\Gamma$ which is a measure theoretic tangent space of Γ. Because of this, $\int_\Gamma \operatorname{div}_\Gamma g\,d\mathcal{H}^n$ for $g \in C_c^1(\mathbb{R}^{n+1};\mathbb{R}^{n+1})$ is a well-defined quantity. For $i \in \{1,\ldots,n+1\}$ and $y \in \mathbb{R}^{n+1}$ fixed, let g be a smooth vector field whose components are all zero except for the i-th component, and let that component be $\Phi_\varepsilon(x-y)$. To motivate the definition, let us assume that Γ is smooth. Use (2) with this g. Writing the i-th component of h_Γ as h_Γ^i, we have

$$\int_\Gamma \sum_{j=1}^{n+1}(T_x\Gamma)_{ij}\frac{\partial\Phi_\varepsilon}{\partial x_j}(x-y)\,d\mathcal{H}^n(x) = -\int_\Gamma h_\Gamma^i(x)\Phi_\varepsilon(x-y)\,d\mathcal{H}^n(x).$$

We then divide both sides by $\varepsilon + \int_\Gamma \Phi_\varepsilon(x-y)\,d\mathcal{H}^n(x)$. When $y \in \Gamma$, due to the property of Φ_ε, the divided right-hand side converges to $-h_\Gamma^i(y)$ as $\varepsilon \to 0+$. When $y \notin \Gamma$, the same quantity converges to 0. Motivated by this observation, one is led to define the approximate mean curvature vector $h_{\varepsilon,\Gamma}$ of Γ as

$$\tilde{h}^i_{\varepsilon,\Gamma}(y) = -\frac{\int_\Gamma \sum_{j=1}^{n+1}(T_x\Gamma)_{ij}\frac{\partial\Phi_\varepsilon}{\partial x_j}(x-y)\,d\mathcal{H}^n(x)}{\varepsilon + \int_\Gamma \Phi_\varepsilon(x-y)\,d\mathcal{H}^n(x)},$$

$$\tilde{h}_{\varepsilon,\Gamma} = (\tilde{h}^1_{\varepsilon,\Gamma},\ldots,\tilde{h}^{n+1}_{\varepsilon,\Gamma}),$$

$$h_{\varepsilon,\Gamma}(y) = (\Phi_\varepsilon * \tilde{h}_{\varepsilon,\Gamma})(y) = \int_{\mathbb{R}^{n+1}} \Phi_\varepsilon(x-y)\tilde{h}_{\varepsilon,\Gamma}(y)\,dy.$$

As long as Γ is $\mathcal{H}^n$ measurable and countably n-rectifiable with locally finite $\mathcal{H}^n$ measure, $\tilde{h}_{\varepsilon,\Gamma}$ and $h_{\varepsilon,\Gamma}$ are well-defined and smooth vector fields. Here, the reason that we take another convolution for $h_{\varepsilon,\Gamma}$ is that we can obtain a set of good estimates (see [6, Section 5.2] for the detail) and having these two similar quantities is essential. One can prove that

$$\sup_{y\in\mathbb{R}^{n+1}}\left\{\varepsilon^2|h_{\varepsilon,\Gamma}(y)|, \varepsilon^4|\nabla h_{\varepsilon,\Gamma}(y)|\right\} \le c(n)(1+\varepsilon\mathcal{H}^n(\Gamma)).$$

Define, for $\Delta t \ll \varepsilon^4$, $f_{\varepsilon,\Gamma}(x) = x + \Delta t\, h_{\varepsilon,\Gamma}(x)$. Then $f_{\varepsilon,\Gamma}$ is a diffeomorphism on $\mathbb{R}^{n+1}$. Then, starting from Γ_0, we may inductively define $\Gamma_{(k+1)\Delta t} = f_{\varepsilon,\Gamma_{k\Delta t}}(\Gamma_{k\Delta t})$ for $k \in \mathbb{N}$. So starting from Γ_0 we move $\Gamma_{k\Delta t}$ by the approximate mean curvature vector $h_{\varepsilon,\Gamma_{k\Delta t}}$ for the time interval of Δt. This seems like a good way to construct an approximate mean curvature flow. In fact, one can estimate the change of $\mathcal{H}^n$ measure of $\Gamma_{k\Delta t}$ and prove that it is decreasing in time with some small errors. However, there are two essential problems with this approach. The first problem is that this does not allow any topological changes for the flow, since $f_{\varepsilon,\Gamma}$ is always a diffeomorphism. For example, for $n = 1$ we may want to split a cross figure into two triple junctions connected by a short line segment. Also some domain should vanish in finite time. With $f_{\varepsilon,\Gamma}$, these changes are not possible. The second problem is that, when we take a limit $\varepsilon \to 0$, we do not have any information on the scale smaller than ε since the motion itself is smoothed out by Φ_ε. For this reason, the actual construction in [6] is different. In each step, before we compute the approximate mean curvature vector, we insert a measure-reducing Lipschitz deformation which has a regularity effect in ε-scale, and also allows desired topological changes. We also work in the framework which guarantees non-triviality of the MCF in the end and define the following notion of open partitions.

Definition 4.1. *A collection of subsets $E_1,\ldots,E_N \subset \mathbb{R}^{n+1}$ is called an open partition of N elements if*

(a) $E_1, \ldots, E_N$ *are open and mutually disjoint,*
(b) $\mathcal{H}^n(\mathbb{R}^{n+1} \setminus \bigcup_{i=1}^N E_i) < \infty$,
(c) $\bigcup_{i=1}^N \partial E_i$ *is countably n-rectifiable.*

If these properties are satisfied, we write $\mathcal{E} := \{E_1, \ldots, E_N\} \in \mathcal{OP}_N$. *We also write* $\partial\mathcal{E} = \bigcup_{i=1}^N \partial E_i$.

Starting from $E_{0,1}, \ldots, E_{0,N}$, which is in fact an open partition of N elements according to the above definition, we move these domains step by step. As stated above, we move these domains by a suitable class of measure-reducing Lipschitz deformation and then by the approximate mean curvature. For this purpose, we define

Definition 4.2. *For a given* $\mathcal{E} = \{E_1, \ldots, E_N\} \in \mathcal{OP}_N$, *a function* $f : \mathbb{R}^{n+1} \to \mathbb{R}^{n+1}$ *is called $\mathcal{E}$-admissible if it is Lipschitz and satisfies the following. First define* $\tilde{E}_i := \operatorname{int} f(E_i)$ *for each* $i = 1, \ldots, N$, *where* $\operatorname{int} f(E_i)$ *is the set of interior points of* $f(E_i) \subset \mathbb{R}^{n+1}$. *Then*

(a) $\tilde{E}_1, \ldots, \tilde{E}_N$ *are mutually disjoint,*
(b) $\mathbb{R}^{n+1} \setminus \bigcup_{i=1}^N \tilde{E}_i \subset f(\bigcup_{i=1}^N \partial E_i)$,
(c) $\sup_{x \in \mathbb{R}^{n+1}} |f(x) - x| < \infty$.

We define $f_\star\mathcal{E} := \{\tilde{E}_1, \ldots, \tilde{E}_N\}$ *and* $\partial f_\star\mathcal{E} := \bigcup_{i=1}^N \partial\tilde{E}_i$.

One can easily prove that $f_\star\mathcal{E} \in \mathcal{OP}_N$ if f is $\mathcal{E}$-admissible. Also, the definition allows some desirable topological changes for the boundary (see [6, Section 4.3] for some typical pictures). To obtain a suitable measure-reducing Lipschitz deformation, we need to introduce a class of test functions. For $j \in \mathbb{N}$ (which eventually goes to ∞), define

$$\mathcal{A}_j := \big\{\phi \in C^2(\mathbb{R}^{n+1}; \mathbb{R}^+) : \phi(x) \le 1,\ |\nabla\phi(x)| \le j\phi(x),\\ \|\nabla^2\phi(x)\| \le j\phi(x) \text{ for all } x \in \mathbb{R}^{n+1}\big\}.$$

Note that $\phi(x) \le \exp(j|x-y|)\phi(y)$ for any x, y, thus ϕ behaves like a constant function within the length scale of $o(1/j)$ while any test function $\phi \in C_c^2(\mathbb{R}^{n+1}; \mathbb{R}^+)$ may be uniformly approximated by a function in $\mathcal{A}_j$ for large j. Using $\mathcal{A}_j$, we next define a restricted class of admissible functions as follows.

Definition 4.3. *For* $\mathcal{E} = \{E_1, \ldots, E_N\} \in \mathcal{OP}_N$ *and* $j \in \mathbb{N}$, *define* $\mathbf{E}(\mathcal{E}, j)$ *to be the set of all $\mathcal{E}$-admissible functions* $f : \mathbb{R}^{n+1} \to \mathbb{R}^{n+1}$ *such that, writing* $\{\tilde{E}_i\}_{i=1}^N = f_\star\mathcal{E}$,

(a) $|f(x) - x| \le 1/j^2$ *for all* $x \in \mathbb{R}^{n+1}$,
(b) $\mathcal{L}^{n+1}((\tilde{E}_i \setminus E_i) \cup (E_i \setminus \tilde{E}_i)) \le 1/j$ *for all* $i = 1, \ldots, N$,
(c) $\mathcal{H}^n\lfloor_{\partial f_\star\mathcal{E}}(\phi) \le \mathcal{H}^n\lfloor_{\partial\mathcal{E}}(\phi)$ *for all* $\phi \in \mathcal{A}_j$.

The condition (c) means that the deformation is measure reducing for all test functions in $\mathcal{A}_j$. Since $1 \in \mathcal{A}_j$ in particular, $\mathcal{H}^n(\partial f_\star \mathcal{E}) \leq \mathcal{H}^n(\partial \mathcal{E})$. It turned out that (c) puts a strong restriction on $\mathbf{E}(\mathcal{E}, j)$ in the sense that, if $\partial\mathcal{E}$ is a C^2 hypersurface with no singularities, then for large enough j, $\mathbf{E}(\mathcal{E}, j)$ consists only of the identity map $f(x) = x$. With this $\mathbf{E}(\mathcal{E}, j)$, define the following quantity which is the indication of how much one may reduce the measure of $\partial\mathcal{E}$ by this restricted class of Lipschitz deformations.

Definition 4.4. *For $\mathcal{E} \in \mathcal{OP}_N$ and $j \in \mathbb{N}$, we define*

$$\Delta_j \|\partial\mathcal{E}\| := \inf_{f \in \mathbf{E}(\mathcal{E}, j)} (\mathcal{H}^n(\partial f_\star \mathcal{E}) - \mathcal{H}^n(\partial\mathcal{E})).$$

In general, $\Delta_j\|\partial\mathcal{E}\| \leq 0$ since $\mathbf{E}(\mathcal{E}, j)$ contains the identity map. If $\partial\mathcal{E}$ is C^2 hypersurface, as mentioned above, $\mathbf{E}(\mathcal{E}, j)$ consists of the identity map for all sufficiently large j and $\Delta_j\|\partial\mathcal{E}\| = 0$. If $\Delta_j\|\partial\mathcal{E}\|$ is close to 0, the idea is that $\partial\mathcal{E}$ is almost measure minimizing for this restricted class of deformations.

With these tools defined, the construction of approximate MCF proceeds as follows. First, we are given $\mathcal{E}_0 := \{E_{0,1}, \ldots, E_{0,N}\} \in \mathcal{OP}_N$ as the initial data. For each $j \in \mathbb{N}$, fix $0 < \varepsilon_j < j^{-6}$ and $0 < \Delta t_j < (\varepsilon_j)^{3n+20}$. The choices of these numbers come from needs to satisfy various estimates in [6], but the key features are that $\varepsilon_j \ll j^{-2}$ and $\Delta t_j \ll \varepsilon_j$. Define $\mathcal{E}_{j,0} := \mathcal{E}_0$. Assume inductively that we have chosen $\mathcal{E}_{j,k} \in \mathcal{OP}_N$. Then, we choose $f \in \mathbf{E}(\mathcal{E}_{j,k}, j)$ such that

$$\mathcal{H}^n(\partial f_\star \mathcal{E}_{j,k}) - \mathcal{H}^n(\partial\mathcal{E}_{j,k}) \leq (1 - j^{-5})\Delta_j\|\partial\mathcal{E}_{j,k}\|.$$

The factor $1 - j^{-5}$ is not so important but we need to choose some f which numerically accounts for the extent of $\partial\mathcal{E}_{j,k}$ being non-minimizing. Recall also that f would be the identity map if $\partial\mathcal{E}_{j,k}$ is C^2 and j is sufficiently large. Set $\mathcal{E}^*_{j,k+1} := f_\star\mathcal{E}_{j,k}$ with this f. Since f is $\mathcal{E}_{j,k}$-admissible, $\mathcal{E}^*_{j,k+1} \in \mathcal{OP}_N$. We then compute the approximate mean curvature of $\partial\mathcal{E}^*_{j,k+1}$, and set $h_{j,k+1} := h_{\varepsilon_j, \partial\mathcal{E}^*_{j,k+1}}$. We move $\mathcal{E}^*_{j,k}$ by the diffeomorphism $\hat{f}(x) := x + h_{j,k+1}\Delta t_j$ and define $\mathcal{E}_{j,k+1} := \hat{f}_\star\mathcal{E}^*_{j,k+1}$, closing the inductive cycle. We repeat this until $k \approx j/\Delta_j t$. We naturally define a piece-wise constant flow $\mathcal{E}_j(t)$ so that $\mathcal{E}_j(k\Delta t_j) = \mathcal{E}_{j,k}$.

One carries out various error estimates for each maps (f and $\hat{f}$ above) and proves (see [6, Section 6] for the precise statements with $c_1 = 0$ there)

$$\sup_{t\in[0,j]} \mathcal{H}^n(\partial\mathcal{E}_j(t)) \le \mathcal{H}^n(\partial\mathcal{E}_0) + \varepsilon_j^{\frac{1}{8}},$$

$$\int_0^j \int_{\mathbb{R}^{n+1}} \frac{1}{4}|\tilde{h}_{\varepsilon_j,\partial\mathcal{E}_j(t)}|^2(\Phi_{\varepsilon_j} * \mathcal{H}^n\lfloor_{\partial\mathcal{E}_j(t)} + \varepsilon_j)\,dxdt$$
$$+ \int_0^j \frac{1-j^{-5}}{\Delta t_j}\big|\Delta_j\|\partial\mathcal{E}_j(t)\|\big|\,dt \le \mathcal{H}^n(\partial\mathcal{E}_0) + \varepsilon_j^{\frac{1}{8}}$$

and also similar estimates weighted by test functions $\phi \in \mathcal{A}_j$. These estimates in turn show that $\partial\mathcal{E}_j(t)$ converges subsequentially to a Brakke flow. The hard part is to prove certain compactness results for varifolds which are analogous to Allard's compactness theorem of integral varifolds. The main such compactness theorem is the following ([6, Theorem 8.6]).

Theorem 4.1. *Suppose that* $\{\mathcal{E}_j\}_{j=1}^\infty \in \mathcal{OP}_N$ *and* $\{\varepsilon_j\}_{j=1}^\infty$ *satisfy*
(a) $\lim_{j\to\infty} j^4\varepsilon_j = 0$,
(b) $\sup_j \mathcal{H}^n(\partial\mathcal{E}_j) < \infty$,
(c) $\liminf_{j\to\infty} \int |\tilde{h}_{\varepsilon_j,\partial\mathcal{E}_j}|^?(\Phi_{\varepsilon_j} * \mathcal{H}^n\lfloor_{\partial\mathcal{E}_j} + \varepsilon_j)\,dx < \infty$,
(d) $\lim_{j\to\infty} j^{2(n+1)}\Delta_j\|\partial\mathcal{E}_j\| = 0$.

Then there exists a converging subsequence $\{\partial\mathcal{E}_{j_l}\}_{l=1}^\infty$ *whose limit is an integral varifold. In particular, the limit measure of* $\{\mathcal{H}^n\lfloor_{\partial\mathcal{E}_{j_l}}\}_{l=1}^\infty$ *is of the form* $\theta\mathcal{H}^n\lfloor_\Gamma$, *where* Γ *is a countably* n*-rectifiable set and* $\theta : \Gamma \to \mathbb{N}$ *is a positive integer-valued* $\mathcal{H}^n$ *measurable function.*

The sequence of approximate flows satisfies the assumptions of above theorem for a.e. $t > 0$ and we can be assured that the limiting measures have integer multiplicities for a.e. $t > 0$, if not for all t. Additionally we need to prove that the limit is a Brakke flow and that we have a set of continuously moving domains $E_i(t)$ for $i = 1, \ldots, N$ and the key ingredients are the relative isoperimetric inequality and Huisken's monotonicity formula.

References

[1] K. Brakke, *The Motion of a Surface by its Mean Curvature*, Math. Notes, **20**, Princeton Univ. Press, Princeton, NJ, 1978.

[2] Y.-G. Chen, Y. Giga and S. Goto, *Uniqueness and existence of viscosity solutions of generalized mean curvature flow equations*, J. Differential Geom., **33** (1991), no. 3, p. 749–786.

[3] L. C. Evans and J. Spruck, *Motion of level sets by mean curvature*, I, J. Differential Geom., **33** (1991), no. 3, p. 635–681.

[4] T. Ilmanen, *Convergence of the Allen-Cahn equation to Brakke's motion by mean curvature*, J. Differential Geom., **38** (1993), no. 2, p. 417–461.

[5] K. Kasai and Y. Tonegawa, *A general regularity theory for weak mean curvature flow*, Calc. Var. PDE., **50** (2014), p. 1–68.

[6] L. Kim and Y. Tonegawa, *On the mean curvature flow of grain boundaries*, Annales de l'Institut Fourier (Grenoble), **67** (2017) no. 1, p. 43–142.

[7] C. Mategazza, M. Novaga, A. Pluda and F. Schulze, *Evolution of networks with multiple junctions*, arXiv:1611.08254.

[8] L. Simon, *Lectures on geometric measure theory*, in Proceedings of the Centre for Mathematical Analysis, Australian National University, vol. 3, 1983.

[9] Y. Tonegawa, *A second derivative Hölder estimate for weak mean curvature flow*, Adv. Cal. Var., **7** (2014), no. 1, p. 91–138.

Department of Mathematics, Tokyo Institute of Technology,
152-8551, Tokyo, Japan
E-mail address: `tonegawa@math.titech.ac.jp`

Advanced Studies in Pure Mathematics 85, 2020
The Role of Metrics in the Theory of Partial Differential Equations
pp. 441–452

Equivalent extensions of partial differential equations on surfaces

Lindsay Martin, Jay Chu and Richard Tsai

Abstract.

This paper gives a general framework for the computation of several classes of partial differential equations (PDEs) on hypersurfaces. The approach is to work in a sufficiently thin narrow band, surrounding the surface, in the embedding Euclidean space and appropriately extend the differential operators so that the solutions on the narrow band and surface are equivalent. The solutions in the narrow band are equivalent to the solutions on the surface as in the former are constant along surface normal extensions of the latter. Consequently, it is possible to use existing (high order) numerical methods developed on grids in Euclidean space to solve PDEs on surfaces, with narrow bands whose widths are a small constant multiple of uniform Cartesian grid spacing. We apply the formulation to PDEs that originate from variational principles defined on surfaces as well as Hamilton-Jacobi-Bellman equations on surfaces that are derived from optimal control problems. This framework also provides a systematic way for solving PDE's on the unstructured point clouds that are sampled from the surface.

§1. Introduction

We present a general framework for the computation of partial differential equations on surfaces. We are motivated by the many applications that require numerical solutions of partial differential equations defined on (moving) interfaces. For example, the modeling of two phase fluids under the presence of surfactant molecules adhering to the fluid interface. Another example comes from solving Hamilton-Jacobi-Bellman

Received March 30, 2019.
Revised July 2, 2019.
2010 *Mathematics Subject Classification.* Primary 70H20, 49J20, 35R01 58E10, 65N06.
Key words and phrases. Hamilton-Jacobi-Bellman equations, obstacle problems.

(HJB) equations on surfaces. The corresponding characteristic curves can be extracted and used in many applications. Some examples include mesh generation [6], path planning [9, 15], and brain mapping [2, 10].

Typically, in order to derive numerical methods to solve PDEs defined on surfaces, one has to deal first with how the surface is defined. There are a large body of works using triangulated surfaces — with such approximation of surfaces, it is necessary to translate the continuum geometrical quantifies, such as curvatures, to consistent approximations involving the adjacent triangles. If the surface is given by one parameterization, one may simply discretize the surface PDEs using the parameterization. However, this is not the case for most surfaces. If the surface is defined by different coordinate patches, then one needs to deal with how discretizations using different patches can be "stitched" together. In this paper, we generalize the approach introduced in [4] to compute a class of optimal control problems on smooth closed surfaces. We also demonstrate that the proposed approach can be applied to discretize and solve obstacle problems defined on smooth closed surfaces.

With the proposed extension approach, one avoids the above problems, and one would be able to discretize the resulting problems defined in the embedding Euclidean space using state-of-the-art numerical schemes, without the need of coordinate patches nor triangulation. The approach also provides a systematic and natural way to derive and implement boundary conditions at the boundaries of the narrow band. Furthermore, independent of the narrow band's thickness (up to what the geometry of the surface allows), the solutions of the extended problems will be the constant along surface normal extension of the solutions to the original problems given on the surface. This is a major difference of the proposed approach to other embedding methods.

§2. Related work

Level set methods and closest point methods are two approaches to solving PDEs on surfaces embedded in Euclidean spaces. In level set methods, e.g., [3, 5], partial derivatives on a surface, Γ, are written in the form of the orthogonal projections of the gradient operator in the Euclidean space. The idea is that the restriction of the projected gradient or divergence operators onto the right level surface will have the values as the surface gradients or divergence operators that one wishes to approximate. Equations in the narrow band are derived using such type of projected differential operators. Extending functions and differential operators defined on surfaces, via a closest point mapping, to the embedding Euclidean space, is used in [11, 13] to compute solutions to a class of PDEs. The methods assume the constant-along-normal extension of quantities defined on the surface and replace the surface gradient operator, ∇_Γ, by the gradient operator, ∇, defined in the embedding Euclidean space. The methods typically require iterations to enforce the constraint that the solutions are constant-along-normal. The solution of the surface PDE is extracted by restricting the solution on the narrow band to Γ and interpolating.

§3. Extension of surface gradient via closest point mappings

Let Γ be a closed C^2-surface in $\mathbb{R}^3$ of co-dimension one. For any $\epsilon > 0$, we define the narrow band of Γ as

$$T_\epsilon = \left\{ z \in \mathbb{R}^3 : \min_{x\in\Gamma} |x - z| < \epsilon \right\}. \tag{1}$$

The closest point mapping $P_\Gamma : T_\epsilon \mapsto \Gamma$ is defined by

$$P_\Gamma(z) := \mathrm{argmin}_{x\in\Gamma} |z - x|. \tag{2}$$

If more than one point of Γ are equidistant to z, randomly assign one of them to be $P_\Gamma(z)$. Since Γ is assumed to be C^2, P_Γ is differentiable while ϵ is small enough. Moreover, if the distant function

$$d_\Gamma(z) := |P_\Gamma(z) - z|$$

is differentiable, then $P_\Gamma(z) = z - d_\Gamma(z)\nabla d_\Gamma(z)$.

Finally, we define the constant normal extension of function on Γ, $u : \Gamma \to \mathbb{R}$, to be $\overline{u} : T_\epsilon \to \mathbb{R}$ given by

$$\overline{u}(z) := u(P_\Gamma z).$$

Again, our goal is to define extended PDEs on the narrow band so that the solution is the constant normal extension to the solution of the desired equation on the surface.

It can be derived easily, see e.g. [4], that the surface gradient $\nabla_\Gamma u$, when embedded as a vector in $\mathbb{R}^3$, satisfies

$$\overline{\nabla_\Gamma u}(z) = B(z;\mu)\nabla_z \overline{u}(z), \tag{3}$$

where $B(z;\mu) = B_0(z) + \mu B_1(z)$,

$$B_0(z) := \sigma_1^{-1} t_1 \otimes t_1 + \sigma_2^{-1} t_2 \otimes t_2,$$
$$B_1(z) := n \otimes n,$$

and t_1, t_2 are the two orthonormal tangent vectors corresponding to the directions that yield the principle curvatures of Γ, n is the unit normal vector, μ can be any real number, and σ_1 and σ_2 are two largest singular values of DP_Γ, where DP_Γ is the Jacobian matrix of P_Γ.

Next, we give two examples of how (3) is used to define equivalent PDEs in the narrow band.

§4. Extension of integro-differential operators

Consider integro-differential operators of the form

$$E_\Gamma[u] := \int_\Gamma L(x, u(x), \nabla_\Gamma u(x))\, dS(x).$$

By extending the integral on T_ϵ and replacing $\nabla_\Gamma u$ by $B\nabla\bar{u}$, we can find an equivalent energy function

$$E_{T_\epsilon}[v] := \int_{T_{\epsilon'}} \bar{L}(v(z), \nabla v(z))\, dz$$

defined on $\mathcal{H}^1(T_\epsilon)$ that satisfies

$$E_{T_\epsilon}[\bar{u}] := E_\Gamma[u] \tag{4}$$

for all $u \in \mathcal{H}^1(\Gamma)$. The integrand $\bar{L}$ is derived from L by

$$\bar{L}(v(z), \nabla v(z)) = K(z) L(v(z), B\nabla v(z)) J(z), \tag{5}$$

where $K(z) = k_\epsilon(d_\Gamma(z))$ with $\int_{-\epsilon}^{\epsilon} k_\epsilon(\eta)\, d\eta = 1$, and J is the Jacobian that comes from the change of variables $z \mapsto P_\Gamma z \in \Gamma$. In fact it is derived in [8] that $J(z) = \sigma_1(z)\sigma_2(z)$. It can be shown that if the energy function $E_\Gamma[u]$ is convex, then the extended energy E_{T_ϵ} is also convex by choosing positive parameter μ to convexify the energy. Let u^* and v^* be, respectively, the minimizer of $E_\Gamma[u]$ and $E_{T_{\epsilon'}}[v]$ in suitable Hilbert spaces. By the uniqueness of minimizers for convex problems, we obtain the equivalence of the minimizers:

$$v^* \equiv u^* \iff v^*(x) = \overline{u^*}(x),\ \forall x \in T_{\epsilon'}.$$

The detail of the construction of the extension E_{T_ϵ} for a class of differentiable energies and the proof of the equivalence of minimizers can be found in [4].

One of the objectives of this paper is to show that the above approach can be generalized to consider a wider class of problems which are not necessarily differentiable or which involve inequalities. The idea is to use numerical optimization techniques to solve for the problems derived from a discretization of E_{T_ϵ}, instead of solving for the Euler-Lagrange equation of E_{T_ϵ}. For this purpose, we consider the obstacle problem defined on the manifold Γ. The energy function E is given by

$$\min_{\varphi_1 \le u \le \varphi_2} E(u) = \frac{1}{2}\int_\Gamma |\nabla_\Gamma u|^2\,dx, \tag{6}$$

where φ_1 and φ_2 are obstacle functions. For non-flat surfaces, it is generally more difficult to find the minimizer in (6) directly. We generalize the approach proposed in [14], where an obstacle problem on a planar domain is solved via the solution of an unconstrained but non-differentiable optimization problem. First (6) is replaced by the relaxed minimization problem:

$$\min_u \hat{E}(u) = \frac{1}{2}\int_\Gamma |\nabla_\Gamma u|^2 dx + \mu_1 \int_\Gamma (\varphi_1 - u)_+ dx + \mu_2 \int_\Gamma (u - \varphi_2)_+ dx \tag{7}$$

with some appropriate parameters $\mu_1, \mu_2 > 0$. Here the notation $(\cdot)_+$ is defined by $(\psi)_+ = \max\{\psi, 0\}$. Then we derive the corresponding extended energy function on T_ϵ

$$\begin{aligned}\min_v \hat{E}_{T_\epsilon}(v) = {} & \frac{1}{2}\int_{T_\epsilon} K(z)|B(z,\mu)\nabla v|^2 J(z)\,dz \\ & + \mu_1 \int_{T_\epsilon} K(z)(\overline{\varphi}_1 - v)_+\,dz + \mu_2 \int_{T_\epsilon} K(z)(v - \overline{\varphi}_2)_+\,dz,\end{aligned} \tag{8}$$

where $\overline{\varphi}_1$ and $\overline{\varphi}_2$ are the constant normal extensions of φ_1 and φ_2, respectively. With positive parameters μ_1 and μ_2, we can see the minimizer of (8) must be the constant normal extension of the minimizer of (7). Now (8) is a variable coefficient version of the one analyzed in [14], and can be solved easily. See Section 6.1.

Another objective is to consider optimization of optimal control problems defined on surfaces. This type of problems involve optimizing functionals of curves, which correspond to solution trajectories of some given control systems. This topic is explored in the next section.

§5. Hamilton-Jacobi-Bellman Equations

Suppose we are interested in trajectories on Γ governed by the dynamical system where $a : [0, \infty) \to A$ is the given control of the system:

$$\frac{dy_x}{dt}(t) = f(y_x(t), a(t))a(t), \ t > 0 \tag{9}$$

$$y_x(0) = x, \tag{10}$$

where $f : \Gamma \times A \to \mathbb{R}$ represents the velocity with $A = S^{n-1} \cap TM(\Gamma)$ where $TM(\Gamma)$ is the tangent bundle of Γ. Then given a running cost function, r, and exit time penalty, g, the value function of the optimal control problem, $u : \Gamma \to \mathbb{R}$, is defined to be the minimal total cost to $\mathcal{T} \subset \Gamma$ starting at x and is given by

$$u(x) = \inf_{a(\cdot)\in\mathcal{A}_x} \left\{ \int_0^T r(y_x(t), a(t))\, dt + g(y_x(T)) \right\}, \tag{11}$$

where $T = \min\{t \mid y_x(t) \in \mathcal{T}\}$ and $\mathcal{A}_x$ is the set of controls where the trajectories remain in Γ.

The corresponding Hamilton-Jacobi-Bellman equation on Γ can be derived via the dynamic programming principle [1] and is given by

$$\min_{a\in A_x} \left\{ r(x,a) + \nabla_\Gamma u(x) \cdot f(x,a)a \right\} = 0, \quad x \in \Gamma \backslash \mathcal{T}, \tag{12}$$

where $A_x := S^{n-1} \cap T_x(\Gamma)$ and $T_x(\Gamma)$ is the collection of tangent vectors to Γ at the point x. The boundary condition is

$$u(x) = g(x), \quad x \in \mathcal{T}. \tag{13}$$

It is a classic result that the value function u coincides with the unique viscosity solution of (12)–(13) under certain assumptions [1].

We can extend the optimal control problem on the surface to one in the narrow band, T_ϵ, as defined in (1).

The extended dynamical system in T_ϵ is given by

$$\frac{dy}{dt}(t) = f(P_\Gamma y(t), a(t))B(y(t), \mu)a(t), \; t > 0 \tag{14}$$

$$y(0) = z, \; z \in T_\epsilon. \tag{15}$$

Now, let $\overline{\mathcal{T}} = \{z \in T_\epsilon \mid P_\Gamma z \in \mathcal{T}\}$. The value function $v : T_\epsilon \to \mathbb{R}$ is the minimal total cost to $\overline{\mathcal{T}}$ starting at z and is given by

$$v(z) = \inf_{a(\cdot)\in\mathcal{A}_z} \left\{ \int_0^{\overline{T}} r(P_\Gamma y_z(t), a(t))\, dt + g(P_\Gamma y_z(\overline{T})) \right\}, \tag{16}$$

where $\overline{T} = \min\{t \mid y_z(t) \in \overline{\mathcal{T}}\}$ and $\mathcal{A}_z$ is the set of controls where the trajectories remain on the parallel surface to which z belongs.

The corresponding Hamilton-Jacobi-Bellman equation on T_ϵ is

$$\min_{a\in A_z} \left\{ r(P_\Gamma z, a) + \nabla v(z) \cdot f(P_\Gamma z, a)B(z,\mu)a \right\} = 0, \quad z \in T_\epsilon \backslash \overline{\mathcal{T}}, \tag{17}$$

and the boundary condition is

$$v(z) = \overline{g}(z), \quad z \in \overline{\mathcal{T}}. \tag{18}$$

Again under certain assumptions, the value function coincides with the unique viscosity solution [1]. It will be shown in a forthcoming paper, [12], that the viscosity solution to (17)–(18), derived in this paper, is the constant normal extension of the viscosity solution of (12)–(13).

§6. Numerical Simulations

We present several examples of the formulation demonstrated on surfaces of co-dimension one in three dimensions. We use a uniform Cartesian grid with step size h, and denote the discretized narrow band as T_ϵ^h. Provides that ϵ is sufficiently small compared to the curvatures of the surface, the choice of it will not influence the solution to the original problem, but it may place some constraint on the order of approximation to the solutions. High order numerical approximations typically require wider stencil, which in turn, require larger ϵ. In the following simulations, we chose $\epsilon = 4h$.

We must take careful consideration of the discretization near the boundary, ∂T_ϵ^h. When approximating the partial derivatives of v, a neighboring grid node, called a ghost node, may lie outside of T_ϵ^h. In our implementation, we provide a boundary closure procedure to enforce the fact that the solution is a constant along normal function. For each ghost node, we perform the following procedure:

(1) Project the node into the narrow band.
(2) The value at the ghost node is then calculated by interpolating grid values surrounding the projected point. Formally, we must use an interpolation scheme of order higher to the discretization of the PDE on T_ϵ^h.

6.1. Example: the obstacle problem on a surface

We consider the obstacle problem on the unit sphere $\{(x, y, z) \in \mathbb{R}^3 | x^2 + y^2 + z^2 = 1\}$ with the following obstacles functions given by

$$\varphi_1(x, y, z) := \begin{cases} 1, & |x + y + z| > 1.7, \\ -2, & \text{otherwise}, \end{cases}$$

and

$$\varphi_2(x, y, z) := \begin{cases} -1, & |-x - y + z| > 1.7 \\ 2, & \text{otherwise}. \end{cases}$$

We approximate the minimizer, u, of (6) on the unit sphere by finding the minimizer of (8) in an extended narrow band of the unit sphere:

$$1 - 4h \leq \sqrt{x^2 + y^2 + z^2} \leq 1 + 4h.$$

The parameters used in the relaxation (8) are chosen by $\mu_1 = \mu_2 = 300$.

A uniform Cartesian grid, $h\mathbb{Z}^3$, is placed over the domain. For the unit sphere, the extension of the surface gradient is simply given by $\overline{\nabla_\Gamma u(z)} = r\nabla \overline{u(z)}$ with $r = \sqrt{x^2 + y^2 + z^2}$ and the Jacobian is r^{-2}. The discretized minimization problem of (8) is

$$\text{(19)} \qquad \min_{u_{ijk}\in\mathbb{R},\ (ih,jh,kh)\in\mathcal{T}_\epsilon^h} \left\{ \sum_{i,j,k} |\nabla_h u_{ijk}|^2 + r_{ijk}^{-2}(\overline{\varphi_1}_{ijk} - u_{ijk})_+ + r_{ijk}^{-2}(u_{ijk} - \overline{\varphi_2}_{ijk})_+ \right\} K_{ijk} h^3.$$

Here ∇_h is the central differencing and the notation $f_{ijk} = f(ih, jh, kh)$, for $i, j, k \in \mathbb{Z}$ is used for any function f. Notice that in the above formula, the matrix B that should appear in the gradient term cancels with the Jacobian. To solve the discretized minimization problem (19), we perform Bregman iterations as in [14] until $\|u^{n+1} - u^n\| < 10^{-4}$. The parameters used in the iterations are chosen by $\lambda_1 = \lambda_2 = 45$ and u is initialized using the obstacles $u^0 = \frac{1}{2}(\varphi_1 + \varphi_2)$. The results are shown in the left of Figure 1. The right plot in Figure 1 shows the decrease of the energy of the numerical solution u^n (estimated by (8)) along the iterations.

6.2. Example: solving HJB equations on surfaces represented as point clouds

We give three examples of computing the solution of an HJB equation on surfaces represented by a point cloud. The Lax-Friedrichs sweeping scheme in [7] to discretize and solve (17)–(18) on a uniform Cartesian grid a grid spacing $h = 1/100$, and the width of the narrow band is $\epsilon = 4h$.

Notice that the closest point projection P_Γ and it's Jacobian matrix are needed to define the extended Hamiltonian. We compute the closest point mapping by the following description: Denote the point cloud by $\Gamma_N \subset \Gamma$, with N indicating the total number of points in the set. We approximate P_Γ using the following strategy:

(1) For each $z_i \in T_\epsilon^h$, estimate Γ_N locally on the grid nodes. Call the local surface associated to the point z_i, $\Gamma_N(z_i)$.

(2) Compute the closest point of z_i on the local surface, $\Gamma_N(z_i)$.

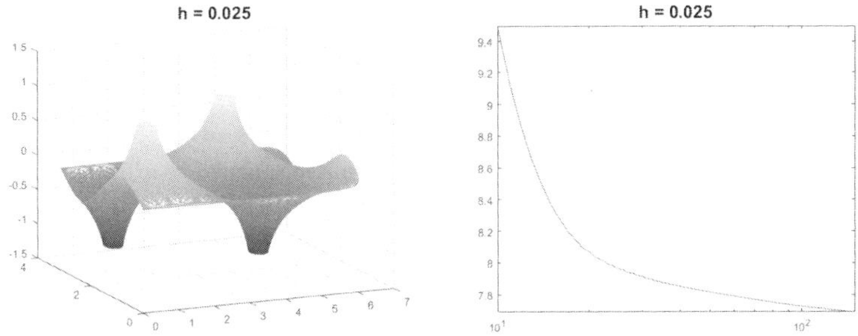

Fig. 1. Left: Numerical solutions to the obstacle problem with step size 0.025. We use the spherical coordinate system to represent points on the unit sphere and plot the minimizers on the (θ, ϕ) plane. Right: Energy of the minimizer versus $log(n)$, where n is the number of iterations.

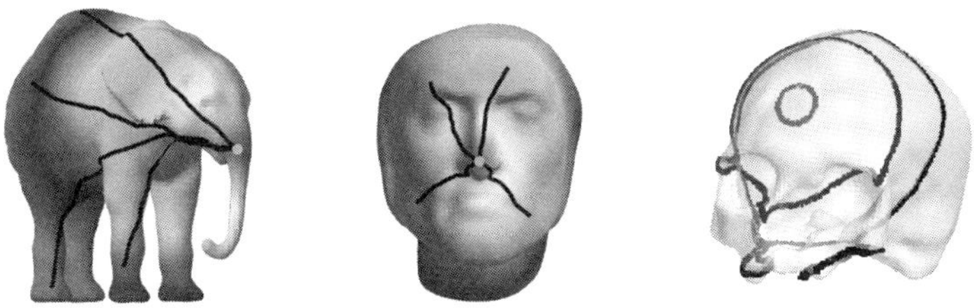

Fig. 2. Left and Center: Distance function to a source point on an elephant and mask with geodesics plotted. Right: Points from the corresponding point cloud of a skull whose distance from the source point lies in the interval (0.045, 0.055), (0.245, 0.255), and (0.395, 0.405).

We use biquadratic interpolation at each z_i to compute $\Gamma_N(z_i)$, and then use Newton's method to obtain the closest point of each z_i on $\Gamma_N(z_i)$. Denote the closest point mappings by $P^{\text{int}}_{\Gamma_N}(z_i)$. Then on each grid node z_i, the equation is discretized as usual, with $P'_\Gamma(z_i)$ being approximated by finite differences of $P^{\text{int}}_{\Gamma_N}(z_i)$. We use a fourth-order finite difference scheme to estimate P'_Γ and use singular value decomposition to obtain the singular values and singular vectors of P'_Γ.

In Figure 2, we show three examples of solving (17)–(18) for different surfaces represented as point clouds. The point clouds for the elephant, mask, and skull contains 65,292 points, 1,199,988 points, and 234,618 points, respectively. In our computation, 105,704, 87,088, and 85,214 number of grid nodes are used for the elephant, mask, and skull examples, respectively. In our approach, we can use the grid spacing to control the resolution of the surface geometry. When the density of the point cloud is high compared to the grid, it will be downsampled/compressed; otherwise, as we observed in the case of the "elephant" point cloud, new points obtained via local interpolation are "inserted" to the point cloud for computation.

We choose $r, f \equiv 1$ so that the HJB equation on the surface reduces to the Eikonal equation, $|\nabla_\Gamma u| = 1$. We set $\mathcal{T} = \{x_0\}$ and $g(x_0) = 0$, then solution is well known to be the distance function on Γ to the point x_0. We also plot some corresponding characteristics, also known as geodesics, of the distance function in Figure 2. The characteristics can be obtained by solving the dynamical system backwards:

$$\begin{aligned} \frac{dy}{dt}(t) &= \frac{-B(y(t),\mu)\nabla v(y(t))}{|B(y(t),\mu)\nabla v(y(t))|},\ t>0,\\ y(0) &= z,\, z\in\Gamma, \end{aligned}$$

where v is the solution to (17)-(18). The formulation also allows us to sort point clouds. The right of Figure 2 displays level "belts" of the point cloud for a skull, i.e., points in the point cloud whose distance from the source point lies in given interval.

§7. Summary

We presented a general formulation for solving PDEs on surfaces. By appropriately extending the differential operators and variational integrals, the solutions in the narrow band are the constant normal extensions of the solutions on the surface. Therefore instead of solving the PDEs on surfaces, we can use a variety of existing robust numerical methods and different meshes to solve the extended PDEs in Euclidean space on a very thin narrow band, coupled with a simple boundary closure procedure. With two rather different examples, we demonstrated that the approach is general and lead to convenient mathematical formulations that are suitable for numerical computation purposes.

§ Acknowledgement

Chu's research is partially supported by Ministry of Science and Technology of Taiwan under research grant MOST-106-2115-M-007 -002. Martin and Tsai's research is partially supported by National Science Foundation grant DMS-1720171. Tsai also thanks National Center for Theoretical Sciences (NCTS) for hosting this collaboration.

References

[1] M. Bardi and I. Capuzzo-Dolcetta, *Optimal control and viscosity solutions of Hamilton-Jacobi-Bellman equations*, Systems & Control: Foundations & Applications. Birkhäuser Boston, Inc., Boston, MA, 1997. With appendices by Maurizio Falcone and Pierpaolo Soravia.

[2] A. Bartesaghi and G. Sapiro, A system for the generation of curves on 3d brain images, *Human Brain Mapping*, 14(1):1–15, 2001.

[3] L.-T. Cheng, P. Burchard, B. Merriman, and S. Osher, Motion of curves constrained on surfaces using a level-set approach, *J. Comput. Phys.*, 175(2):604–644, 2002.

[4] J. Chu and R. Tsai, Volumetric variational principles for a class of partial differential equations defined on surfaces and curves, *Res. Math. Sci.*, 5(2):Paper No. 19, 38, 2018.

[5] G. Dziuk and C. M. Elliott, Finite element methods for surface PDEs, *Acta Numer.*, 22:289–396, 2013.

[6] P. Hoch and M. Rascle, Hamilton-Jacobi equations on a manifold and applications to grid generation or refinement, *SIAM J. Sci. Comput.*, 23(6):2055–2073, 2002.

[7] C. Y. Kao, S. Osher and J. Qian, Lax-Friedrichs sweeping scheme for static Hamilton-Jacobi equations, *J. Comput. Phys.*, 196(1):367–391, 2004.

[8] C. Kublik and R. Tsai, Integration over curves and surfaces defined by the closest point mapping, *Res. Math. Sci.*, 3:Paper No. 3, 17, 2016.

[9] A. Kumar and A. Vladimirsky, An efficient method for multiobjective optimal control and optimal control subject to integral constraints, *J. Comput. Math.*, 28(4):517–551, 2010.

[10] C. Lenglet, E. Prados, J.-P. Pons, R. Deriche and O. Faugeras, Brain connectivity mapping using Riemannian geometry, control theory, and PDEs, *SIAM J. Imaging Sci.*, 2(2):285–322, 2009.

[11] C. B. Macdonald and S. J. Ruuth, The implicit closest point method for the numerical solution of partial differential equations on surfaces, *SIAM J. Sci. Comput.*, 31(6):4330–4350, 2009/10.

[12] L. Martin and R. Tsai, Equivalent extensions of Hamilton-Jacobi-Bellman equations on hypersurfaces, In preparation.

[13] S. J. Ruuth and B. Merriman, A simple embedding method for solving partial differential equations on surfaces, *J. Comput. Phys.*, 227(3):1943–1961, 2008.

[14] G. Tran, H. Schaeffer, W. M. Feldman and S. J. Osher, An L^1 penalty method for general obstacle problems, *SIAM J. Appl. Math.*, 75(4):1424–1444, 2015.

[15] J. N. Tsitsiklis, Efficient algorithms for globally optimal trajectories, *IEEE Trans. Automat. Control*, 40(9):1528–1538, 1995.

Department of Mathematics, The University of Texas at Austin,
TX 78712, USA.
E-mail address: lmartin@math.utexas.edu

Department of Mathematics, National Tsing Hua University,
101 Section 2 Kuang-Fu Road, Hsinchu, Taiwan.
E-mail address: ccchu@math.nthu.edu.tw

Department of Mathematics and Oden Institute for
Computational Engineering and Sciences,
The University of Texas at Austin,
TX 78712, USA and KTH Royal Institute of Technology, Sweden.
E-mail address: tsai@kth.se

Advanced Studies in Pure Mathematics 85, 2020
The Role of Metrics in the Theory of Partial Differential Equations
pp. 453–462

Quasi-local mass and isometric embedding with reference to a static spacetime

Mu-Tao Wang

Abstract.

The mathematical theory of isometric embedding is applied to study the notion of quasilocal mass in general relativity. In particular, I shall report some recent progress of quasilocal mass with reference to a cosmological spacetime, such as the de Sitter or the Anti-de Sitter spacetime, or a blackhole spacetime, such as the Schwarzschild spacetime. This article is based on joint work with Po-Ning Chen, Ye-Kai Wang, and Shing-Tung Yau.

§1. Isometric embedding into $\mathbb{R}^3$

A good starting point of the theory of surface isometric embedding is Weyl's problem [39]: Given a Riemannian metric σ of positive Gauss curvature on the two-sphere S^2, does there exist a closed convex surface in $\mathbb{R}^3$ whose first fundamental form is σ? In a local coordinate system (u, v) on S^2,

$$\sigma = E(u,v)\,du^2 + 2F(u,v)\,dudv + G(u,v)\,dv^2$$

with $EG - F^2 > 0$, we ask if there exists an embedding $X : S^2 \to \mathbb{R}^3$ such that

$$\begin{cases} \left\langle \frac{\partial X}{\partial u}, \frac{\partial X}{\partial u} \right\rangle = E \\ \left\langle \frac{\partial X}{\partial u}, \frac{\partial X}{\partial v} \right\rangle = F \\ \left\langle \frac{\partial X}{\partial v}, \frac{\partial X}{\partial v} \right\rangle = G. \end{cases}$$

Received January 11, 2019.
Revised May 30, 2019.
2010 *Mathematics Subject Classification.* 53Cxx, 83Cxx.
Key words and phrases. general relativity, quasi-local mass, isometric embedding, gravitational energy.

When the Gauss curvature is positive, this is a nonlinear PDE system of "elliptic" type and the problem was solved independently by Nirenberg [30] and Pogorelov [33]. The uniqueness of the solution was proved by Cohn-Vossen [13] in the analytic case and by Herglotz [19] in the smooth case. Therefore, Weyl's problem always has a unique solution, i.e. any two solutions differ by a rigid motion of $\mathbb{R}^3$.

§2. Quasilocal mass in general relativity

How is the surface isometric embedding problem related to the notion of quasilocal mass in general relativity? A fundamental difficulty in general relativity is that, unlike any other physical theory, there is no mass/energy density for gravitation. This is a direct consequence of Einstein's equivalence principle. The familiar formula that mass is the bulk integral of mass density is simply not available. In the case of Newtonian gravity, which is the predecessor of the theory of general relativity, there is the formula $\Delta\Phi = 4\pi\rho$, where ρ is the mass density. For a domain $\Omega \subset \mathbb{R}^3$, by the divergence theorem, the total mass $\int_\Omega \rho$ is the same as the flux integral

$$\frac{1}{4\pi}\int_{\partial\Omega}\frac{\partial\Phi}{\partial\nu}$$

on the boundary surface $\partial\Omega$.

In 1982, Penrose [31] proposed a list of major unsolved problems in classical general relativity, and the first one is: Find a suitable definition of quasi-local mass for a spacelike domain Ω in a general physical spacetime, as an integral over the boundary surface $\partial\Omega = \Sigma$.

There are many different approaches to this problem (Hawking [17], Penrose [32], Bartnik [3], Brown-York [5], Dougan-Mason [14], etc.). A very promising one is the Hamilton-Jacobi analysis of gravitational action [5, 18] which indeed leads to another fundamental difficulty in general relativity: there is no canonical reference system.

Applying the Hamilton-Jacobi analysis to the Hilbert-Einstein action $\int \mathfrak{R}$, where $\mathfrak{R}$ is the scalar curvature of spacetime, Brown and York [5] derived a surface Hamiltonian (see also Hawking-Horowitz [18], Kijowski [21]) on a two-dimensional spacelike surface Σ. The expression depends on t^μ, a timelike unit vector field restricting to Σ, and u^μ, a timelike unit normal vector field along Σ. One considers a spacelike domain Ω bounded by Σ and let v^μ be the outward spacelike unit normal of Σ that is tangent to Ω. u^ν can then be taken to be the timelike normal that is orthogonal to v^μ. Along Σ, decompose

$$t^\mu = Nu^\mu + N^\mu,$$

where N is called the lapse function and N^μ is called the shift vector. The surface Hamiltonian (up to a factor of $\frac{1}{8\pi}$) is

$$\mathcal{H}(t^\mu, u^\mu) = \int_\Sigma [Nk - N^\mu P_{\mu\nu} v^\nu],$$

where k is the mean curvature of Σ in Ω, and $P_{\mu\nu}$ is the second fundamental form of Ω in spacetime. It turns out the expression depends only on the geometric data (σ, H) where σ is the induced metric on Σ, and H is the mean curvature vector field of Σ, in addition to the gauge choice of (t^μ, u^ν) along Σ. The mean curvature vector field H is the unique normal vector field of Σ which characterizes the area expansion or variation of Σ.

To obtain the *energy*, one needs a reference Hamiltonian to define:

$$energy = \text{physical Hamiltonian} - \text{reference Hamiltonian}.$$

The reference Hamiltonian should correspond to a surface term in a reference spacetime. The theory of surface isometric embeddings is applied in finding the reference surface. One also needs to specify t^μ and u^μ.

Brown and York [5] fixed Ω and took $t^\mu = u^\mu$, which corresponds to $N = 1$ and $N^\mu = 0$. The reference surface was taken to be an isometric embedding of σ into $\mathbb{R}^3$. The Brow-York mass is defined to be

$$\frac{1}{8\pi} \int_\Sigma (H_0 - k),$$

where H_0 is the mean curvature of the image surface of the isometric embedding of σ. The expression does converge to the ADM mass [1]under standard asymptotically flat conditions. Another important justification of the Brown-York mass is the positivity/rigidity theorem of Shi-Tam [34]. Yau proposed that a spacetime version is needed to study the dynamics of Einstein's equation. Liu-Yau [25] took the spacetime mean curvature vector field H (assumed to be spacelike) of Σ and defined the Liu-Yau mass [25, 26] (see also Kijowski [21], Booth-Mann [4], Epp [15]) to be

$$\frac{1}{8\pi} \int_\Sigma (H_0 - |H|),$$

and proved the positivity. However, these definitions turn out to be "too positive" as there are surfaces (O'Murchadha et al [29]) in the Minkowski spacetime $\mathbb{R}^{3,1}$ with strictly positive Brown-York or Liu-Yau mass.

At this point, it appeared clear that one should look for a quasilocal mass definition that uses the Minkowski spacetime as a reference spacetime. An important criterion is that the quasilocal mass of a surface in the Minkowski spacetime should be zero, as there is no matter or gravitation.

§3. Isometric embeddings into $\mathbb{R}^{3,1}$

Consider an isometric embedding of σ as $X : S^2 \to \mathbb{R}^{3,1}$ with components $X = (X_0, X_1, X_2, X_3)$ such that

$$-(dX_0)^2 + \sum_{i=1}^{3}(dX_i)^2 = \sigma.$$

Denoting the image surface $X(\Sigma)$ by Σ_0, we also consider the projection $\widehat{\Sigma}$ given by $(X_1, X_2, X_3) : S^2 \to \mathbb{R}^3$. The induced metric on $\widehat{\Sigma}$ is $\sigma + (dX_0)^2$. The above observation motivates the following procedure of constructing isometric embeddings. Take any function τ on (S^2, σ), $\sigma + (d\tau)^2$ gives another Riemannian metric. As long as the Gauss curvature of $\sigma + (d\tau)^2$ is positive, one can first isometrically embed $\sigma + (d\tau)^2$ into $\mathbb{R}^3$ with image $\widehat{\Sigma}$, and then take the graph of τ over $\widehat{\Sigma}$ in $\mathbb{R}^{3,1}$. The space of isometric embedding into $\mathbb{R}^{3,1}$ is essentially parametrized by such time function τ. But which τ is the best?

§4. The definition of quasilocal mass with $\mathbb{R}^{3,1}$ reference

It turns out the surface Hamiltonian satisfies the following gravitational conservation law that can be used to relate surface Hamiltonians on Σ_0 and $\widehat{\Sigma}$.

$$\int_{\widehat{\Sigma}} \widehat{H} = \int_{\Sigma_0} \left[|H_0|\sqrt{1+|\nabla\tau|^2}\cosh\theta_0 - \nabla\theta_0 \cdot \nabla\tau - \alpha_{H_0}(\nabla\tau) \right], \tag{4.1}$$

where θ_0 is given by

$$\sinh\theta_0 = \frac{-\Delta\tau}{\sqrt{1+|\nabla\tau|^2}\,|H_0|}.$$

Here H_0 is the mean curvature vector field of Σ_0 in $\mathbb{R}^{3,1}$, and α_{H_0} is the connection one-form determined by H_0.

The identity (4.1) provides a way to motivate the quasilocal energy definition of Wang-Yau [36, 37]. Give a spacelike 2-surface Σ in a physical spacetime with the data $(\sigma, |H|, \alpha_H)$. Let X be an isometric embedding of σ into $\mathbb{R}^{3,1}$ with time function τ. Let $\widehat{\Sigma}$ be the projection of $\Sigma_0 = X(\Sigma)$, the quasilocal energy is defined to be:

$$8\pi E(\Sigma, X, \tau) = \int_{\widehat{\Sigma}} \widehat{H} - \int_{\Sigma} \left[|H|\sqrt{1+|\nabla\tau|^2}\cosh\theta - \nabla\theta\cdot\nabla\tau - \alpha_H(\nabla\tau) \right]$$

where θ is given by

$$\sinh\theta = \frac{-\Delta\tau}{\sqrt{1+|\nabla\tau|^2}\,|H|}. \tag{4.2}$$

For a surface in the Minkowski spacetime, the above conservation law (4.1) implies the corresponding quasilocal energy is zero. After the positivity [36, 37] is established, one can minimize among the time functions to define the quasilocal mass. The Euler-Lagrange equation is called the optimal isometric embedding equation:

$$\begin{aligned}(\widehat{H}\hat{\sigma}^{ab} - \hat{\sigma}^{ac}\hat{\sigma}^{bd}\hat{h}_{cd})&\frac{\nabla_b\nabla_a\tau}{\sqrt{1+|\nabla\tau|^2}} \\ &= \operatorname{div}_\sigma\left(\frac{\nabla\tau}{\sqrt{1+|\nabla\tau|^2}}\cosh\theta|H| - \nabla\theta - \alpha_H\right)\end{aligned} \tag{4.3}$$

with $\sinh\theta$ given by (4.2).

Coupling the last equation with the isometric embedding equation into $\mathbb{R}^{3,1}$, we obtain four equations for the four unknown coordinate functions (τ, X_1, X_2, X_3). The solution gives the optimal isometric embedding and produces a surface in $\mathbb{R}^{3,1}$ that is supposed to be the "best match" for the physical surface [9]. We can then define other quasilocal conserved quantities such as quasilocal angular momentum based on this reference surface [10].

§5. The definition of quasilocal mass with static reference

There have been recent progresses [7, 11] to replace the reference Minkowski spacetime by a static reference spacetime. A static spacetime is a time-oriented Lorentzian 4-manifold (with possibility nonempty smooth boundary) such that there exists a static chart (t, x^1, x^2, x^3) under which the metric takes the form

$$\check{g} = -V^2(x^1,x^2,x^3)\,dt^2 + \bar{g}_{ij}(x^1,x^2,x^3)\,dx^i dx^j, \tag{5.1}$$

where the static potential V satisfies $V > 0$ on the interior and $V = 0$ on the boundary.

A static slice (t = constant) is a Riemannian 3-manifold with the induced metric $\bar{g} = \bar{g}_{ij}$. For simplicity, we assume the spacetime is vacuum with possibly nonzero cosmological constant and thus V satisfies the static equation

$$\bar{\Delta}V\bar{g} - \bar{\nabla}^2 V + V\bar{Ric} = 0.$$

Isometric embeddings of 2-surfaces into our examples such as the de Sitter, Anti-de Sitter, and Schwarzschild spacetimes have been studied by Chang-Xiao [6], Lin-Wang [24], Guan-Lu [16], etc.

Again for a 2-surface Σ in a physical spacetime with geometric data (σ, H), we consider an isometric embedding X of σ into a reference static spacetime time with time function τ (with respect to $T_0 = \frac{\partial}{\partial t}$) and the projection $\hat{\Sigma}$ on a static slice (with mean curvature $\hat{H}$), the quasilocal energy $E(\Sigma, X, T_0)$ [7, 11] is defined to be

$$\frac{1}{8\pi}\int_{\hat{\Sigma}} V\widehat{H} - \frac{1}{8\pi}\int_{\Sigma}\Bigg[\sqrt{(1+V^2|\nabla\tau|^2)|H|^2V^2 + \operatorname{div}(V^2\nabla\tau)^2} \\ - \operatorname{div}(V^2\nabla\tau)\sinh^{-1}\frac{\operatorname{div}(V^2\nabla\tau)}{V|H|\sqrt{1+V^2|\nabla\tau|^2}} - V^2\alpha_H(\nabla\tau)\Bigg].$$

We note that when $V = 1$ this is exactly the expression for the quasilocal energy with respect to Minkowski reference. For a surface in the static spacetime, the expression vanishes by a gravitational conservation law similar to (4.1).

One can check that for any spacelike surface in the reference static spacetime, the variation of E vanishes. When the reference isometric embedding X lies in a static slice (i.e. $\tau = 0$), the quasilocal mass expression becomes

$$\frac{1}{8\pi}\int V(H_0 - |H|)\, d\Sigma.$$

The Brown-York-Liu-Yau mass corresponds to the special case with $\mathbb{R}^3$ reference when $V = 1$. Combing with the quasi-spherical construction of Bartnik-Shi-Tam [2, 34], the positivity of $\int V(H_0 - H)$ with respect to the $\mathbb{H}^3$ reference is proved by Wang-Yau [38] and Shi-Tam [35]. Recently, Lu-Miao [28] established a positivity theorem with respect to the Schwarzschild reference that leads to the proof of a quasilocal Penrose's inequality. In retrospect, such a consideration is natural as the model spacetime for Penrose's inequality in which the equality holds is exactly the Schwarzschild spacetime.

§6. Rigidity properties

I would like to discuss the rigidity property in the rest of this article. The following theorem corresponds to the simplest version, which can be interpreted as a spacetime infinitesimally rigidity theorem of isometric embedding.

Theorem 1 ([9]). *Let $(\sigma, |H|, \alpha_H)$ be the data of a spacelike surface Σ in the Minkowski spacetime. Suppose the projection of Σ onto* *the orthogonal complement of T_0 is convex. Then the kernel of the linearized optimal isometric embedding system consists precisely of Lorentz motions. Moreover, the second variation of the quasilocal energy $E(\Sigma, \tau)$ is non-negative definite.*

Comparing with Cohn-Vossen's rigidity theorem of isometric embedding of convex surfaces in $\mathbb{R}^3$, the theorem implies that surfaces with prescribed $(\sigma, |H|, \alpha_H)$ are infinitesimally rigid in $\mathbb{R}^{3,1}$. The identification of the kernel of the linearized optimal embedding system enables the solution of the optimal isometric embedding system in a perturbative configuration by an inverse function argument.

The above theorem remains true for a convex surface in a static slice of de Sitter or Anti-de Sitter spacetime, see the next theorem. However, a new phenomenon occurs for the Schwarzschild spacetime.

Definition 1. *An isometric embedding into a static slice is infinitesimally rigid if the kernel of the linearized isometric embedding equation consists of the restriction of the Killing vector fields of the static slice to the surface.*

Remark 1. *Every convex surface in $\mathbb{R}^3$, $\mathbb{S}^3$, or $\mathbb{H}^3$ is infinitesimally rigid. However, this is no longer true for convex surfaces in a static slice of the Schwarzschild spacetime. In fact, even a round sphere of symmetry admits isometric deformations that are not induced by any ambient isometry, see Li-Wang* [23]. *On the other hand, there are rigidity theorems for isometric embeddings into a static slice of the Schwarzschild spacetime by Chen-Zhang* [12] *and Li-Miao-Wang* [22] *under an additional condition.*

Isometric embedding theorems into the Schwarzschild spacetime were established by the work of Lu [27], Guan-Lu [16], and Li-Wang [23].

Since convex surfaces in $\mathbb{S}^3$ and $\mathbb{H}^3$ are infinitesimally rigid, we have the following spacetime rigidity theorem for the de Sitter and Anti-de Sitter spacetimes.

Theorem 2 ([7]). *The spacetime infinitesimal rigidity theorem holds true for convex surfaces in a static slice of the de Sitter or Anti-de Sitter spacetime.*

Namely $\delta^2 E \geq 0$ and the kernel of the linearized optimal isometric embedding consists of the global isometry of the ambient spacetime. Such a rigidity theorem should be enough to anchor a unique solution of the optimal isometric embedding system for a perturbative configuration of the background spacetime.

This material is based upon work supported by the National Science Foundation under Grants No. DMS-1405152 and No. DMS-1810856 (Mu-Tao Wang).

References

[1] R. Arnowitt, S. Deser and C. W. Misner, *The dynamics of general relativity*, in Gravitation: An introduction to current research, 227–265, Wiley, New York.

[2] R. Bartnik, *Quasi-spherical metrics and prescribed scalar curvature*, J. Differential Geom., **37** (1993), no. 1, 31–71.

[3] R. Bartnik, *New definition of quasi-local mass*, Phys. Rev. Lett., **62** (1989), no. 20, 2346–2348.

[4] I. S. Booth and R. B. Mann, *Moving observers, nonorthogonal boundaries, and quasilocal energy*, Phys. Rev. D (3), **59** (1999), 064021.

[5] J. D. Brown and J. W. York, *Quasi-local energy and conserved charges derived from the gravitational action*, Phys. Rev. D (3), **47** (1993), no. 4, 1407–1419.

[6] J.-E. Chang and L. Xiao, *The Weyl problem with nonnegative Gauss curvature in hyperbolic space*, Canad. J. Math., **67** (2015), no. 1, 107–131.

[7] P.-N. Chen, M.-T. Wang, Y.-K. Wang and S.-T. Yau, *Quasi-local energy with respect to a static spacetime*, Adv. Theor. Math. Phys., **22** (2018), no. 1, 1–23, arXiv:1604.02983.

[8] P.-N. Chen, M.-T. Wang and S.-T. Yau, *Evaluating quasi-local energy and solving optimal embedding equation at null infinity*, Comm. Math. Phys., **308** (2011), no. 3, 845–863.

[9] P.-N. Chen, M.-T. Wang and S.-T. Yau, *Minimizing properties of critical points of quasi-local energy*, Comm. Math. Phys., **329** (2014), no. 3, 919–935.

[10] P.-N. Chen, M.-T. Wang and S.-T. Yau, *Conserved quantities in general relativity: from the quasi-local level to spatial infinity*, Comm. Math. Phys., **338** (2015), no. 1, 31–80.

[11] P.-N. Chen, M.-T. Wang and S.-T. Yau, *Quasi-local energy with respect to de Sitter/Anti-de Sitter reference*, to appear in Comm. Anal. Geom., arXiv:1603.02975.

[12] P.-N. Chen and X. Zhang, *A rigidity theorem for surfaces in Schwarzschild manifold*, to appear in Int. Math. Res. Not., arXiv:1802.00887.

[13] E. Cohn-Vossen, *Zwei Sätze über die Starrheit der Eiflächen*, in Nachrichten von der Gesellschaft der Wissenschaften zu Göttingen, Mathematish-Physikalische Klasse, 1927, pp. 125–134.

[14] A. J. Dougan and L. J. Mason, *Quasilocal mass constructions with positive energy*, Phys. Rev. Lett., **67** (1991), no. 16, 2119–2122.

[15] R. J. Epp, Phys. Rev. D 62, 124018 (2000).

[16] P. Guan and S. Lu, *Curvature estimates for immersed hypersurfaces in Riemannian manifolds*, Invent. Math., **208** (2017), no. 1, 191–215.

[17] S. W. Hawking, *Gravitational radiation in an expanding universe*, J. Math. Phys., **9**, 598 (1968).

[18] S. W. Hawking and G. T. Horowitz, *The gravitational Hamiltonian, action, entropy and surface terms*, Classical Quantum Gravity, **13** (1996), no. 6, 1487–1498.

[19] G. Herglotz, *Über die Starrheit der Eiflächen*, (German), Abh. Math. Sem. Hansischen Univ., **15** (1943). 127–129.

[20] G. T. Horowitz and M. J. Perry, *Gravitational energy cannot become negative*, Phys. Rev. Lett., **48** (1982), no. 6, 371–374.

[21] J. Kijowski, *A simple derivation of canonical structure and quasi-local Hamiltonians in general relativity*, Gen. Relativity Gravitation, **29** (1997), no. 3, 307–343.

[22] C. Li, P. Miao and Z. Wang, *Uniqueness of isometric immersions with the same mean curvature*, J. Funct. Anal., **276** (2019) no. 9, 2831–2855, arXiv:1802.04244.

[23] C. Li and Z. Wang, *The Weyl problem in warped product space*, to appear in J. Differential Geom., arXiv:1603.01350.

[24] C.-Y. Lin and Y.-K. Wang, *On Isometric embeddings into Anti-de Sitter Space-times*, Int. Math. Res. Not., (2015), no. 16, 7130–7161.

[25] C.-C. M. Liu and S.-T. Yau, *Positivity of quasilocal mass*, Phys. Rev. Lett., **90** (2003), no. 23, 231102.

[26] C.-C. M. Liu and S.-T. Yau, *Positivity of quasi-local mass* II, J. Amer. Math. Soc., **19** (2006), no. 1, 181–204.

[27] S. Lu, *On Weyl's embedding problem in Riemannian manifolds*, to appear in Int. Math. Res. Not..

[28] S. Lu and P. Miao, *Minimal hypersurfaces and boundary behavior of compact manifolds with non-negative scalar curvature*, to appear in J. Differential Geom..

[29] N. Ó Murchadha, L. B. Szabados and K. P. Tod, Comment on: "Positivity of quasi-local mass", Phys. Rev. Lett., **92** (2004), no. 25, 259001, 1 p.

[30] L. Nirenberg, *The Weyl and Minkowski problems in differential geometry in the large*, Comm. Pure Appl. Math., **6** (1953), 337–394.

[31] R. Penrose, *Some unsolved problems in classical general relativity*, Seminar on Differential Geometry, pp. 631–668, Ann. of Math. Stud., **102**, Princeton Univ. Press, Princeton, N.J., 1982.

[32] R. Penrose, *Quasi-local mass and angular momentum in general relativity*, Proc. Roy. Soc. London Ser. A, **381** (1982), no. 1780.

[33] A. V. Pogorelov, *Regularity of a convex surface with given Gaussian curvature*, (Russian), Mat. Sbornik N.S., **31** (73), (1952), 88–103.

[34] Y. Shi and L.-F. Tam, *Positive mass theorem and the boundary behaviors of compact manifolds with nonnegative scalar curvature*, J. Differential Geom., **62** (2002), no. 1, 79–125.

[35] Y. Shi and L.-F. Tam, *Rigidity of compact manifolds and positivity of quasi-local mass*, Classical Quantum Gravity, **24** (2007), no. 9, 2357–2366.

[36] M.-T. Wang and S.-T. Yau, *Quasi-local mass in general relativity*, Phys. Rev. Lett., **102** (2009), no. 2, no. 021101.

[37] M.-T. Wang and S.-T. Yau, *Isometric embeddings into the Minkowski space and new quasi-local mass*, Comm. Math. Phys., **288** (2009), no. 3, 919–942.

[38] M.-T. Wang and S.-T. Yau, *A generalization of Liu-Yau's quasi-local mass*, Comm. Anal. Geom., **15** (2007), no. 2, 249–282.

[39] H. Weyl, *Über die Bestimmung einer geschlossenen konvexen Fläche durch ihr Linienelement*, Vierteljahrschrift Naturforsch. Gesell. Zürich, 3 : 2 (1916) pp. 40–72.

Department of Mathematics, Columbia University,
2990 Broadway, New York, NY 10027
E-mail address: mtwang@math.columbia.edu

Advanced Studies in Pure Mathematics 85, 2020
The Role of Metrics in the Theory of Partial Differential Equations
pp. 463–470

Quasi-periodic solutions to nonlinear PDEs

Wei-Min Wang

Abstract.

We present recent developments in the theory of quasi-periodic solutions to nonlinear PDEs, such as the nonlinear Schrödinger and the nonlinear Klein-Gordon equations. These solutions hold in arbitrary dimensions, and the quasi-periodicity can be either in time or space. The method hinges on multi-scale analysis, harmonic analysis and algebraic analysis.

§1. Introduction

We consider the nonlinear Schrödinger equation (NLS) posed on $\mathbb{R}^d$:

$$-i\frac{\partial}{\partial t}u - \Delta u + |u|^{2p}u = 0, \tag{1}$$

where $p \in \mathbb{N}$ and $p \geq 1$ is *arbitrary*, and u is a complex valued function; and the nonlinear Klein-Gordon equation (NLKG)posed on $\mathbb{R}^d$:

$$\frac{\partial^2 u}{\partial t^2} - \Delta u + u + u^{2p+1} = 0, \tag{2}$$

where $p \in \mathbb{N}$ and $p \geq 1$ (arbitrary), and u is real-valued.[1]

In Section 2, we discuss solutions which are periodic in space $\mathbb{R}^d$ and quasi-periodic ("periodic" with several frequencies) in time. In Section 3, we proceed to discuss solutions which are quasi-periodic *in space* and periodic in time, the so called standing wave solutions. One of the

Received August 20, 2018.
Revised December 4, 2018.
2010 *Mathematics Subject Classification.* 35M20; 42A16.
Key words and phrases. quasi-periodic solutions, Newton iteration, semi-algebraic geometry.

[1]For brevity, we have written plus signs in front of the nonlinear terms in (1) and (2), however, since we seek small solutions, the results in this paper hold for both signs.

main differences of these two types of solutions is that for the latter, the "dynamical" variables are the space variables and are multi-dimensional. As we will see, however, our method provides a unified approach to these two different problems.

§2. Time quasi-periodic solutions

We identify periodic solutions to (1) and (2) on $\mathbb{R}^d$ to solutions on the torus $\mathbb{T}^d = \mathbb{R}^d/(2\pi\mathbb{Z})^d$. Our aim in this section is to present the ideas for the proof of the following result.

Theorem 1. *There is a set of global solutions to the NLS and NLKG in* (1), (2), *in arbitrary dimension d and for arbitrary nonlinearity p. These are time quasi-periodic solutions, which are Gevrey functions in space and time.*

Remark 1. This set of global solutions has positive Hausdorff measure in the following sense: if it is quasi-periodic of b basic frequencies in time, then the b-dimensional Hausdorff measure is positive. For more precise statements and proofs, see [11], [12].

Below we give an idea of the approach.

2.1. NLS

We start from the linear Schrödinger equation on the torus:

$$i\frac{\partial}{\partial t}u + \Delta u = 0, \tag{3}$$

with a point of view toward *time quasi-periodic* solutions. Let $j \in \mathbb{Z}^d$, the spectrum of the Laplacian $-\Delta$ consists of the integers:

$$j_1^2 + j_2^2 + \cdots + j_d^2 := j^2, \quad j_1, j_2, \ldots, j_d \in \mathbb{Z}.$$

The eigenfunctions to (3) are all of the form

$$e^{-ij^2t}e^{ij\cdot x}, \quad j \in \mathbb{Z}^d. \tag{4}$$

They generate the space-time Fourier series:

$$u = \sum a_j e^{-ij^2t}e^{ij\cdot x}, \quad j \in \mathbb{Z}^d,$$

which are, in turn, solutions to (3).

We seek solutions to the NLS in (1) of finite number of basic frequencies, b frequencies, in time. It is useful to add a dimension for each frequency in time and view $u^{(0)}$ as a function on $\mathbb{T}^b \times \mathbb{T}^d := \mathbb{T}^{b+d} \supset \mathbb{T}^d$. We therefore seek u of the form:

$$u(t,x) = \sum_{(n,j)\in\mathbb{Z}^{b+d}} \mathfrak{a}(n,j)e^{in\cdot\omega t}e^{ij\cdot x}, \tag{5}$$

where $\mathfrak{a}$ and ω are to be determined. We express the NLS in (1), using the ansatz in (5). By analogy with the standard Fourier series we write $\hat{u}$ for $\mathfrak{a}$ and define $\hat{\bar{u}}$ to be $\hat{\bar{u}}(n,j) = \bar{\mathfrak{a}}(-n,-j)$ for all $(n,j) \in \mathbb{Z}^{b+d}$. To simplify notations we write $\hat{v}$ for $\hat{\bar{u}}$. Equation (1) can then be written as a nonlinear (infinite) matrix equation:

$$\operatorname{diag}(n\cdot\omega + j^2)\hat{u} + (\hat{u} * \hat{v})^{*p} * \hat{u} = 0, \tag{6}$$

where diag $\cdot$ denotes a diagonal matrix, $\omega \in \mathbb{R}^b$ is to be determined.

We use a Newton scheme to solve (6), by seeking solutions which are close to *appropriate* linear solutions $u^{(0)}$ of b frequencies:

$$u^{(0)} = \sum_{k=1}^{b} a_k e^{-ij_k^2 t}e^{ij_k\cdot x}, \quad j_k \in \mathbb{Z}^d, \tag{7}$$

with small $a_k \in \mathbb{R}$, $k = 1, 2, \ldots, b$. Here appropriate means

(1) the j_k, $k = 1, 2, \ldots, b$, satisfy certain non-resonance conditions;
(2) the a_k, $k = 1, 2, \ldots, b$, satisfy certain small-divisor conditions.

The conditions in (1) are geometric, given by algebraic equations in j_k. The solutions $u^{(0)}$ that satisfy (1) are called *generic*. This is the algebraic analysis component developed in [11], cf. also [7], [10]. After fixing the generic j_k, $k = 1, 2, \ldots, b$, we need to choose the a_k, $k = 1, 2, \ldots, b$. Contrary to (1), which comprises of "one step", so to speak, (2) is iterative, based on multi-scale analysis and harmonic analysis, as in [3] and Chap. 19 in [1]. The a_k, $k = 1, 2, \ldots, b$, in (2), generally speaking, belong to a Cantor set, albeit of positive measure, because of the requirement to avoid small divisors coming from the Newton iterations.

These appropriate linear solutions $u^{(0)}$ then bifurcate leading to the solutions u to the NLS in Theorem 1.

2.2. NLKG

For the NLKG in (2), since we seek real solutions, in lieu of (5), the ansatz is

$$u(t,x) = \sum_{(n,j)\in\mathbb{Z}^{b+d}} \hat{u}(n,j)\cos(n\cdot\omega t + j\cdot x). \tag{8}$$

Equation (2) then leads to the nonlinear matrix equation:

$$\text{diag}\left[-(n\cdot\omega)^2+j^2+1\right]\hat{u}+\hat{u}^{*(2p+1)}=0. \tag{9}$$

The linear Klein-Gordon equation is the following:

$$\frac{\partial^2 u}{\partial t^2}-\Delta u+u=0. \tag{10}$$

Define the wave operator to be

$$D=\sqrt{-\Delta+1}. \tag{11}$$

Then the spectrum of D on the torus $\mathbb{T}^d$, consists of square roots of integers:

$$\sqrt{j_1^2+j_2^2+\cdots+j_d^2+1}:=\sqrt{j^2+1},\quad j_1,j_2,\ldots,j_d\in\mathbb{Z}. \tag{12}$$

For $d\geq 2$, the spacing of non-equal eigenvalues tends to 0. This is a major complication compared to the spectrum of the Laplacian, which consists of integers. The spacings of non-equal eigenvalues for the Laplacian are uniformly bounded below by 1.

As for the NLS in (1), we seek solutions u close to

$$u^{(0)}(t,x)=\sum_{k=1}^{b}a_k\cos\left(-\left(\sqrt{j_k^2+1}\right)t+j_k\cdot x\right),$$

which are appropriate solutions to the linear equation (10). Being appropriate here means the same as before, namely, $j_k\in\mathbb{Z}^d$, $k=1,2,\ldots,b$, satisfy non-resonance conditions and a_k, $k=1,2,\ldots,b$, satisfy small-divisor conditions. The idea of the proof is therefore similar to that for NLS. However, due to the presence of square roots, aside from geometric conditions on j_k, $k=1,2,\ldots,b$, we also need arithmetic conditions, namely we require that j_k^2+1, $k=1,2,\ldots,b$, are square-free integers [12]. (A recent paper [8] indicates that such square-free integers have positive density in $\mathbb{Z}^d$.) The non-resonance conditions can again be written as algebraic equations as for NLS. The Newton iteration proceeds as for NLS, leading to small-divisor conditions on a_k. Subsequently these appropriate linear solutions $u^{(0)}$ bifurcate and become solutions u to the NLKG in Theorem 1.

We conclude by noting that *no* global (in time) solutions are known in general for NLS (1) and NLKG (2) in higher dimensions ($d\geq 3$). This is largely due to resonances produced by the spectra of the linear operators. Theorem 1 produces a set of global solutions by controlling the resonances using the geometric non-resonance and the small-divisor conditions, cf. [11], [12].

2.3. This new method

At the origin of this new method lie three ingredients, namely, the multiscale analysis of Fröhlich-Spencer [6], the harmonic analysis of Bourgain-Goldstein-Schlag [3] and Bourgain, Chap. 18–20 [1] and the algebraic analysis of [11], [12]. It works for *both* limit-elliptic equations such as NLS, and hyperbolic equations such as NLKG. The latter is perhaps an important point.

We note that for NLS, KAM-type techniques are applicable as in Eliasson-Kuksin [5] and Procesi-Procesi [10], providing linearly stable solutions.

§3. Space quasi-periodic solutions

Here it is conventional to write NLS as

$$-i\frac{\partial}{\partial t}u - \Delta u - |u|^{2p}u = 0. \tag{13}$$

(The difference with (1) is the minus sign in front of the nonlinear term.) We seek standing wave solutions to (13) of the form:

$$u(t,x) = e^{-iEt}v(x), \tag{14}$$

where $E \in \mathbb{R}$, and v is *even* and *quasi-periodic* in each x_k, $k = 1, 2, \ldots, d$, given by a quasi-periodic cosine series:

(15)

$$v(x) = v(x_1, x_2, \ldots, x_d) = \sum_{j_1, j_2, \ldots, j_d} \hat{v}(j_1, j_2, \ldots, j_d) \prod_{k=1}^{d} \cos(j_k \cdot \lambda_k)x_k,$$

where for each $k \in \{1, 2, \ldots, d\}$, $j_k \in \mathbb{Z}^2$ and $\lambda_k \in (1/2, 3/2)^2$. The λ_k's are the *parameters* in the problem, and are assumed to be irrational, satisfying

$$\|j_k \cdot \lambda_k\|_{\mathbb{T}} \neq 0,$$

for all $j_k \neq 0$, where $\|\cdot\|_{\mathbb{T}}$ denotes distance to the integers.

We note that the quasi-periodic series (15) reduces to a periodic cosine series if j_k and λ_k were one dimensional: $j_k \in \mathbb{Z}$ and $\lambda_k \in (1/2, 3/2)$ for $k = 1, 2, \ldots, d$. For example, setting $\lambda_k = 1$ for all k, leads to a periodic series with period 2π in each directions. This reverts to the setting in Theorem 1, where solutions with more general time dependence, namely, the time quasi-periodic solutions, are shown to exist.

Substituting the ansatz (14) into (1) yields the following stationary, nonlinear elliptic problem:

$$-\Delta v - |v|^{2p} v = Ev.$$

For $v \in H^1(\mathbb{R}^d)$ with a *fixed* $L^2(\mathbb{R}^d)$ norm, there is a well established variational structure under appropriate conditions on p: E is a Lagrange multiplier and (1.3) are the minimizers for the energy functional:

$$\mathcal{E}(v) = \int_{\mathbb{R}^d} dx \left[\frac{1}{2}\|\nabla v\|^2 - \frac{1}{2p+2}|v|^{2p+2}\right].$$

Localized standing wave solutions are well known from the works of, for example, Cazenave and Lions [4]. (Cf. also the references therein.) The v's given by (15), however, are only in L^∞. We are not aware of a variational approach, instead, we use a Newton scheme and the semi-algebraic geometry technique developed in [2] to construct the nonlinear (generalized) eigenfunctions.

To simplify notations, define

$$(j \cdot \lambda)^2 := \sum_{k=1}^{d} (j_k \cdot \lambda_k)^2,$$

where $j = (j_1, j_2, \ldots, j_d) \in \mathbb{Z}^{2d}$ and $\lambda = (\lambda_1, \lambda_2, \ldots, \lambda_d) \in (1/2, 3/2)^{2d}$. Let

$$\tilde{u} = ae^{-i(\tilde{j}\cdot\lambda)^2 t} \prod_{k=1}^{d} \cos(\tilde{j}_k \cdot \lambda_k)x_k, \tag{16}$$

where $a \in \mathbb{R}$ and $\tilde{j} \in \mathbb{Z}^{2d}$. (If $\tilde{j} = 0$, $\tilde{u} = ae^{i|a|^{2p}t}$ trivially solves (1).) Then $\tilde{u}$ satisfies the linear equation:

$$i\frac{\partial}{\partial t}\tilde{u} = -\Delta \tilde{u}; \tag{17}$$

and

$$\tilde{u} = a \prod_{k=1}^{d} \cos(\tilde{j}_k \cdot \lambda_k)x_k$$

satisfies

$$-\Delta \tilde{u} = \tilde{E}\tilde{u},$$

with

$$\tilde{E} = (\tilde{j} \cdot \lambda)^2.$$

Our main result of the section is

Theorem 2. [13] *For every solution to the linear equation* (17) *in the form* (16):

$$\tilde{u} = ae^{-i(\tilde{j}\cdot\lambda)^2 t} \prod_{k=1}^{d} \cos(\tilde{j}_k \cdot \lambda_k)x_k,$$

where $a \in \mathbb{R}$ *and* $\tilde{j} \in \mathbb{Z}^{2d}$, *there is a set in* λ, $\Lambda \subset (1/2, 3/2)^{2d}$ *satisfying*

$$meas\ \Lambda \geq 1 - \mathcal{O}(|a|^{p/2}),$$

provided $|a| \ll 1$. *If* $\lambda \in \Lambda$, *then there is a solution* u, *bifurcating from* $\tilde{u}$, *to the nonlinear equation* (1) *in the form* (14), (15):

$$u(t,x) = e^{-i[(\tilde{j}\cdot\lambda)^2 + \mathcal{O}(|a|^{2p})]t} \left[a \prod_{k=1}^{d} \cos(\tilde{j}_k \cdot \lambda_k)x_k + \mathcal{O}(|a|^p) \right]. \tag{18}$$

The nonlinear eigenvalue E, *as a function in* λ, *is* C^1 *on* $(1/2, 3/2)^{2d}$.

The solutions in (18) are nonlinear eigenfunctions, i.e., solutions which are periodic in time with the eigenvalue as the frequency; while spatially they are quasi-periodic in $\mathbb{R}^d$. To our knowledge, Theorem 2 is the first such result on global in time, non-decaying solutions u in arbitrary dimensions, which do not have an underlying translation symmetry group. (Cf. Moser [9] for an iterative method in the space periodic setting, i.e., on the quotient space $L^2(\mathbb{R}^d/(2\pi\mathbb{Z})^d) := L^2(\mathbb{T}^d)$.)

References

[1] J. Bourgain, Green's function estimates for latttice Schrödinger operators and applications, Ann. of Math. Studies, **158**, 2005; Princeton University Press.

[2] J. Bourgain, Anderson localization for quasi-periodic lattice Schrödinger operators on $\mathbb{Z}^d$, d arbitrary, Geom. Funct. Anal., **17** (2007), 682–706.

[3] J. Bourgain, M. Goldstein and W. Schlag, Anderson localization for Schrödinger operators on $\mathbb{Z}^2$ with quasi-periodic potential, Acta Math., **188**, (2002), 41–86.

[4] T. Cazenave and P. L. Lions, Orbital stability of standing waves for some nonlinear Schrödinger equations, Commun. Math. Phys., **85**, (1982), 549–561.

[5] L. H. Eliasson and S. E. Kuksin, KAM for the nonlinear Schrödinger equation, Ann. of Math. (2), **172** (2010), 371–435.

[6] J. Fröhlich and T. Spencer, Absence of diffusion in the Anderson tight binding model for large disorder or low energy, Commun. Math. Phys., **88** (1983), 151–184.

[7] J. Geng, X. Xu and J. You, An infinite dimensional KAM theorem and its application to the two dimensional cubic Schrödinger equation, Adv. in Math. no. 6 (2011), 5361–5402.

[8] K. Lapkova and S. Y. Xiao, Density of power-free values of polynomials, Mathematika, **65(4)**, (2019), 1038–1050.

[9] J. Moser, A rapidly convergent iteration method and nonlinear partial differential equations I, Ann. Scuola Norm. Sup. Pisa (3), **20(2)** (1966), 265–315.

[10] C. Procesi and M. Procesi, A KAM algorithm for the resonant non-linear Schrödinger equation, Adv. in Math., **272** (2015), 399–470.

[11] W.-M. Wang, Energy supercritical nonlinear Schrödinger equations: Quasi-periodic solutions, Duke Math. J., **165** (2016), no. 6, 1129–1192.

[12] W.-M. Wang, Quasi-periodic solutions for nonlinear Klein-Gordon equations, arXiv:1609.00309v2, (2017), 66 pp.

[13] W.-M. Wang, Space quasi-periodic standing waves for nonlinear Schrödinger equations, Commun. Math. Phys., **378** (2020), 783–806.

CNRS and Department of Mathematics, Université Cergy-Pontoise,
95302 Cergy-Pontoise Cedex, France
E-mail address: wei-min.wang@math.cnrs.fr

Advanced Studies in Pure Mathematics 85, 2020
The Role of Metrics in the Theory of Partial Differential Equations
pp. 471–478

Simons cone and saddle solutions of the Allen-Cahn equation

Kelei Wang

Abstract.

Simons cone is a minimal cone in $\mathbb{R}^8$, which provides a counterexample to the Bernstein problem. Corresponding objects for the Allen-Cahn equation are saddle solutions. They are conjectured to be minimal in dimension 8. We discuss some progress on this problem, in particular, the existence of minimizing solutions to Allen-Cahn equation in $\mathbb{R}^8$.

This note is based on the author's joint work with Yong Liu and Juncheng Wei, [13]. We are mainly concerned with the existence of minimizing solutions to the Allen-Cahn equation

$$-\Delta u = u - u^3 \quad \text{in } \mathbb{R}^n, \ |u| < 1. \tag{1}$$

This problem arise in the study of De Giorgi conjecture [7], which states that for $n \leq 8$, if a solution to (1) is monotone in one direction, then its level sets are all hyperplanes.

§1. Simons cone and Bernstein problem

There is an intricate connection between Allen-Cahn equation and minimal surfaces. De Giorgi conjecture is parallel to the Bernstein theorem in minimal surface theory.

Theorem 1.1. *Suppose $u \in C^2(\mathbb{R}^n)$ is a solution to the minimal surface equation*

$$\operatorname{div}\left(\frac{\nabla u}{\sqrt{1+|\nabla u|^2}}\right) = 0. \tag{2}$$

If $n \leq 7$, then u is an affine function.

Received October 22, 2018.
2010 *Mathematics Subject Classification.* 35J25, 35J20, 35B33, 35B40.
Key words and phrases. Allen-Cahn equation, De Giorgi conjecture, Minimal cone, Global minimizer.

There is a corresponding version for area minimizing hypersurfaces.

Theorem 1.2. *Suppose Σ is an area minimizing minimal hypersurface in $\mathbb{R}^n$. If $n \le 7$, then Σ is a hyperplane.*

Starting from Bernstein, many people including Fleming, De Giorgi and Almgren contributed to the solution of this conjecture. The above two theorems are finally proved by Simons in [17]. The proof relies on the following classification of stable minimal cones.

Theorem 1.3. *Suppose Σ is a stable minimal cone in $\mathbb{R}^n$. If $n \le 7$, then Σ is a hyperplane.*

The dimension bound is sharp. Simons showed that

$$\left\{x_1^2 + x_2^2 + x_3^2 + x_4^2 = x_5^2 + x_6^2 + x_7^2 + x_8^2\right\} \subset \mathbb{R}^8$$

is a stable minimal cone. Clearly it has a singular point, the origin. This cone is the Simons cone.

Simons also conjectured that this cone minimizes area. This was proved by Bombieri-De Giorgi-Giusti in [3] In fact, they established the following result.

Theorem 1.4. *There exist two smooth minimal hypersurfaces $\Sigma_\pm$, which lie on two sides of the Simons cone. Moreover, $\cup\{\lambda\Sigma_\pm, \lambda > 0\}$ forms a foliation of $\mathbb{R}^8$.*

Because of the $O(4) \times O(4)$ symmetry of Simons cone, they find these minimal hypersurfaces in the class with the same symmetry. Let $r = (x_1^2 + \cdots + x_4^2)^{1/2}$, $s = (x_5^2 + \cdots + x_8^2)^{1/2}$. If a curve $(r(t), s(t))$ satisfies the second order ODE

$$r''s' - r's'' + 3\left[(r')^2 + (s')^2\right]\left(\frac{r'}{s} - \frac{s'}{r}\right) = 0,$$

then it represents a minimal hypersurface in $\mathbb{R}^8$. They established the existence of $\Sigma_\pm$ by solving this ODE.

In the same paper Bombieri-De Giorgi-Giusti also proved that.

Theorem 1.5. *There exists a non-planar minimal graph in $\mathbb{R}^9$.*

§2. De Giorgi conjecture and saddle solutions

As for the De Giorgi conjecture, it has been proven to be true for $n = 2$ (Ghoussoub-Gui [10]), $n = 3$ (Ambrosio-Cabré [2]), and for $4 \le n \le 8$ (Savin [15]) under an additional condition

$$\lim_{x_n \to \pm\infty} u(x', x_n) = \pm 1. \tag{3}$$

This condition together with the monotone property implies that u is a global minimizer in the sense that, for any smooth bounded domain $\Omega \subset \mathbb{R}^n$,

$$\mathcal{J}(u) \leq \mathcal{J}(u+\phi), \text{ for all } \phi \in C_0^\infty(\Omega),$$

where

$$\mathcal{J}(u) := \int_\Omega \left[\frac{1}{2}|\nabla u|^2 + \frac{(u^2-1)^2}{4}\right],$$

see Alberti-Ambrosio-Cabré [1] and Savin [15].

On the other hand, it turns out that for $n \geq 9$, there do exist monotone solutions which are not one dimensional. These nontrivial examples have been constructed in Del Pino-Kowalczyk-Wei [8] using the machinery of *infinite dimensional Lyapunov-Schmidt reduction method.* The nodal sets of these solutions are actually close to the Bombieri-De Giorgi-Giusti minimal graphs in Theorem 1.5.

Regarding solutions which are not necessarily monotone, Savin [15] also proved that if u is a global minimizer in $\mathbb{R}^n$ and $n \leq 7$, then u is one dimensional. While the monotone solutions of Del Pino-Kowalczyk-Wei provide examples of nontrivial global minimizers in dimension $n \geq 9$, it is still not known whether there are nontrivial global minimizers for $n = 8$. Due to the connection with minimal surface theory, one expects nontrivial global minimizers should exist in dimension 8 and higher. This was pointed out in [11, Section 1.3] and posed as an open question in [6]. The existence of global minimizers is our main focus in [13].

There are analogous objects as Simons cone in the theory of Allen-Cahn equation. They are the so-called saddle-shaped solutions, which are solutions in $\mathbb{R}^{2m}$ of (1) vanishing exactly on Simons cone (Cabré-Terra [4, 5] and Cabré [6]). In [4, 5] it is proved that for $2 \leq m \leq 3$, saddle solutions are unstable, while for $m \geq 7$, they are stable ([6]). It is also conjectured in [6] that for $m \geq 4$, the saddle solution should be a global minimizer. This problem is still open. In fact, even if they are stable for $4 \leq m \leq 6$ is still open.

Recall that Σ_λ, $\lambda \in (-\infty, +\infty)$, is a foliation of $\mathbb{R}^8$, where each leave is a minimal hypersurface and all except Σ_0 (Simons cone) are smooth. Pacard and Wei in [14], by using a perturbative argument, proved

Theorem 2.1. *For any $\lambda \in (-\infty, \Lambda) \cup (\Lambda, +\infty)$ where Λ is a large constant, there exists a solution U_λ to (1) with $\{U_\lambda = 0\}$ close to Σ_λ. Moreover, U_λ is continuous and increasing in λ.*

By differentiating in λ, we get

$$-\Delta \frac{\partial U_\lambda}{\partial \lambda} = \left(1 - 3U_\lambda^2\right) \frac{\partial U_\lambda}{\partial \lambda}.$$

Since $\frac{\partial U_\lambda}{\partial \lambda} > 0$, this implies U_λ is stable. However, these solutions are quite different from saddle solutions.

§3. Global minimizers in $\mathbb{R}^8$

Recall that for each $\lambda \in \mathbb{R}$, Σ_λ is asymptotic to Simons cone at the rate λr^{-2} as r tends to infinity. Here is the main result of [13].

Theorem 3.1. *For each $\lambda \in \mathbb{R}$, there exists a global minimizer U_λ of the Allen-Cahn equation* (1) *which is invariant under $O(4) \times O(4)$ and has the following property: outside a compact set, the nodal set of U_λ is represented by $s = F_\lambda(r)$, where*

$$F_\lambda(r) = r + \lambda r^{-2} + o(r^{-2}), \quad as \ r \to +\infty.$$

Remark 3.2. *If one could show that this family of solutions U_λ depends continuously on λ and is ordered: $U_{\lambda_1} < U_{\lambda_2}$ for $\lambda_1 < \lambda_2$, then the saddle-shaped solution belongs to this family and it will be a global minimizer. However, our variational construction gives no information on the continuous dependence on λ.*

Remark 3.3. *This result is readily generalized to high dimensional case $\mathbb{R}^{2m}$ for any $m \geq 4$ as well as Lawson cones (see* [12]*).*

As we will see later on, the existence of at least one global minimizer is an easy consequence of the result of Pacard-Wei, Theorem 2.1. Recall that a result of Jerison and Monneau [11] tells us that the existence of a nontrivial global minimizer in $\mathbb{R}^8$ which is even in all of its variables implies the existence of a family of counter-examples for the De Giorgi conjecture in $\mathbb{R}^9$. Hence an immediate corollary of Theorem 3.1 is the following

Corollary 3.4. *There is a family of monotone solutions to the Allen-Cahn equation* (1) *in $\mathbb{R}^9$, which is not one-dimensional and having $O(4) \times O(4)$ symmetry in the first* 8 *variables.*

This corollary could be regarded as a parallel result to Simon's existence result on entire minimal graphs [16]. Using other Lawson cones we can get many counterexamples to De Giorgi conjecture in $\mathbb{R}^9$.

§4. Proof

In this section we outline the main steps in the proof of Theorem 3.1. We refer the reader to [13] again for more details.

Step 1. Construction of solutions in bounded domains

Take two solutions, $u^{-\Lambda} < u^{\Lambda}$, constructed in Theorem 2.1 as barriers. Construct minimizing solutions u_R^λ under the constraint $\{u^{-\Lambda} < u < u^{\Lambda}\}$ and suitable Dirichlet boundary values in expanding bounded domains Ω_R, $R \to +\infty$.

Then we will let $R \to +\infty$. It is readily verified that $u_R^\lambda \to U_\lambda$ in $C^2_{loc}(\mathbb{R}^8)$ and the limit U_λ is a minimizing solution of (1). The problem is to show that the nodal set of U_λ is very close to Σ_λ. To this end, we need to show that the nodal set of u_R^λ, $\{u_R^\lambda = 0\}$, is very close to Σ_λ in a uniform way. Hence the boundary value of u_R^λ is taken to be $g(\mathrm{dist}_{\Sigma_\lambda})$ (with a small perturbation), where g is the one dimensional solution of (1) and $\mathrm{dist}_{\Sigma_\lambda}$ is the signed distance to Σ_λ.

Step 2. Refined asymptotics

Outside a large ball B_{R_0}, $\{u_R^\lambda = 0\}$ is a graph over Simons cone Σ_0, $\{z = h_R^\lambda(\ell)\}$. Here Simons cone is viewed as the ray $\{x_1 = x_2\}$ in $\mathbb{R}^2$ and $\ell = (x_1^2 + x_2)^{1/2}$, $z = \mathrm{dist}_{\Sigma_0}$ is the signed distance to Simons cone.

The next step is to establish the refined asymptotics for h_R^λ. Such a refined asymptotics is possible by the fact that h_R^λ is almost a Jacobi field on Simons cone

$$\frac{d^2 h_R^\lambda(\ell)}{d\ell^2} + \frac{6}{\ell}\frac{dh_R^\lambda(\ell)}{d\ell} + \frac{6}{\ell^2} h_R^\lambda(\ell) = O(\ell^{-5}). \tag{4}$$

By this ODE, there exists $R_1 > R_0$ (independent of R) such that

$$R_1^2 h_R^\lambda(R_1) = \ell^2 h_R^\lambda(\ell) + O(\ell^{-1}), \quad \forall \ell \in [R_1, R].$$

Recall that we have prescribed a boundary value at $\ell = R$, $h_R^\lambda(R) = \lambda R^{-2}$. Therefore we have

$$h_R^\lambda(\ell) = \lambda \ell^{-2} + O(\ell^{-3}), \quad \forall \ell \in [R_1, R].$$

With this refined asymptotics in hand, we can let $R \to +\infty$ and conclude that the nodal set of U_λ has the same refined asymptotics as $\ell \to +\infty$.

Step 3. Infinite dimensional Lyapunov-Schmidt reduction

Equation (4) is derived by using the *infinite dimensional Lyapunov-Schmidt reduction method.* This method was previously used by Del Pino, Kowalczyk and Wei to construct solutions of (1) starting from some good minimal hypersurfaces. We use this method *reversely*: from a solution of (1), u_R^λ, we show that its nodal set $\{u_R^\lambda = 0\}$ is *uniformly* close to a minimal hypersurface Σ_λ. Furthermore, since the nodal set of u^λ is close to Simons cone and it is represented as a graph on this cone, we get an almost linearized equation, the almost Jacobi field equation (4).

Now we detail this reduction method. In Fermi coordinates with respect to Simons cone (ℓ, z), where $\ell \in \Sigma_0$ and z is the signed distance to Σ_0, we take the decomposition

$$u_R^\lambda(\ell, z) = g\big(z - \widetilde{h}_R^\lambda(\ell)\big) + \phi(\ell, z). \tag{5}$$

Here it is necessary to choose $\widetilde{h}_R^\lambda$ so that two goals are achieved at the same time:

- ϕ is a *small* error;
- $\widetilde{h}_R^\lambda$ is a Jacobi field with *controlled* error.

Note that $\widetilde{h}_R^\lambda$ is not exactly h_R^λ, but it can be shown that their difference decays at the order $O(\ell^{-3})$ as $\ell \to +\infty$.

By the decomposition (5), we get the equation

$$\Delta_{\mathbb{R}^n}\phi = W''(g_*)\phi + g'\big(z - \widetilde{h}_R^\lambda(y)\big)\left[\Delta_{\Sigma_0}\widetilde{h}_R^\lambda - |A_{\Sigma_0}|^2\widetilde{h}_R^\lambda\right] + h.o.t. \tag{6}$$

where h.o.t. are higher order terms quadratic in $\widetilde{h}_R^\lambda$ and ϕ.

The function $\widetilde{h}_R^\lambda$ is chosen to satisfy the orthogonal condition:

$$\int \phi(\ell, z) g'\big(z - \widetilde{h}_R^\lambda(\ell)\big)\, dz = 0, \quad \forall \ell. \tag{7}$$

Then (6) can be decoupled into two equations:

- the parallel part (to g') gives the equation for $\widetilde{h}_R^\lambda$, the Jacobi field equation with error of higher order, which follows by multiplying (6) by g' and integrating in z and applying (7);
- the orthogonal part (to g') gives a good equation for ϕ, and consequently a good estimate on it, which follows by multiplying (6) by ϕ and integrating in z.

Step 4. The decoupling: a model case

Let us look at a model case. Suppose $\phi \in L^\infty(\mathbb{R}^n)$ solves

$$\begin{cases} \Delta_{\mathbb{R}^{n-1}}\phi + \partial_{zz}\phi = W''(g(z))\phi + a(y)g'(z) + E, \\ \displaystyle\int_{-\infty}^{+\infty} \phi(y, z) g'(y, z)\, dz = 0, \quad \forall y \in \mathbb{R}^{n-1}. \end{cases} \tag{8}$$

Here $E \in L^\infty(\mathbb{R}^n)$ is a given function, a point in $\mathbb{R}^n$ is denoted by (y, z), where $y \in \mathbb{R}^{n-1}$ and $z \in \mathbb{R}$. Then we claim that both ϕ and a are controlled by E, see [9, Proposition 4.1]. In other words, the single equation (8) implies two equations, one for ϕ and the other one for a.

To this end, the main tool we need on the one dimensional solution is its nondegeneracy, i.e. kernel of the linearized operator at g is spanned by g'. Furthermore, there exists a spectral gap. Recall that g' corresponds to eigenvalue 0. The spectral gap says that the second eigenvalue of this linearized operator is positive.

By this spectral gap, multiplying (8) by ϕ, integrating in z and using the orthogonal condition in (8), we get

$$\Delta_{\mathbb{R}^{n-1}} \int_{-\infty}^{+\infty} \phi(y,z)^2\, dz \geq \mu \int_{-\infty}^{+\infty} \phi(y,z)^2\, dz + \int_{-\infty}^{+\infty} \phi(y,z)E(y,z)\, dz.$$

Here μ is a positive constant related to the spectral gap. From this differential inequality we get an estimate of ϕ in term of E.

On the other hand, multiplying (8) by g', integrating in z and using the orthogonal condition in (8) once again, we get

$$a(y) \int_{-\infty}^{+\infty} g'(z)^2 dz = -\int_{-\infty}^{+\infty} E(y,z)g'(z)dz.$$

Hence a is also controlled by $E^{\perp}$.

Coming back to our original setting, in (6), $E^{\perp} = O(\|\phi\|^2) + O(\|\widetilde{h}_R^{\lambda}\|^2)$, so if at first we know $\|\phi\| = o(1)$ and $\|\widetilde{h}_R^{\lambda}\| = o(1)$, then these bounds can be improved after an iteration.

Remark 4.1. *In fact, there is a technical issue: we can only show that $\phi = O(\ell^{-2})$. Therefore in the ODE for $\widetilde{h}_R^{\lambda}$, the error is only of the order $O(\ell^{-4})$. This is not sufficient to deduce the required expansion of $\widetilde{h}_R^{\lambda}$. To get an $O(\ell^{-5})$ bound, we take a further decomposition of ϕ into an ℓ^{-2} decay component and other ones of higher order decay, see* [13] *for more details.*

References

[1] G. Alberti, L. Ambrosio and X. Cabré, On a long-standing conjecture of E. De Giorgi: symmetry in 3D for general nonlinearities and a local minimality property. Special issue dedicated to Antonio Avantaggiati on the occasion of his 70th birthday, *Acta Appl. Math.*, 65 (2001), no. 1–3, 9–33.

[2] L. Ambrosio and X. Cabré, Entire solutions of semilinear elliptic equations in $\mathbb{R}^3$ and a conjecture of De Giorgi, *J. Amer. Math. Soc.*, 13 (2000), no. 4, 725–739.

[3] E. Bombieri, E. De Giorgi and E. Giusti, Minimal cones and the Bernstein problem, *Invent. Math.*, 7 (1969), 243–268.

[4] X. Cabré and J. Terra, Saddle-shaped solutions of bistable diffusion equations in all of $\mathbb{R}^{2m}$, *J. Eur. Math. Soc. (JEMS)*, 11 (2009), no. 4, 819–843.
[5] X. Cabré and J. Terra, Qualitative properties of saddle-shaped solutions to bistable diffusion equations, *Comm. Partial Differential Equations*, 35 (2010), no. 11, 1923–1957.
[6] X. Cabré, Uniqueness and stability of saddle-shaped solutions to the Allen-Cahn equation, *J. Math. Pures Appl.* (9), 98 (2012), no. 3, 239–256.
[7] E. De Giorgi, Convergence problems for functionals and operators, Proc. Int. Meeting on Recent Methods in Nonlinear Analysis (Rome, 1978), 131–188, Pitagora, Bologna (1979).
[8] M. del Pino, M. Kowalczyk and J. Wei, On De Giorgi's conjecture in dimension $N \geq 9$, *Ann. of Math.* (2), 174 (2011), no. 3, 1485–1569.
[9] M. del Pino, M. Kowalczyk, J. Wei and J. Yang, Interface foliation near minimal submanifolds in Riemannian manifolds with positive Ricci curvature, *Geom. Funct. Anal.*, 20 (2010), no. 4, 918–957.
[10] N. Ghoussoub and C. Gui, On a conjecture of De Giorgi and some related problems, *Math. Ann.*, 311 (1998), no. 3, 481–491.
[11] D. Jerison and R. Monneau, Towards a counter-example to a conjecture of De Giorgi in high dimensions, *Ann. Mat. Pura Appl.* (4), 183 (2004), no. 4, 439–467.
[12] B. Lawson, The equivariant Plateau problem and interior regularity, *Trans. Amer. Math. Soc.*, 173 (1972).
[13] Y. Liu, K. Wang and J. Wei, Global minimizers of Allen-Cahn equation in dimensions $n \geq 8$, *J. Math. Pure Appl.*, 108 (2017), no. 6, 818–840.
[14] F. Pacard and J. Wei, Stable solutions of the Allen-Cahn equation in dimension 8 and minimal cones, *J. Funct. Anal.*, 264 (2013), no. 5, 1131–1167.
[15] O. Savin, Regularity of flat level sets in phase transitions, *Ann. of Math.* (2), 169 (2009), no. 1, 41–78.
[16] L. Simon, Entire solutions of the minimal surface equation, *J. Differential Geometry*, 30 (1989), 643–688.
[17] J. Simons, Minimal varieties in Riemannian manifolds, *Ann. of Math.* (2), 88 (1968), 62–105.

School of Mathematics and Statistics &
Computational Science Hubei Key Laboratory,
Wuhan University,
Wuhan 430072, China.
E-mail address: wangkelei@whu.edu.cn

Advanced Studies in Pure Mathematics 85, 2020
The Role of Metrics in the Theory of Partial Differential Equations
pp. 479–489

Harmonic maps with prescribed singularities and applications in general relativity

Gilbert Weinstein

Abstract.

This paper presents a general existence and uniqueness result for harmonic maps with prescribed singularities into non-positively curved targets, and surveys a number of applications to general relativity. It is based on a talk delivered by the author at *The* 11*th Mathematical Society of Japan Seasonal Institute, The Role of Metrics in the Theory of Partial Differential Equations.*

§1. Introduction

We begin with a brief review of harmonic maps. Let (M, g) and (N, h) be Riemannian manifolds and let $\varphi\colon M \to N$ be a smooth map. Then $d\varphi\colon M \to TM \otimes \varphi^{-1}TN$ is a cross-section of the tangent bundle of M tensor-product with the pull back bundle of TN under φ. In local coordinates we have:

$$d\varphi = \frac{\partial \varphi^\mu}{\partial x^i}\, dx^i \otimes \frac{\partial}{\partial y^\mu}.$$

There is a canonical positive definite inner product on this bundle with the norm expressed in local coordinates:

$$|d\varphi|^2 = \frac{\partial \varphi^\mu}{\partial x^i}\,\frac{\partial \varphi^\nu}{\partial x^j}\, g^{ij}\, h_{\mu\nu}.$$

From this we can form an energy:

$$E_\Omega(\varphi) = \int_\Omega |d\varphi|^2\, dV_g.$$

Received February 4, 2019.
Revised June 13, 2019.
Key words and phrases. Harmonic maps, black holes.
The author wishes to express his thanks to the organizers of the MSJ-SI for their invitation to participate in this conference, and for their hospitality in Sapporo.

Definition 1.1. A map $\varphi \colon M \to N$ is *harmonic* if for each $\Omega \subset\subset M$ it is a critical point of E_Ω. The map φ is an *local energy minimizer* if for every $\Omega \subset\subset M$, $E_\Omega(\varphi) \leq E_\Omega(\psi)$ for all ψ which agrees with φ on the boundary of Ω. It is an energy minimizer if the same is true with $\Omega = M$.

Clearly every local energy minimizer is a harmonic map. The Euler-Lagrange equations read

$$\tau^\mu(\varphi) = \Delta_g \varphi^\mu + \Gamma^\mu_{\alpha\beta}(\varphi)\, g(\nabla\varphi^\alpha, \nabla\varphi^\beta) = 0, \tag{1}$$

where $\Gamma^\mu_{\alpha\beta}(\varphi)$ are the Christoffel symbols of N evaluated along φ, and $\Delta_g = \operatorname{tr} \nabla^2$ is the Laplace-Beltrami operator on M. This forms a system of semi-linear elliptic PDEs on M with the nonlinearity being quadratic in the first derivatives of φ. For an arbitrary map φ, the left-hand side τ is a cross section of $\varphi^{-1}TN$ called the *tension* of φ.

Familiar examples of harmonic maps are the two cases: (i) $\dim N = 1$, then $\Gamma = 0$ and the harmonic map problem reduces to the linear Laplace equation; (ii) $\dim M = 1$, then the harmonic map problem becomes the ODE problem for geodesics. The first significant non-linear PDE result concerning harmonic maps was obtained by Eells-Sampson [6]: *If the sectional curvature of N satisfies $\kappa_N \leq 0$, and $\varphi_0 \colon M \to N$ is any smooth map, then there is a harmonic map φ homotopic to φ_0.* Without the assumption on κ_N, harmonic maps and even energy minimizer can develop singularities, and the best one can do is control the size of the singular set [18].

§2. Harmonic maps with prescribed singularities

Consider first the linear case. Let (M, g) be a complete Riemannian manifold[1] and let $\Gamma \subset M$ be a submanifold of co-dimension ≥ 2. Let $\rho \colon \Gamma \to \mathbb{R}^+$ be a positive continuous function, and let $G(x, y)$ be the Green function for Δ_g on M, then

$$u(x) = \int_\Sigma \rho(y)\, G(x, y)\, dv_\Sigma(y),$$

is a harmonic function on $M \setminus \Gamma$ which blows up on Γ, $\Delta_g u = \rho\delta_\Gamma$. Now if γ is a geodesic in N, then $\gamma \circ u$ is a harmonic map into N which blows up on Γ. Before we state and prove a nonlinear analog of this result, we need two definitions.

[1] A similar result can be stated for a compact manifold with boundary.

Definition 2.1. A *model map* is a map $\varphi_0 \colon M \setminus \Gamma \to N$ such that (i) $|\tau(\varphi_0)|$ is bounded; and (ii) $|\tau(\varphi_0)|$ decays "fast enough", in the sense that there is a function $v > 0$ with $v \to 0$ at infinity and $|\tau(\varphi_0)| \leq -\Delta_g v$.

It turns out for instance that for $M = \mathbb{R}^3$, the case of interest for our applications, it is sufficient to require $|\tau(\varphi_0)| = O(r^{-2-\epsilon})$ with $\epsilon > 0$.

Definition 2.2. Two maps $\varphi_1, \varphi_2 \colon M \setminus \Gamma \to N$ are *asymptotic* if $\mathrm{dist}_N(\varphi_1, \varphi_2)$ is (i) bounded; and (ii) falls off to 0 at infinity.

Part (i) of this definition is inspired by the analogous definition for geodesics [5].

Theorem 1. *Let $\varphi_0 \colon M \setminus \Gamma \to N$ be a model map, and let either $-b^2 < \kappa_N < -a^2$, or N be a symmetric space of non-compact type and rank ≤ 2, then there is a unique harmonic map $\varphi \colon M \setminus \Gamma \to N$ which is asymptotic to φ_0.*

Proof. For completeness, we give some of the details of the proof. We need the following Lemma [23].

Lemma 1. *Let $\varphi_1, \varphi_2 \colon M \setminus \Gamma \to N$ be smooth maps, and suppose $\kappa_N \leq 0$, then*

$$\Delta_g \sqrt{1 + \mathrm{dist}_N(\varphi_1, \varphi_2)^2} \geq -(|\tau(\varphi_1)| + |\tau(\varphi_2)|).$$

The Hessian H of the distance function $d_N \colon N \times N \to \mathbb{R}$ is positive semi-definite, thanks to the hypothesis $\kappa_N \leq 0$ [19], hence denoting $d_p(\cdot) = d_N(p, \cdot)$, we have

$$\begin{aligned}\Delta_g d_N(\varphi_1, \varphi_2) &= \nabla d_{\varphi_1} \cdot \tau(\varphi_2) + \tau(\varphi_1) \cdot \nabla d_{\varphi_2} + H(d\varphi_1 + d\varphi_2, d\varphi_1 + d\varphi_2) \\ &\geq -(|\tau(\varphi_1)| + |\tau(\varphi_2)|),\end{aligned}$$

wherever $d_N(\varphi_1, \varphi_2) > 0$. Now, we get

$$\begin{aligned}\Delta \sqrt{1 + d_N^2} &\geq \frac{d_N}{\sqrt{1 + d_N^2}} \Delta d_N \geq -\frac{d_N}{\sqrt{1 + d_N^2}} (|\tau(\varphi_1)| + |\tau(\varphi_2)|) \\ &\geq -(|\tau(\varphi_1)| + |\tau(\varphi_2)|),\end{aligned}$$

when $d_N > 0$, and since $\sqrt{1 + d_N^2} \geq 0$ has a global minimum on the set $\{d_N = 0\}$, it follows trivially that $\Delta \sqrt{1 + d_N^2} \geq 0$ also there. This completes the proof of the Lemma.

We now return to the proof of the Theorem. We first establish uniqueness. If $\varphi_1, \varphi_2 \colon M \setminus \Gamma \to N$ are both harmonic and asymptotic to φ_0, then φ_1 is asymptotic to φ_2 and by Lemma 1, $u = d_N(\varphi_1, \varphi_2)$ is a bounded subharmonic function on $M \setminus \Gamma$ which tends to 0 at infinity.

Since Γ is of co-dimension ≥ 2, it follows from Lemma 8 in [22] that u is weakly harmonic on M. Hence by the maximum principle $u = 0$ and $\varphi_1 = \varphi_2$.

We now prove existence. Fix some point $x_0 \in M \backslash \Gamma$, pick $\delta > 0$, and let $M_\delta = B_{1/\delta}(x_0) \cap \{x \in M \colon d_M(x, \Gamma) > \delta\}$. Since $\overline{M}_\delta$ is compact and φ_0 is smooth on $\overline{M}_\delta$, there exists a harmonic map $\varphi_\delta \colon M_\delta \to N$ such that $\varphi_\delta = \varphi_0$ on ∂M_δ. Let $u_\delta = \sqrt{1 + d_N(\varphi_0, \varphi_\delta)^2}$, then by Lemma 1 and Definition 2.1, we have

$$\Delta_g(u_\delta - v) \geq 0,$$

on M_δ, where v is the function in Definition 2.1, hence $v > 0$ and $v \to 0$ at infinity. Also $u_\delta - v \leq 1$ on ∂M_δ, thus by the maximum principle it follows immediately that $u_\delta \leq C$ independent of $\delta > 0$, and $d_N(\varphi_0, \varphi_\delta) \to 0$ at infinity.

We now consider the maps φ_δ restricted to a fixed set M_{δ_0}. These map into a fixed ball $B_R \subset N$, and hence we can obtain local pointwise a priori bounds uniform in δ not just for φ_δ but also for any number of derivatives. From here it is straightforward to obtain a sequence $\delta_i \to$ such that φ_{δ_i} converges uniformly on compact subsets of $M \setminus \Gamma$ together with two derivatives, and hence the limit φ will be harmonic and asymptotic to φ_0.

To achieve this, the first step is to obtain a priori local energy bounds. These follow from the pointwise bound above and require a particular coordinate system on the target N of the harmonic map:

$$ds^2 = \sum_{j=1}^{k} dr_j^2 + Q(\xi, \xi)$$

where $1 \leq k \leq 2$ is either 1 or the rank of N, and Q is a quadratic form on $\mathbb{R}^{l-k}$ satisfying

$$2a\, Q(\xi, \xi) \leq \frac{d}{dr_j} Q(\xi, \xi) \leq 2b\, Q(\xi, \xi), \quad \forall \xi \in \mathbb{R}^{l-k},$$

for some constants $0 < a < b$. Here where l is the dimension of N. The existence of such a coordinate system for symmetric spaces of non-compact type and rank 2 was proved in [12]. However, we conjecture the same holds true for any rank. In this coordinate system, k of the harmonic map equations read

$$\Delta_g r_j = \frac{\partial}{\partial r_j} Q(\nabla \xi, \nabla \xi) \geq 2a. \tag{2}$$

Multiplying by $\chi^2 r_j$, where $\chi \in C_0^\infty(M_{\delta_0})$ is a cut-off function, and integrating by parts, we obtain

$$\int_{M_{\delta_0}} \chi^2 |\nabla r_j|^2 \leq 4\left(\sup_{M_{\delta_0}} r_j^2\right) \int_{M_{\delta_0}} |\nabla \chi|^2.$$

Now going back to Equation(2), multiplying by χ^2 and integrating by parts, we obtain

$$\begin{aligned}\int_{M_{\delta_0}} \chi^2 Q(\nabla\xi, \nabla\xi) &\leq \frac{1}{a}\int_{M_{\delta_0}} \chi\nabla\chi\nabla r_j \\ &\leq \frac{1}{a}\left(\int_{M_{\delta_0}} |\nabla r_j|^2\right)^{1/2}\left(\int_{M_{\delta_0}} |\nabla\chi|^2\right)^{1/2}.\end{aligned}$$

Combining these two inequalities gives the required local energy estimate $E_\Omega(\varphi_\delta) \leq C$ where C depends on $\Omega \Subset M_{\delta_0}$ but is independent of δ.

We can now establish a pointwise bound on the derivatives using the Bochner identity:

$$\Delta_g |d\varphi_\delta|^2 = |\nabla d\varphi_\delta|^2 + \mathrm{Ric_M}(d\varphi_\delta, d\varphi_\delta) - \mathrm{Riem_N}(d\varphi_\delta, d\varphi_\delta, d\varphi_\delta, d\varphi_\delta).$$

Since $\mathrm{Riem_N} \leq 0$, it follows

$$\Delta_g |d\varphi_\delta|^2 \geq -C|d\varphi_\delta|^2,$$

where $C = \sup_{M_{\delta_0}} |\mathrm{Ric_M}|$, and standard elliptic theory immediately gives an L^∞ on $|d\varphi_\delta|^2$ from the L^1 bound already obtained. One can now bootstrap using the harmonic map equations (1) to get pointwise bounds on any number of derivatives independent of δ. Q.E.D.

§3. Applications in general relativity

The applications to general relativity of harmonic maps with prescribed singularities stem from the following fact or variations thereof. Under the hypothesis of *stationarity* and $(D-3)$-*axisymmetry*, i.e. admitting an isometric action of the group $\mathbb{R} \times U(1)^{D-3}$, the Einstein vacuum equations reduce to a harmonic map problem $\varphi\colon \mathbb{R}^3 \setminus \Gamma \to N$ where $\Gamma \subset z$-axis, and $N = SL(D-2,\mathbb{R})/SO(D-2)$. Note that N is a symmetric space of non-compact type and rank $D-3$. We recall that the *rank* is the maximal dimension of a *flat*, i.e. a totally geodesic submanifold with vanishing curvature. For the sake of brevity, we will not expound here further on this reduction; see [12, Section 2] for more details. We only comment on two essential features: the quotient space is 2 dimensional, and the Einstein equations decouple so that the system becomes semi-linear.

3.1. Non-degenerate vacuum black holes in 4-d gravity

In this case, the target is the hyperbolic plane $N = SL(2,\mathbb{R})/SO(2) = \mathbb{H}^2$ [23]. This line of investigation, which begun in the static linear case by Bach-Weyl [1], was pursued further in a series of papers by this author starting with [20], culminating with [23], in which an earlier version of Theorem 1 was proved.

The metrics considered are of the form

$$g = -f^{-1}\rho^2 dt^2 + f(d\phi + v dt)^2 + f^{-1}e^{2\sigma}(d\rho^2 + dz^2), \tag{3}$$

and are asymptotically flat. The *twist potential* is defined by $d\omega = *(\xi \wedge d\xi)$, and is related to the metric function v by $dv = -\rho f^{-2} *_2 d\omega$, where $*_2$ is the Hodge dual in the flat ρz-plane. The one form ω measures how far ξ is from being hypersurface orthogonal. It vanishes in the *static* case, indicating the absence of rotation, in which case the Einstein equations reduce to one linear equation[2]. The map blows up on the axis Γ. The singular boundary conditions can be used to specify the number of black holes, their mass and their angular momenta. Once that data is prescribed, one constructs a model map φ_0, and there is, by Theorem 1, a unique solution to the harmonic map problem.

Also as pointed out in [23], given such a harmonic map, one can reconstruct a solution of the original Einstein vacuum equations. The metric (3) thus constructed can be extended across Γ if and only if there are no conical singularities, and if the metric components are sufficiently smooth up to the axis. The latter condition, which we call *analytic regularity*, was established in [20]; see also generalizations in [15, 16]. The angle deficiency $e^{\beta} = \lim_{\rho\to 0} \rho^2 e^{2\sigma} f^{-2}$ is constant along any connected component of the axis, and it in fact vanishes on the two unbounded component [13]. With two black holes, it is related to the gravitational force acting between them $F = 1/4(e^{-\beta/2} - 1)$. In the static linear case, Bach and Weyl first calculated $F = m_1 m_2/(R^2 - (m_1 + m_2)^2)$, where m_1 and m_2 is a measure of the masses of the two bodies, and $R - m_1 - m_2$ is a measure of their separation. It is conjectured that it is impossible to balance several rotating black holes. This is equivalent to a conical singularity always being present. Some results were obtained in [14, 21].

These results were generalized to the Einstein-Maxwell equations, and also later further generalized to the Einstein-Abelian-Yang-Mills equations, in which case $N = \mathbb{H}^2_{\mathbb{C}}$, and $\mathbb{H}^k_{\mathbb{C}}$ respectively [23].

[2]This is the case studied by Bach and Weyl in 1917.

3.2. Degenerate black holes in 4-d gravity

The generalization of these results to the *degenerate* horizon case, sometimes also called the *extreme* case, is primarily motivated by an idea of S. Dain [4] which allows one to prove lower bounds on the ADM mass of axisymmetric data in terms of angular momentum and charge. Dain showed that the ADM mass is bounded below by a renormalized harmonic map energy and that this renormalized energy is minimized by the harmonic maps corresponding to extreme black holes. In particular, Dain [4] proved the inequality $m \geq |J|$ where m is the ADM mass and J the angular momentum. This was generalized to multiple black holes where we proved $m \geq \mathcal{F}$, and $\mathcal{F}$ is the harmonic map renormalized energy for the corresponding extreme black hole [3]. Dain's result was later generalized to the charged single black hole case [2] $m^2 \geq \frac{1}{2}\left(q + \sqrt{q^2 + 4J^2}\,\right)$, and we again generalized this to the multiple black hole case [11].

A key ingredient in these results is the construction of the minimizer corresponding to multiple black holes. The existence result for harmonic maps with prescribed singularities described in Section 2 is used to achieve this goal.

In the multiple black hole case, both in vacuum and in the charged case, the lower bound is not explicit, but it is conjectured that it is the same as in the single black hole case with q and J denoting the total charge and angular momentum.

3.3. Black holes in 5-d gravity

In this latest application of harmonic maps with prescribed singularities, we considered the problem of vacuum black holes in 5-d with non-spherical topology. In 4-d, a result of Hawking [9] shows that the cross section of any connected component of the event horizon in an asymptotically flat (AF) stationary spacetime satisfying the dominated energy condition (DEC), has positive Euler characteristic, and hence must be topologically a 2-sphere. However, the higher dimensional analogous result by Galloway and Schoen [8] only shows that black hole boundaries are $(n-2)$-dimensional Riemannian manifolds with positive Yamabe invariant. In particular, in 5-d this allows prime 3-d manifolds of positive Yamabe type, S^3, $S^2 \times S^1$, and the lens spaces $L(p, q)$. Some explicit examples with $S^2 \times S^1$ horizon topology, referred to as Black Rings, are known [7, 17].

The target for the harmonic map problem is now the symmetric space $N = SL(3, \mathbb{R})/SO(3)$. In [12], we constructed harmonic maps corresponding to AF spacetimes, and in [13] solutions corresponding to asymptotically locally Euclidean (ALE) spacetimes, i.e. whose spacelike infinity is modeled on $(a,\infty)\times L(p,q)$, as well as asymptotically Kaluza-Klein (AKK), i.e. with spacelike infinity modeled on $(a,\infty)\times S^1\times S^2$. Here too, we are able to construct multiple black hole solutions, provided the necessary topological obstructions are avoided.

The metrics we consider are now of the form

$$g = f^{-1}e^{2\sigma}(d\rho^2+dz^2) - f^{-1}\rho^2 dt^2 + f_{ij}(d\phi^i + v^i dt)(d\phi^j + v^j dt), \tag{4}$$

where ρ^2 is minus the determinant of the metric on the orbits of $\mathbb{R}\times U(1)^2$, f_{ij} is the metric on the orbits of $U(1)^2$ and f is the determinant of f_{ij}. Also as before, the resulting metrics can be extended across the axis if and only if analytic regularity holds, and there are no conical singularities.

The topology of the horizon is prescribed via the *rod structure*, a pair of relatively prime integers (m,n) assigned to each *axis rod* $\subset\Gamma$. Rods are subintervals of Γ on which the rank of f_{ij} is either 1 on axis rods, or 2 on horizon rods. The end points of horizon rods are *poles*, points separating axis rods are called *corners*. The rod structure (m,n) of an axis rod is defined to be such that $m\partial_{\phi_1}+n\partial_{\phi_2}$ is the generator of $U(1)^2$ which degenerates on that rod. The topology of the horizon is determined by the rod structures of the two adjacent rods. Some simple examples of rod structure configurations are given in Figure 1. It turns out that the potentials v^i are constant on each component of Γ. The difference between the two constant values of v^i adjacent to any horizon determines the angular momentum associated with that generator for that black hole. The data consists of the rod structure configuration and these constants. Given any data satisfying a mild admissibility condition, we showed in [12] that a model map could be constructed, and hence a unique harmonic map exists corresponding to this data.

In [10], we analyzed the topology of the domain of outer communications of these solutions using *linear plumbing*. Plumbing is a construction used to glue together two disk bundles by identifying the bundles over two neighborhoods, crisscrossing base and fiber. Utilizing this technique we were able, given any rod structure configuration, to determine that the topology of the $4D$ slice in our solutions is a sequence of disk bundle plumbings.

§4. Conclusion

In this paper, we sketched the proof of an existence and uniqueness result for harmonic maps with prescribed singularities and surveyed a number of applications to general relativity, particularly to black hole existence and uniqueness problems in dimensions 4 and 5, both with

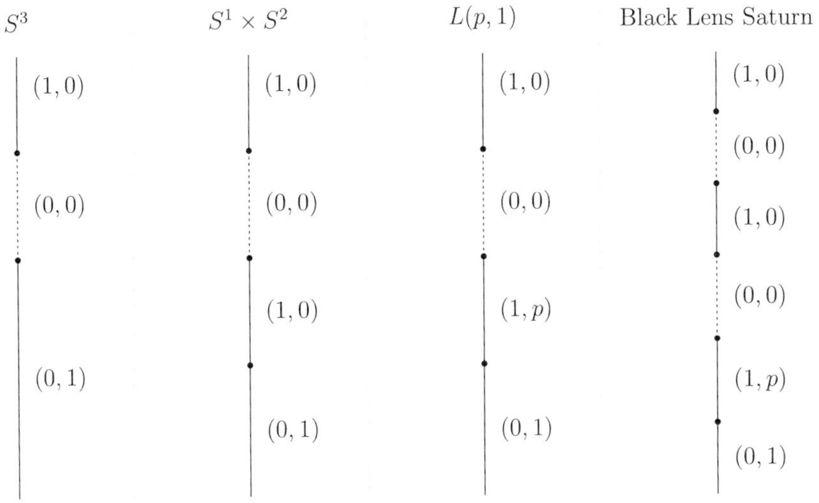

Fig. 1. Some examples of rod structure configurations.

non-degenerate and degenerate horizons. In particular, the applications in dimension 5 provided a glimpse into an interesting interface between general relativity and low dimensional topology.

We believe there are many more applications of this versatile method which should yield answers to a number of interesting unanswered questions.

References

[1] Rudolf Bach and Hermann Weyl, *Republication of: New solutions to Einstein's equations of gravitation*, B. Explicit determination of static, axially symmetric fields. By Rudolf Bach. With a supplement on the static two-body problem. By H. Weyl, General Relativity and Gravitation, **44**, (2012Mar), no. 3, 817–832.

[2] Piotr T Chruściel and João Lopes Costa, *Mass, angular-momentum and charge inequalities for axisymmetric initial data*, Classical and Quantum Gravity, **26** (2009), no. 23, 235013.

[3] Piotr T. Chruściel, Yanyan Li and Gilbert Weinstein, *Mass and angular-momentum inequalities for axi-symmetric initial data sets.* II. *Angular momentum*, Ann. Physics, **323** (2008), no. 10, 2591–2613.

[4] Sergio Dain, *Proof of the angular momentum-mass inequality for axisymmetric black holes*, J. Differential Geom., **79** (2008), no. 1, 33–67, MR2414749 (2009j:83007).

[5] P. Eberlein and B. O'Neill, *Visibility manifolds*, Pacific J. Math., **46** (1973), no. 1, 45–109.

[6] James Eells and J. H. Sampson, *Harmonic Mappings of Riemannian Manifolds*, American Journal of Mathematics, **86** (1964), no. 1, 109–160.

[7] Roberto Emparan and Harvey S. Reall, *Black holes in higher dimensions*, Living Reviews in Relativity, **11** (2008Sep), no. 1, 6.

[8] Gregory J. Galloway and Richard Schoen, *A Generalization of Hawking's Black Hole Topology Theorem to Higher Dimensions*, Communications in Mathematical Physics, **266** (2006), no. 2, 571–576.

[9] S. W. Hawking, *Black holes in general relativity*, Comm. Math. Phys., **25** (1972), no. 2, 152–166.

[10] Marcus Khuri, Yukio Matsumoto, Gilbert Weinstein and Sumio Yamada, *Plumbing Constructions and the Domain of Outer Communication for 5-Dimensional Stationary Black Holes*, (2018), available at `1807.03452`.

[11] Marcus Khuri and Gilbert Weinstein, *The Positive Mass Theorem for Multiple Rotating Charged Black Holes*, Calculus of Variations and Partial Differential Equations, **55** (April 2016-04), no. 2, 1–29, available at `1502.06290`.

[12] Marcus Khuri, Gilbert Weinstein and Sumio Yamada, *Stationary Vacuum Black Holes in 5 Dimensions*, to appear in Communications in PDE, (2017), available at `http://arxiv.org/abs/1711.05229`.

[13] ———, *Asymptotically locally euclidean/Kaluza-Klein stationary vacuum black holes in five dimensions*, Progress of Theoretical and Experimental Physics, **2018** (2018), no. 5, 053E01, available at `/oup/backfile/content_public/journal/ptep/2018/5/10.1093_ptep_pty052/3/pty052.pdf`.

[14] Yan Yan Li and Gang Tian, *Nonexistence of axially symmetric, stationary solution of Einstein vacuum equation with disconnected symmetric event horizon*, Manuscripta mathematica, **73** (1991), 83–90.

[15] ———, *Regularity of harmonic maps with prescribed singularities*, Comm. Math. Phys., **149** (1992), no. 1, 1–30.

[16] Luc Nguyen, *Singular Harmonic Maps and Applications to General Relativity*, Communications in Mathematical Physics, **301** (2011), no. 2, 411–441.

[17] A. A. Pomeransky and R. A. Sen'kov, *Black ring with two angular momenta*, (2006), available at `hep-th/0612005`.

[18] Richard Schoen and Karen Uhlenbeck, *A regularity theory for harmonic maps*, J. Differential Geom., **17** (1982), no. 2, 307–335.

[19] SRichard choen and Shing Tung Yau, *Compact group actions and the topology of manifolds with non-positive curvature*, Topology, **18** (1979), no. 4, 361–380.

[20] Gilbert Weinstein, *On rotating black holes in equilibrium in general relativity*, Comm. Pure Appl. Math., **43** (1990), no. 7, 903–948, MR1072397 (91h:83063).

[21] ———, *On the force between rotating co-axial black holes*, Trans. Amer. Math. Soc., **343** (1994), no. 2, 899–906, MR1214787 (94h:83087).

[22] ———, *On the Dirichlet problem for harmonic maps with prescribed singularities*, Duke Math. J., **77** (1995), no. 1, 135–165, MR1317630 (96a:58073).

[23] ———, *Harmonic maps with prescribed singularities into Hadamard manifolds*, Math. Res. Lett., **3** (1996), no. 6, 835–844, MR1426540 (98b:58049).

Department of Mathematics & Department of Physics
Ariel University, Ariel, Israel 40700
E-mail address: gilbertw@ariel.ac.il

Advanced Studies in Pure Mathematics 85, 2020
The Role of Metrics in the Theory of Partial Differential Equations
pp. 491–511

Regularity of minimal and CMC stable hypersurfaces: a brief overview

Neshan Wickramasekera

§1. Introduction

This article is largely based on the material covered in the author's lecture at the 11th Mathematical Society of Japan Seasonal Institute held in Sapporo in July 2018. Our goal here, as was in the lecture, is to provide a brief discussion of the regularity theory of minimal and constant-mean-curvature (CMC) stable hypersurfaces of Riemannian manifolds, focusing mainly on the recent developments ([BCW], [BelWic-1], [Wic14]). We refer the reader to the articles [Wic14b] and [Wic] for more detailed surveys of the results presented here.

We shall begin by presenting, in Section 2, basic notions and definitions necessary for the statements of the results in later sections. Section 3 contains an overview of the work [Wic14] which provides a regularity and compactness theory for mass bounded stable, stationary (i.e. zero generalized mean curvature) hypersurfaces of a given $(n+1)$-dimensional smooth Riemannian manifold N. Once regularity is known, these hypersurfaces are critical points of the functional $\mathcal{A} : M \mapsto \mathcal{H}^n(M)$ on hypersurfaces $M \subset N$, where $\mathcal{H}^n$ is the n-dimensional Hausdorff measure on N induced by the metric on N. The work [Wic14] can be viewed as providing a complete regularity theory for stable minimal hypersurfaces for the embedded case. It builds on the earlier fundamental work of Schoen–Simon ([SchSim81]) that treated a special case, giving primarily compactness for locally uniformly area bounded stable minimal hypersurfaces with small singular sets. The embeddedness criterion provided by the work [Wic14], namely the absence of *classical singularities* (see the definition in Section 3 below), is readily checked for

Received June 7, 2019.
Revised July 8, 2019.
2010 *Mathematics Subject Classification.* Primary 53-02; Secondary 35-02.
Key words and phrases. integral varifolds, generalised mean curvature, minimal hypersurfaces, CMC hypersurfaces, Allen–Cahn equation.

locally area minimizing integral currents, so the work [Wic14] subsumes also the well-known regularity theory for locally area minimizing integral currents established by the combined pioneering work of De Giorgi ([DeG61]), Federer ([Fed70]) and Simons ([JSim68]) in the decade 1960-1970, with intermediate advances due to Reifenberg ([Rei60]), Fleming ([Fle62]) and Almgren ([Alm66]).

As regards the question of existence of critical points of $\mathcal{A}$ in compact Riemannian manifolds, there are now two basic variational approaches producing minimal hypersurfaces that are embedded away from a possible closed singular set of Hausdorff dimension no more than $n-7$. Both these approaches yield in the first instance a critical point not of $\mathcal{A}$ but of the *mass* functional, i.e. Hausdorff measure weighted by an integer valued multiplicity function, on the space of *integral n-varifolds* —a space of generalized hypersurfaces of N equipped with a topology of measure-theoretic convergence. The original approach to the existence problem, dating back to the work of Almgren ([Alm65]) and Pitts ([Pit81]) in 1965 and 1977 respectively, and revisited and streamlined recently by De Lellis–Tasnady ([DeLTas13]), involves direct construction of a min-max varifold that is a critical point of mass. For regularity conclusions this method relies both on the regularity theory for codimension 1 locally area minimizing integral currents and the Schoen–Simon compactness theory.

In contrast to this highly non-trivial, direct geometric construction, which is carried out in a non-linear space (of varifolds) with little structure, recently the work [Wic14] has made way for a simpler PDE-theoretic alternative in which the min-max construction is transferred to the classical setting of a functional satisfying the Palais–Smale compactness criterion on a Hilbert space. Specifically, this new approach exploits the long known approximability property of the mass functional by the Allen–Cahn functional on the Hilbert space $W^{1,2}(N)$, and uses more elementary, standard PDE min-max theorems to produce, in the first instance, bounded solutions to (elliptic) Allen–Cahn equations in a 1-parameter family, with a uniform Morse index upper bound and upper and lower Allen–Cahn energy bounds (Guaraco ([Gua18]), Gaspar–Guaraco ([GasGua1])). In the singular limit as the parameter value tends to zero, the Allen–Cahn energies of these solutions concentrate on a stationary integral n-varifold V on N (Hutchinson–Tonegawa ([HutTon00])). Regularity of V is deduced, after verifying stability of its regular part in small ambient balls (Tonegawa, [Ton05]), by applying the main regularity theorem of [Wic14] in its full strength in a fairly direct way (Tonegawa-Wickramasekera ([TonWic12])). We shall give a more detailed discussion of this PDE approach in Section 4.

Finally in Section 5 we shall discuss some newer results. These extend the results of [Wic14], furnishing a regularity and compactness theory (joint work of the author with C. Bellettini ([BelWic-1])) applicable to a much larger class of integral codimension 1 varifolds. This class consists of codimension 1 integral n-varifolds with generalized mean curvature locally integrable to a power $p > n$, with no classical singularities, with appropriately controlled behaviour near touching singularities (see definition in Section 5) and satisfying two variational hypotheses on their regular parts: namely, stationarity of the C^1-embedded orientable portions with respect to the functional $\mathcal{J}_\lambda : M \mapsto \mathcal{A}(M) + \lambda vol\,(M)$ for some constant $\lambda \in \mathbb{R}$, where *vol* denotes the "enclosed volume" functional, and weak stability with respect to $\mathcal{J}_\lambda$ of the C^2-immersed part (which is automatically a classical CMC immersion if $\lambda \neq 0$ by virtue of the fact that the C^1 embedded part is stationary with respect to $\mathcal{J}_\lambda$). We shall end the article by mentioning, also in Section 5, the recently established curvature estimates for weakly stable CMC hypersurfaces (joint work of the author with C. Bellettini and O. Chodosh ([BCW])). These estimates extend the earlier curvature estimates of Schoen–Simon–Yau ([SSY75]) and of Schoen–Simon ([SchSim81]) for strongly stable hypersurfaces. (Even more recently, in [BelWic-2] the codimension 1 regularity theory has been extended to hypersurfaces whose orientable regular parts have mean curvature prescribed by a general ambient $C^{1,\alpha}$ function g, and whose Morse index (with respect to the relevant functional determined by g) is locally finite, but we shall not discuss these latest developments here.)

Acknowledgement. It is a pleasure to thank the organisers of the 11th MSJ Seasonal Institute, Professors Yoshikazu Giga, Hideo Kubo and Tohru Ozawa, for their invitation to deliver a lecture, and for their hospitality during the event.

§2. Basic notions: first variation, mean curvature and varifolds

2.1. First variation and stationarity

Let N be a smooth, $(n+1)$-dimensional Riemannian manifold, $k \in \{1, 2, \ldots, n\}$ and let M be an embedded k-dimensional C^1 submanifold of N with locally finite mass in N (i.e. with $\mathcal{H}^k\,(M \cap K) < \infty$ for each compact set $K \subset N$). To define the notion of (unconstrained) critical point of the area functional $\mathcal{A}$, we start with a smooth compactly supported vector field X on N, and note that we can use X to deform M in the following way: choose a smooth map $\varphi : I \times N \to N$, where

$I \subset \mathbb{R}$ is an interval containing zero, such that: (i) $\varphi(0,x) = x$ for every $x \in N$; (ii) $\varphi(t,x) = x$ for every $x \in N \setminus \operatorname{spt} X$ and $t \in I$; (iii) $\frac{d}{dt}\big|_{t=0} \varphi(t,x) = X(x)$ for every $x \in N$; (iv) the map $\varphi_t : N \to N$ given by $\varphi_t(x) = \varphi(t,x)$ is a diffeomorphism of N for every $t \in I$. Given X, there are many such maps φ, each giving rise to a 1-parameter family of submanifolds $M_t = \varphi_t(M)$, $t \in I$ with $M_0 = M$. The *first variation* of M with respect to $\mathcal{A}$ in the direction X, denoted $\delta_X(M)$, is defined by

$$\delta_X(M) = \frac{d}{dt}\bigg|_{t=0} \mathcal{A}(M_t).$$

A calculation using the area formula (see e.g. [Sim84]) shows that

$$\delta_X(M) = \int_M \operatorname{div}_M X \, d\mathcal{H}^k \tag{1}$$

where div_M denotes the tangential divergence on M, given, at a point $x \in M$, by

$$\operatorname{div}_M X(x) = \sum_{j=1}^{k} < \tau_j, \nabla_{\tau_j} X(x) > \tag{2}$$

for any orthonormal basis $\{\tau_1, \tau_2, \ldots, \tau_k\}$ for the tangent space $T_x M$ (and is independent of the choice of the orthonormal basis). The identity (1) is known as the *first variation formula* and it shows that the first variation $\delta_X(M)$ depends only on X and not on the choice of the family of diffeomorphisms φ_t as above. The submanifold M is said to be *stationary* in N if

$$\delta_X(M) = 0$$

for every compactly supported vector field X on N.

Now suppose that M is C^2 and $\overline{M}$ is a C^2 submanifold-with-boundary. For each $x \in M$, we can write $X(x) = X^T(x) + X^\perp(x)$, the sum of components of $X(x)$ tangential and normal to M at x. Since for the normal part $X^\perp$ we have

$$\operatorname{div}_M X^\perp = - < H_M, X >$$

where H_M is the mean curvature vector of M, we get by the divergence theorem that

$$\int_M \operatorname{div}_M X \, d\mathcal{H}^k = -\int_M < H_M, X > \, d\mathcal{H}^k + \int_{\partial M} < X, \nu > d\mathcal{H}^{k-1} \tag{3}$$

where ν is the outward pointing unit co-normal to ∂M. Thus for C^2 submanifolds M such that $\overline{M}$ is a C^2 submanifolds-with-boundary, the *variational property* of being stationarity in N with respect to the area functional (in the sense described above) is characterized by the absence of manifold boundary together with the simple *geometric property* of being minimal, i.e. having vanishing mean curvature—a fact that sparks a degree of surprise at first sight!

2.2. Varifolds and generalized mean curvature

While the classical notion of mean curvature, and hence the right hand side of the formula (3), requires the submanifold M to be of class C^2, the left hand side of this formula (i.e. the first variation $\delta_X(M)$) makes sense when M is of class C^1. So stationarity can be defined for C^1 submanifolds M. A closer look at the stationarity condition in fact reveals that a useful form of stationarity can be obtained under far less regularity assumptions on M than C^1: all that is necessary is that M is a $\mathcal{H}^k$ measurable subset with locally finite $\mathcal{H}^k$ measure, and that a unique, measure-theoretic k-dimensional tangent plane T_xM to M at x exists for $\mathcal{H}^k$ a.e. $x \in M$. Such a set M, called a *k-rectifiable subset* of N, can of course be extremely singular and in fact need not correspond to a C^1 submanifold anywhere. On the other hand, it turns out that relaxing in the first instance the regularity requirements on M in this way helps with the task of establishing existence of generalized (rectifiable) stationary submanifolds.

The well known compactness theorem of Federer–Fleming ([FedFle60]) implies the existence of stationary (in fact area minimizing) integer multiplicity currents in non-trivial homology classes of compact Riemannian manifolds. Subsequently Almgren ([Alm65]) developed a general min-max method for finding unstable critical points of area, which gives existence of stationary submanifolds (technically stationary integral varifolds —see below) in arbitrary compact Riemannian manifolds regardless of whether homology is non-trivial. Such a min-max approach requires working in a space (of generalised submanifolds) with a topology with respect to which the functional (i.e. area) is continuous, in addition to having a meaningful notion of first variation of a general element in the space. The space of k-rectifiable subsets M of N admitting, for $\mathcal{H}^k$ a.e. $x \in M$, a (measure-theoretic) tangent space T_xM with a positive integer multiplicity $\theta(x)$ such that $\theta(\cdot)$ is locally $\mathcal{H}^k$ integrable on M, taken with the topology of an appropriate form of weak convergence (see below), meets these requirements, provided the area functional is extended to be the *mass* (or *weighted area*) functional $V = (M,\theta) \mapsto \|V\|(N)$ defined by $\|V\|(A) = \int_{M\cap A} \theta \, d\mathcal{H}^k$ for $A \subset N$. A pair $V = (M,\theta)$ where M is k-rectifiable and $\theta : M \to \mathbb{N}^+$ is a locally $\mathcal{H}^k$-integrable multiplicity function is called an *integral k-varifold*, and the Radon measure

$\|V\|$ on N defined as above is called the *weight* (*or mass*) *measure* associated with V. (Strictly speaking V is an equivalence class of pairs (M, θ) where $(M, \theta) \sim (M', \theta')$ if $\mathcal{H}^k((M \setminus M') \cup (M' \setminus M)) = 0$ and $\theta(x) = \theta'(x)$ for $\mathcal{H}^k$-a.e. $x \in M \cap M'$). We now briefly outline the basic facts about integral and more general varifolds necessary for our purpose of describing the existence and regularity theorems in the next sections. (For a comprehensive treatment of the theory of varifolds see [Sim84] or [All72].)

In analogy with the classical setting described above, the first variation of an integral varifold $V = (M, \theta)$ with respect to mass is defined by $\delta_V(X) = \frac{d}{dt}\big|_{t=0} \|\varphi_{t,\#} V\|(\operatorname{spt} X)$, where for X a given compactly supported C^1 vector field on N, the deformation of V by X is given by $\varphi_{t\,\#} V = (\varphi_t(M), \theta \circ \varphi_t^{-1})$ where φ_t is a 1-parameter family of C^1 diffeomorphisms of N generated by X as described in the preceding section. Similar to the case of C^1 submanifolds described in the preceding section, we have the formula $\delta_V(X) = \int_M \operatorname{div}_M X(x)\theta(x)\, d\mathcal{H}^k(x)$, where for (the full-measure set of) points $x \in M$ at which the measure-theoretic tangent space $T_x M$ exists, $\operatorname{div}_M X(x)$ is given by the same expression (2) as in the C^1 case.

How should we take limits of sequences of integral k-varifolds V under uniform mass bounds? There is an obvious topology, namely that of weak convergence of Radon measures on N (restricted to the weight measures $\|V\|$ associated with V), but there is no clear analytically useful way to extend the first variation formula to the limits obtained in this way. A natural fix for this is to consider integral varifolds as elements of the space $\mathcal{V}_k(N)$ of Radon measures on the k-dimensional Grassmanian bundle

$$G_k(N) = \big\{(x, S) : x \in N,\ S \text{ is a } k\text{-dimensional subspace of } T_x N\big\}$$

(equipped with the metric induced by the usual product metric on $N \times G(k, n)$ where $G(k, n)$ is the space of k-dimensional subspaces of $\mathbb{R}^{n+1}$), via the lifting that takes $V = (M, \theta)$ to the element $v(V) \in \mathcal{V}_k(N)$ defined (by duality) by

$$v(V)(\varphi) = \int_M \varphi(x, T_x M)\theta\, d\mathcal{H}^k$$

for each $\varphi \in C^0_c(G_k(N))$.

Accordingly, extending the notion of integral k-varifolds, we simply define a *k-varifold* on N to be a Radon measure on $G_k(N)$, and associated to a k-varifold V on N its weight (or mass) measure to be the Radon measure $\|V\|$ on N given by $\|V\|(A) = V(\{(x, S) \in G_k(N) : x \in A\})$ for

$A \subset N$. Given a compactly supported C^1 vector field X on N, the first variation $\delta_V(X)$ of a k-varifold V with respect to mass in the direction of X is defined, in a manner analogous to the above, by considering push forwards $\varphi_{t\,\#} V$ of V by 1-parameter ambient space diffeomorphisms φ_t generated by X, where $\varphi_{t\,\#} V$ is defined by extending the area formula (see [Sim84]). We then have the first variation formula $\delta_V(X) = \int_{G_k(N)} \operatorname{div}_S X(x)\, dV(x,S)$, with $\operatorname{div}_S X(x)$ given by (2) for any orthonormal basis $\{\tau_1, \tau_2, \ldots, \tau_k\}$ for the k-dimensional subspace S of $T_x N$. These definitions and the first variation formula are, of course, consistent with the corresponding ones in the case of integral k-varifolds. Varifold convergence means simply the convergence as Radon measures on $G_k(N)$, with respect to which not only the mass functional but also the first variation (for a fixed vector field X) is continuous.

In analogy with (3), we say that an $L^1_{\text{loc}}(\|V\|)$ vector field H_V on N is the *generalized mean curvature* of the varifold V if $\int_{G_k(N)} \operatorname{div}_S X\, dV(x,S) = -\int_N < H_V(x), X(x) > d\|V\|(x)$ for each compactly supported C^1 vector field X on N. This formula identifies a sub class of varifolds (namely, those that admit generalized mean curvature). If V admits generalized mean curvature H_V, then V is stationary in N if and only if $H_V = 0$.

Definition. For a k-varifold V defined as above, we let $\operatorname{reg}_1 V$ be the set of points $p \in \operatorname{spt}\|V\|$ near which $\operatorname{spt}\|V\|$ is a C^1 embedded k-dimensional submanifold, and define the singular set $\operatorname{sing} V$ by $\operatorname{sing} V = (\operatorname{spt}\|V\| \setminus \operatorname{reg}_1 V)$. We let $\operatorname{reg} V$ be the set of points $p \in \operatorname{spt}\|V\|$ near which $\operatorname{spt}\|V\|$ is a C^∞ embedded k-dimensional submanifold.

To return to the discussion of integral varifolds, let V be an integral k-varifold with generalized mean curvature H_V (so $V = v(M, \theta)$ with M, θ as above). To have any hope of regularity of $\operatorname{spt}\|V\|$, we need $H_V \in L^p_{loc}(\|V\|)$ for some $p \geq k$. In the absence of this hypothesis $\operatorname{spt}\|V\|$ need not even be k-dimensional as can be seen by taking V to be the multiplicity 1 varifold associated with the union of suitable countably many concentric two-dimensional spheres around rational points in $\mathbb{R}^3$, chosen so as to ensure that $\operatorname{spt}\|V\|$ contains each rational point and hence is equal to the whole ambient space. If on the other hand $H_V \in L^p_{loc}(\|V\|)$ for some $p > k$, then $\operatorname{spt}\|V\|$ is k-rectifiable, $\mathcal{H}^k((\operatorname{spt}\|V\| \setminus M) \cup (M \setminus \operatorname{spt}\|V\|)) = 0$, the area density $\Theta(\|V\|, p) = \lim_{\rho \to 0} \frac{\|V\|(B^N_\rho(p))}{\omega_k \rho^k}$ exists for every $p \in N$, and $\operatorname{spt}\|V\| = \{p \in N : \Theta(\|V\|, p) \geq 1\}$. Here ω_k denotes the volume of the unit ball in $\mathbb{R}^k$, and $B^N_\rho(p)$ is the open ball in N with centre p and radius ρ. Moreover, in this case we have an embryonic level of regularity, namely, that

$\operatorname{reg}_1 V$ is a (relatively open) dense subset of $\operatorname{spt}\|V\|$; in fact $\operatorname{reg}_1 V$ is of class $C^{1,1-\frac{k}{p}}$ if $k < p < \infty$. Additionally, any subset of integral varifolds with locally uniformly bounded mass and locally uniformly L^p-bounded mean curvature for some $p > 1$ is compact in the varifold topology. These facts, established in the landmark work of Allard ([All72]) that extended and streamlined the fundamental work of Almgren ([Alm65]), make the space of integral k-varifolds with mean curvature summable to a power $p > k$ a "good" space of generalized k-dimensional submanifolds for many purposes.

The hypothesis $H_V \in L^p_{\text{loc}}(\|V\|)$ for $p > k$ alone does not imply a.e. regularity of V, i.e. that $\operatorname{sing} V$ has zero k-dimensional Hausdorff measure. There is a well-known example due to Brakke ([Bra78]), refined recently by Kolasiński–Menne ([KolMen15]), of an integral 2-varifold W in $\mathbb{R}^3$ with the property that $H_W \in L^\infty_{\text{loc}}(\|V\|)$ and $\mathcal{H}^2(\operatorname{sing} W) > 0$. There is no such known example with $H_W = 0$: it is a major open question whether the singular set of a stationary integral k-varifold must have zero k-dimensional Hausdorff measure. However, for several important closed subfamilies of integral varifolds (stationary and more general), sharp upper bounds on the Hausdorff dimension of the (lower dimensional) singular set are known. We shall describe some of these results in Sections 3 and 5 below.

§3. Regularity and compactness theory for stable stationary codimension 1 varifolds

A stationary k-varifold V on N is said to be stable if it has non-negative second variation with respect to the mass functional, in the sense that $\frac{d^2}{dt^2}\big|_{t=0}\|\varphi_{t\,\#} V\|(\operatorname{spt} X) \geq 0$ for every compactly supported C^1 vector field X on N, where φ_t is as in Section 2. In the regularity theory discussed below, we require the stability assumption on the orientable regions of *only the regular part* $\operatorname{reg} V$ in ambient normal coordinate balls and with respect to normal variations (i.e. with respect to those vector fields X with $\operatorname{spt} X \subset (N \setminus \operatorname{sing} V) \cap B$ for some ambient normal coordinate ball B and with $\operatorname{spt} X \cap \operatorname{reg} V$ orientable and $X(x)$ normal to $\operatorname{reg} V$ for each $x \in \operatorname{reg} V$).

To describe this theory, and also its generalisation to the CMC setting in Section 5, we need to identify a very special type of singularity as follows:

Definition. A *classical singularity* of a set $Y \subset N$ is a point $p \in Y$ such that for some $\rho > 0$, $Y \cap B^N_\rho(p)$ is the union of three or more embedded $C^{1,\alpha}$ hypersurfaces-with-boundary meeting only along their common boundary Γ, with $p \in \Gamma$ and with at least one pair meeting transversely. A classical singularity of a varifold V on N is a classical singularity of $\operatorname{spt} \|V\|$.

The main content of the codimension 1 regularity and compactness theory for stable stationary varifolds is contained in the following theorem. This theorem extends and unifies two earlier theories: compactness for mass bounded stable minimal hypersurfaces with small singular sets ([SchSim81]) and the regularity theory for locally are minimizing integral currents ([DeG61], [Fed70], [JSim68]):

Theorem 3.1 ([Wic14]). *Let N be a smooth $(n+1)$-dimensional Riemannian manifold and let V be a stationary integral n-varifold on N. If each point $p \in N$ has a normal coordinate ball $B^N_\rho(p)$ such that every orientable open subset $U \subset \operatorname{reg} V \cap B^N_\rho(p)$ is stable with respect to area (i.e. n-dimensional Hausdorff measure) for normal deformations of U leaving a neighborhood of its boundary ∂U fixed, and if V has no classical singularities, then* $\operatorname{sing} V = \emptyset$ *if* $n \leq 6$, $\operatorname{sing} V$ *is discrete if* $n = 7$ *and* $\dim_{\mathcal{H}} (\operatorname{sing} V) \leq n - 7$ *for general n. Moreover, if $\mathcal{V}$ is a locally uniformly mass bounded family of stationary integral n-varifolds on N with no classical singularities and if there is a collection $\mathcal{B}$ of normal coordinate balls of N with $N = \cup \mathcal{B}$ such that the above stability property holds with each $B^N_\rho(p) \in \mathcal{B}$ and each $V \in \mathcal{V}$, then $\mathcal{V}$ is compact in the topology of varifold convergence; additionally, the convergence of any sequence in $\mathcal{V}$ is with respect to the C^k norm for any $k \geq 1$ (and with multiplicity ≥ 1) away from the singular set of the limit.*

Remark. Obviously, it suffices in applications of Theorem 3.1 to verify the no-classical-singularities hypothesis away from any subset $Z \subset \operatorname{spt} \|V\|$ of zero $(n-1)$-dimensional Hausdorff measure (regardless of whether Z is closed). For if $\operatorname{spt} \|V\|$ has a classical singularity then it follows, by the definition of classical singularity, that $\operatorname{spt} \|V\|$ has classical singularities on a set of positive $(n-1)$-dimensional Hausdorff measure, and hence if no point $\in \operatorname{spt} \|V\| \setminus Z$ is a classical singularity, then no point of $\operatorname{spt} \|V\|$ is a classical singularity. In particular, it suffices to verify that no point $y \in \operatorname{spt} \|V\| \setminus Z$ has a classical tangent cone, i.e. a tangent cone supported on three or more half-hyperplanes meeting along an $(n-1)$-dimensional subspace.

The special case of the (compactness part) of this theorem where the varifolds are assumed to have singular sets of Hausdorff dimension $\leq n - 7$, which follows from [SchSim81], serves as an important step in the proof the general theorem. We refer the reader to the survey article [Wic14b] for a discussion of the key aspects of the proof of Theorem 3.1.

Applications of Theorem 3.1 include a strong maximum principle and a unique continuation theorem for stationary codimension 1 integral varifolds under a sharp hypothesis on the size of the allowable singular sets ([Wic14a]). A further nice application, described in the next section, is a new, PDE theoretic approach to the existence theory for minimal hypersurfaces in compact Riemannian manifolds (Theorem 4.1).

§4. Existence theory for minimal hypersurfaces: old and new

A theorem due to Almgren [Alm65], dating back to 1965, asserts that every compact $(n+1)$-dimensional Riemannian manifold contains a k-dimensional "boundaryless" stationary integral k-varifold V for each $k \in \{1, \ldots, n\}$. However, we know very little about regularity of these varifolds except in the one dimensional case where it is known that $\text{spt}\,\|V\|$ at every point consists of a finite number of embedded geodesic arcs coming together in a regular manner. Our understanding of the existence problem in the hypersurface case is also a lot better, and in fact the following theorem has long been known:

Theorem 4.1. *Every compact, smooth $(n+1)$-dimensional Riemannian manifold N without boundary contains a stationary integral n-varifold V whose singular set* $\text{sing}\, V$ *is empty if $n \leq 6$, consists of isolated points if $n = 7$ and satisfies*

$$\dim_{\mathcal{H}}(\text{sing}\, V) \leq n - 7$$

in general dimensions.

There are now two different proofs of this theorem, with fundamental differences in their points of view.

Original proof. This is based on: (i) a geometric min-max argument for the mass functional carried out in a space of integral varifolds, due to Pitts ([Pit81]) that builds on the earlier work of Almgren ([Alm65]); (ii) regularity theory for locally area minimizing rectifiable currents, and (iii) compactness for stable minimal hypersurface with bounded area and singular sets of Hausdorff dimension $\leq n-7$, due to Schoen–Simon–Yau for $n \leq 5$ ([SSY75]), and Schoen–Simon for general n ([SchSim81]).

More recent work of De Lellis–Tasnady ([DeLTas13]) streamlined Pitts' argument in the case of 1-parameter varifold min-max construction.

Recent proof. This requires the sharper regularity theory of Theorem 3.1 (in place of the Schoen–Simon version), but its advantage is that it replaces the Almgren–Pitts construction with a considerably simpler

PDE argument. The idea of this method is to obtain the minimal hypersurface as the varifold limit of nodal sets (or, equivalently, the energy concentration set) of solutions to (elliptic) Allen–Cahn equations in a certain 1-parameter family.

To describe this method in more detail, consider, for $\epsilon \in (0,1)$, the Allen–Cahn functionals $E_\epsilon(u) = \int_N \frac{\epsilon |\nabla u|^2}{2} + \frac{W(u)}{\epsilon}\, d\mathcal{H}^{n+1}$, $u \in W^{1,2}(N)$ where $W : \mathbb{R} \to \mathbb{R}$ is a fixed smooth non-negative double-well potential with precisely two non-degenerate minima at ± 1 with $W(\pm 1) = 0$. (e.g. $W(t) = \frac{1}{4}(1-t^2)^2$).

For proving existence of minimal hypersurfaces, the starting point is a result due to Hutchinson–Tonegawa ([HutTon00]), adapted in a straightforward manner to Riemannian ambient spaces (described in [Gua18]) as follows: *If $\epsilon_j \to 0^+$, $u_j \in W^{1,2}(N)$ is a critical point of E_{ϵ_j} (so $-\epsilon_j \Delta u_j + \epsilon_j^{-1} W'(u_j) = 0$) and $E_{\epsilon_j}(u_j) + \sup_N |u_j| \le c$ for fixed c and all j, then*

$$\sigma^{-1}\left(\epsilon_{j'} |\nabla u_{j'}|^2 + \epsilon_{j'}^{-1} W(u_{j'})\right) d\mathcal{H}^{n+1} \to \theta\, d\mathcal{H}^n$$

where $\sigma = \sigma(W)$ is an explicit constant and $\theta : N \to \mathbb{N} \cup \{0\}$. If $M \equiv \{x \in N : \theta(x) > 0\}$ then M is n-rectifiable and $V_{ac} = (M, \theta)$ is a stationary integral varifold in N. This theorem is the elliptic version of the pioneering work of Ilmanen ([Ilm93]) in the parabolic case.

By a result of Tonegawa ([Ton05]) (adapted again as described in [Gua18] to the Riemannian setting), we also have the following: *if u_j are stable, i.e. if*

$$\left.\frac{d^2}{dt^2}\right|_{t=0} E_{\epsilon_j}(u_j + t\varphi) \equiv \int_N \epsilon_j |\nabla \varphi|^2 + \epsilon_j^{-1} W''(u_j)\varphi^2 \ge 0 \quad \forall\, \varphi \in C_c^1(N),$$

then $\operatorname{reg} V_{ac}$ *is stable with respect to the area functional, i.e. the stability hypothesis of Theorem* 3.1 *holds with V_{ac} in place of V.*

Applying Theorem 3.1 in its full strength we obtain the sharp regularity conclusion for V_{ac} as follows ([TonWic12]): *if V_{ac} as above corresponds to a sequence of stable solutions u_j, then* $\dim(\operatorname{sing} V_{ac}) \le n - 7$ *and* $\operatorname{reg} V_{ac}$ *is an embedded stable minimal hypersurface of N.*

There are two reasons why the full strength of Theorem 3.1 is needed in the proof of this result in [TonWic12]:

(1) There is no a priori control of the size of the set T of singularities of V_{ac} where one tangent cone is planar. Convergence of level sets of u_j is not known to be strong enough to give any information about T. (In case $\dim N = 3$ however, recently Chodosh–Mantoulidis ([ChoMan]) proved, with additional work building on the results of Wei–Wang ([WanWei])

and of Ambrosio–Cabŕe ([AmbCab00]), $C^{2,\alpha}$ convergence of the level sets.)

(2) The second fundamental forms of the level sets of u_j may concentrate near some set $Z \subset \operatorname{spt}\|V_{ac}\|$ (in the sense of positive $(n-2)$-dimensional upper density for the limit measure associated with the second fundamental forms; see [TonWic12, Section 4]). An easy covering argument shows that $\dim_{\mathcal{H}}(Z) \leq n-2$, but otherwise no control on Z is available.

Both these difficulties are automatically handled by Theorem 3.1. Indeed, Theorem 3.1 requires no control of T whatsoever, and in its full strength, the theorem yields full regularity conclusions upon ruling out classical tangent cones away from Z (see the remark following Theorem 3.1); the latter is fairly straightforward to do since by definition of Z, the second fundamental form does not concentrate near any point of $\operatorname{spt}\|V_{ac}\| \setminus Z$ but on the other hand if there is a classical tangent cone at a point in $\operatorname{spt}\|V_{ac}\|$ then the level sets must undergo significant bending in small balls about that point forcing the second fundamental form to concentrate.

The above results reduce the existence proof for minimal hypersurfaces to a construction of appropriate solutions to the ϵ Allen–Cahn equations. Recent work of Guaraco ([Gua18]) achieves precisely that, by establishing the following: *For each* $\epsilon \in (0,1)$, *there is a min-max critical point* u_ϵ *of* E_ϵ *such that*: (i) *Morse index of* $u_\epsilon \leq 1$; (ii) $\sup_N |u_\epsilon| \leq 1$ *and* (iii) $0 < \liminf_{\epsilon\to 0^+} E_\epsilon(u_\epsilon) \leq \limsup_{\epsilon\to 0^+} E_\epsilon(u_\epsilon) < \infty$.

It is an elementary observation that because $\operatorname{Index}(u_\epsilon) \leq 1$, for any ambient open ball B, u_ϵ is stable either in B or in $N \setminus \overline{B}$. In view of this, the above regularity result of [TonWic12] can be applied to the limit varifold, giving the desired conclusion ([Gua18]): *if* u_ϵ *are as in the construction of* [Gua18] *described above, any sequence* $\epsilon_j \to 0^+$ *has a subsequence* $\epsilon_{j'}$ *such that* $\sigma^{-1}\big(\epsilon_{j'}|\nabla u_{\epsilon_{j'}}|^2 + \epsilon_{j'}^{-1}W(u_{\epsilon_{j'}})\big)\, d\mathcal{H}^{n+1} \to \|V_{ac}\|$ *for some non-trivial stationary integral varifold* V_{ac} *with* $\dim_{\mathcal{H}}(\operatorname{sing} V_{ac}) \leq n-7$.

A comparison with the Almgren–Pitts approach.

(1) The key difference between the two approaches is that the Allen–Cahn method makes way for a substantial gain in simplicity on the construction side of the theorem, by reducing the construction to a PDE min-max question and shifting the overall burden more to the regularity side. Indeed, Guaraco's elegant proof of the existence of the Allen–Cahn solutions u_ϵ as above is more elementary than the geometric varifold min-max construction of Almgren–Pitts (or of De Lellis–Tasnady), and is obtained by applying a standard PDE mountain pass lemma taking

advantage of the fact that the Allen–Cahn functional satisfies the Palais–Smale compactness criterion. As such, and crucially, one obtains uniform control of the Morse index of u_ϵ (independently of ϵ).

(2) For regularity conclusions in either proof, some form of local stability of the constructed varifold is necessary. In Almgren–Pitts, this follows from an "almost minimizing" property built into the min-max construction. In the new proof, stability needs to be checked only on the regular part of V_{ac} in any small ambient ball—a less onerous task accomplished as a very direct consequence of the uniform index bound on the Allen–Cahn critical points and the above theorem of Tonegawa ([Ton05]).

(3) The Morse index bounds for the resulting minimal hypersurfaces have also been established: for the Almgren–Pitts method by Marques–Neves ([MarNev16]) and for the Allen–Cahn approach by Hiesmayr ([Hie]) for two-sided limits, and also by Gaspar ([Gas]) based on work of Le ([Le11], [Le15]) for the general case. The simplicity of the argument in the Allen–Cahn setting is again in striking contrast with that in the Almgren–Pitts setting.

In recent years, both these basic approaches have been used and extended in various ways to show that the space of minimal hypersurfaces in compact manifolds is extremely rich. See e.g. [GasGua1], [GasGua2], [IMN18], [MarNev14], [Son].

§5. Regularity and compactness theory for stable CMC codimension 1 varifolds

The above regularity and compactness theory for stable minimal hypersurfaces can be embedded in a more general theory applicable to codimension 1 integral n-varifolds V whose generalised mean curvature $H_V \in L^p_{\text{loc}}(\|V\|)$ for some $p > n$ and: (i) whose C^1-embedded, orientable parts are stationary and C^2-immersed part is stable for area with respect to volume preserving deformations; (ii) whose support admits no classical singularities and satisfies one other (also necessary) structural condition. This work is described in Theorems 5.1 and 5.2 below, proved in the author's recent joint work with C. Bellettini ([BelWic-1]).

We will assume that the ambient space is an open subset U of $\mathbb{R}^{n+1}$. The generalization of the results below to Riemannian ambient spaces (as well as to the setting of more generally prescribed mean curvature) can be found in [BelWic-2].

Since in these theorems the variational hypotheses only concern the regular parts of the varifold, let us first consider the case of an embedded, oriented C^1 hypersurface $M \subset U$ with $\partial M \cap U = \emptyset$ and with ν a continuous choice of unit normal on M. For an open set $\mathcal{O} \subset\subset U$, write

$$\mathcal{A}_{\mathcal{O}}(M) = \mathcal{H}^n(M \cap \mathcal{O}) \text{ and,}$$
$$vol_{\mathcal{O}}(M) = \frac{1}{n+1}\int_{M\cap\mathcal{O}} x \cdot \nu(x)\, d\mathcal{H}^n(x).$$

Note that by the divergence theorem, $vol_{\mathcal{O}}(M)$ is the volume enclosed by M in case M is the boundary of a bounded open set $\Omega \subset \mathcal{O}$ and ν is the outward pointing unit normal to M.

Definition. M is *stationary in U with respect to the area functional for volume preserving deformations* if $\frac{d}{dt}\big|_{t=0} \mathcal{A}_{\mathcal{O}}(\varphi_t(M)) = 0$ for each open $\mathcal{O} \subset\subset U$ and each smooth map $\varphi : U \times (-\epsilon, \epsilon) \to U$, $\epsilon > 0$, with: (i) $\varphi_t = \varphi(\cdot, t) : U \to U$ a diffeomorphism for each $t \in (-\epsilon, \epsilon)$, (ii) $\varphi_0 = identity$, (iii) $\varphi_t|_{U\setminus\overline{\mathcal{O}}} = identity|_{U\setminus\overline{\mathcal{O}}}$ for each $t \in (-\epsilon, \epsilon)$ and (iv) $vol_{\mathcal{O}}(\varphi_t(M)) = vol_{\mathcal{O}}(M)$ for each $t \in (-\epsilon, \epsilon)$.

Write H_M for the mean curvature vector of M whenever M is of class C^2. Let $\lambda \in \mathbb{R}$ be a constant, and let

$$\mathcal{J}_{\mathcal{O}}(M) = \mathcal{A}_{\mathcal{O}}(M) + \lambda\, vol_{\mathcal{O}}(M).$$

It is well-known, and is straightforward to verify, that the following statements are equivalent (see e.g. [BarDoc84]):

(a) M is of class C^2 and is a CMC hypersurface (i.e. $H_M \cdot \nu$ is constant).
(b) M is stationary in U with respect to area for volume preserving deformations.
(c) For some constant $\lambda \in \mathbb{R}$ and each open $\mathcal{O} \subset\subset U$, M is stationary with respect to $\mathcal{J}_{\mathcal{O}}(\cdot)$ for arbitrary deformations (i.e. for φ_t as above but not necessarily with $vol_{\mathcal{O}}(\varphi_t(M)) = vol_{\mathcal{O}}(M)\ \forall t$).

Now consider an integral n-varifold V on U with generalized mean curvature vector H_V and associated weight measure $\|V\|$, and suppose that $H_V \in L^p_{\text{loc}}(\|V\|)$ for some $p > n$. Note that as mentioned at the end of Section 2.2 above the Brakke example ([Bra78], [KolMen15]) shows that a.e. regularity (i.e. the fact that $\mathcal{H}^n(\text{sing}\, V) = 0$) cannot follow from the assumption that $H_V \in L^p_{\text{loc}}(\|V\|)$, $p > n$ alone.

Let $\text{sing}_C\, V$ be the set of classical singularities of spt $\|V\|$, defined in the preceding section.

We need to identify a second special type of singularity that will play a key role in the main theorems below.

Definition. A point $p \in \text{spt}\, \|V\| \cap U$ is a *touching singularity* of V if $p \notin \text{sing}_C\, V \cup \text{reg}_1\, V$ and there are $\sigma > 0$, an affine hyperplane L

containing p, $\alpha \in (0,1)$ and two $C^{1,\alpha}$ functions $u_1, u_2 : L \to L^{\perp}$ such that $\operatorname{spt} \|V\| \cap B_\sigma^{n+1}(p) = (\operatorname{graph} u_1 \cup \operatorname{graph} u_2) \cap B_\sigma^{n+1}(p)$. (Note that it follows from this definition that u_1, u_2 are distinct, and that $u_1(p) = u_2(p)$, $Du_1(p) = Du_2(p)$.)

Definition. If p is a touching singularity of V and σ, u_1, u_2 are as above, the *coincidence set of* V *in* $B_\sigma^{n+1}(p)$ is the set $\operatorname{graph} u_1 \cap \operatorname{graph} u_2 \cap B_\sigma^{n+1}(p)$.

Let $\operatorname{sing}_T V$ be the set of touching singularities of V.

Note that near either a classical singularity or a touching singularity, by definition the varifold consists of pieces of hypersurfaces of class $C^{1,\alpha}$ coming together in a regular fashion. Thus these singularities appear in parts of the varifold that are quite regular, and hence their presence, and other local properties of the varifold near them, are relatively easy to check.

The second variation assumption (i.e. stability) in our theorems below will be required on the smoothly immersed part of the varifold. It is convenient to use the following terminology in stating this assumption:

Definition. A point $X \in \operatorname{spt} \|V\|$ is a *generalized regular point of* V if either (i) $X \in \operatorname{reg} V$ or (ii) $X \in \operatorname{sing}_T V$ and we may choose *smooth* functions u_1 and u_2 corresponding to X (as in the definition above) such that $u_1 \geq u_2$.

Let gen-reg V denote the set of generalised regular points of V.

The main regularity and compactness theorems can now be given as follows:

Theorem 5.1 (CMC regularity theorem [BelWic-1]). *Let V be an integral n-varifold on an open subset $U \subset \mathbb{R}^{n+1}$, $n \geq 2$, with $H_V \in L^p_{loc}(\|V\|)$ for some $p > n$. Suppose that V satisfies the following:*

Structural Hypotheses:

(a) $\operatorname{sing}_C V = \emptyset$;

(b) *For each $p \in \operatorname{sing}_T V$, there is $\sigma > 0$ such that the coincidence set of V in $B_\sigma^{n+1}(p)$ has zero n-dimensional Hausdorff measure.*

Variational Hypotheses:

(c) *stationarity: for each open $\mathcal{O} \subset\subset U \setminus \operatorname{sing} V$ such that $\operatorname{reg}_1 V \cap \mathcal{O}$ is orientable, the varifold $W \equiv V \llcorner \mathcal{O}$ is stationary in $\mathcal{O}$ with respect to the mass functional $\overline{\mathcal{A}}_{\mathcal{O}}(W) = \|W\|(\mathcal{O})$ for ambient deformations that leave the region outside a compact subset of $\mathcal{O}$ fixed and preserve $\overline{vol}_{\mathcal{O}}(W) = \frac{1}{n+1}\int x \cdot \nu \, d\|W\|$, where ν is a choice of continuous unit normal on $\operatorname{reg}_1 V \cap \mathcal{O}$;*

(d) *weak stability*: *gen-reg* V *is stable (as a CMC immersion) with respect to the (multiplicity* 1*) area functional for volume preserving normal deformations.*[1]

Then there exists a closed set $\Sigma \subset \operatorname{spt}\|V\|$ *with* $\dim_{\mathcal{H}}(\Sigma) \leq n-7$ *such that* $\operatorname{spt}\|V\| \setminus \Sigma$ *locally near each point is either a smoothly embedded disk or the union of precisely two smoothly embedded disks intersecting tangentially along a set contained in a smooth* $(n-1)$*-dimensional submanifold; moreover, either* $H_V = 0$ *everywhere, or* $\operatorname{spt}\|V\| \setminus \Sigma$ *is an orientable smooth immersion and there is a constant* $\lambda \in \mathbb{R} \setminus \{0\}$ *such that* $H_V = \lambda\nu$ *on* $\operatorname{spt}\|V\| \setminus \Sigma$ *where* ν *is a choice of orientation on* $\operatorname{spt}\|V\| \setminus \Sigma$.

Remark. (1) Hypothesis (a) cannot be dropped, as shown by a piece of two intersecting unit spheres or cylinders.
(2) If hypothesis (b) is dropped, then C^2 regularity as asserted by the theorem cannot be guaranteed. This is shown by the example $\Gamma \times \mathbb{R}^{n-1}$ where Γ is the following 1-dimensional varifold in a suitable open subset of $\mathbb{R}^2$:

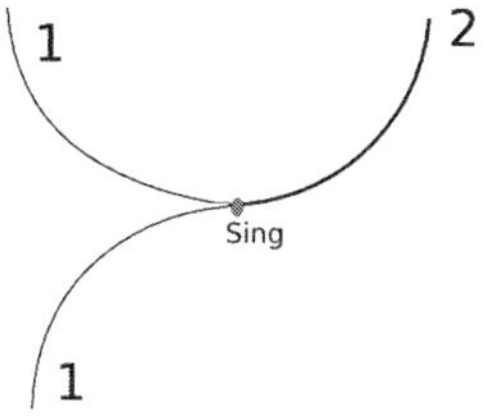

In this picture, each arc is a piece of a unit circle, and the numbers 1, 2 denote the multiplicity on an arc. The singular point is a touching singularity (not a classical singularity, since no pair of arcs meet transversely). The only hypothesis of the theorem not satisfied by this example is (b). The support of the varifold is not the union of two C^2 graphs (it is however the union of a C^2 graph and a $C^{1,1}$ graph).

In view of De Giorgi's structure theorem for Caccioppoli sets ([DeG54], [DeG55]; also [Sim84]), the following is an immediate consequence of Theorem 5.1:

Corollary 5.1 ([BelWic-1]). *If* V *is the multiplicity* 1 *varifold associated with the boundary of a Caccioppoli set, and if the hypotheses* (a), (c), (d) *of Theorem* 5.1 *hold, then* V *satisfies the conclusions of the Theorem* 5.1.

[1] In fact it suffices that stability holds with respect to volume preserving normal deformations "induced by ambient functions" in a certain specific sense; see [BelWic-1] for the precise requirement.

See [BelWic-1] for another corollary for Caccioppoli sets ([BelWic-1, Corollary 2.2]) that says that hypothesis (a) can be dropped, the stability assumption can be weakened to require it only on the embedded part, and that the conclusion regarding the touching singularities can be strengthened, provided a stronger (and yet natural) stationarity condition involving deformation through Caccioppoli sets is assumed.

The estimates required to prove Theorem 5.1 also imply the following compactness result:

Theorem 5.2 (CMC compactness theorem [BelWic-1]). *Let (V_j) be a sequence of integral n-varifolds in open $U \subset \mathbb{R}^{n+1}$, $n \geq 2$, satisfying $H_{V_j} \in L^{p_j}_{loc}(\|V_j\|)$ for some $p_j > n$ and* (a)–(d) *(as in the above theorem) with $V = V_j$. If*

$$\limsup_{j\to\infty} \|V_j\|(K) < \infty$$

for each compact $K \subset U$ and $\limsup_{j\to\infty} |H_{V_j}| < \infty$ (note that $|H_{V_j}|$ is constant for each j by the above theorem), then there is an integral n-varifold V in U satisfying the conclusions of Theorem 5.1, *and a subsequence $\{j'\}$ such that $V_{j'} \to V$ as varifolds in U.*

Finally, we mention the following effective versions of Theorem 5.2 giving curvature estimates for weakly stable CMC hypersurfaces. These results, proved in the author's joint work with C. Bellettini and O. Chodosh ([BCW]), generalize the curvature estimates proved by Schoen–Simon–Yau ([SSY75]) (corresponding to Theorem 5.3) and Schoen–Simon ([SchSim81]) (corresponding to Theorem 5.4) for strongly stable hypersurfaces.

Theorem 5.3 (Pointwise curvature estimate in low dimension [BCW]). *For each $H_0 > 0$ and $\Lambda \geq 1$, there exists $C = C(H_0, \Lambda)$ such that the following holds: Let $2 \leq n \leq 6$ and let $\Sigma \subset B_R(0) \subset \mathbb{R}^{n+1}$ be a smooth immersed hypersurface with $(\overline{\Sigma}\setminus\Sigma)\cap B_R(0) = \emptyset$, $\mathcal{H}^n(\Sigma) \leq \Lambda R^n$ and with constant scalar mean curvature H such that $|H| \leq H_0 R^{-1}$. Assume that Σ is weakly stable as a CMC immersion. If $n = 6$ suppose additionally either that Σ contains no point of transverse self-intersection (or equivalently, by the maximum principle, for each point $p \in \Sigma$ where Σ is not embedded, there is $\rho > 0$ such that $\Sigma \cap B^{n+1}_\rho(p)$ is the union of at two embedded smooth CMC hypersurfaces intersecting only tangentially), or that $\Lambda = 3 - \delta$ for some $\delta \in (0, 1)$.*

Then

$$\sup_{x\in\Sigma\cap B_{R/2}(0)} |A_\Sigma|(x) \leq CR^{-1},$$

where A_Σ denotes the second fundamental form of Σ.

Remark. The reason, in case $n = 6$, for the additional restrictions in Theorem 5.3 (that either Σ has no transverse points or $\Lambda = 3 - \delta$) is that it is not known if a pointwise curvature estimate holds for 6-dimensional immersed strongly stable minimal hypersurfaces satisfying an arbitrary mass bound; such an estimate is only known to hold if the minimal hypersurface is either embedded ([SchSim81]) or is immersed and satisfies a mass bound corresponding to $\Lambda = 3-\delta$ for some $\delta \in (0,1)$ ([Wic08]).

In general dimensions, we have the following:

Theorem 5.4 (Sheeting Theorem [BCW]). *Let $\Lambda, H_0 > 0$ and $n \geq 2$. Suppose that $\Sigma \subset B_R(0) \subset \mathbb{R}^{n+1}$ is an immersed hypersurface with $\mathcal{H}^n(\Sigma) \leq \Lambda R^n$, with constant scalar mean curvature H such that $|H| \leq H_0 R^{-1}$ and with $\mathcal{H}^{n-7+\alpha}((\overline{\Sigma} \setminus \Sigma) \cap B_R(0)) = 0$ for all $\alpha > 0$. Suppose that Σ contains no point of transverse self-intersection (or equivalently, by the maximum principle, for each $p \in \Sigma$ where Σ is not embedded there is $\rho > 0$ such that $\Sigma \cap B^{n+1}_\rho(p)$ is the union of exactly two embedded smooth CMC hypersurfaces intersecting only tangentially), and that Σ is weakly stable as a CMC immersion. There exists $\delta_0 = \delta_0(n, H_0, \Lambda)$ so that if additionally*

$$\Sigma \subset \left\{|x^{n+1}| \leq \delta_0 R\right\}$$

then $\overline{\Sigma} \cap \left(B^n_{R/2}(0) \times \mathbb{R}\right)$ is equal to the union of the graphs of finitely many ordered functions $u_1 \leq \cdots \leq u_k$ defined on $B^n_{R/2}(0) := B_{R/2}(0) \cap \{x^{n+1} = 0\}$ and satisfying

$$\sup_{B^n_{R/2}(0)} \left(|Du_i| + R|D^2 u_i|\right) \leq C \left(R^2 H^2 + R^{-n-2} \int_{B_R(0)\cap\Sigma} (x^{n+1})^2\right)^{\frac{1}{2}}$$

for $i = 1, \ldots, k$, where $C = C(n)$; moreover, each u_i is a smooth CMC graph.

It is interesting to note the following: Consider a CMC hypersurface Σ immersed in $\mathbb{R}^{n+1}$ with mean curvature H (possibly equal to zero). Recall that the *Morse index* $I(\Sigma)$ of Σ is the number of negative Dirichlet eigenvalues (counting with multiplicity) of the Jacobi operator $L : C^2(\Sigma) \to \mathbb{R}$ given by $L\phi = -\Delta\,\phi - |A_\Sigma|^2\phi$. It is readily checked that if Σ is weakly stable, then $I(\Sigma) \leq 1$. Theorem 5.4, and hence also Theorem 5.3 above, are *false* in every dimension ≥ 2 if we replace "Σ is weakly stable" with "Σ satisfies $I(\Sigma) = 1$." This can be seen by considering rescalings of the higher-dimensional Catenoid (the unique non-flat rotationally symmetric minimal hypersurface in $\mathbb{R}^{n+1}$) which converge weakly to a hyperplane with multiplicity two, but do not have uniformly bounded curvature (and hence do not satisfy the conclusion of Theorem 5.4). In the context of Theorem 5.3 and Theorem 5.4, the crucial difference between "weakly stable" and "$I(\Sigma) = 1$" is that complete weakly stable surfaces cannot have two ends at infinity while index one surfaces can (e.g., the Catenoid). Indeed, the one-endedness of weakly stable entire minimal hypersurfaces, a result due to Chen, Cheung and Zhou ([CCZ08]), is a key ingredient of our proofs of Theorem 5.3 and Theorem 5.4.

References

[All72] W. K. Allard, *On the first variation of a varifold*, Ann. of Math. **95** (1972), 417–491.

[Alm65] F. J. Almgren, Jr, *The theory of varifolds*, Mimeographed notes, Princeton (1965).

[Alm66] F. J. Almgren, Jr, *Some interior regularity theorems for minimal surfaces and an extension of Bernstein's theorem*, Ann. of Math. **84** (1966), 277–292.

[AmbCab00] L. Ambrosio and X. Cabŕe, *Entire solutions of semilinear elliptic equations in $\mathbb{R}^3$ and a conjecture of De Giorgi*, Jour. Amer. Math. Soc., **13** (4) (2000), 725–739.

[Bra78] K. Bakke, *The motion of a surface by its mean curvature*, Princeton University press, Princeton (1978).

[BarDoc84] J. Barbosa and M. do Carmo, *Stability of hypersurfaces with constant mean curvature*, Math. Z., **185** (1984), 339–353.

[BelWic-1] C. Bellettini and N. Wickramasekera, *Regularity and compactness for stable codimension* 1 *CMC varifolds*, arXiv (2018).

[BelWic-2] C. Bellettini and N. Wickramasekera, *Stable prescribed-mean-curvature integral varifolds of codimension* 1: *regularity and compactness*, arXiv (2019).

[BCW] C. Bellettini, O. Chodosh and N. Wickramasekera, *Curvature estimates and sheeting theorems for weakly stable CMC hypersurfaces*, arXiv (2018), to appear in Adv. Math..

[CCZ08] X. Cheng, L.-F. Cheung and D. Zhou, *The structure of weakly stable constant mean curvature hypersurfaces*, Tohoku Math. J. (2) **60** (2008), no. 1, 101–121.

[ChoMan] O. Chodosh and C. Mantoulidis, *Minimal surfaces and the Allen–Cahn equation on* 3 *manifolds*: *index, multiplicity and curvature estimates*, arXiv (2018).

[DeG54] E. De Giorgi, *Su una teoria generale della misura* $(r-1)$-*dimensionale in uno spazio ad* r *dimensioni*, Ann. Mat. Pura Appl. (4), **36** (1954), 191–213.

[DeG55] E. De Giorgi, *Nuovi teoremi relativi alle misure* $(r-1)$-*dimensionali in uno spazio ad* r *dimensioni*, Ricerche Mat., **4** (1955), 95–113.

[DeG61] E. De Giorgi, *Frontiere orientate di misura minima*, Sem. Mat. Scoula Norm. Sup. Pisa, (1961), 1–56.

[DeLTas13] C. De Lellis and D. Tasnady, *The existence of embedded minimal hypersurfaces*, J. Differential Geom., **95** (2013), no. 3, 355–388.

[Fed70] H. Federer, *The singular sets of area minimizing rectifiable currents with codimension one and area minimizing flat chains modulo two with arbitrary codimension*, Bull. Amer. Math. Soc., **76** (1970), 767–771.

[FedFle60] H. Federer and W. Fleming, *Normal and integral currents*, Ann. of Math., **72** (1960), 458–520.

[Fle62] W. H. Fleming, *On the oriented Plateau problem*, Rend. Circ. Mat. Palermo (2), **11** (1962), 69–90.

[Gas] P. Gaspar, *The second inner variation of energy and the Morse index of limit interfaces*, arXiv (2017).

[GasGua1] P. Gaspar and M. Guaraco, *The Allen–Cahn equation on closed manifolds*, Calc. Var. & PDE, **57** (2018).

[GasGua2] P. Gaspar and M. Guaraco, *The Weyl Law for the phase transition spectrum and density of limit interfaces*, arXiv (2018), to appear in GAFA.

[Gua18] M. Guaraco, *Min-max for phase transitions and the existence of embedded minimal hypersurfaces*, Jour. Diff. Geom., **108** (2018), 91–133.

[Hie] F. Hiesmayr, *Spectrum and index of two sided Allen–Cahn minimal hypersurfaces*, Comm. PDE, (to appear), arXiv (2017).

[HutTon00] J. Hutchinson and Y. Tonegawa, *Convergence of phase interfaces in the van der Waals–Cahn–Hilliard theory*, Calc. Var. PDE, **10** (2000), 49–84.

[Ilm93] T. Ilmanen, *Convergence of the Allen–Cahn equation to Brakke's motion by mean curvature flow*, Jour. Diff. Geom., **38** (1993), 417–461.

[IMN18] K. Irie, F. C. Marques and A. Neves, *Density of minimal hypersurfaces for generic metrics*, Ann. of Math. (2), **187** (2018), no. 3, 963–972.

[KolMen15] S. Kolasiński and U. Menne, *Decay rates for the quadratic and super-quadratic tilt-excess of integral varifolds*, Nonlinear Diff. Equations and Appl. NoDEA, **24** (2015).

[Le11] N. Le, *On the second inner variation of the Allen–Cahn functional and its applications*, Indiana Univ. Math. Jour., **60** (2011), 1843–1856.

[Le15] N. Le, *On the second inner variation of Allen–Cahn type energies and applications to local minimizers*, Jour. des Mat. Pures et Appl., **103** (2015), 1317–1345.

[MarNev14] F. C. Marques and A. Neves, *Existence of infinitely many minimal hypersurfaces in positive Ricci curvature*, Invent. Math., **209** (2017), no. 2, 577–616.

[MarNev16] F. C. Marques and A. Neves, *Morse index and multiplicity of min-max minimal hypersurfaces*, Cambridge Jour. Math., **4** (4) (2016), 463–511.

[Pit81] J. Pitts, *Existence and regularity of minimal surfaces on Riemannian manifolds*, Princeton University press, Princeton, (1981).

[Rei60] R. Reifenberg, *Solution of the Plateau problem for m-dimensional surfaces of varying topological type*, Acta Math., **104** (1960), 1–92.

[SchSim81] R. Schoen and L. Simon, *Regularity of stable minimal hypersurfaces*, Comm. Pure & Appl. Math., **34** (1981), 741–797.

[SSY75] R. Schoen, L. Simon and S. T. Yau, *Curvature estimates for minimal hypersurfaces*, Acta Math., **134** (1975), 276–288.

[Sim84] L. Simon, *Lectures on Geometric Measure Theory*, Centre for Mathematical Analysis, Australian National University, Canberra, (1984).

[JSim68] J. Simons, *Minimal varieties in Riemannian manifolds*, Ann. of Math., **88** (1968), 62–105.

[Son] A. Song, *Existence of infinitely many minimal hypersurfaces in closed manifolds*, arXiv (2018).

[Ton05] Y. Tonegawa, *On stable critical points for a singular perturbation problem*, Comm. Anal. Geom., **13** (2005), 439–459.

[TonWic12] Y. Tonegawa and N. Wickramasekera, *Stable phase interfaces in the van der Waals–Cahn–Hilliard theory*, J. Reine Ang. Mat., (2012), 191–210.

[WanWei] K. Wang and J. Wei, *Finite Morse index implies finite ends*, Comm. Pure Appl. Math., (to appear), arXiv (2017).

[Wic08] N. Wickramasekera, *A regularity and compactness theory for immersed stable minimal hypersurfaces of multiplicity at most* 2, J. Differential Geom., **80** (2008), no. 1, 79–173.

[Wic14] N. Wickramasekera, *A general regularity theory for stable codimension* 1 *integral varifolds*, Ann. of Math., **179** (2014), 843–1007.

[Wic14a] N. Wickramasekera, *A sharp strong maximum principle and a sharp unique continuation theorem for stationary codimension* 1 *integral varifolds*, Calc. Var. PDE, **51** (2014), 799–812.

[Wic14b] N. Wickramasekera, *Regularity of stable minimal hypersurfaces: Recent advances in the theory and applications*, Surveys in Diff. Geom., **19** (2014), 231–301.

[Wic] N. Wickramasekera, *Regularity and compactness for stable codimension* 1 *CMC varifolds*, Current Developments in Mathematics 2017 (to appear), arXiv (2019).

University of Cambridge DPMMS,
Centre for Mathematical Sciences
Wilberforce Road Cambridge, CB3 0WB United Kingdom
E-mail address: n.wickramasekera@dpmms.cam.ac.uk

Advanced Studies in Pure Mathematics 85, 2020
The Role of Metrics in the Theory of Partial Differential Equations
pp. 513–523

A novel integral transform approach to solving hyperbolic equations in the curved space-times

Karen Yagdjian

Abstract.

This is a survey of a novel integral transform and its applications to solving second-order hyperbolic equations in the curved space-times.

§1. Introduction. Description of integral transform

The integral transform of this paper is aimed to generate representation formulas for the solutions to the equation

$$u_{tt} - a^2(t)A(x, \partial_x)u - M^2u = f, \qquad t \in (0, T), \ x \in \Omega, \tag{1}$$

where $M \in \mathbb{C}$ is a parameter, $a \in C(0, T)$, and $A(x, \partial_x)$ is negative uniformly elliptic operator

$$A(x, \partial_x) = \sum_{|\alpha| \le p} a_\alpha(x)\partial_x^\alpha,$$

with smooth coefficients $a_\alpha(x) \in C^\infty(\Omega)$ in the domain in $\Omega \subseteq \mathbb{R}^n$. In this section we give the outline of the definitions without any precise statements which will be presented in the next section. For the given smooth function $f = f(x, t)$ defined in $\Omega \times [0, T]$ consider a solution $v = v(x, t; b)$ of the problem

$$\begin{cases} v_{tt} - A(x, \partial_x)v = 0, & t \in (0, T_1), \quad x \in \Omega, \\ v(x, 0; b) = f(x, b), & v_t(x, 0; b) = 0, \quad x \in \Omega, \end{cases} \tag{2}$$

Received November 20, 2018.
Revised June 23, 2019.
2010 *Mathematics Subject Classification.* Primary 35C15; Secondary 35Q05, 35Q40, 35Q75.
Key words and phrases. Klein-Gordon Equation, de Sitter Space-time, Fundamental Solutions, Huygens' Principle, Ehrenfest Question, Higuchi Bound, Sign Changing Solutions, Higgs Boson Equation.

with the parameter $b \in (0,T)$ and $0 < T_1 \le \infty$. We denote the resolving operator $\mathcal{E}\mathcal{E}_A : f \longmapsto v$ by $v := \mathcal{E}\mathcal{E}_A[f]$. Next we introduce the integral operator

$$\mathcal{K} : v \longmapsto u, \tag{3}$$

which maps function $v = v(x,t;b)$ into solution of the equation (1). The integral operator $\mathcal{K}$ is defined by

$$u(x,t) = \mathcal{K}[v](x,t) := \int_0^t db \int_0^{|\phi(t)-\phi(b)|} K(t;r,b;M)v(x,r;b)\,dr,$$

$x \in \Omega$, $t \in (0,T)$. Here $\phi(t) = \displaystyle\int_0^t a(\tau)\,d\tau$ is a distance function produced by $a = a(t)$ and $\phi(t) \le T_1$ for all $t \in (0,T)$. We note that the integral transform $\mathcal{K}$ of (3) is point-wise in x and non-local in time. This integral transform is applicable to the distributions and fundamental solutions as well. In fact, $u = u(x,t)$ takes the initial values

$$u(x,0) = 0, \; u_t(x,0) = 0, \quad x \in \Omega.$$

Thus, the operator $\mathcal{K}$ generates solutions of different well-known equations of the form (1). (See [25], [26].) A class of operators for which we have obtained representation formulas for the solutions includes $a(t) = t^\ell$, $\ell \in \mathbb{R}$, $a(t) = e^{\pm t}$. If we have a resolving operator $\mathcal{E}\mathcal{E}_A$ of the problem (2), then, by applying integral transform, we obtain the solution operator $\mathcal{G}_A : f \longmapsto u$, $u := \mathcal{G}_A[f]$ hence $\mathcal{G}_A = \mathcal{K} \circ \mathcal{E}\mathcal{E}_A$. The equations of the form (1) are not necessarily hyperbolic. Note, if $a(t) \equiv 1$, $M = 0$ and $A(x,\partial_x) = \Delta$ is the Laplacian, then $K(t;r,b;M) \equiv 1$ and the integral transform becomes a formula that is well known as the Duhamel's principle.

§2. The linear Klein-Gordon equation in the de Sitter spacetime

Now we give precise formulations for one important application of the our integral transform approach. For $a(t) = e^{-t}$ define a *chronological future* $D_+(x_0,t_0)$ and a *chronological past* $D_-(x_0,t_0)$ of the point (x_0,t_0), $x_0 \in \mathbb{R}^n$, $t_0 \in \mathbb{R}$: $D_\pm(x_0,t_0) := \{(x,t) \in \mathbb{R}^{n+1}; |x - x_0| \le \pm(e^{-t_0} - e^{-t})\}$. Then, for $(x_0,t_0) \in \mathbb{R}^n \times \mathbb{R}$, $M \in \mathbb{C}$, we define the function

$$
\begin{aligned}
(4)\qquad & E(x,t;x_0,t_0;M)\\
&:= 4^{-M}e^{M(t_0+t)}\Big((e^{-t_0}+e^{-t})^2-(x-x_0)^2\Big)^{M-\frac{1}{2}}\\
&\quad\times F\left(\frac{1}{2}-M,\frac{1}{2}-M;1;\frac{(e^{-t_0}-e^{-t})^2-(x-x_0)^2}{(e^{-t_0}+e^{-t})^2-(x-x_0)^2}\right),
\end{aligned}
$$

where $(x,t)\in D_+(x_0,t_0)\cup D_-(x_0,t_0)$ and $F(a,b;c;\zeta)$ is the hypergeometric function. When no ambiguity arises, we use the notation $x^2:=|x|^2$ for $x\in\mathbb{R}^n$. Thus, if $r^2=(x-x_0)^2$, then $E(x,t;x_0,t_0;M)=E(r,t;0,t_0;M)$. The kernels $K_0(z,t;M)$ and $K_1(z,t;M)$ are defined by

$$
(5)\qquad K_0(z,t;M):=-\left[\frac{\partial}{\partial b}E(z,t;0,b;M)\right]_{b=0},
$$

$$
(6)\qquad K_1(z,t;M):=E(z,t;0,0,M).
$$

For the equation (1) with $a(t)=e^{-t}$ we have the following result.

Theorem 1 ([25]). *For $f\in C(\Omega\times I)$, $I=[0,T]$, $0<T\le\infty$, and $\varphi_0,\ \varphi_1\in C(\Omega)$, let the function $v_f(x,t;b)\in C^{p,2,0}_{x,t,b}(\Omega\times[0,1-e^{-T}]\times I)$ be a solution to the problem*

$$
(7)\qquad \begin{cases} v_{tt}-A(x,\partial_x)v=0, & x\in\Omega,\quad t\in[0,1-e^{-T}],\\ v(x,0;b)=f(x,b), & v_t(x,0;b)=0,\quad b\in I,\quad x\in\Omega,\end{cases}
$$

and the function $v_\varphi(x,t)\in C^{p,2}_{x,t}(\Omega\times[0,1-e^{-T}])$ be a solution of the problem

$$
\begin{cases} v_{tt}-A(x,\partial_x)v=0, & x\in\Omega,\quad t\in[0,1-e^{-T}],\\ v(x,0)=\varphi(x), & v_t(x,0)=0,\quad x\in\Omega.\end{cases}
$$

Then the function $u=u(x,t)$ defined by

$$
\begin{aligned}
&u(x,t)\\
&=2\int_0^t db\int_0^{\phi(t)-\phi(b)} v_f(x,r;b)E(r,t;0,b;M)\,dr+e^{\frac{t}{2}}v_{\varphi_0}(x,\phi(t))\\
&\quad+2\int_0^{\phi(t)} v_{\varphi_0}(x,s)K_0(s,t;M)\,ds+2\int_0^{\phi(t)} v_{\varphi_1}(x,s)K_1(s,t;M)\,ds,
\end{aligned}
$$

where $\phi(t):=1-e^{-t}$, solves the problem

$$
(8)\qquad \begin{cases} u_{tt}-e^{-2t}A(x,\partial_x)u-M^2u=f, & x\in\Omega,\ t\in I,\\ u(x,0)=\varphi_0(x), & u_t(x,0)=\varphi_1(x),\quad x\in\Omega.\end{cases}
$$

Here the kernels E, K_0 and K_1 have been defined in (4), (5) *and* (6), *respectively.*

We note that the operator $A(x, \partial_x)$ is of arbitrary order, that is, the equation of (8) can be an evolution equation. Then, (7) and (8) can be a mixed initial-boundary value problem with boundary condition.

The Huygens' principle. Recall that a hyperbolic equation is said to satisfy Huygens' principle if the solution vanishes at all points which cannot be reached from the support of initial data by a null geodesic. The Huygens' principle has a wide variety of applications including a derivation of the sharp rate time decay of solutions of wave equations [1]. Let Δ be the Laplace operator in $\mathbb{R}^n$. By the change $\psi = ue^{-nt/2}$ the equation (9) of the next theorem can be written in the form (8).

Theorem 2 ([23]). *The equation*

$$\psi_{tt} + n\psi_t - e^{-2t}\Delta\psi + m^2\psi = 0, \tag{9}$$

obeys the Huygens' principle if and only if $m = \sqrt{n^2-1}/2$ and $n \geq 3$ is an odd number.

Definition 3 ([23]). *We say that the equation* (9) *obeys the incomplete Huygens' principle with respect to the first initial datum, if it obeys the Huygens' principle provided that the second datum vanishes.*

Theorem 4 ([23]). *Suppose that equation* (9) *does not obey the Huygens' principle. Then, it obeys the incomplete Huygens' principle with respect to the first initial datum, if and only if the equation is massless, $m = 0$, and either $n = 1$ or $n = 3$.*

Corollary 5. [23] *Assume that the equations*

$$\psi_{tt} + n\psi_t - e^{-2t}\Delta\psi + m_1^2\psi = 0,$$
$$\psi_{tt} + n\psi_t - e^{-2t}\Delta\psi + m_2^2\psi = 0$$

describe two fields with $m_1 \neq m_2$. Then they obey the incomplete Huygens' principle if and only if the dimension n of the spatial variable x is 3 and $m_1 = 0$, $m_2 = \sqrt{2}$.

Paul Ehrenfest in 1917 [8] raised the question: *In that way does it become manifest in the fundamental laws of physics that space has three dimensions?*. In that article he addressed the question: "*Why has our space just three dimensions?*" or in other words: "*By which singular characteristics do geometries and physics in R_3 distinguish themselves from those in the other R_n's?*". Thus the last corollary sheds light on this question.

In quantum field theory the interval $(0,\sqrt{2})$ is known as the so-called *Higuchi bound.* Higuchi bound $(0,\sqrt{2})$ is the forbidden mass range for spin-2 field theory in de Sitter space-time because negative probability appears if one introduce interactions [15].

The positivity of the kernel functions E, K_0 and K_1 for the real M allows us to discuss the maximum principle and the sign changing solutions of the equation with the Higgs potential [24]

$$\psi_{tt} - e^{-2t}\Delta\psi + n\psi_t = \mu^2\psi - \lambda\psi^3,$$

with $\lambda > 0$ and $\mu > 0$. Denote $M := (n^2/4 - m^2)^{1/2}$.

Proposition 6. [24] *Assume that $M \geq 0$. Then*

$$E(r,t;0,b;M) > 0, \quad \textit{for all } 0 \leq b \leq t,\ r \leq e^{-b} - e^{-t},\ t \in [0,\infty),$$
$$K_1(r,t;M) > 0 \quad \textit{for all} \quad r \leq 1 - e^{-t}, \quad t \in [0,\infty).$$

If we assume that $M > 1$, then

$$K_0(r,t;M) > 0 \ \textit{for all } r \leq 1 - e^{-t} \ \textit{ and for all } \ t > -\ln(1 - 1/M).$$

The sign changing solutions to the Higgs boson equation are investigated in [28].

§3. Semilinear Klein-Gordon equation. Global existence of the self-interacting scalar field in the de Sitter universe

In this section we consider the equation

$$\psi_{tt} + n\psi_t - e^{-2t}A(x,\partial_x)\psi + m^2\psi = F(x,\psi)\,, \tag{10}$$

where $A(x,\partial_x) = \sum_{|\alpha|\leq 2} a_\alpha(x)\partial_x^\alpha$ is a second order negative uniformly elliptic operator with coefficients $a_\alpha \in \mathcal{B}^\infty$, where $\mathcal{B}^\infty$ is the space of all $C^\infty(\mathbb{R}^n)$ functions with uniformly bounded derivatives of all orders. We also assume that the mass m can be a complex number, $m^2 \in \mathbb{C}$.

In physical terms this equation describes a local self-interaction for a scalar particle. The special case of the equation (10) if $A(x,\partial_x) = \Delta$ is the covariant Klein-Gordon equation in the de Sitter space-time.

In [13], [18], [21], [29] the global existence of small data solutions of the Cauchy problem for the semilinear Klein-Gordon equation in the de Sitter space-time was proved. The nonlinearity F was assumed Lipschitz continuous with exponent $\alpha > 0$ (see definition below) while $m \in (0,\sqrt{n^2-1}/2] \cup [n/2,\infty)$. The proof of the global existence in [21] is based on the integral transform and representations of Section 1 and $L^p - L^q$ estimates. Later on, in [13] this result for the same range of the parameters n, m and the same nonlinearity was extended on the equation (10), that is, from the spatially flat de Sitter space-time to the de Sitter space-time with the time slices being, in particular, the Riemannian manifolds. The case of $m \in (\sqrt{n^2-1}/2, n/2)$ was left open in [13]. Another interesting and important case of the complex-valued mass m also was not discussed in [13]. The last case contains the Klein-Gordon model of the Higgs boson equation.

Now there is an increasing interest to the semilinear wave and Klein-Gordon equations in some cosmological models with curved space-time (see, e.g., [2], [3], [16] and references therein).

In [29] paper we generalized and completed the small data global existence result of [13].

Condition ($\mathcal{L}$). *The smooth in x function $F = F(x,\psi)$ is said to be Lipschitz continuous with exponent $\alpha \geq 0$ in the space $H_{(s)}(\mathbb{R}^n)$ if there is a constant $C \geq 0$ such that*

$$\|F(x,\psi_1(x)) - F(x,\psi_2(x))\|_{H_{(s)}} \leq C\|\psi_1-\psi_2\|_{H_{(s)}}\left(\|\psi_1\|^\alpha_{H_{(s)}} + \|\psi_2\|^\alpha_{H_{(s)}}\right)$$

for all $\psi_1, \psi_2 \in H_{(s)}$.

The polynomials in ψ are Lipschitz continuous with some exponent α in the space $H_{(s)}(\mathbb{R}^n)$ when $s > n/2$. Moreover, the exponent α is independent of s. Another interesting functions $F(x,\psi) = \pm|\psi|^{\alpha+1}, \pm|\psi|^\alpha\psi$ are important examples of the Lipschitz continuous with exponent $\alpha > 0$ in the Sobolev space $H_{(s)}(\mathbb{R}^n)$, $s > n/2$, functions provided that α agrees with s and n. Define also the metric space

$$\begin{aligned}(11)\qquad & X(R,s,\gamma) \\ & := \left\{\psi \in C([0,\infty); H_{(s)}) \;\middle|\; \|\psi\|_X := \sup_{t\in[0,\infty)} e^{\gamma t}\|\psi(\cdot,t)\|_{H_{(s)}} \leq R\right\},\end{aligned}$$

where $\gamma \in \mathbb{R}$, with the metric $d(\psi_1,\psi_2) := \sup_{t\in[0,\infty)} e^{\gamma t}\|\psi_1(\cdot,t) - \psi_2(\cdot,t)\|_{H_{(s)}}$.

We study the Cauchy problem for the equation (10) with the initial conditions

$$(12)\qquad \psi(x,0) = \varphi_0(x), \quad \psi_t(x,0) = \varphi_1(x),$$

through the integral equation. To define that integral equation we appeal to the operator

$$\mathcal{G}_A := \mathcal{K} \circ \mathcal{E}\mathcal{E}_A$$

($\mathcal{E}\mathcal{E}_A$ stands for the resolving operator of the evolution equation) as follows. For the function $f(x,t)$ we define

$$v(x,t;b) := \mathcal{E}\mathcal{E}_A[\,f\,](x,t;b),$$

where the function $v(x,t;b)$ is a solution to the Cauchy problem

$$\begin{aligned} &\partial_t^2 v - A(x,D)v = 0, \quad x \in \mathbb{R}^n, \quad t \geq 0, \\ &v(x,0;b) = f(x,b), \quad v_t(x,0;b) = 0, \quad x \in \mathbb{R}^n, \end{aligned}$$

while $\mathcal{K}$ is introduced by

$$\begin{aligned} &\mathcal{K}[v](x,t) \\ &\quad := 2e^{-\frac{n}{2}t} \int_0^t db \int_0^{e^{-b}-e^{-t}} dr\, e^{\frac{n}{2}b} v(x,r;b) E(r,t;0,b;M). \end{aligned} \tag{13}$$

Here $M := (n^2/4 - m^2)^{1/2}$. The kernel $E(r,t;0,b;M)$ is defined by (4) (see [20], [21]). Hence,

$$\begin{aligned} &\mathcal{G}_A[\,f\,](x,t) \\ &\quad = 2e^{-\frac{n}{2}t} \int_0^t db \int_0^{e^{-b}-e^{-t}} dr\, e^{\frac{n}{2}b}\, \mathcal{E}\mathcal{E}_A[\,f\,](x,r;b) E(r,t;0,b;M). \end{aligned} \tag{14}$$

Thus, the Cauchy problem (10) and (12) leads to the following integral equation

$$\psi(x,t) = \psi_0(x,t) + \mathcal{G}_A[F(\cdot,\psi)](x,t). \tag{15}$$

Every solution to the Cauchy problem (10) and (12) solves also the last integral equation with some function $\psi_0(x,t)$, which is a solution for the problem for the linear equation without source term. We define a solution of the Cauchy problem (10) and (12) via integral equation (15). Here we consider the case of the complex mass $m \in \mathbb{C}$ that includes, in particular, the Higgs boson equation. The principal square root $M := (n^2/4 - m^2)^{\frac{1}{2}}$ is the parameter that controls estimates and solvability.

Theorem 7 ([29]). *Assume that the nonlinear term $F(x,\psi)$ is a Lipschitz continuous in the space $H_{(s)}(\mathbb{R}^n)$, $s > n/2 \geq 1$, $F(x,0) \equiv 0$, and $\alpha > 0$.*
(i) *Assume also that $0 < \Re M < 1/2$. Then, there exists $\varepsilon_0 > 0$ such that, for every given functions $\varphi_0, \varphi_1 \in H_{(s)}(\mathbb{R}^n)$, such that*

$$\|\varphi_0\|_{H_{(s)}(\mathbb{R}^n)} + \|\varphi_1\|_{H_{(s)}(\mathbb{R}^n)} \leq \varepsilon, \qquad \varepsilon < \varepsilon_0, \tag{16}$$

there exists a solution $\psi \in C([0,\infty); H_{(s)}(\mathbb{R}^n))$ *of the Cauchy problem* (10), (12). *The solution* $\psi(x,t)$ *belongs to the space* $X(2\varepsilon, s, \frac{n-1}{2})$, *that is,*

$$\sup_{t\in[0,\infty)} e^{\frac{n-1}{2}t}\|\psi(\cdot,t)\|_{H_{(s)}(\mathbb{R}^n)} \le 2\varepsilon.$$

(ii) *Assume that* $1/2 \le \Re M < n/2$ *and* $\gamma \in (0, \frac{1}{\alpha+1}(\frac{n}{2} - \Re M))$. *Then there exists* $\varepsilon_0 > 0$ *such that for every given functions* $\varphi_0, \varphi_1 \in H_{(s)}(\mathbb{R}^n)$, *such that* (16), *there exists a solution* $\psi \in X(2\varepsilon, s, \gamma)$ *of the Cauchy problem* (10)–(12).
(iii) *If* $\Re M > n/2$, *then the lifespan* T_{ls} *of the solution can be estimated from below as follows*

$$T_{ls} \ge -\frac{1}{\Re M - \frac{n}{2}} \ln\left(\|\varphi_0\|_{H_{(s)}(\mathbb{R}^n)} + \|\varphi_1\|_{H_{(s)}(\mathbb{R}^n)}\right) - C(m,n,\alpha)$$

with some constant $C(m,n,\alpha)$.

The main tools to study the problem (10) and (12) are the integral transform from [25] and the energy estimate for the finite time interval for the strictly hyperbolic equation. The integral transforms makes possible to reduce the problem with infinite time to the problem for the hyperbolic equation with time independent coefficients and with the finite time interval due to the fact that the de Sitter space-time has permanently bounded domain of influence. Thus the integral transform allows us to push forward the estimates provided that some integrals of the kernel functions lead to the proper estimates. The proof of the estimates for the kernel functions consists of a long sequence of estimates of integrals involving hypergeometric functions.

The finite time blowup of the solutions is proved in [22] for the Cauchy problem (10) and (12) with the wide class of semilinear term F that is contained in one assumed in (iii) of Theorem 7. In particular, the blow up occurs if $M > n/2$ and $F(x,\psi) = |\psi|^\alpha$ with $\alpha > 0$. The derivation and the proof are also based on the integral transform (13), (14) that leads to representation formulas. That class of F includes sign preserving solutions of the Higgs boson equation (see, e.g., [9, Ch.17])

$$\psi_{tt} - e^{-2t}A(x,D)\psi + n\psi_t = \mu^2\psi - \lambda\psi^3, \tag{17}$$

with $\lambda > 0$, $\mu > 0$, $A(x,D) = \Delta$, and $n = 3$. In fact, the application of the last assertion of Theorem 7 to the Higgs boson equation gives an estimate from below of the lifespan of the Higgs boson. The mathematical proof of the finiteness of the lifespan of solution, which is not necessarily sign preserving, to the equation (17) is an interesting and difficult problem that requires special technique (see, e.g., [22]) and will be published in a forthcoming paper. In the existing extensive literature on the physics of the Higgs boson one can find that the lifetime of the Higgs boson is approximately $10^{-22}sec.$ (see, e.g., [5]) that points at the boundedness of the lifespan of nontrivial solution to the equation (17).

The estimates derived for the linear equation in [29] include equations of scalar fields considered in [4], [10] with $m^2 < 0$ living on the de Sitter universe. The Klein-Gordon scalar quantum fields on the de Sitter manifold with imaginary mass $m^2 = -k(k+n)$, $k = 0, 1, 2, \dots$, present a family of tachyonic quantum fields. Epstein and Moschella [10] give a complete study of such family of linear scalar tachyonic quantum fields. The corresponding linear equation is $\psi_{tt} + n\psi_t - e^{-2t}\Delta\psi + m^2\psi = 0$, for which the kernel of the integral transform $\mathcal{K}$ is $E(x, t; x_0, t_0; M)$, where $M = k + \frac{n}{2}$, $k = 0, 1, 2, \dots$. For an odd n, the mass m takes value at knot points set in the sense of [23]. Theorem 7 contains an estimate for the lifespan of a self-interacting tachyonic field.

The equation (10) in the case of $A(x, \partial_x) = \Delta$ is amenable to the analysis via the Fourier transform and the Bessel functions [11]. For the equation with $A(x, \partial_x) = \Delta$ and $F(x, \Phi) = |\Phi|^p$ the global existence in the energy classes was proved in [7] under some restrictions on p.

Unlike to the case of the operator $A(x, \partial_x) = \Delta$, the linear part of the equation (10) is not invariant with respect to de Sitter group [14]. Nevertheless, the mass intervals $(0, \sqrt{n^2-1}/2)$, $[\sqrt{n^2-1}/2, n/2)$, $[n/2, \infty)$ appear and play important role also in this case.

Finally we mention that it will be interesting to apply the integral transform approach to the derivation of the Strichartz estimates and of the Price's law for the corresponding cosmological models [6], and to the investigation of the gravitational waves in the curved space-times [17].

References

[1] B. Abbasi and W. Craig, On the initial value problem for the wave equation in Friedmann-Robertson-Walker space-times, Proc. R. Soc. Lond. Ser. A Math. Phys. Eng. Sci., **470** (2169), (2014), 20140361.

[2] A. Bachelot, On the Klein-Gordon equation near a De Sitter Brane, J. Math. Pures Appl., (9) **105** (2016), no. 2, 165–197.

[3] D. Baskin, Strichartz Estimates on Asymptotically de Sitter Spaces, Annales Henri Poincaré, **14**, Issue 2, (2013) 221–252.

[4] J. Bros, H. Epstein and U. Moschella, Scalar tachyons in the de Sitter universe, Lett. Math. Phys., **93** (2010), no. 2, 203–211.

[5] CMS Collaboration, Limits on the Higgs boson lifetime and width from its decay to four charged leptons, Phys. Rev. D, **92** (2015), 072010.

[6] J. L. Costa, A. Alho and J. Natário, Spherical linear waves in de Sitter spacetime, J. Math. Phys., **53** (5), (2012), 052501, 9 pp.

[7] M. R. Ebert and M. Reissig, Regularity theory and global existence of small data solutions to semi-linear de Sitter models with power non-linearity, Nonlinear Anal. Real World Appl., **40** (2018), 14–54.

[8] P. Ehrenfest, In what way does it become manifest in the fundamental laws of physics that space has three dimensions? Proceedings Koninklijke Akademie van Wetenschappen, **20** (1918), 200–209.

[9] E. Elbaz, Quantum. The quantum theory of particles, fields, and cosmology, Texts and Monographs in Physics, Springer-Verlag, Berlin, 1998.

[10] H. Epstein and U. Moschella, de Sitter tachyons and related topics, Comm. Math. Phys., **336** (1), (2015), 381–430.

[11] A. Galstian, $L_p - L_q$ decay estimates for the wave equations with exponentially growing speed of propagation, Appl. Anal., **82** (2003), no. 3, 197–214.

[12] A. Galstian and K. Yagdjian, Global solutions for semilinear Klein-Gordon equations in FLRW spacetimes, Nonlinear Anal., **113** (2015), 339–356.

[13] A. Galstian and K. Yagdjian, Global in Time Existence of the Self-Interacting Scalar Field in de Sitter Spacetimes, Nonlinear Anal. Real World Appl., **34** (2017), 110–139.

[14] K. Goto, Wave Fields in de Sitter Space, Progress of Theoretical Physics, **12** (1954), no. 3, 311–354.

[15] A. Higuchi, Forbidden mass range for spin-2 field theory in de Sitter spacetime, Nuclear Phys. B, **282** (2), (1987), 397–436.

[16] P. Hintz and A. Vasy, Semilinear wave equations on asymptotically de Sitter, Kerr-de Sitter and Minkowski spacetimes, Anal. PDE, **8** (8), (2015), 1807–1890.

[17] J. Näf, P. Jetzer, M. Sereno, On gravitational waves in spacetimes with a nonvanishing cosmological constant, Phys. Rev. D, **79**, 024014 (2009).

[18] M. Nakamura, The Cauchy problem for semi-linear Klein-Gordon equations in de Sitter spacetime, J. Math. Anal. Appl., **410** (1), (2014), 445–454.

[19] A. Palmieri and M. Reissig, A competition between Fujita and Strauss type exponents for blow-up of semi-linear wave equations with scale-invariant damping and mass, J. Differential Equations, in press.

[20] K. Yagdjian and A. Galstian, Fundamental Solutions for the Klein-Gordon Equation in de Sitter Spacetime, Comm. Math. Phys., **285** (2009), 293–344.

[21] K. Yagdjian, Global existence of the scalar field in de Sitter spacetime, J. Math. Anal. Appl., **396** (1), (2012), 323–344.

[22] K. Yagdjian, The semilinear Klein-Gordon equation in de Sitter spacetime, Discrete Contin. Dyn. Syst. Ser. S, **2** (3), (2009), 679–696.

[23] K. Yagdjian, Huygens' Principle for the Klein-Gordon equation in the de Sitter spacetime, J. Math. Phys., **54** (9), (2013), 091503.

[24] K. Yagdjian and A. Balogh, The maximum principle and sign changing solutions of the hyperbolic equation with the Higgs potential, J. Math. Anal. Appl., **465** (2018), no. 1, 403–422.

[25] K. Yagdjian, Integral transform approach to solving Klein-Gordon equation with variable coefficients, Mathematische Nachrichten, **288** (17-18), (2015), 2129–2152.

[26] K. Yagdjian, Integral transform approach to generalized Tricomi equations, J. Differential Equations, **259** (2015), 5927–5981.

[27] K. Yagdjian, A note on the fundamental solution for the Tricomi-type equation in the hyperbolic domain, J. Differential Equations, **206** (2004), 227–252.

[28] K. Yagdjian, On the global solutions of the Higgs boson equation, Comm. Partial Differential Equations, **37** (3), (2012), 447–478.

[29] K. Yagdjian, Global existence of the self-interacting scalar field in the de Sitter universe, J. Math. Phys., **60** (2019), 051503.

School of Mathematical and Statistical Sciences,
University of Texas RGV, 1201 W. University Drive,
Edinburg, TX 78539, USA
E-mail address: karen.yagdjian@utrgv.edu

Advanced Studies in Pure Mathematics 85, 2020
The Role of Metrics in the Theory of Partial Differential Equations
pp. 525–532

Harmonic maps and the Einstein equation

Sumio Yamada

Abstract.

In 1917, very shortly after Einstein had announced his master equation on the general relativity and Schwarzschild had discovered the first explicit solution to the equation, H. Weyl characterized the Schwarzschild metric, by a harmonic function. Since then, the solutions to the Einstein equation with a certain set of symmetries are identified with elliptic variational problems, in particular the harmonic map equation. In collaboration with Marcus Khuri, Yukio Matasumoto and Gilbert Weinstein, we constructed a new set of stationary solutions to the 5-dimensional vacuum Einstein equation, which contains non-spherical event horizons. The higher dimensional spacetime exhibit a wider range of topological structures, compared to our 4-dimensional physical spacetime, and those stationary solutions are thus geometrically interesting.

§1. Introduction

Spacetime is a Lorentzian $n+1$-dimensional manifold $(\mathcal{N}^{n+1}, \mathbf{g})$ satisfying the Einstein equations

$$R_{\mu\nu} - \frac{1}{2} R_{\mathbf{g}}\, \mathbf{g}_{\mu\nu} = T_{\mu\nu}$$

where $R_{\mu\nu}$ is the Ricci curvature of the metric $\mathbf{g}$, $R_{\mathbf{g}}$ the scalar curvature of the Lorentzian metric $\mathbf{g}$, and $T_{\mu\nu}$ is the energy-momentum-stress tensor of the matter fields. When vacuum; $T_{\mu\nu} = 0$, taking the trace of the Einstein vacuum equations we get $R_{\mathbf{g}} = 0$, and hence the vacuum Einstein equation is equivalent to $R_{\mu\nu} = 0$.

Received December 25, 2018.
Revised June 2, 2019.
2010 *Mathematics Subject Classification.* Primary 83C05; Secondary 35Q76, 53C43.
Key words and phrases. Einstein equation, harmonic map, 4+1 dimension, stationary solution, axial symmetry.

Now we make a brief exposition of harmonic map. Let (M, g) and (N, h) be Riemannian manifolds, and let $\varphi\colon M \to N$ be a continuously differentiable map. The Dirichlet energy density of φ is

$$|d\varphi|^2 = h_{ij}\, g^{\mu\nu}\, \partial_\mu \varphi^i\, \partial_\nu \varphi^j .$$

If $\Omega \subset\subset M$ then the *energy* of φ over Ω is $E_\Omega(\varphi) = \int_\Omega |d\varphi|^2\, dV_g$.

A map φ is *harmonic* if it is critical with respect to E_Ω for every $\Omega \subset\subset M$. The harmonic map equations are the vanishing of the tension field:

$$\tau^i = g^{\mu\nu}\left(\partial_\mu \partial_\nu \varphi^i - {}^M\Gamma^\sigma_{\mu\nu}\partial_\sigma \varphi^i + {}^N\Gamma^i_{jk}\partial_\mu \varphi^j \partial_\nu \varphi^k\right) = 0,$$

which is a cross section $\tau(\varphi) = \tau^i \partial_i\colon M \to TN$ of the pull back bundle $\varphi^{-1}TN$. When the domain (M, g) is $\mathbb{R}$, the image of the harmonic map is a geodesic in (N, h). When the target (N, h) is $\mathbb{R}$, the map is a harmonic function.

§2. The Einstein Equation according to H. Weyl

H. Weyl wrote down (1917) the Schwarzschild metric (1916)

$$ds^2 = -\left(1 - \frac{m}{r}\right) dt^2 + \left(1 - \frac{m}{r}\right)^{-1} dr^2 + r^2 d\omega^2_{S^2},$$

as

$$ds^2 = -V^2\, dt^2 + V^{-2}\{\rho^2 d\phi^2 + e^{2\sigma}(d\rho^2 + dz^2)\} \tag{1}$$

and when $V = e^u$, $R_{ij} = 0$ becomes

$$\triangle_{\mathbb{R}^3} u = 0, \quad \sigma_z = 2\rho u_z u_\rho, \quad \sigma_\rho = \rho(u_\rho^2 - u_z^2)$$

and the harmonicity is the integrability condition for the quadrature

$$d\sigma = \frac{1}{2}\rho\, (\mathrm{Re}|\partial \mathrm{u}|^2)\, \mathrm{d}\xi \qquad (\xi = \rho + \mathrm{iz}).$$

In other words, the VEE satisfying the Schwarzschild metric written in the form of Equation (1) with two Killing vector fields $\frac{\partial}{\partial t}$ and $\frac{\partial}{\partial \phi}$, is reduced to finding the harmonic function u in ρ, z. The function u determines the $\mathbf{g}$-norm of $\frac{\partial}{\partial \phi}$. We formulate this as follows: With the global coordinate system

$$[\mathbb{R} \times U(1)] \times \{(\rho, z) \mid \rho > 0,\ z \in \mathbb{R}\}$$

the Einstein metric is written as

$$\mathbf{g} = G_{ij}(\rho, z)\, dx^i \otimes dx^j + e^{2\sigma(\rho,z)}(d\rho^2 + dz^2)$$

where

$$\rho = \sqrt{|\det G_{ij}|}, \quad G_{ij} = \begin{pmatrix} -V^2 & 0 \\ 0 & \rho^2 V^{-2} \end{pmatrix}$$

$$G_{ij} = \begin{pmatrix} -V^2 & 0 \\ 0 & \rho^2 V^{-2} \end{pmatrix} = \begin{pmatrix} -e^{2u_0} & 0 \\ 0 & e^{2u_1} \end{pmatrix}$$

where

$$u_0 = \log\left[\frac{-\mu - z + \sqrt{(\mu+z)^2 + \rho^2}}{\mu - z + \sqrt{(\mu - z)^2 + \rho^2}}\right], \qquad u_1 = \log\rho - \log u_0$$

with $\mu = \mu(m)$. In particular when $\rho = 0$ (z-axis)

$$u_0 = 0 \quad (-\mu \le z \le \mu), \quad u_1 = 0 \quad (|z| > \mu)$$

$$V_0 = \begin{pmatrix} 1 \\ 0 \end{pmatrix} \text{ for } |z| \le \mu, \quad V_1 = \begin{pmatrix} 0 \\ 1 \end{pmatrix} \text{ for } |z| > \mu$$

are, respectively, a Killing, null, zero-eigen-vector of G_{ij} ($GV = 0$) generating $\mathbb{R}$-symmetry (stationarity) and a Killing, spacelike, zero-eigen-vector of G_{ij} generating the $U(1)$-symmetry. The rod $\{|z| \le \mu\}$ in z-axis is called horizon rod, and is a minimal surface in $\{t = 0\}$. The rod $\{|z| \ge \mu\}$ is called axis rod, representing the $U(1)$ fixed points, where the norm of the Killing vector $(0,1)^t$ degenerates.

Recall that the Minkowski metric is the trivial Schawarzschild metric ($\mu = 0$)

$$ds^2 = (dt, d\phi)\begin{pmatrix} -1 & 0 \\ 0 & \rho^2 \end{pmatrix}\begin{pmatrix} dt \\ d\phi \end{pmatrix} + d\rho^2 + dz^2$$

so that $u_0 = 0$ and $u_1 = \log\rho$, with no horizon rod, and the entire z-axis is the axis rod. Note that u_1 is Green's function on $\mathbb{R}^2$ with its set of poles aligned on the z-axis, when $\mathbb{R}^3$ is seens as $\mathbb{R}^2 \times \mathbb{R}$. (Perhaps the starting point for Weyl's thoughts.)

In other words, the u_1 for the Schwarzschild metric is Green's function whose poles are evenly distributed over the axis-rods.

§3. Ernst Reduction

Having seen the contribution by Weyl, who identified the vacuum Einstein equation with a decoupled system of a Laplace equation and a quadrature in a static and axisymmetric setting, it is reasonable to expect that stationary solutions are also obtained in the context of elliptic variational problems. In this light, we introduce the so-called Ernst reduction scheme in $3+1$ dimension. A comprehensive survey on this subject can be found, for example, in [1]. Let $(\mathcal{N}^{3+1}, \mathbf{g})$ be asymtptptically flat, stationary and axially symmetric, and let G be the symmetry group ($\cong \mathbb{R} \times U(1)$). We further define X: the (Lorentzian) norm (> 0) of the Killing field generator $\xi = \partial/\partial\phi$, and Y: the potential of the *twist* $*(\xi \wedge d\xi)$. Then the two quantities $\varphi = (X, Y)\colon \mathcal{M}/G \to \mathbb{H}^2$ form a (weighted) harmonic map, i.e. a critical point of the energy

$$\int \left(\frac{|\nabla X|^2 + |\nabla Y|^2}{X^2} \right) \rho \, d\mu_g,$$

where g is the quotient metric on $\mathcal{M}/G$ and ρ is the area element of the orbits of symmetry. Then there are the following observations.

Fact 0: ρ is harmonic, and $\nabla\rho \neq 0$. Let z be a conjugate harmonic function.
Fact 1: $g = e^{2\nu}(d\rho^2 + dz^2)$ on $\mathcal{M}/G =$ the ρz-half plane ($\rho > 0$)
Fact 2: $X \approx \rho^2$ near $\Gamma \subset \{z - \text{axis}\}$, $\{X = 0\} \sim$ the physical *axis*.

(X, Y) completely determines the 2×2 metrix G_{ij}, and vice versa of the Lorentz metric with the symmetries

$$\mathbf{g} = G_{ij}(\rho, z)\, dx^i \otimes dx^j + e^{2\sigma(\rho,z)}(d\rho^2 + dz^2).$$

To see this, it should be noted that Y determines ω up to a constant, as they are related via a differential relation. Namely; the Weyl-Papapetrou form for $3+1$ dimension is

$$\mathbf{g} = \left[-\frac{\rho^2}{X} dt^2 + X(d\phi + \omega dt)^2 \right] + e^{2\nu}(d\rho^2 + dz^2).$$

The map φ is identified with an θ-axisymmetric harmonic map (B. Carter)

$$\varphi : \mathbb{R}^3 \setminus \Gamma \to \mathbb{H}^2 \ni (X, Y),$$

with $\varphi(\rho, \theta, z) = (X, Y)$. (Note $d\mu_{\mathbb{R}^3} = [\rho d\rho dz] d\theta$ with θ is a dummy variable.) Furthermore, Y is constant on any component of Γ, and the difference between constants is the angular momentum induced by the existence of the horizon between the two components. The *data* consists of Γ together with a constant for each component of Γ.

And $\mathbf{g}$ satisfies the VEE is equivalent to the fact (B. Carter) that

$$(\rho, z, \phi) \mapsto (X, Y)$$
$$[G] = \frac{1}{X}\begin{pmatrix} 1 & Y \\ Y & X^2 + Y^2 \end{pmatrix}$$

is a harmonic map $\mathbb{R}^3 \setminus \{\rho = 0\} \to SL(2, \mathbb{R})/SO(2)$, being a critical point of

$$\int \left(\frac{|\nabla X|^2 + |\nabla Y|^2}{X^2} \right) d\mu_{\mathbb{R}^3}.$$

Two maps $\varphi_1, \varphi_2 \colon M \to N$ are *asymptotic* if $\mathrm{dist}_N(\varphi_1, \varphi_2)$ is bounded. We define that $\varphi_0 \colon \mathbb{R}^3 \setminus \Gamma \to \mathbb{H}^2$ is a *model map* for a given data if $\tau(\varphi_0)$ is $O(r^{-2-\epsilon})$ at infinity, $|\tau(\varphi_0)|$ is bounded and φ_0 assumes the given constants on Γ. The asymptotics of a harmonic map is encoded in the model map, and conversely given the data, there exists a unique harmonic map (Weinstein [14]). When reconstructing the spacetime metric $\mathbf{g}$, conical singularities can appear on (bounded components of) Γ. (Bach-Weyl 1921, Li-Tian 1991, Weinstein [14])

§4. Ernst Reduction in 4 + 1 dimension

We now seek stationary solutions in the higher dimeisnion, namely find spacetimes $(\mathcal{N}^{4+1}, \mathbf{g})$ of the Einstein vacuum equations with three mutually commuting Killing fields $\frac{\partial}{\partial t}$, $\frac{\partial}{\partial \phi_1}$, $\frac{\partial}{\partial \phi_2}$, the generators of an isometry group $\mathbb{R} \times U(1)^2$ acting on N.

If $(\mathcal{N}^{4+1}, \mathbf{g})$ has an event horizon, its connected components are diffeomorphic to 3-manifolds Σ^3 of positive Yamabe type, i.e. each component admits a metric of positive scalar curvature (Galloway-Schoen [3]). Under the additional symmetry condition above, the list of topological types (Hollands-Yazadjiev [4]) is further restricted to S^3, $S^1 \times S^2$, $\mathbb{R}P^3$, $L(n, m)$. Examples of solutions include

- No horizon: Minkowski space: $\mathbb{R}^{4,1}$ with its metric
- S^3: Myers-Perry ([11])— 4+1 analog of Kerr solution, vacuum.
- $S^2 \times S^1$: Emperan-Reall ([2]) Pomeransky-Senkov([12])— *Black Ring*, vacuum.
- $L(2, 1)$: Kunduri-Lucietti ([9])— *Black Lens*, minimal supergravity.
- $L(n, 1)$: Nozawa-Tomizawa ([13])— minimal supergravity.

Following the 3+1-dimensional cases, solutions $\mathbf{g}$ on $(\mathcal{N}^{4+1} \setminus \{\text{axis}\})$ can be modeled by the Weyl-Papapetrou Coordinates

$$\left(\mathbb{R} \times U(1)^2\right) \times \{(\rho, z) \colon z \in \mathbb{R},\ \rho > 0\}$$

with

$$\mathbf{g} = G_{ij}\, dx^i\, dx^j + e^{2\nu}(d\rho^2 + dz^2) \quad (y^1 = \phi^1,\ y^2 = \phi^2,\ y^3 = t)$$

and

$$\rho = \sqrt{|\det G_{ij}|}, \quad \nu = \nu(\rho, z), \quad G_{ij} = G_{ij}(\rho, z).$$

Note: ρ is harmonic, and thus a coordinate function, i.e. $\nabla\rho \neq 0$, and Let z be its conjugate harmonic function.

The Einstein vacuum equations now reduce to a *harmonic map* ([10])

$$\varphi\colon \mathbb{R}^3 \setminus \Gamma \to SL(3,\mathbb{R})/SO(3) := X, \text{ with}$$

$$G_{ij}\, dx^i\, dx^j = -\frac{\rho^2}{f}\, dt^2 + \sum_{1\le i,j,\le 2}^{2} f_{ij}(d\phi^i + \omega^i\, dt)(d\phi^j + \omega^j\, dt),$$

where $f = \det(f_{ij})$, define a positive definite symmetric matrix with $\det = 1$:

$$\Phi = \frac{1}{f}\begin{pmatrix} 1 & -v_1 & -v_2 \\ -v_1 & f\,f_{11} + (v_1)^2 & f\,f_{12} + v_1 v_2 \\ -v_2 & f\,f_{21} + v_2 v_1 & f\,f_{22} + (v_2)^2 \end{pmatrix}$$

representing a point $\Psi \in SL(3,\mathbb{R})/SO(3)$ $(\Phi = \Psi^T\Psi)$. Here v_i is the Ernst potential for the Killing field $\frac{\partial}{\partial\phi^i}$. We recall here that the symmetric space X is the space of ellipsoids of unit volume in $\mathbb{R}^3$:

$$X = \{x \in GL(3,\mathbb{R}) \mid x^t = x,\ x > 0,\ \det x = 1\}$$

and $SL(3)$ acts on X by $g \cdot x := gxg^t$. The left-invariant Riemannian metric on X at $x_0 = \mathrm{Id}_3$ is

$$ds^2 = \langle Y, Z\rangle := \mathrm{tr}(Y \cdot Z^t)$$

for $Y, Z \in T_{x_0}X$ with $Y^t = Y$ and $\mathrm{tr}Y = 0$.

On the z-axis $\{\rho = \sqrt{|\det G|} = 0\}$, we have $\dim\ker(G(0,z)) = 1$ except at isolated values $\{p_i\}_{i=1}^N$ on the z-axis (corners). Let $(0,z)$ be a non-corner point, and let $0 \neq V \in \ker G$. There are two cases:

- V *null* $\Rightarrow$ the rod $p_i \le z \le p_{i+1}$ corresponds to *a horizon rod.*
- V *spacelike* $\Rightarrow$ the rod $p_j \le z \le p_{j+1}$ corresponds to *an axis rod.*

In the second case, one can scale V so that $V = n\frac{\partial}{\partial\phi^1} + m\frac{\partial}{\partial\phi^2}$ and $n, m \in \mathbb{Z}$ with $gcd(n,m) = 1$.

Definition. The *rod structure* consists of the set of axis rods $\{\Gamma_i\}_{i=1}^N$ together with the integers (n_i, m_i) for each axis rod, plus the values of Ernst potentials v_1, v_2 for each axis rod Γ_i.

The precise definition of v_1, v_2 is stated in eqn. (2.10) of [6]. They measure the deviations of the Killing vector fields $\partial/\partial\phi^i$ being hypersurface-orthogonal.

The rod structure, defined as above in [4] for the first time, encodes the topological types of the horizons, as the three manifolds on our lists have singular torus foliations whose singular fibers are S^1's, appearing at one of the axis rods.

We present our main theorems. Denote by Γ the set of axis rods $\{\Gamma_i\}$.

Theorem 1 (Khuri-Weinstein-Yamada [6, 7]). *Given any rod structure* $\Gamma = \cup\Gamma_i$, *and any set of axis rod constants* $v_1 = a_i$, $v_2 = b_i$ *for* Γ_i, *there exists* a unique harmonic map $\varphi\colon \mathbb{R}^3 \setminus \Gamma \to SL(3,\mathbb{R})/SO(3)$ *with singularities on* Γ *whose asymptotic behavior near* Γ *and spacetime infinity corresponds to the correct rod structure and to asymptotic flatness/asymptotic Kaluxa-Klein-ness/asymptitic locally-Euclidean-ness respectively.*

When reconstructing the spacetime, there may be *conical singularities* along the (bounded) axes. Also, the uniqueness of the stationary solution given a rod structure had been shown in [4].

Open problem. Can these be removed by adjusting the parameters? (Clearly such adjustments are possible in some cases; Myers-Perry, Emperan-Reall.) The angular momentum should have the effect of counteracting the gravitational attraction, as observed by Emperan-Reall blackring configuration [2].

The topological classification of the domain of outer communications in 5 dimension is given by the following statement utilizing the language of disc bundles over S^2, and their so-called plumbing construction. We should remark that there are previously known results such as [5] which complement the theorem below.

Theorem 2 (Khuri-Matsumoto-Weinstein-Yamada [8]). *Given the rod structure, the domain of the outer communication (DOC) of the spacetime (quotient-ed by the time-symmetry* $\mathbb{R}$*) is represented by a 4-dimensional manifold with boundary obtained by a plumbing construction of a series of disc bundles over* S^2 *minus the black hole region(s). In particular the DOC is diffeomrophic to one of the list*

$$S^4, \#(S^2 \times S^2), (\#n\mathbb{C}P^2)\#(\#\ell\overline{\mathbb{C}P^2})$$

minus the interior of the horizons and the end.

In [8], we give the explicit plumbing constructions of some of the known blackhole spacetimes.

References

[1] P. T. Chruściel, J. L. Costa and M. Heusler, Living Rev. Relativ., (2012) **15**: 7, https://doi.org/10.12942/lrr-2012-7.

[2] R. Emparan and H. Reall, *A rotating black ring in five dimensions*, Phys. Rev. Lett., **88** (2002) 101101, arXiv:hep-th/0110260.

[3] G. Galloway and R. Schoen, *A generalization of Hawking's black hole topology theorem to higher dimensions*, Commun. Math. Phys., **266** (2006), no. 2, 571–576.

[4] S. Hollands,and S. Yazadjiev, *A uniqueness theorem for stationary Kaluza-Klein black holes*, Commun. Math. Phys., **302** (2011), no. 3, 631–674, arXiv:0812.3036.

[5] S. Hollands, J. Holland and A. Ishibashi, Ann. Henri Poincaré, (2011) **12**: 279, https://doi.org/10.1007/s00023-011-0079-2.

[6] M. Khuri, G. Weinstein and S. Yamada, *Stationary vacuum black holes in 5 dimensions*, Comm. Part. Diff. Eqn., **43** (2018), 1205–1241, arXiv:1711.05229.

[7] M. Khuri, G. Weinstein and S. Yamada, *Asymptotically locally Euclidean/ Kaluza-Klein stationary vacuum black holes in 5 dimensions*, PTEP. Prog. Theor. Exp. Phys., (2018), no. 5, 053E01, 13pp, arXiv:1802.02457.

[8] M. Khuri, Y. Matsumoto, G. Weinstein and S. Yamada, *Plumbing Constructions and the Domain of Outer Communication for 5-Dimensional Stationary Black Holes*, Trans. Amer. Math. Soc., (2019), https://doi.org/10.1090/tran/7812.

[9] J. Lucietti and H. Kunduri, *A supersymmetric black lens*, Phys. Rev. Lett., **113** (2014), 211101. arXiv:1408.6083.

[10] D. Maison, *Ehlers-Harrison-type transformations for Jordan's extended theory of gravitation*, Gen. Relativity Gravitation, **10** (1979), no. 8, 717–723.

[11] R. Myers and M. Perry, *Black holes in higher dimensional space-times*, Ann. Physics, **172** (1986), no. 2, 304–347.

[12] A. Pomeransky and R. Sen'kov, *Black ring with two angular momenta*, preprint, arXiv:hep-th/0612005.

[13] S. Tomizawa and M. Nozawa, *Supersymmetric black lenses in five dimensions*, Phys. Rev. D, **94** (2016), 044037. arXiv:1606.06643.

[14] G. Weinstein, *Harmonic maps with prescribed singularities into Hadamard manifolds*, Mathematical Research Letters, **3** (1996), no. 6, 835–844.

Department of Mathematics, Gakushuin University,
1-5-1 Mejiro, Toshima, Tokyo, Japan 171-8588
E-mail address: yamada@math.gakushuin.ac.jp

Advanced Studies in Pure Mathematics 85, 2020
The Role of Metrics in the Theory of Partial Differential Equations
pp. 533–543

Global regularity for Einstein-Klein Gordon system with $U(1) \times \mathbb{R}$ isometry group

Haoyang Chen and Yi Zhou

Abstract.

This is an announcement of [3] and [4] which study the global regularity of the 3+1 dimensional Einstein-Klein Gordon system with a $U(1) \times \mathbb{R}$ isometry group. We reduce the Cauchy problem of the Einstein-Klein Gordon system to a 2+1 dimensional system. Next, the energy of this system cannot concentrate near the first possible singularity, and hence is small. Then, we show that the global regularity holds for the reduced 2+1 system with initial data of small energy.

§1. Introduction and previous results

Let $({}^{(4)}M, {}^{(4)}g)$ be a 3+1 dimensional globally hyperbolic Lorentzian manifold which satisfies the following Einstein-scalar field equations:

$$(1) \quad \begin{cases} {}^{(4)}G_{\mu\nu} = {}^{(4)}T_{\mu\nu} = \partial_\mu\phi\partial_\nu\phi - \frac{1}{2}{}^{(4)}g_{\mu\nu}\partial^\lambda\phi\partial_\lambda\phi - {}^{(4)}g_{\mu\nu}V(\phi) \\ \Box_{{}^{(4)}g}\phi = V'(\phi) \end{cases}$$

where ${}^{(4)}g$ is the Lorentzian metric, ${}^{(4)}G_{\mu\nu}$ is the Einstein tensor of ${}^{(4)}g$, ${}^{(4)}T_{\mu\nu}$ is the stress-energy tensor given as above, ϕ is the scalar field, and we take the potential $V(\phi) = \frac{1}{2}m^2\phi^2$.

Then, the equation that the scalar field satisfies is a Klein-Gordon equation, which makes the system an Einstein-Klein Gordon system. The Einstein-Klein Gordon system (1) is equivalent to the following equations,

$$(2) \quad \begin{cases} {}^{(4)}R_{\mu\nu} = {}^{(4)}\rho_{\mu\nu} \triangleq \partial_\mu\phi\partial_\nu\phi + {}^{(4)}g_{\mu\nu}\frac{1}{2}m^2\phi^2 \\ \Box_{{}^{(4)}g}\phi = m^2\phi \end{cases}$$

Received March 5, 2019.
Revised June 18, 2019.
2010 *Mathematics Subject Classification.* 35Q76; 35L70.
Key words and phrases. Einstein-Klein Gordon system, Cauchy problem, global regularity, non-concentration of energy.

where ${}^{(4)}R_{\mu\nu}$ is the Ricci curvature tensor.

One fundamental open problem in the field of general relativity is the cosmic censorship conjectures by Penrose. Roughly speaking, the weak cosmic censorship may be formulated as follows: For generic asymptotically flat Cauchy data, of the vacuum equations or suitable Einstein-matter systems, the maximal development possesses a complete future null infinity. While the strong cosmic censorship states that thc maximal Cauchy development is inextendible for generic initial data. This question is partially related to the study of the formulation of event horizons. Although still open in general, there is a series of results in spherical symmetric case by Christodoulou for the Einstein equations coupled with a massless scalar field. On the other hand, of our interest, it is related to the global well-posedness of the Cauchy problem in large.

We aim to show the global regularity of the above 3+1 Einstein-Klein Gordon systems with a $U(1) \times \mathbb{R}$ isometry group. Our study is motivated by research on the vacuum Einstein equations with spacelike Killing vector fields which is related to the study of wave map systems. We first review some related results on the vacuum case and the wave map systems.

For the 3+1 dimensional vacuum Einstein equations with one spacelike Killing field, as we can see in [7], the Einstein equations can be reduced to a 2+1 dimensional Einstein-wave map system on a 2+1 dimensional Lorentzian manifold where the target manifold is the hyperbolic space $\mathbb{H}^2$. Yet the global existence problem of the Einstein-wave map system is still open. As a first step towards this global existence conjecture, Andersson, Gudapati and Szeftel proved that the global regularity holds for the equivariant case in [1], where, of our concern, the 2+1 dimensional spacetime admits $U(1)$ symmetry. This results from some pioneering work on equivariant wave maps.

Shatah and Tahvildar-Zadeh have proved the global regularity for 2+1 equivariant wave maps with the target geodesically convex in [15]. This condition was later laxed by Grillakis to include a certain class of nonconvex targets, see [10]. Their proof of regularity was also simplified later by Shatah and Struwe in [14]. They gave several more results on equivariant wave maps in the areas of existence and uniqueness, regularity, asymptotic behavior, development of singularities, and weak solutions, see [16]. Further, as an improvement of these above results, for target manifolds that do not admit nonconstant harmonic spheres, global existence of smooth solutions to the Cauchy problem for corotational wave maps with smooth equivariant data was shown by Struwe in [17].

Then, for the 3+1 vacuum Einstein equations with G_2 symmetry, the symmetry we consider, it was shown in [2] that the system reduce to a spherically symmetric wave map $u : \mathbb{R}^{2+1} \to \mathbb{H}^2$, where $\mathbb{R}^{1+2}$ is the 2+1 dimensional Minkowski spacetimes and the target $\mathbb{H}^2$ is the hyperbolic space. Thus the global regularity can be proved by the work of Christodoulou and Tahvildar-Zadeh [8] on 2+1 dimensional spherically symmetric wave maps. In [8], the range of the wave map u should be contained in a convex part of the target N. This restriction was later shown unnecessary by Struwe in [19] as the target is the standard sphere. Further, Struwe give a more general result in [18], where the target is any smooth, compact Riemannian manifold without boundary. We refer to [9] for more results and references of wave maps.

§2. The 3+1 spacetime with $U(1) \times \mathbb{R}$ isometry group

In our case, the Lorentzian manifold we study is of the form ${}^{(4)}M = \mathbb{R} \times \mathbb{R}^2 \times \mathbb{R}$ with a Lorentzian metric ${}^{(4)}g$ on it, and we consider the case where the two spacelike Killing fields of $({}^{(4)}M, {}^{(4)}g)$ are hypersurface orthogonal. Then, with the existence of the translational Killing vector, the metric can be written in the following form,

$$ {}^{(4)}g = e^{-2\gamma}\, {}^{(3)}g + e^{2\gamma}(dx^3)^2 $$

where ∂_{x^3} is the translational spacelike Killing vector field.

As we mentioned before in [7], the 3+1 vacuum Einstein equations with spacelike Killing field reduce to a 2+1 Einstein-wave map system with the target manifold $\mathbb{H}^2$. We will give a similar reduction to a 2+1 dimensional Einstein-wave-Klein Gordon system in Section 3, where γ in ${}^{(4)}g$ satisfies a wave equation and the scalar field ϕ satisfies a Klein Gordon equation.

In the vacuum case with G_2 symmetry, the wave maps equations reduced from the Einstein equations is a semilinear wave equations, for instance, special solutions of this case are the Einstein-Rosen waves, see [13] and references therein. While one major difficulty in our problem is that the wave equations are coupled with Einstein equations, which make the system quasilinear. However, we develop a new way to solve this problem in 2+1 dimension with symmetry.

Now we assume that the reduced spacetime $({}^{(3)}M, {}^{(3)}g)$ is a globally hyperbolic 2+1 dimensional spacetime with Cauchy surface diffeomorphic to $\mathbb{R}^2$, on which the reduced Einstein-wave-Klein Gordon system is radially symmetric. And we assume that the $U(1)$ action on M is generated by a hypersurface orthogonal Killing field ∂_θ. In particular, we write the metric ${}^{(3)}g$ in the following form in this paper

$$
\begin{aligned}
{}^{(3)}g &= -e^{2\alpha(t,r)}\,dt^2 + e^{2\beta(t,r)}\,dr^2 + r^2\,d\theta^2 \\
&\triangleq \breve{g} + r^2\,d\theta^2,
\end{aligned} \tag{3}
$$

where $\breve{g}$ is a metric on the orbit space $\mathcal{Q} = M/\mathbb{S}^1$ and r is the radius function, defined such that $2\pi r(p)$ is the length of the $\mathbb{S}^1$ orbit through p.

§3. Reduction to the 2+1 Einstein-wave-Klein Gordon system

Consider the Einstein equations (2), where the spacetime $({}^{(4)}M, {}^{(4)}g)$ admits a spacelike translational Killing vector field. The Einstein-Klein Gordon system can be reduced to a 2+1 Einstein-wave-Klein Gordon system in a similar way as we know in [5].

With the metric ${}^{(3)}g$ taking the form (3), we write in [3] the 2+1 radially symmetric Einstein-wave-Klein Gordon system in local coordinates,

$$
\beta_r = re^{2\beta-2\alpha}{\gamma_t}^2 + r{\gamma_r}^2 + \frac{r}{2}e^{2\beta-2\alpha}{\phi_t}^2 + \frac{r}{2}{\phi_r}^2 + \frac{m^2}{2}re^{2\beta-2\gamma}\phi^2 \tag{4}
$$

$$
\beta_t = 2r\gamma_t\gamma_r + r\phi_t\phi_r \tag{5}
$$

$$
\alpha_r = re^{2\beta-2\alpha}{\gamma_t}^2 + r{\gamma_r}^2 + \frac{r}{2}e^{2\beta-2\alpha}{\phi_t}^2 + \frac{r}{2}{\phi_r}^2 - \frac{m^2}{2}re^{2\beta-2\gamma}\phi^2 \tag{6}
$$

$$
\begin{aligned}
&e^{-2\beta}\alpha_{rr} - e^{-2\alpha}\beta_{tt} + e^{-2\beta}\alpha_r(\alpha_r - \beta_r) + e^{-2\alpha}\beta_t(\alpha_t - \beta_t) \\
&= -\frac{m^2}{2}e^{-2\gamma}\phi^2 + e^{-2\alpha}\left({\gamma_t}^2 + \frac{1}{2}{\phi_t}^2\right) - e^{-2\beta}\left({\gamma_r}^2 + \frac{1}{2}{\phi_r}^2\right)
\end{aligned} \tag{7}
$$

$$
\begin{aligned}
\Box_{{}^{(3)}g}\gamma &= -e^{-2\alpha}(\gamma_{tt} + (\beta_t - \alpha_t)\gamma_t) \\
&\quad + e^{-2\beta}\left(\gamma_{rr} + \frac{\gamma_r}{r} + (\alpha_r - \beta_r)\gamma_r\right) \\
&= -\frac{m^2}{2}e^{-2\gamma}\phi^2
\end{aligned} \tag{8}
$$

$$
\begin{aligned}
\Box_{{}^{(3)}g}\phi &= -e^{-2\alpha}(\phi_{tt} + (\beta_t - \alpha_t)\phi_t) \\
&\quad + e^{-2\beta}\left(\phi_{rr} + \frac{\phi_r}{r} + (\alpha_r - \beta_r)\phi_r\right) \\
&= m^2e^{-2\gamma}\phi.
\end{aligned} \tag{9}
$$

§4. The Cauchy data

As we mentioned before, to study the Cauchy problem of the 3+1 dimensional Einstein-Klein Gordon system, we can equivalently consider the Cauchy problem of the reduced 2+1 Einstein-wave-Klein Gordon system.

Now we introduce the definition of the Cauchy data set for the 2+1 Einstein-wave-Klein Gordon systems with $U(1)$ isometry group as follows,

Definition 4.1 (Cauchy data set for the 2+1 Einstein-wave-Klein Gordon system with $U(1)$ isometry group). *A Cauchy data set for the 2+1 dimensional Einstein-wave-Klein Gordon system with a $U(1)$ isometry group is a 7-tuple $(M_0, g_0, K, \gamma_0, \gamma_1, \phi_0, \phi_1)$ consisting of a Remannian 2-manifold (M_0, g_0) with a spacelike rotational Killing vector field and a 2-tensor K which is the second fundamental form and symmetric under the same action, γ_0, γ_1 are initial data for the wave equation that γ satisfies, ϕ_0, ϕ_1 are initial data for the Klein Gordon equation that the scalar field satisfies. g_0, K are functions of r only and the following constraints equations hold:*

$$R_0 - K_{\alpha\beta}K^{\alpha\beta} + (trK)^2 = 2^{(3)}T_{\alpha\beta}n^\alpha n^\beta \tag{10}$$

$$D^\beta K_{\alpha\beta} - D_\alpha K^\beta_\beta = {}^{(3)}T_{\alpha\mu}n^\mu \tag{11}$$

where $n^\alpha(t,r)$ is the future directed unit normal, R_0 is the scalar curvature on M_0, D_α is the intrinsic covariant derivative on M_0, and ${}^{(3)}T_{\alpha\beta}$ is the stress-energy tensor for the reduced system.

For a smooth solution of the 2+1 dimensional Einstein-wave-Klein Gordon system with $U(1)$ symmetry, it must hold that $\alpha(t,r), \beta(t,r)$ are even functions of r. And we give some normalisation of the metric functions $\alpha(t,r)$ and $\beta(t,r)$ on the axis. It must hold that $\beta(t,0) = 0$, in order to avoid a conical singularity at the axis Γ, which means that the perimeter of a circle of radius r grows like $2\pi cr$ at the axis, instead of $2\pi r$ in the Euclidean metric. This condition can be realized by appropriately choosing the Cauchy data such that $\beta(0,0) = 0$, then, by (5), we have $\beta(t,0) = 0$ on Γ. Further, $\alpha(t,0)$ is determined only up to a choice of time parametrization. We shall choose a time coordinate such that $\alpha(t,0) = 0$.

Finding solutions to the constraint equations is a research area in itself. Note that Cecile has proved the existence of such constraint equations in vacuum with translational Killing vector field in [11], [12]. In our case, we will just briefly show that such data exists, without going further into the study of the constraints.

We will construct an asymptotically flat[1] Cauchy data set which satisfies the following initial conditions

$$\begin{cases} \gamma_t(0,r) = \gamma_1(r) \geq 0, \\ \gamma(0,r) = \gamma_0(r) \geq 0, \quad \gamma_r(0,r) = \gamma_0'(r) > -\dfrac{1}{2}r^{-1}. \end{cases} \tag{12}$$

Remark 1. *The following initial condition in* (12)

$$\gamma_r(0,r) = \gamma_0'(r) > -\frac{1}{2}r^{-1}$$

can be replaced by

$$\gamma_r(0,r) = \gamma_0'(r) > -Cr^{-1}$$

with C an arbitrary positive constant. Here, we take $C = \frac{1}{2}$ just for simplicity.

For metrics of the form (3), it can be calculated that the second fundamental form K of a Cauchy t-level takes the following form

$$K = K_{rr}dr^2,$$

where $K_{rr} = e^{-\alpha+2\beta}\beta_t$.

For the reduced 2+1 dimensional Einstein-wave-Klein Gordon system (see Section 2), the constraint equations (10), (11) take the following form in local coordinates,

$$\begin{cases} 2e^{-2\beta}r^{-1}\beta_r \\ \quad = 2\left(e^{-2\alpha}{\gamma_t}^2 + e^{-2\beta}{\gamma_r}^2 + \dfrac{1}{2}e^{-2\alpha}{\phi_t}^2 + \dfrac{1}{2}e^{-2\beta}{\phi_r}^2 + \dfrac{m^2}{2}e^{-2\gamma}\phi^2\right), \\ r^{-1}e^{-\alpha}\beta_t = 2e^{-\alpha}\gamma_t\gamma_r + e^{-\alpha}\phi_t\phi_r. \end{cases}$$

If we take $\beta_t = 0$ and $\gamma_t = 0$, the constraint equations become

$$\begin{cases} 2e^{-2\beta}r^{-1}\beta_r = 2\left(e^{-2\beta}{\gamma_r}^2 + \dfrac{1}{2}e^{-2\alpha}{\phi_t}^2 + \dfrac{1}{2}e^{-2\beta}{\phi_r}^2 + \dfrac{m^2}{2}e^{-2\gamma}\phi^2\right), \\ e^{-\alpha}\phi_t\phi_r = 0. \end{cases}$$

Remark 2. *Noting that $\beta_t = 0$ means that M_0 is a totally geodesic submanifold. Moreover, what is more interesting, a direct calculation shows that the additional condition $\gamma_t = 0$ makes the corresponding Cauchy hypersurface of the 3+1 Einstein-Klein Gordon system a totally geodesic submanifold.*

[1]We say the Cauchy data is 'asymptotically flat' in the sense of Andersson [1] here and hereafter.

Noting that all quantities above are functions of r only. Thus, we can properly choose the initial data sets for γ and ϕ which is compatible with (12) such that the second equation holds. Here, we set $\phi_t = 0$. What left to be done is solving the first equation, which is just an ordinary differential equation. Noting that it is equivalent to the following equation

$$-\partial_r(e^{-2\beta}) = 2r\left(e^{-2\beta}{\gamma_r}^2 + \frac{1}{2}e^{-2\beta}{\phi_r}^2 + \frac{m^2}{2}e^{-2\gamma}\phi^2\right)$$

which is a first order linear differential equation of $e^{-2\beta}$, and with the condition $\beta(0,0) = 0$, the solution takes the form

$$\beta(r) = \frac{1}{2}e^{\int_0^r(2\xi{\gamma_r}^2+\xi{\phi_r}^2)d\xi} - \frac{1}{2}\ln\left(1 - \int_0^r \xi m^2 e^{-2\gamma}\phi^2 e^{\int_0^\xi(2\eta{\gamma_r}^2+\eta{\phi_r}^2)d\eta}\,d\xi\right).$$

Thus, properly choose the initial data sets for γ and ϕ such that

$$\int_0^r \xi m^2 e^{-2\gamma}\phi^2 e^{\int_0^\xi(2\eta{\gamma_r}^2+\eta{\phi_r}^2)d\eta}d\xi < 1, \tag{13}$$

and

$$\phi = \mathcal{O}(r^{-\frac{11}{8}}), \quad \gamma = \mathcal{O}(r^{-\frac{11}{8}}) \tag{14}$$

as $r \to \infty$. The first condition (13) guarantees that the initial energy is finite. The second condition (14) on asymptotic behavior guarantees the asymptotic flatness, which is to say that

$$\beta = \beta_\infty + \tilde{\beta}.$$

Here, β_∞ is a constant and $(\tilde{\beta}, K) \in H_\delta^{s+1} \times H_{\delta+1}^s$ with H_δ^{s+1} and $H_{\delta+1}^s$ the weighted Sobolev space (see [1], [11]) where we take $\delta = -\frac{1}{2}$ and $s > 1$. Thus, we finally give a non-trivial solution of the constraint equations.

Remark 3. *If we eliminate the condition $\gamma_t = 0$, it is also easy to give a construction in a similar way by properly choosing γ_t and ϕ_t which have the same asymptotic behavior as above.*

§5. The problem of global well-posedness

The proof by Choquet-Bruhat and Geroch (see [5], [6]) of existence and uniqueness of maximal solutions to the Cauchy problem for the vacuum Einstein equations, together with the equivalence of the Cauchy data, can be generalized to our case as follows, which guaranteed the local well-posedness.

Theorem 4. *Let $(M_0, g_0, k, \gamma_0, \gamma_1, \phi_0, \phi_1)$ be the Cauchy data set for the 2+1 Einstein-wave-Klein Gordon system with $U(1)$ isometry group. Then there is a unique, maximal Cauchy development $({}^{(3)}M, {}^{(3)}g, \gamma, \phi)$ satisfying the 2+1 Einstein-wave-Klein Gordon system.*

Our goal is to prove the following theorem about the global well-posedness,

Theorem 5. *Let $({}^{(3)}M, {}^{(3)}g)$ be the maximal Cauchy development of a regular Cauchy data set aforementioned in Section 1.3 for the 2+1 dimensional Einstein-wave-Klein-Gordon system. Then there is a global in time smooth solution for the Cauchy problem of the equations.*

Remark 6. *In our present work [3] and [4], we construct solutions where $({}^{(4)}M, {}^{(4)}g)$ satisfies the Einstein-scalar field equations equivalently.*

For our further consideration, it is worthwhile to study the asymptotic behaviour of the solution obtained in Theorem 5, which is related with the future causal geodesic completeness of the 3+1 dimensional spacetime $({}^{(4)}M, {}^{(4)}g)$.

The outline of the proof for Theorem 5 will be given in the following part.

§6. Outline of proof for Theorem 5

Firstly, using vector field methods as in [1], we can get energy estimates of the system. This is because the 2+1 dimensional spacetime we study shares same symmetry with what is considered in [1], while the coupled matter fields are quite different. Then, we construct a global null coordinate system with respect to which the 3-dimensional manifold (M, g) admits a coordinate system (u, v, θ) such that g takes the form

$$g = -e^{2\lambda}(u, v)dudv + r^2(u, v)d\theta^2$$

where now $d\theta^2$ is the line element on the $\mathbb{S}^1$ symmetry orbit. Then we can define

$$T = \frac{u+v}{2}, \quad R = \frac{v-u}{2}.$$

We study the Einstein-wave-Klein Gordon system in these coordinates from now on.

In null coordinate system, we show in [3] that γ has a lower bound under the given initial conditions (12) by a bootstrap argument, i.e.

$$\gamma \geq -1.$$

Next, the first possible singularity must occur on the axis. This is equivalent to show that the Cauchy problem admits a regular solution away from the axis, see [3].

Proposition 7. *For any $(\bar{u}, \bar{v})$ away from the axis, the solution of the Einstein-wave-Klein Gordon system is regular at $(\bar{u}, \bar{v})$.*

Then, we show in [4] the non-concentration of the energy inside the causal past $J^-(O)$ of O,

$$E^O(t) := \int_{\Sigma_t \cap J^-(O)} \mathbf{e}\bar{\mu}_q$$

with O the first possible singularity on the axis.

Theorem 8. *Let (M, g) be the maximal Cauchy development of the Cauchy data set in Theorem 5. Then, the energy of the Einstein-wave-Klein Gordon system (4)–(9) cannot concentrate, i.e. $E^O(t) \to 0$ as $t \to 0$, where O is the first possible singularity.*

What left to be done is to give a proof of small energy implying global regularity, which is the technically difficult part in our work.

Theorem 9. *There exists an $\varepsilon > 0$, such that for any regular Cauchy data of energy $E_0 < \varepsilon$, the Cauchy problem for the Einstein-wave-Klein Gordon system admits an unique globally smooth solution.*

Remark 10. *The Cauchy data set we mentioned in the above theorem is a surface close enough to the first possible singularity.*

This can be shown by setting up a bootstrap argument as Christodoulou did in [8]. However, as we mentioned before, one major difficulty come from the fact that the wave equations are coupled with Einstein equations which make our system quasilinear, while the wave map system in [8] is semilinear. Moreover, different from the 2+1 dimensional equivariant Einstein-wave map system that Andersson studied in [1], there is not such a condition that $U = (\gamma, \phi)$ vanishes at the axis Γ, which brings difficulties in studying the regularity near the first possible singularity on the axis.

We give a uniform estimate on the full gradient of our system, which means that the solution can be extended beyond the first possible singularity. This is performed in [4] as follows. Let

$$X \triangleq \sup_{(T,R) \in K_0} r^\delta |U_v|$$

where $\frac{1}{2} < \delta < \frac{2}{3}$ and K_0 a small region sufficiently close to the first possible singularity. We show that X is uniformly bounded firstly. Then, in a similar but rather technically complicated way, we get higher regularity by dealing with the equations for $V = \partial_R U$, where we establish a Morawetz type energy estimate. Estimates for r and λ can be obtained during the above procedure. It follows that the full gradients of all unknowns are bounded. Then the higher regularity of the unknowns following using energy estimates. This implies the global existence of smooth solutions.

References

[1] L. Andersson, N. Gudapati and J. Szeftel, Global regularity for the 2+1 dimensional equivariant Einstein-wave map system, *Ann. PDE*, **3** (2017), 142 pp.

[2] B. K. Berger, P. T. Chruściel and V. Moncrief, On the "Asymptotically Flat" Space-Times with G_2-invariant Cauchy Surfaces, *Ann. of Phys.*, **237** (1995), 322–354.

[3] H. Chen and Y. Zhou, Global regularity for Einstein-Klein-Gordon system with $U(1) \times \mathbb{R}$ isometry group, I, *preprint*, (2019).

[4] H. Chen and Y. Zhou, Global regularity for Einstein-Klein-Gordon system with $U(1) \times \mathbb{R}$ isometry group, II, *preprint*, (2019).

[5] Y. Choquet-Bruhat, General Relativity and the Einstein Equations, *Oxford Science Press*, (2009).

[6] Y. Choquet-Bruhat and R. Geroch, Global aspects of the Cauchy problem in general relativity, *Comm. Math. Phys.*, **14** (1969), 329–335.

[7] Y. Choquet-Bruhat and V. Moncrief, Future Global in Time Einsteinian Spacetimes with U(1) Isometry Group, *Ann. Henri Poincaré*, **2** (2001), 1007–1064.

[8] D. Christodoulu and A. S. Tahvildar-Zadeh, On the regularity of spherically symmetric wave maps, *Comm. Pure Appl. Math.*, **46**, (1993), 1041–1091.

[9] D. Geba and M. Grillakis, An Introduction to the theory of wave maps and related geometric problems, *New Jersey: World Scientific*, (2017).

[10] M. Grillakis, Classical slutions for the equivariant wave maps in 1+2 dimensions, *preprint*, (1991).

[11] C. Huneau, Constraint equations for 3+1 vacuum Einstein equations with a translational space-like Killing field in the asymptotically flat case, *Ann. Henri Poincaré*, **17(2)** (2016), 271–299.

[12] C. Huneau, Constraint equations for 3+1 vacuum Einstein equations with a translational space-like Killing field in the asymptotically flat case II, *Asymptot. Anal.*, **96(1)** (2015), 51–89.

[13] C. Huneau, Stability in exponential time of Minkowski space-time with a translation space-like Killing field, *Ann. PDE*, **2(1)** (2016), 151 pp.

[14] J. Shatah and M. Struwe, Geometric Wave Equations, Volume 2 of Courant Lecture Notes in Mathematics.

[15] J. Shatah and A. S. Tahvildar-Zadeh, Regularity of harmonic maps from the Minkowski space into rotationally symmetric manifolds, *Comm. Pure Appl. Math.*, **45(8)** (1992), 947–971.
[16] J. Shatah and A. S. Tahvildar-Zadeh, On the Cauchy problem for equivariant wave maps, *Comm. Pure Appl. Math.*, **47(5)** (1994), 719–754.
[17] M. Struwe, Equivariant wave maps in two space dimensions, *Comm. Pure Appl. Math.*, **56** (2003), 0815–0823.
[18] M. Struwe, Radially symmetric wave maps form (1+2)-dimensional Minkowski space to general targets, *Calc. Var.*, **16**, (2003), 431–437.
[19] M. Struwe, Radially symmetric wave maps from (1+2)-dimensional Minkowski space to the sphere, *Math. Z.*, **242**, (2002), 407–414.

School of Mathematical Sciences, Fudan University,
Shanghai, China.
E-mail address: 15110180018@fudan.edu.cn
E-mail address: yizhou@fudan.edu.cn

Advanced Studies in Pure Mathematics

A SERIES OF UP-TO-DATE GUIDES OF LASTING INTEREST TO ADVANCED MATHEMATICS

Volume 1 Algebraic Varieties and Analytic Varieties.
Edited by S. Iitaka. February, 1983

Volume 2 Galois Groups and their Representations.
Edited by Y. Ihara. December, 1983

Volume 3 Geometry of Geodesics and Related Topics.
Edited by K. Shiohama. June, 1984

Volume 4 Group Representations and Systems of Differential Equations.
Edited by K. Okamoto. March, 1985

Volume 5 Foliations.
Edited by I. Tamura. February, 1986

Volume 6 Algebraic Groups and Related Topics.
Edited by R. Hotta. March, 1985

Volume 7 Automorphic Forms and Number Theory.
Edited by I. Satake. February, 1986

Volume 8 Complex Analytic Singularities.
Edited by T. Suwa and P. Wagreich. February, 1987

Volume 9 Homotopy Theory and Related Topics.
Edited by H. Toda. February, 1987

Volume 10 Algebraic Geometry, Sendai, 1985.
Edited by T. Oda. July, 1987

Volume 11 Commutative Algebra and Combinatorics.
Edited by M. Nagata and H. Matsumura.
October, 1987

Volume 12 Galois Representations and Arithmetic Algebraic Geometry.
Edited by Y. Ihara. November, 1987

Volume 13 Investigations in Number Theory.
Edited by T. Kubota. March, 1988

Volume 14 Representations of Lie Groups, Kyoto, Hiroshima, 1986.
Edited by K. Okamoto and T. Oshima. February, 1989

Volume 15 Automorphic Forms and Geometry of Arithmetic Varieties.
Edited by K. Hashimoto and Y. Namikawa. July, 1989

Volume 16 Conformal Field Theory and Solvable Lattice Models.
Edited by M. Jimbo, T. Miwa and A. Tsuchiya.
June, 1988

Volume 17 Algebraic Number Theory—in honor of K. Iwasawa.
Edited by J. Coates, R. Greenberg, B. Mazur and I. Satake. August, 1989

Volume 18 I—Recent Topics in Differential and Analytic Geometry.
Edited by T. Ochiai. November, 1990
II—Kähler Metric and Moduli Spaces.
Edited by Edited by T. Ochiai. November, 1990

Volume 19 Integrable Systems in Quantum Field Theory and Statistical Mechanics.
Edited by M. Jimbo, T. Miwa and A. Tsuchiya.
November, 1989

Volume 20 Aspects of Low Dimensional Manifolds.
Edited by Y. Matsumoto and S. Morita.
December, 1992

Volume 21 Zeta Functions in Geometry.
Edited by N. Kurokawa and T. Sunada.
December, 1992

Volume 22 Progress in Differential Geometry.
Edited by K. Shiohama. September, 1993

Volume 23 Spectral and Scattering Theory and Applications.
Edited by K. Yajima. February, 1994

Volume 24 Progress in Algebraic Combinatorics.
Edited by E. Bannai and A. Munemasa. May, 1996

Volume 25 CR-Geometry and Overdetermined Systems.
Edited by T. Akahori, G. Komatsu, K. Miyajima, M. Namba and K. Yamaguchi. July, 1997

Volume 26 Analysis on Homogeneous Spaces and Representation Theory of Lie Groups, Okayama-Kyoto.
Edited by M. Kashiwara, T. Kobayashi, T. Matsuki, K. Nishiyama and T. Oshima. April, 2000

Volume 27 Arrangements – Tokyo 1998.
Edited by M. Falk and H. Terao. August, 2000

Volume 28 Combinatorial Methods in Representation Theory.
Edited by M. Kashiwara, K. Koike, S. Okada, I. Terada, and H.-F. Yamada. December, 2000

Volume 29 Singularities – Sapporo 1998.
Edited by J.-P. Brasselet and T. Suwa.
December, 2000

Volume 30 Class Field Theory – Its Centenary and Prospect.
Edited by K. Miyake. April, 2001

Volume 31 Taniguchi Conference on Mathematics Nara '98.
Edited by M. Maruyama and T. Sunada. May, 2001

Volume 32 Groups and Combinatorics — in memory of Michio Suzuki.
Edited by E. Bannai, H. Suzuki, H. Yamaki and T. Yoshida. October, 2001

Volume 33 Computational Commutative Algebra and Combinatorics.
Edited by T. Hibi. February, 2002

Volume 34 Minimal Surfaces, Geometric Analysis and Symplectic Geometry.
Edited by K. Fukaya, S. Nishikawa and J. Spruck.
June, 2002

Volume 35 Higher Dimensional Birational Geometry.
Edited by S. Mori and Y. Miyaoka. July, 2002

Volume 36 Algebraic Geometry 2000, Azumino.
Edited by S. Usui, M. Green, L. Illusie, K. Kato, E. Looijenga, S. Mukai and S. Saito. November, 2002

Volume 37 Lie Groups, Geometric Structures and Differential Equations — One Hundred Years after Sophus Lie —.
Edited by T. Morimoto, H. Sato and K. Yamaguchi.
December, 2002

Volume 38 Operator Algebras and Applications.
Edited by H. Kosaki. January, 2004

Volume 39 Stochastic Analysis on Large Scale Interacting Systems.
Edited by T. Funaki and H. Osada. March, 2004

Volume 40 Representation Theory of Algebraic Groups and Quantum Groups.
Edited by T. Shoji, M. Kashiwara, N. Kawanaka, G. Lusztig and K. Shinoda. March, 2004

Volume 41 Stochastic Analysis and Related Topics in Kyoto
In honour of Kiyosi Itô.
Edited by H. Kunita, S. Watanabe and Y. Takahashi.
March, 2004

Volume 42 Complex Analysis in Several Variables
– Memorial Conference of Kiyoshi Oka's Centennial Birthday, Kyoto/Nara 2001.
Edited by K. Miyajima, M. Furushima, H. Kazama, A. Kodama, J. Noguchi, T. Ohsawa, H. Tsuji and T. Ueda. June, 2004

Volume 43 Singularity Theory and Its Applications.
Edited by S. Izumiya, G. Ishikawa, H. Tokunaga, I. Shimada and T. Sano. January, 2007

Volume 44 Potential Theory in Matsue.
Edited by H. Aikawa, T. Kumagai, Y. Mizuta and N. Suzuki. August, 2006

Volume 45 Moduli Spaces and Arithmetic Geometry (Kyoto, 2004).
Edited by S. Mukai, Y. Miyaoka, S. Mori, A. Moriwaki and I. Nakamura. January, 2007

Volume 46 Singularities in Geometry and Topology 2004.
Edited by J.-P. Brassele and T. Suwa. February, 2007

Volume 47 Asymptotic Analysis and Singularities.
1—Hyperbolic and dispersive PDEs and fluid mechanics.
2—Elliptic and parabolic PDEs and related problems.
Edited by H. Kozono, T. Ogawa, K. Tanaka, Y. Tsutsumi and E. Yanagida. October, 2007

Volume 48 Finsler Geometry, Sapporo 2005
— In Memory of Makoto Matsumoto.
Edited by S. V. Sabau and H. Shimada.
November, 2007

Volume 49 Probability and Number Theory — Kanazawa 2005.
Edited by S. Akiyama, K. Matsumoto, L. Murata and H. Sugita. December, 2007

Volume 50 Algebraic Geometry in East Asia — Hanoi 2005.
Edited by K. Konno and V. Nguyen-Khac.
February, 2008

Volume 51 Surveys on Geometry and Integrable Systems.
Edited by M. Guest, R. Miyaoka and Y. Ohnita.
November, 2008

Volume 52 Groups of Diffeomorphisms
in honor of Shigeyuki Morita
on the occasion of his 60th birthday.
Edited by R. Penner, D. Kotschick, T. Tsuboi, N. Kawazumi, T. Kitano and Y. Mitsumatsu.
November, 2008

Volume 53 Advances in Discrete Dynamical Systems.
Edited by S. Elaydi, K. Nishimura, M. Shishikura and N. Tose. June, 2009

Volume 54 Algebraic Analysis and Around
in honor of Professor Masaki Kashiwara's 60th birthday.
Edited by T. Miwa, A. Matsuo, T. Nakashima and Y. Saito. February, 2009

Volume 55 Noncommutativity and Singularities
Proceedings of French–Japanese symposia held at IHÉS in 2006.
Edited by J.-P. Bourguignon, M. Kotani, Y. Maeda and N. Tose. July, 2009

Volume 56 Singularities — Niigata–Toyama 2007.
Edited by J.-P. Brasselet, S. Ishii, T. Suwa and M. Vaquie. November, 2009

Volume 57 Probabilistic Approach to Geometry.
Edited by M. Kotani, M. Hino and T. Kumagai.
March, 2010

Volume 58 Algebraic and Arithmetic Structures of Moduli Spaces (Sapporo 2007).
Edited by I. Nakamura and L. Weng. June, 2010

Volume 59 New Developments in Algebraic Geometry, Integrable Systems and Mirror Symmetry (RIMS, Kyoto, 2008).
Edited by M.-H. Saito, S. Hosono and K. Yoshioka.
August, 2010

Volume 60 Algebraic Geometry in East Asia — Seoul 2008.
Edited by J. Keum, S. Kondō, K. Konno and K. Oguiso.
September, 2010

Volume 61 Exploring New Structures and Natural Constructions in Mathematical Physics.
Edited by K. Hasegawa, T. Hayashi, S. Hosono and Y. Yamada. October, 2011

Volume 62 Arrangements of Hyperplanes—Sapporo 2009.
Edited by H. Terao and S. Yuzvinsky. March, 2012

Volume 63 Galois–Teichmüller Theory and Arithmetic Geometry.
Edited by H. Nakamura, F. Pop, L. Schneps and A. Tamagawa. October, 2012

Volume 64 Nonlinear Dynamics in Partial Differential Equations.
Edited by S.-I. Ei, S. Kawashima, M. Kimura and T. Mizumachi. April, 2015

Volume 65 Algebraic Geometry in East Asia — Taipei 2011.
Edited by J. A. Chen, M. Chen, Y. Kawamata and J. Keum. April, 2015

Volume 66 Singularities in Geometry and Topology 2011.
Edited by V. Blanlœil and O. Saeki. May, 2015

Volume 67 Variational Methods for Evolving Objects.
Edited by L. Ambrosio, Y. Giga, P. Rybka and Y. Tonegawa. June, 2015

Volume 68 School on Real and Complex Singularities in São Carlos, 2012.
Edited by O. Saeki, R. N. Araújo dos Santos, T. Nishimura and V. H. Jorge Pérez. July, 2016

Volume 69 Development of Moduli Theory—Kyoto 2013.
Edited by O. Fujino, S. Kondō, A. Moriwaki, M.-H. Saito and K. Yoshioka. July, 2016

Volume 70 Minimal Models and Extremal Rays (Kyoto, 2011).
Edited by J. Kollár, O. Fujino, S. Mukai and N. Nakayama. September, 2016

Volume 71 Schubert Calculus — Osaka 2012.
Edited by H. Naruse, T. Ikeda, M. Masuda and T. Tanisaki. December, 2016

Volume 72 Geometry, Dynamics, and Foliations 2013
In honor of Steven Hurder and Takashi Tsuboi on the occasion of their 60th birthdays.
Edited by T. Asuke, S. Matsumoto and Y. Mitsumatsu. August, 2017

Volume 73 Hyperbolic Geometry and Geometric Group Theory.
Edited by K. Fujiwara, S. Kojima and K. Ohshika. September, 2017

Volume 74 Higher Dimensional Algebraic Geometry
in honour of Professor Yujiro Kawamata's sixtieth birthday.
Edited by K. Oguiso, C. Birkar, S. Ishii and S. Takayama. November, 2017

Volume 75 Algebraic Varieties and Automorphism Groups.
Edited by K. Masuda, T. Kishimoto, H. Kojima, M. Miyanishi and M. Zaidenberg. December, 2017

Volume 76 Representation Theory, Special Functions and Painlevé Equations — RIMS 2015.
Edited by H. Konno, H. Sakai, J. Shiraishi, T. Suzuki and Y. Yamada. June, 2018

Volume 77 The 50th Anniversary of Gröbner Bases.
Edited by T. Hibi. August, 2018

Volume 78 Singularities in Generic Geometry.
Edited by S. Izumiya, G. Ishikawa, M. Yamamoto, K. Saji, T. Yamamoto and M. Takahashi. September, 2018

Volume 79 Mathematics of Takebe Katahiro and History of Mathematics in East Asia.
Edited by T. Ogawa and M. Morimoto.
December, 2018

Volume 80 Operator Algebras and Mathematical Physics.
Edited by M. Izumi, Y. Kawahigashi, M. Kotani, H. Matui and N. Ozawa. May, 2019

Volume 81 Asymptotic Analysis for Nonlinear Dispersive and Wave Equations.
Edited by K. Kato, T. Ogawa and T. Ozawa.
October, 2019

Volume 82 Differential Geometry and Tanaka Theory — Differential Systems and Hypersurface Theory—.
Edited by T. Shoda and K. Shibuya. November, 2019

Volume 83 Primitive Forms and Related Subjects — Kavli IPMU 2014.
Edited by K. Hori, C. Li, S. Li and K. Saito.
December, 2019

Volume 84 Various Aspects of Multiple Zeta Functions — in honor of Professor Kohji Matsumoto's 60th birthday.
Edited by H. Mishou, T. Nakamura, M. Suzuki and Y. Umegaki. March, 2020

Volume 85 The Role of Metrics in the Theory of Partial Differential Equations.
Edited by Y. Giga, N. Hamamuki, H. Kubo, H. Kuroda and T. Ozawa. December, 2020

TO BE CONTINUED